D1383144

Nuclear Spectroscopy and Reactions

PART C

Nuclear Spectroscopy and Reactions

PART C

Edited by *JOSEPH CERNY*

Department of Chemistry
and Lawrence Berkeley Laboratory
University of California
Berkeley, California

ACADEMIC PRESS   New York and London   1974

A Subsidiary of Harcourt Brace Jovanovich, Publishers

ACADEMIC PRESS, INC.
111 Fifth Avenue, New York, New York 10003

United Kingdom Edition published by
ACADEMIC PRESS, INC. (LONDON) LTD.
24/28 Oval Road, London NW1

Library of Congress Cataloging in Publication Data

Cerny, Joseph.
    Nuclear spectroscopy and reactions.

    (Pure and applied physics, v. 40-C)
    Includes bibliographies.
    1.   Nuclear spectroscopy.   2.   Nuclear reactions.
I.   Title.   II.   Series.
QC454.N8C47   543'.085   72-13606
ISBN 0−12−165203−3 (pt. C)

PRINTED IN THE UNITED STATES OF AMERICA

CONTENTS

VII   GAMMA-RAY SPECTROSCOPY

VII.A   Coulomb Excitation

F. K. McGowan and P. H. Stelson

VII.B   Coulomb Reorientation

O. Häusser

VII.G   Angular Correlation Methods

A. J. Ferguson

VII.H   Lifetime Measurements

D. B. Fossan and E. K. Warburton

VIII   OTHER TOPICS

VIII.A   Photonuclear Reactions

B. L. Berman

VIII.B   Nuclear Spectroscopy from Delayed Particle Emission

J. C. Hardy

VIII.C In-Beam Atomic Spectroscopy

Indrek Martinson

VIII.D Effects of Extranuclear Fields on Nuclear Radiations

D. A. Shirley

VIII.E A Guide to Nuclear Compilations

F. Ajzenberg-Selove

LIST OF CONTRIBUTORS

Numbers in parentheses indicate the pages on which the authors' contributions begin.

F. AJZENBERG-SELOVE, Department of Physics, University of Pennsylvania, Philadelphia, Pennsylvania (551)

B. L. BERMAN, Lawrence Livermore Laboratory, University of California, Livermore, California (377)

E. CHEIFETZ,† Lawrence Berkeley Laboratory, University of California, Berkeley, California (229)

A. J. FERGUSON, Nuclear Physics Branch, Chalk River Nuclear Laboratories, Atomic Energy of Canada Limited, Chalk River, Ontario, Canada (277)

D. B. FOSSAN, State University of New York, Stony Brook, New York (307)

O. HÄUSSER, Chalk River Nuclear Laboratories, Atomic Energy of Canada Limited, Chalk River, Ontario, Canada (55)

J. C. HARDY, Physics Division, Chalk River Nuclear Laboratories, Atomic Energy of Canada Limited, Chalk River, Ontario, Canada (417)

A. E. LITHERLAND, University of Toronto, Toronto, Canada (143)

F. K. McGOWAN, Oak Ridge National Laboratory, Oak Ridge, Tennessee (3)

INDREK MARTINSON, Research Institute for Physics, Stockholm, Sweden (467)

J. O. NEWTON, Department of Nuclear Physics, Institute of Advanced Studies, Australian National University, Canberra, Australia (185)

E. RECKNAGEL, Physics Department, Free University of Berlin, and Nuclear and Radiation Physics Department, Hahn–Meitner Institute for Nuclear Research, Berlin, Germany (93)

† Present address: Weizmann Institute of Science, Rehovoth, Israel.

ix

C. Rolfs, University of Toronto, Toronto, Canada (143)

D. A. Shirley, Department of Chemistry and Lawrence Berkeley Laboratory, University of California, Berkeley, California (513)

P. H. Stelson, Oak Ridge National Laboratory, Oak Ridge, Tennessee (3)

E. K. Warburtcn, Brookhaven National Laboratory, Upton, New York (307)

J. B. Wilhelmy,[†] Lawrence Berkeley Laboratory, University of California, Berkeley, California (229)

[†] Present address: Los Alamos Scientific Laboratory, Los Alamos, New Mexico.

PREFACE

This work presents a survey of the development of a substantial fraction of the field of nuclear spectroscopy and reactions, with an emphasis on in-beam spectroscopy. It basically attempts to follow the spirit of "Nuclear Spectroscopy," edited by Fay Ajzenberg-Selove, which was published more than a decade ago. Hopefully, by bringing together at one time some 48 related chapters by 59 experts dealing with particular subdivisions of this active research area, the user will be able to acquire a broad, contemporary perspective of this field.

As before, these volumes have been primarily designed for use by graduate students engaging in experimental studies in nuclear spectroscopy and reactions as well as by specialists interested in ideas or techniques developed in other areas of research. Each author was requested to make his contribution accessible to a student who has completed graduate-level courses in nuclear physics and quantum mechanics.

This work is organized into four parts, each of which attempts to present a coherent area within the field. Part C contains an extensive coverage of gamma-ray spectroscopy as well as chapters on important related or recently unreviewed topics and a guide to nuclear compilations.

Due to the large number of contributors, inevitable problems of duplication, variable depth of coverage, and differing notation have arisen. Insofar as possible these problems have been minimized, but retaining the flavor of the original contribution as well as the notation employed in a particular research specialty were often overriding concerns.

As the editor I have profited both from many discussions with colleagues concerning the nature and substance of the work as well as from the pleasant interactions with the contributing authors. In particular I would like to express my deep appreciation to F. Ajzenberg-Selove, B. G. Harvey, D. L. Hendrie, and A. M. Poskanzer for much valuable advice.

CONTENTS OF OTHER PARTS

VII
Gamma-Ray Spectroscopy

VII.A COULOMB EXCITATION

F. K. McGowan and *P. H. Stelson*

OAK RIDGE NATIONAL LABORATORY [†]
OAK RIDGE, TENNESSEE

[†] Operated by Union Carbide for the U.S. AEC.

I. Introduction

When two nuclei approach each other but do not come sufficiently close to allow nuclear forces to act, the nuclei may still undergo transitions to excited states generated by the Coulomb forces acting between the nuclei. Such a process is known as Coulomb excitation. That nuclei can be excited by this process is not so important per se. What has proved to be enormously fruitful is the fact that the cross sections for this process can be accurately related to electromagnetic properties of nuclei. These electromagnetic properties of nuclei, such as the reduced electromagnetic transition probabilities $B(E2)$, $B(M1)$, $B(E3)$, and $B(E4)$ and the static electric quadrupole and magnetic dipole moments of excited states of nuclei, have provided bases for both new nuclear models and continuing important tests of all nuclear models.

The field of experimental Coulomb excitation is now about 20 years old. One might think that the field has been fairly well worked over by now and, yet, our impression is that the number of research papers continues to increase. Advances in technology have played an important role. During the first decade, the de-excitation γ rays following Coulomb excitation were detected with NaI scintillation detectors. The second decade witnessed the discovery and development of the Ge(Li) γ-ray detector. The Ge(Li) detector has such superior energy resolution [typically 2 keV width for Ge(Li) versus 50 keV width for NaI] that it was usually quite rewarding to repeat previous measurements done with NaI detectors. In addition to sharpening the traditional experimental results, such as γ-ray yields and angular distributions, the Ge(Li) detectors brought to fruition important new techniques for lifetime determinations of Coulomb-excited states by the observation of Doppler shifts.

The increased emphasis on heavy-ion reactions in nuclear physics has stimulated Coulomb excitation research. As we shall see, the Coulomb excitation mechanism becomes much more interesting in the case of highly charged projectiles. The increasing availability of energetic heavy-ion beams makes experiments practical. Of equal importance is the pervasive and influential role played by the Coulomb interaction in our understanding of heavy-ion nuclear reactions. Knowledge of pure Coulomb excitation theory and experiments should be valuable for research in the field of heavy-ion nuclear reactions.

As examples of the influence of the Coulomb interaction in nuclear reactions produced by heavy ions we cite three recent experiments:

(a) Diamond *et al.* (1968) reported an experiment which determined the cross sections for proton transfer between ^{12}C projectiles and ^{197}Au target nuclei. They found the surprising result that the cross section for transfer of a proton out of ^{12}C into ^{197}Au was about 100 times larger than the cross section for transfer of a proton out of ^{197}Au into the ^{12}C nucleus. They proposed that the Coulomb interaction between the two nuclei was largely responsible for this behavior.

(b) Vitoux *et al.* (1971) measured cross sections for excitation of the first 2^+ state of ^{24}Mg by inelastic scattering of ^{16}O ions. At lower bombarding energies, the cross section varied as predicted by Coulomb excitation theory. However, at somewhat higher energies, the cross section dropped rapidly below that predicted. This behavior is believed to result from the destructive interference between the repulsive Coulomb force and the attractive nuclear force which comes into play for the closer collisions at the higher bombarding energies.

(c) When two heavy nuclei approach each other in a collision, the Coulomb interaction is so strong that distortions in the shapes of the nuclei may occur (Thomas, 1968). This would cause an increase of the height of the Coulomb barrier. Recent work by Gauvin *et al.* (1972) suggests that these distortions are observable for ^{84}Kr ions bombarding ^{116}Cd.

A.   Simple Picture of the Coulomb Excitation Mechanism

The theory of pure Coulomb excitation is well understood but it is mathematically complicated. The modern approach is to use a computer program to generate specific theoretical results applicable to a particular experimental situation. For these reasons it may take considerable time and effort to develop a good "feel" for the process. We therefore believe it is valuable to consider a simplified picture.

We will take the case of the Coulomb excitation of a ^{238}U nucleus by a 140-MeV ^{40}Ar projectile which is scattered through an angle of 160°. This situation is shown in Fig. 1. A familiar parameter of Coulomb excitation theory is η, where

$$\eta = \frac{a}{\lambdabar} = \frac{Z_1 Z_2 e^2}{\hbar v}, \qquad a = \frac{Z_1 Z_2 e^2}{M_0 v^2}, \qquad M_0 = \frac{M_1 M_2}{M_1 + M_2} \tag{1}$$

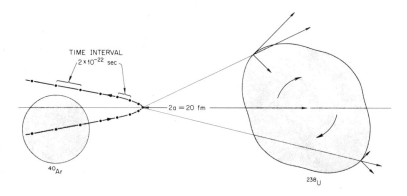

Fig. 1. Simple picture of the Coulomb excitation of a ^{238}U nucleus by the scattering of a 140-MeV ^{40}Ar projectile through an angle of 160°. To a very good approximation, the ^{40}Ar projectile travels along the classical hyperbolic orbit shown. Also shown are characteristic distances and time intervals for the collision. The electrical forces between the two nuclei exert a torque on the ^{238}U nucleus which gives it a rotational excitation.

The quantity $2a$ is the distance of closest approach in a head-on collision and M_1, Z_1, M_2, Z_2 are the mass and charge numbers of projectile and target nucleus, respectively, v is the velocity of the particle at large distances, and M_0 is the reduced mass. In the present case a is 10 fm and λ is 0.07 fm, so that the ratio is 140. Thus we have quite a good approximation to a classical system with the incident ^{40}Ar moving along a hyperbolic orbit since the wavelength of the projectile is so short compared to the characteristic distances of the collision.

The ^{238}U nucleus is known to have a "football" shape as shown in Fig. 1. The lowest excitations of this nucleus consist of collective end-over-end tumbling motions. Such a rotational spectrum is shown at the right of Fig. 2. At the point of closest approach in Fig. 1 we show the electrostatic forces exerted by the ^{40}Ar nucleus on the two tips of the ^{238}U nucleus. These forces are then resolved into normal and tangential components which show a net "couple" that tries to orient the nucleus so that a line through the tips of the nucleus is made perpendicular to the line connecting the centers of the projectile and target nuclei. To say this in another way, the ^{238}U nucleus tries to keep its "belly" facing the charged projectile. It is this twisting action which causes the ^{238}U nucleus to be excited by the incident ^{40}Ar projectile.

Several features of Coulomb excitation are evident from this simple picture. First, the larger the charge on the projectile is, the larger will be the

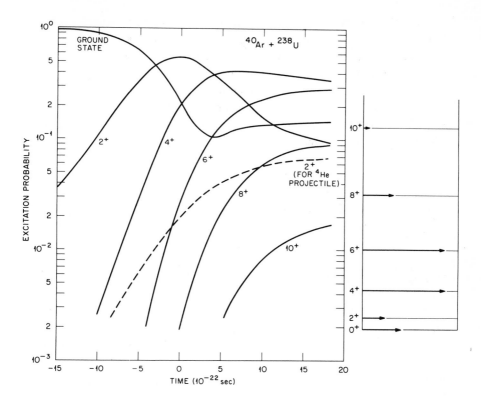

Fig. 2. Probabilities for excitation of the ground-state rotational band of ^{238}U as a function of time for the case of a 140-MeV ^{40}Ar projectile scattered through 160°. The time scale can be related to the position of the ^{40}Ar projectile by noting the dots along the orbit shown in Fig. 1. These dots are separated by 2×10^{-22} sec. The final relative probabilities for excitation of the different rotational states are shown by the length of the arrows in the level diagram on the right of the figure. The dashed curve shows the excitation probability for the 2$^+$ state excited by a ^{4}He projectile traveling the same orbit as the ^{40}Ar projectile.

torque on the nucleus. Second, the closer the projectile approaches the nucleus, the larger will be the torque. We will discuss this feature in more detail below. Third, a little reflection about the situation shown in Fig. 1 suggests that the more deformed the nucleus is, the larger the torque will be. It is the quantification of this feature that yields information about the shapes of nuclei from Coulomb excitation cross sections.

Another very important aspect of the Coulomb excitation mechanism is the time interval during which appreciable torque acts on the nucleus and how this time compares with the period of the nuclear state being excited. For the example shown in Fig. 1, the first excited rotational state (2^+) of the ^{238}U nucleus occurs at 45 keV. Using the fact that $\tau = h/\Delta E$, we find the period for rotation of the nucleus in the 2^+ state is 920×10^{-22} sec. Along the hyperbolic path of Fig. 1 is shown a series of dots which indicate the position of the ^{40}Ar projectile at constant time intervals of 2×10^{-22} sec. When the ^{40}Ar projectile is at the position indicated, the torque on the tips of the ^{238}U nucleus is only about $\frac{1}{3}$ of the value at the point of closest approach. Therefore most of the torque will be exerted during a time interval of about 30×10^{-22} sec. It is customary to take the effective time interval for the collision to be $2\pi a/[v \sin(\theta/2)]$, which, for the present example, is 23×10^{-22} sec. A useful parameter is $\xi(\theta)$, which is the ratio of this collision time to the nuclear period:

$$\xi(\theta) = (a \, \Delta E)/[\hbar v \sin(\theta/2)] \qquad (2)$$

Often one refers simply to ξ, which is taken to be the value $\xi(\theta = 180°)$. For the present example, ξ has the quite small value of 0.025.

There is an optimum probability for achieving Coulomb excitation when the collision time is much less than the nuclear period ($\xi \ll 1$). If $\xi = 1$, the probability for excitation has already decreased by about a factor of 30 from the $\xi = 0$ case (for E2 mode of excitation). For larger values of ξ, the probability continues to decrease exponentially; thus as ξ increases from 1.0 to 1.15, the probability for Coulomb excitation decreases by a factor of two. The most effective Coulomb excitation is achieved under the "sudden shock" condition rather than under the adiabatic condition.

The fact that Coulomb excitation probabilities decrease quite rapidly when $\xi \gtrsim 1$ imposes a strong limit on the excitation energies in nuclei which are accessible to Coulomb excitation. A 2-MeV excited state in ^{238}U would have a ξ value of 1.2 for 140-MeV ^{40}Ar ions. Therefore, although the ^{40}Ar projectile may have a high probability for losing some energy (large Coulomb excitation cross sections for low-lying states), it cannot lose a very significant amount of its total energy to the target nucleus and its path is very little changed by the energy given to the ^{238}U nucleus. It then follows that the total differential cross section for scattering (elastic plus inelastic) will be very nearly given by the well-known Rutherford cross section

$$d\sigma/d\Omega = (a^2/4) \sin^{-4}(\theta/2) \qquad (3)$$

Corrections for the small energy loss to the Coulomb excitation of the target can be made by using "symmetrized parameters" to be discussed.

The situation shown in Fig. 1 was deliberately chosen to emphasize the importance of large Coulomb excitation probabilities and the concept of multiple excitation. By multiple excitation, one means excitation of a final level by excitation through several successive intermediate states during a single collision. In such a situation the best approach is the direct numerical integration of the coupled differential equations which describe the amplitudes of the nuclear levels during the course of the collision. To illustrate this situation, we have used the Winther–de Boer computer program (Winther and de Boer, 1966) together with the theoretical rotational matrix elements and known rotational levels for ^{238}U to calculate the excitation probabilities for the rotational levels as a function of time. These probabilities are shown in Fig. 2.

B. The Problem of the Coulomb Barrier

We have pointed out the strong dependence of the Coulomb excitation probability on the parameter ξ when $\xi \gtrsim 1$. Recalling that ξ is the ratio of the collision time to the nuclear period, we conclude that for a given nuclear excited state, the only way to decrease ξ and therefore increase the probability for excitation is to try to decrease the collision time. It is instructive to rewrite the expression for ξ as a product of several factors, namely

$$\xi = a^{3/2} (M_0/Z_1)^{1/2} (2\pi \, \Delta E/heZ_2^{1/2}) \qquad (4)$$

For a given projectile (second factor) and a given nuclear state (third factor), we see that the only way to decrease ξ is to decrease the quantity a, which is half the distance of closest approach in a head-on collision. But this course of action brings on the distinct danger of invalidating the basic assumption of Coulomb excitation theory, namely that the influence of nuclear forces is negligible. For this reason, we find a strong concern about the nature of the Coulomb barrier of nuclei among experimental physicists engaged in Coulomb excitation.

To discuss this problem in quantitative terms, we will make the (incorrect) assumption that nuclei are spherical. Then we have that the "distance of closest approach" D is

$$D = a[1 + \csc(\theta/2)] = R_1 + R_2 + S = r_0(A_1^{1/3} + A_2^{1/3}) + S \qquad (5)$$

where θ is the CM scattering angle, R_1 and R_2 are the radii of the projectile and target nuclei, and S is the separation of the two nuclear surfaces. We have assumed that $R = r_0 A^{1/3}$. What value to take for r_0 is not clear. In dealing with this particular problem, it has become somewhat customary to take $r_0 = 1.25$ fm. With this value of r_0, the early prescription of de Boer and Eichler (1968) was that S should never be less than 3 fm. However, abundant recent evidence has clearly shown that 3 fm is too small. At the present time, the experimental evidence suggests that S should be at least 5 fm and possibly as large as 6 fm to ensure pure Coulomb excitation. This information on the Coulomb barrier is of fundamental importance to our understanding of nuclear physics and the continuing development of our knowledge of Coulomb barriers for collisions between heavy ions is clearly of central importance for Coulomb excitation.

C. METHODS FOR DETECTING COULOMB EXCITATION

Coulomb excitation can be measured either by the direct analysis of the energy spectrum of the scattered projectiles or by analysis of the radiation (γ rays, internal conversion electrons) subsequently emitted by the excited nuclei. Both methods are extensively used. We will briefly point out the virtues and drawbacks of the two methods.

The chief advantage deriving from a direct measurement of the spectrum of scattered projectiles is the good accuracy with which Coulomb excitation cross sections can be determined. The total cross section (elastic plus inelastic) is the known Rutherford cross section which provides a built-in calibration. The usual experimental problems associated with uncertainties in detection efficiency, target thickness, incident particle flux, stopping powers, etc., do not enter; one only needs simple ratios of peak areas. Coulomb excitation cross sections having an accuracy of 1% have been obtained from such experiments (Cline *et al.*, 1969; Ford *et al.*, 1971). Such highly accurate cross sections are valuable for determining such nuclear properties as static quadrupole moments, hexadecapole moments, and possible centrifugal stretching. They also provide useful information for the interpretation of inelastic electron scattering measurements and for comparison with results from muonic X rays.

The disadvantages of the direct measurement of inelastic projectile spectra (in comparison to γ-ray detection) are the low counting rates, the limited energy resolution, and the rather restricted nuclear information obtained from this type of measurement. The method requires (a) projectiles with very good energy homogeneity, (b) quite thin targets, and (c) a high-resolution charged-particle spectrometer. The heavier the projectiles are, the more stringent are the experimental requirements. For example, to measure the Coulomb excitation of ^{114}Cd with ^{84}Kr requires typically ^{84}Kr projectiles with 250 MeV energy. A beam resolution of 10^{-3} and a spectrometer resolution of 10^{-3} will result in 350 keV of width. A target of 5 $\mu g/cm^2$ will contribute an additional spread of 200 keV for backward scattering. The total energy spread is therefore about 400 keV, whereas the first 2^+ state in ^{114}Cd is only 558 keV above the ground state.

The detection of γ rays following Coulomb excitation offers the experimental advantages of excellent energy resolution [with Ge(Li) detectors] and relatively good counting rates. Most importantly, these advantages are largely preserved as one proceeds to heavier projectiles. An exception to this statement is the Doppler broadening which occurs when the nuclear lifetime is sufficiently short. Doppler broadening becomes increasingly pronounced for heavy projectiles and can produce substantial increases in peak widths.

Charged-particle spectrometers and Ge(Li) γ-ray detectors have roughly comparable overall detection efficiencies. However, since γ-ray detection allows the use of targets which are 10^2–10^3 times thicker than those used for charged-particle spectrometry, one gains a large factor in the quantity excitations/incident particle. There is also no stringent requirement on the energy homogeneity of the incident projectiles for the γ-ray measurements.

Probably the strongest motivation for measuring the γ rays following Coulomb excitation is the increased scope of the nuclear information which results from such experiments. The angular distributions of γ rays provide information both on the spins of excited states and on the multipolarity mixing ratios of the emitted γ rays. Thus, in addition to $B(E2)$ values, one also obtains $B(M1)$ values. The γ rays provide valuable information on the different modes of decay of a Coulomb-excited state, and from these results we obtain $B(E\lambda)$ and $B(M\lambda)$ values for transitions between excited states. The Doppler shift of the γ-ray energies provides useful information on the total lifetime of the Coulomb-excited state. Perturbed angular correlations of the Coulomb-excited γ rays can be measured to determine magnetic moments of excited states of nuclei.

D. Relevance of Coulomb Excitation to Inelastic Electron Scattering and to Muonic X-Ray Spectra

At the present time, the effort to obtain highly accurate values for $B(E\lambda)$ can scarcely be justified from the point of view of comparisons with predictions of nuclear models—probably 10% accuracy would suffice for this purpose. However, the availability of precise $B(E\lambda)$ values has produced a significant impact on the neighboring research fields of inelastic electron scattering and muonic X-ray spectra. Both of these fields share with Coulomb excitation the virtue of pure electromagnetic interactions, but they differ from Coulomb excitation in that the extracted $B(E\lambda)$ values are dependent on the assumed nuclear model. This model dependence, although possibly regarded initially as a drawback, eventually becomes a virtue because it offers the possibility of providing useful new information about nuclei.

The $B(E\lambda)$ values do not provide complete tests of the electromagnetic aspects of nuclear models since they measure only the λth radial moment of the transition density. Inelastic electron scattering can measure higher radial moments of the transition densities and thus provide more detailed information. In practice, one usually assumes a nuclear model and then compares a series of predicted inelastic cross sections for different amounts of transferred momentum with the experiment values. One then varies the models or the parameters of a given model until the best overall agreement is obtained (Curtis et al., 1969; Itoh et al., 1970; Horikawa et al., 1971; De Jager and Boeker, 1972). An important test of the whole procedure is provided by comparing the resultant $B(E\lambda)$ with the model-independent value from Coulomb excitation. The nuclear models tested so far have not provided completely satisfactory fits to the electron inelastic scattering data. There are also differences of 10–20% between the $B(E\lambda)$ obtained from these fits and the model-independent $B(E\lambda)$ value.

The analysis of muonic X-ray spectra provides extensive and detailed information on the charge distribution of nuclei (see Anderson and Jenkins, Chapter V.A, and Wu and Wilets, 1969). For permanently deformed nuclei such as the rare earth or actinide nuclei, one can extract accurate values for the intrinsic quadrupole moment Q_0. A number of years ago, it was noticed by the Columbia group (Wu, 1967) that the Q_0 values from muonic X rays for the tungsten nuclei disagreed by about 10% with the Q_0 values deduced from the $B(E2)$ values obtained from Coulomb excitation. The experimental uncertainties in both cases were only about 1%. In an attempt to reconcile the Q_0 values, it was assumed that the nuclear skin thickness t was a function of the polar angle. By making the surface layer appreciably more diffuse in the equatorial plane, one could bring the Q_0 values into agreement. However, the magnitude of the skin thickness parameter then differed strongly from the value obtained from electron scattering.

Recently, Chen (1970) resolved this situation by pointing out that the nuclear polarization effect implies that the muonic X-ray spectra measure an "effective" Q_0 which is in fact several percent larger than the true Q_0 value. When this correction is applied to the muonic X-ray spectra, one no longer needs to assume a variable nuclear skin thickness and, furthermore, the value for the nuclear skin thickness now agrees with the value from electron scattering. To illustrate the fact that the muonic X-ray spectra measure an "effective" Q_0, we show recent results for ^{232}Th and ^{238}U in Fig. 3. Three recent values for Q_0 from muonic X rays are shown which were not corrected for nuclear polarization (McKee, 1969; Cote *et al.*, 1969; De Wit *et al.*, 1967).

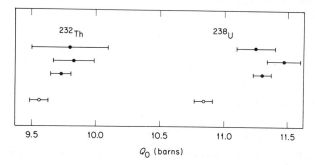

Fig. 3. The comparison of the Q_0 values from Coulomb excitation (•) for ^{232}Th and ^{238}U with three different experimental determinations of the "effective" Q_0 values obtained from muonic X-ray spectra (∘). Nuclear polarization can account for this difference in Q_0 values.

We also show recent Q_0 values from Coulomb excitation (Ford *et al.*, 1971). It is apparent that the "effective" Q_0 values from muonic X rays are a few percent higher than those from Coulomb excitation.

II.   Relevant Aspects of the Theory of Coulomb Excitation

Many of the modern accelerators (tandem Van de Graaffs, cyclotrons, and heavy-ion linear accelerators) are capable of accelerating a variety of projectiles. Therefore, a brief review is given of the pertinent treatments of the Coulomb excitation process which are useful for the analysis of Coulomb excitation experiments. For many experimental conditions, the applicability of the semiclassical treatment of Coulomb excitation is well justified and this approximation simplifies the theoretical evaluation of the process. The semiclassical theory makes the assumption that the nuclear levels are excited by the time-dependent Coulomb field of the impinging projectile acting on the nucleus while the projectile moves along a classical hyperbolic orbit.

Let us for simplicity assume that only the target nucleus can be excited and that the probability of finding this nucleus in a state f after the collision is

$P_{i \to f}$. Then the differential cross for inelastic scattering to the state f is given by

$$d\sigma_{i \to f} = P_{i \to f} \, d\sigma_R \tag{6}$$

where $d\sigma_R$ is the classical Rutherford differential cross section in the CM system [see Eq. (3)].

If the excitation probabilities P are small compared to unity, a first-order perturbation treatment of the excitation process is adequate. For larger values of P, second-order processes may be important and in some instances, an adequate treatment may be obtained by carrying the perturbation approach to second order in the charge of the projectile. Finally, for strong interactions leading to probabilities comparable to unity, the multiple Coulomb excitation theory must be used.

A. FIRST-ORDER PERTURBATION THEORY

When light ions, such as protons and ^{4}He ions, are used as projectiles, it is possible to extract from Coulomb excitation experiments rather accurate matrix elements of the multipole operators between the different nuclear states using the first-order perturbation theory. A detailed discussion of the relevant theory is given in a comprehensive review (Alder *et al.*, 1956) and we simply quote equations from this reference. The semiclassical differential and total cross sections may be written in the form

$$d\sigma_{E\lambda} = (Z_1 e / \hbar v)^2 a^{-2\lambda+2} B(E\lambda, J_i \to J_f) \, df_{E\lambda}(\theta, \xi) \tag{7}$$

$$\sigma_{E\lambda} = (Z_1 e / \hbar v)^2 a^{-2\lambda+2} B(E\lambda, J_i \to J_f) f_{E\lambda}(\xi) \tag{8}$$

The $B(E\lambda, J_i \to J_f)$ is the reduced transition probability

$$B(E\lambda, J_i \to J_f) = (2J_i + 1)^{-1} |\langle J_f \| \mathcal{M}(E\lambda) \| J_i \rangle|^2 \tag{9}$$

where $\mathcal{M}(E\lambda)$ is the reduced nuclear 2^λ-pole matrix element of Bohr and Mottelson (1969) and of Alder *et al.* (1956).

In the semiclassical description of Coulomb excitation, the effect of the energy loss on the projectile motion is neglected. From a WKB and a quantum mechanical treatment of the Coulomb excitation process in first-order perturbation theory, Alder *et al.* (1956) found an essential improvement of the semiclassical description by the following symmetrization of the parameters a and ξ given by

$$a = Z_1 Z_2 e^2 / M_0 v_i v_f \tag{10}$$

and

$$\xi = \eta_f - \eta_i = \frac{Z_1 Z_2 e^2}{\hbar} \left(\frac{1}{v_f} - \frac{1}{v_i} \right) \tag{11}$$

It is readily seen that expressions (10) and (11) for a and ξ are equal to the simpler expressions (1) and (2) to lowest order in $\Delta E/E$. The symmetrized excitation cross sections are

$$d\sigma_{E\lambda} = (Z_1 e/\hbar v_i)^2 a^{-2\lambda+2} B(E\lambda, J_i \rightarrow J_f) \, df_{E\lambda}(\theta, \xi) \tag{12}$$

$$\sigma_{E\lambda} = (Z_1 e/\hbar v_i)^2 a^{-2\lambda+2} B(E\lambda, J_i \rightarrow J_f) \, f_{E\lambda}(\xi) \tag{13}$$

The quantum mechanical cross sections for $\lambda = 1$ and 2 have the same form as Eqs. (12) and (13) except that the $df_{E\lambda}$ and $f_{E\lambda}$ depend on η_i. In the limit $\eta_i \rightarrow \infty$, the quantum mechanical cross sections reduce to the semiclassical cross sections.

The semiclassical $df_{E\lambda}$ and $f_{E\lambda}$ have been evaluated numerically and tabulated by Alder *et al.* (1956) and by Alder and Winther (1956) for $\lambda = 1, 2, 3$, and 4. The quantal functions $f_{E1}(\eta_i, \xi)$ and $f_{E2}(\eta_i, \xi)$ are also tabulated. The semiclassical function $f_{E2}(\xi)$ differs from the quantal function $f_{E2}(\eta_i, \xi)$ by less than 3% for $\eta_i \gtrsim 4$. A corresponding accuracy for the semiclassical differential cross section (Griffy and Biedenharn, 1962) is achieved for E2 excitation at $\eta_i \gtrsim 4$ and $\theta \gtrsim 120°$. The availability of these calculations for finite η_i permits the extraction of accurate values of $B(E2, J_i \rightarrow J_f)$ from measured total cross sections with light ions, e.g., $\eta_i \approx 4$ for 3-MeV protons on $Z_2 = 44$.

The angular distribution of the de-excitation γ rays for the usual case in which the scattered projectiles are not observed is of the form

$$W(\theta_\gamma) = \sum_\nu A_\nu a_\nu^{E\lambda}(\eta_i, \xi) \, P_\nu(\cos \theta_\gamma) \tag{14}$$

where the summation extends over even Legendre polynomials and θ_γ is the angle between the direction of the incident beam and the γ ray. The coefficients A_ν are the γ–γ directional correlation coefficients tabulated by Biedenharn and Rose (1953) and by Ferentz and Rosenzweig (1955) for the decay sequence $J_1(L_1) J_2(L_2) J_3$, where the J's are the spins of the target nucleus, the Coulomb-excited state, and the final state after γ-ray emission, respectively. The particle parameters $a_\nu^{E\lambda}(\eta_i, \xi)$ depend on the Coulomb excitation process.

In contrast to the total cross section, the semiclassical limit $(\eta_i \rightarrow \infty)$ is less good as an approximation for the evaluation of the particle parameters which enter into Eq. (14) for the angular distribution of the de-excitation γ rays. For finite η_i the quantum mechanical results deviate considerably from the semiclassical results because the particle parameters are sensitive to the phases of the amplitudes even though the absolute values of the semiclassical and quantum mechanical excitation amplitudes nearly coincide. The particle parameters $a_v^{E2}(\eta_i, \xi)$ have been calculated numerically and tabulated by Biedenharn et al. (1956) and Alder et al. (1956). The semiclassical particle parameters $a_2^{E2}(\eta_i \to \infty, \xi)$ differ from the quantal $a_2^{E2}(\eta_i, \xi)$ by less than 5% for $\eta_i \gtrsim 20$. Numerical results (Thaler et al., 1956; Alder and Winther, 1955) have also been tabulated for $a_2^{E1}(\eta_i, \xi)$.

Finally, the angular distribution of the deexcitation γ rays following other multipole excitations, as well as the angular distributions for a specified direction of the inelastically scattered projectile, may be obtained in the semiclassical approximation from Eqs. (75) and (76) of Chapter II.A of Alder et al. (1956) for $a_v^{E\lambda}(\eta_i \to \infty, \xi)$ using the tabulated values (Alder and Winther, 1956) of the orbital integrals $I_{\lambda\mu}$ for $\lambda = 1, 2, 3,$ and 4. Although the use of semiclassical particle parameters $a_v^{E3}(\eta_i \to \infty, \xi)$ for the interpretation of γ-ray angular distributions following E3 Coulomb excitation for experimental conditions $\eta_i = 13–20$ has limited accuracy, it is possible to obtain definite spin assignments for Coulomb excited states (see Section IV).

B. SECOND-ORDER PERTURBATION THEORY

When the first-order probability is not too large, an adequate treatment of the Coulomb excitation reaction may be obtained by carrying the perturbation approach to second order in the projectile charge. The parameter $\chi_{i\to f}^{(\lambda)}$ is introduced which is a useful measure of the applicability of the perturbation treatment:

$$\chi_{i\to f}^{(\lambda)} = \frac{(16\pi)^{1/2}(\lambda + 1)!}{(2\lambda + 1)!!} \frac{Z_1 e}{\hbar(v_i v_f)^{1/2}} \frac{\langle J_f \| \mathscr{M}(E\lambda) \| J_i \rangle}{a_{if}^{\lambda}(2J_i + 1)^{1/2}} \tag{15}$$

where $2a_{if}$ is the symmetrized distance of closest approach in a head-on collision. The $\chi_{i\to f}^{(\lambda)}$ is a measure of the strength of the interaction with which state J_i is coupled to state J_f by multipole order λ. For some experimental conditions where $\chi_{i\to f}^{(2)} < 1$, the perturbation treatment to second order is useful for the analysis of data. For instance, states which are not accessible by direct E2 excitation may be populated by double E2 excitation. Also, the first-order excitation amplitude may be altered by second-order terms such as the Coulomb "reorientation" effect due to the interaction from static moments (see Häusser, Chapter VII.B). In many experiments, the perturbation treat-

ment of the Coulomb excitation reaction is also a useful guide for the physical understanding of the process and for selecting suitable experimental
conditions.

Alder *et al.* (1956) have presented the general formulation of the semiclassical theory to second order

$$dσ = dσ^{(1)} + dσ^{(1, 2)} + dσ^{(2)} \tag{16}$$

The first term in Eq. (16) is the first-order excitation cross section, Eq. (7).
The second term represents the interference between first- and second-order
excitations. The third term is the second-order excitation cross section. In
general, there are additional terms of similar order of magnitude originating
from the interference between first- and third-order amplitudes.

The exact quantal treatment (Smilansky, 1968; Alder and Pauli, 1969)
of the multiple excitation process leads to a set of coupled differential equations for the radial wave functions which must be solved. The WKB method
has been used to make a few computations of the quantum corrections to the
"reorientation" effect. More recently, the second-order quantum mechanical
theory of Coulomb excitation has been parameterized in a convenient form
(Alder *et al.*, 1972) from which quantum corrections may be extracted. These
quantum corrections may then be applied to results obtained from data which
have been analyzed by the semiclassical treatment of the Coulomb excitation
reaction to second order. Numerical computations are tabulated (Alder
et al., 1972) for a number of interesting experimental conditions of the Coulomb excitation reaction:

(1) The relative excitation probabilities for pure double E2 excitation of
states $J_f = 0, 1, 2, 3$, and 4 via an initial $0 \rightarrow 2$ transition are given for $θ = 180°$,
160°, and 140° and for $η_i = \infty, 8$, and 4. These relative excitation probabilities are also applicable for pure double E2 excitation of states with arbitrary
spins of the initial, intermediate, and final states designated by J_i, J_z, and
J_f.

(2) E2 excitation of a state J_f [the dominant contribution is from the E2
transition matrix element in $χ_{i \rightarrow f}^{(2)}$] is considered where the term $dσ^{(1,2)}$ in
Eq. (16) includes the influence of the static electric quadrupole moments of
the initial and final states and the influence of double E2 excitation via an
intermediate state J_z on the excitation probability. The relative excitation
probabilities are given for $θ = 180°$, 160°, and 140° and for $η_i = \infty, 8$, and 4.

(3) Finally, the relative excitation probabilities are tabulated for the
case of pure double E2 excitation and direct E4 excitation of a state J_f for an

arbitrary spin sequence $J_i \rightarrow J_z \rightarrow J_f$ and again for the conditions $\theta = 180°$, $160°$, and $140°$ and $\eta_i = \infty, 8$, and 4.

For many experimental conditions, the use of the semiclassical theory to second order is certainly justified. In many other cases, this approach can serve as a useful guide. However, each experimental situation must be carefully assessed. For example, when ${}^4\text{He}$ ions are used as a projectile, the parameter η_i is of the order of ten and quantum mechanical corrections in second-order processes can be important. This is especially true when accurate values of nuclear structure properties are extracted from experimental data or when the nuclear structure property to be determined makes a relatively small contribution to the measured quantity in a Coulomb excitation experiment. For instance, the quantum mechanical correction for pure double E2 excitation of a 4^+ state in even-A deformed nuclei with ${}^4\text{He}$ ions is 3–5%. Therefore, values of the reduced hexadecapole moments $\langle 4 \| \mathcal{M}(\text{E}4) \| 0 \rangle$ deduced from experimental measurements (Stephens *et al.*, 1971; McGowan *et al.*, 1971) of the differential cross section of 4^+ states are very sensitive to these quantum mechanical corrections (see Section VII.E).

It turns out that the quantum mechanical corrections (Alder and Pauli, 1969) are of the order of $1/\bar{\eta}$ compared with the semiclassical results, where $\bar{\eta} = (\eta_i \eta_f)^{1/2}$. Thus, quantum corrections for other values of η_i can be obtained by the $1/\bar{\eta}$ rule from the available tables (Alder *et al.*, 1972).

C. Multiple Coulomb Excitation Theory

When the interaction strength parameters $\chi_{i \rightarrow f}^{(\lambda)}$ become large, $\chi \gtrsim 1$, the Coulomb excitation process involves many nuclear states. This condition exists in many experiments when energetic heavy ions are used. The parameter η_i is large ($\eta_i \gtrsim 25$) for heavy ions with $Z_1 \geq 8$. Furthermore, the projectile loses only a small fraction of its energy in a collision. Excitation energies of nuclear states populated by Coulomb excitation are usually less than 2.5 MeV, while projectile energies are typically 25–200 MeV. Consequently, the semiclassical approximation of hyperbolic projectile orbits is an excellent one.

In multiple Coulomb excitation, a particular nuclear state can be populated by virtual excitation through several intermediate states. To describe the reaction accurately, one must solve the set of coupled differential equations describing the population of nuclear states during the collision. An excellent computer program, developed by Winther and de Boer (1966) for E2 excitation and modified by Holm (1969) to include E1, E3, and E4 excitation, is available for multiple Coulomb excitation in the semiclassical approxima-tion. The excitation amplitudes are evaluated by direct numerical integration of the coupled differential equations for the time-dependent amplitudes. Input quantities required by the program are the nuclear charges and masses, spins and energies of the nuclear levels, projectile energy and scattering angle, and all reduced $E\lambda$ matrix elements. It is assumed that the phases of nuclear wave functions are chosen in such a way that the reduced matrix elements are real. The input matrix elements then satisfy the following symmetry relation:

$$\langle J_f \| i^\lambda \mathscr{M}(E\lambda) \| J_i \rangle = (-1)^{\lambda + J_i - J_f} \langle J_i \| i^\lambda \mathscr{M}(E\lambda) \| J_f \rangle \qquad (17)$$

The program computes the excitation probabilities, differential cross sections, and the angular distributions of the de-excitation γ rays.

   This program is independent of any specific nuclear model; however, a knowledge of the reduced $E\lambda$ matrix elements is required. In practice many of the $E\lambda$ matrix elements are known from experiments done with lighter ions where the analysis of the data with either first- or second-order perturbation theory is appropriate, and from lifetime measurements (see Fossan and Warburton, Chapter VII.H). This is a primary motive for continuing to per-form Coulomb excitation experiments with light ions, i.e., to build up an accurate base of matrix elements which may be used in the analysis of ex-perimental data from multiple Coulomb excitation. This is especially true of experiments designed to determine the static E2 moments of excited nuclear states. In instances where some of the input $E\lambda$ matrix elements for a compu-tation are not known, one must resort to a computation of these from ap-propriate nuclear model parameters. For example, in the limit of the rigid-rotor model, the reduced $E\lambda$ matrix elements (Alder *et al.*, 1956) are given by

$$M_{if}(E\lambda) \equiv \langle J_f \| \mathscr{M}(E\lambda) \| J_i \rangle$$
$$= (2J_i + 1)^{1/2} [(2\lambda + 1)/16\pi]^{1/2} Q_{\lambda 0} \langle J_i \lambda K0 \mid J_i \lambda J_f K \rangle \qquad (18)$$

where the intrinsic quadrupole moment Q_{20} for a prolate nuclear shape can be obtained from the experimental value of $B(E2, 0 \to 2)$.

   The computer program approach also has the advantage that the computed probabilities include contributions from all processes allowed by the input matrix elements. The excitation probabilities may be calculated to any desired accuracy, subject only to the limitation of the semiclassical approximation and the time allowed for the calculation. For four levels (0^+, 2^+, 4^+, and 6^+) with an input of six E2 and seven E4 matrix elements, the time required for numerical integration of the coupled differential equations on a IBM System/360 Model 91 is 6.6 sec. The computer program is useful to make estimates of higher-order effects (higher than second order) which may be used to apply corrections to results deduced from data analyzed with first- or second-order perturbation theory.

III.   Gamma-Ray Spectroscopy

The most extensively used method for determination of Coulomb excitation cross sections is the measurement of thick-target yields of de-excitation γ rays by either direct γ-ray spectroscopy, or γ–γ coincidence spectroscopy, or particle–γ coincidence spectroscopy. The technique of measuring γ-ray yields with a Ge(Li) spectrometer or a NaI scintillation spectrometer is especially suitable for de-excitation γ rays following Coulomb excitation of most nuclei. The γ rays have convenient energies; internal conversion is small (notable exceptions are the low-energy transitions from ground-state rotational states in deformed nuclei); and isotopically enriched samples are available from which targets may be prepared.

A.   GAMMA-RAY SPECTROMETERS

Prior to the Ge(Li) detector, γ-ray yields from the Coulomb excitation reaction were measured with a NaI scintillation detector. A distinct advantage of this detector is the large total efficiency which can be computed by numerical integration with high accuracy for the case of small sources placed at various distances from the crystal face along the axis of the cylinder (Neiler and Bell, 1965). The dimensions of the cylindrical crystal are typically 3 in. in diameter and 3 in. in length. Under the most favorable condition, the absolute γ-ray yield has been determined to an accuracy of $\pm 4\%$ with this detector (Stelson and McGowan, 1958). The main disadvantage of the NaI detector is the poor energy resolution compared to the Ge(Li) detector.

The combination of energy proportionality, high efficiency, and short pulse duration makes the scintillation detector ideally suited to γ–γ and particle–γ coincidence spectrometry. For the latter, the de-excitation γ rays are generally observed in coincidence with projectiles backscattered into an annular silicon surface barrier detector. A typical arrangement of the annular detector corresponds to detecting projectiles scattered between $153°$ and $171°$ ($\Omega_L \approx 0.6$ sr).

By the use of large (30–60 cm^3) Ge(Li) γ-ray detectors, the quality of direct γ-ray spectra is greatly improved compared to NaI scintillation spectra. Uncertainties in background subtraction and in spectral decomposition are reduced with the better energy resolution. The location of nuclear states may be determined with a high degree of accuracy using energy calibration standards (Greenwood et al., 1970; Helmer et al., 1971) from radioactive sources in the energy range 100–2000 keV. Typical uncertainties for the standards are 0.004–0.030 keV. Great care must, of course, be taken to measure the nonlinearity of the amplifier–analyzer system with a precision pulser or preferably with a set of radioactive sources with known γ-ray energies.

The Ge(Li) detector does have the disadvantage that the absolute efficiency is small compared to the NaI scintillation detector. Typically the efficiency is 3–12%, depending on the size of the Ge(Li) detector, relative to a 3 × 3 in. NaI detector. This figure of merit is the ratio of the areas of the full-energy peak measured at $E_\gamma = 1.332$ MeV with a ^{60}Co source. Although the smaller detection efficiency of the Ge(Li) detector is not a disadvantage for most direct γ-ray spectroscopy experiments, it does restrict the types of coincidence spectroscopy experiments which are practical. However, with the larger Ge(Li) detectors, absolute γ-ray yield measurements are becoming feasible for particle–γ coincidence spectroscopy (Casten et al., 1969) and for γ–γ coincidence spectroscopy using a combination of NaI and Ge(Li) detectors.

The absolute efficiency of Ge(Li) can be determined to an accuracy of ±4% over the energy range 100–2000 keV. This is usually accomplished with a set of calibrated γ-ray sources (^{241}Am, ^{57}Co, ^{203}Hg, ^{137}Cs, ^{54}Mn, ^{60}Co, ^{22}Na, ^{88}Y) obtained from the International Atomic Energy Agency Laboratories IAEA, Vienna, Austria. These source strengths are calibrated to about 1% accuracy. Additional sources can be calibrated to an accuracy of ±4% with a standard NaI scintillation spectrometer. There are several radioactive sources which provide a convenient set of relative efficiencies (Jardine, 1971), i.e., ^{182}Ta, ^{180}Hf, ^{226}Ra. As an example, Fig. 4 shows an efficiency calibration for a 30-cm^3 Ge(Li) detector obtained with a set of IAEA sources. The turnover of the absolute full-energy peak efficiency at low γ-ray energies is due to absorptive materials between γ-ray source and the detector. For comparison purposes, the absolute full-energy peak efficiency of a 7.62 × 7.62 cm NaI detector is shown in Fig. 4 along with the energy resolution for both detectors. The ability to observe a full-energy γ-ray peak above the Compton distributions of other γ rays in a pulse-height spectrum depends on the energy resolution and the peak-to-total ratio, which is the ratio of the number of events in the full-energy peak to the number of events in the entire pulse-height distribution of a single γ-ray energy.

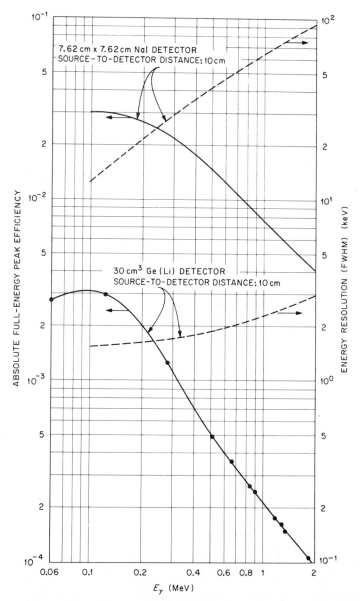

Fig. 4. Absolute efficiency and energy resolution of a 30-cm³ coaxial Ge(Li) detector and of a 7.62 × 7.62 cm NaI detector as a function of γ-ray energy. The absolute full-energy peak efficiency is defined as the product of the full-energy peak efficiency and the detector solid angle. The energy resolution is full-width at half-maximum.

B. Targets and Contaminant γ Rays

Targets which are thick to the range of the impinging projectiles are, in general, more convenient than thin targets for γ-ray spectroscopy from the Coulomb excitation reaction. Metallic targets of normal and isotopically enriched abundance may be prepared by electrodeposition, by sintering metallic powders into foils, or by rolling of metals to produce foils. Nickel backings are useful for preparation of targets by electrodeposition and for mounting sintered and rolled foils. The foils are mounted on the nickel target backing by forming a sandwich of the nickel backing, the foil, and a nickel collar; the collar is then spot-welded to the backing. Normally, targets range between 30 and 150 mg/cm^2 in thickness by 0.5 in. in diameter. Approximately 100 mg of an isotopically enriched sample is sufficient for sintering at room temperature under a pressure of 25 tons/in.^{2}

Gamma rays from nuclear reactions produced by bombardment of low-Z target impurities are sometimes a source of troublesome background. It is found that preparation of targets by electrodeposition eliminates impurities to a considerable extent. Some of the most common impurity γ rays are:

$$^{14}\text{N}(\alpha,\text{p})^{17}\text{O } E_\gamma = 871 \text{ keV}$$
$$^{18}\text{O}(\alpha,\text{n})^{21}\text{Ne } E_\gamma = 350 \text{ keV}$$
$$^{19}\text{F}(\text{p},\text{p}')^{19}\text{F and } ^{19}\text{F}(\alpha,\alpha')^{19}\text{F } E_\gamma = 110 \text{ and } 197 \text{ keV}$$
$$^{19}\text{F}(\alpha,\text{n})^{22}\text{Na } E_\gamma = 583 \text{ and } 891 \text{ keV}$$
$$^{19}\text{F}(\alpha,\text{p})^{22}\text{Ne } E_\gamma = 1274.5 \text{ keV}$$
$$^{23}\text{Na}(\text{p},\text{p}')^{23}\text{Na and } ^{23}\text{Na}(\alpha,\alpha')^{23}\text{Na } E_\gamma = 440 \text{ keV}$$
$$^{23}\text{Na}(\alpha,\text{n})^{26}\text{Al } E_\gamma = 417 \text{ keV}$$
$$^{27}\text{Al}(\text{p},\text{p}')^{27}\text{Al } E_\gamma = 843 \text{ and } 1014 \text{ keV}$$
$$^{28}\text{Si}(\text{p},\text{p}')^{28}\text{Si } E_\gamma = 1779 \text{ keV}$$
$$^{12}\text{C}(^{16}\text{O},\alpha)^{24}\text{Mg } E_\gamma = 1368.5 \text{ keV}$$
$$^{12}\text{C}(^{16}\text{O},\text{pn})^{26}\text{Al } E_\gamma = 417 \text{ keV}$$
$$^{12}\text{C}(^{16}\text{O},\alpha\text{p})^{23}\text{Na } E_\gamma = 440 \text{ keV}$$

and $E_\gamma = 511.0$ keV from β^+ activities produced in the targets.

C. Thick-Target Yields and Angular Distributions

The thick-target yield Y is given by

$$Y\left(\frac{\text{exc}}{\text{particle}}\right) = N \int_0^{x(E=0)} \sigma(x)\, dx = N \int_0^{E_p} \frac{\sigma(E)\, dE}{dE/dx} \qquad (19)$$

where N is the number of target nuclei per cm^3 and E_p denotes the incident projectile energy. It is convenient to define the stopping power of the projectile in the target material as

$$S(E) = dE/d\varrho x \qquad (20)$$

measured in units MeV-cm^2/mg. If the projectile charge state is denoted by z, the thick-target yield zY (exc/μC) in excitations per μC of particles is

$$Y \text{ (exc/}\mu\text{C)} = (3.759 \times 10^{33}/zA_2') \int_0^{E_p} [\sigma(E)/S(E)] \, dE \qquad (21)$$

where A_2' is the atomic weight of the normal target element.

Inserting numerical values into the symmetrized excitation cross sections (13) from first-order perturbation theory, the $B(E\lambda)$ for the interesting cases of direct E2 and E3 Coulomb excitation are given by

$$B(E2, J_i \rightarrow J_f) = \frac{5.55 \times 10^{-59} A_2'(CZ_2)^2 zY \text{ (exc/}\mu\text{C)}}{A_1 I_{E2}} \, e^2 \text{ cm}^4 \qquad (22)$$

$$B(E3, J_i \rightarrow J_f) = \frac{2.88 \times 10^{-85} A_2'(CZ_2)^4 Z_1{}^2 zY \text{ (exc/}\mu\text{C)}}{A_1 I_{E3}} \, e^2 \text{ cm}^6 \qquad (23)$$

where the thick-target integrals $I_{E\lambda}$ are defined by

$$I_{E2} = \int_0^{E_p} \frac{(E - C \, \Delta E) f_{E2}(\eta_i, \xi)}{S(E)} \, dE \qquad \frac{\text{MeV mg}}{\text{cm}^2} \qquad (24)$$

$$I_{E3} = \int_0^{E_p} \frac{E(E - C \, \Delta E)^2 f_{E3}(\xi)}{S(E)} \, dE \qquad \frac{\text{MeV}^3 \text{ mg}}{\text{cm}^2} \qquad (25)$$

A_1 and A_2 are the masses of the projectile and target in amu, E_p is the incident projectile energy in MeV, ΔE is the nuclear excitation energy in MeV, and $C = 1 + A_1/A_2$. The thick-target integrals are normally evaluated numerically with computer programs using the tabulated values of $f_{E2}(\eta_i, \xi)$ and $f_{E3}(\xi)$. Either a function of the form $aE^{(b + cE)}$ or a table of numbers has been used to represent the stopping power $S(E)$.

In the analysis of thick-target γ-ray angular distributions, the particle parameters $a_v^{E\lambda}(\eta_i, \xi)$ in Eq. (14) must be replaced by the thick-target particle parameters

$$[a_v^{E\lambda}(E_p)]_t = \int_0^{E_p} \frac{\sigma_{E\lambda}(E) \, a_v^{E\lambda}(\eta_i, \xi)}{S(E)} \, dE \bigg/ \int_0^{E_p} \frac{\sigma_{E\lambda}(E)}{S(E)} \, dE \qquad (26)$$

These thick-target particle parameters are also normally evaluated numerically with computer programs using the tabulations of $a_v^{E\lambda}(\eta_i, \xi)$. The factors for finite angular resolution of the γ-ray detectors which are required in the analysis of angular distribution data are tabulated for NaI detectors (Yates, 1965) and for Ge(Li) detectors (Camp and Van Lehn, 1969).

For the analysis of thick-target yields from multiple Coulomb excitation experiments, the computer program of Winther and de Boer (1966) and of Holm (1969) must be augmented to provide integrated results over angle and energy. Sayer (1968) has given a detailed description of a modified version of the Winther–de Boer computer program. Integration may be performed over E and θ to calculate total cross sections, multiple Coulomb excitation yields for a thick target, and the Coulomb excitation probabilities $P(J)$ and statistical tensors (Winther and de Boer, 1966) for a thick target and finite solid angle of a particle detector. The latter is needed for the analysis of γ-ray angular distributions.

The time required for calculation of total cross sections and thick-target γ-ray yields with the Winther–de Boer computer program is not necessarily small. A typical calculation of a thick-target γ-ray yield requires about 100 solutions (11 angles and nine energies) of the coupled differential equations for a given set of input conditions. The total computation time for a four- or five-state calculation of E2 excitation processes on an IBM system/360 Model 91 is 200–400 sec. There are instances where the Winther–de Boer computer program need not be used since the use of second-order perturbation theory is quite adequate. An important application of second-order excitation is the pure double E2 Coulomb excitation of $J_f = 0$ and 4 via an initial $0 \rightarrow 2$ transition. Douglas (1963) has given explicit expressions for the cross section to second order. The procedure to extract $B(E2, 2 \rightarrow J_f)$ from thick-target γ-ray yields has been described in detail by McGowan et al. (1965).

D. STOPPING POWER $S(E)$

The $B(E\lambda)$ values extracted from absolute thick-target γ-ray yields depend directly on the stopping power $S(E)$. An excellent early survey of the experimental data concerning the energy loss of charged particles, principally hydrogen and helium ions, has been made by Whaling (1958) and Bichsel (1963). Recently, Andersen et al. (1967, 1968, 1969) have made precision measurements of the stopping power of a number of elements for 2.25–12 MeV protons with an accuracy of $\pm 0.3\%$. Accurate $S(E)$ values for ^{4}He ions are generally obtained indirectly by taking advantage of the fact that in the same stopping material, $S(E)$ for different particles is a function only of the effective charge and velocity of the moving particle. Thus, the stopping power of α particles is

$$S(E)_\alpha = 4S(E/3.972)_\text{proton} \tag{27}$$

This relation has been checked to within 1% for 28-MeV α particles and 7.05-MeV protons (Andersen *et al.*, 1967). At α-particle energies of 6–18 MeV, which is the important energy region for Coulomb excitation, the effective charge of the α particle differs from 2.0 by less than 0.1%. Therefore, the α-particle stopping powers based on the proton stopping power values are accurate to ±2%.

For numerical evaluation of the thick-target integrals, it is convenient to represent $S(E)$ for protons and α particles by a function of the form $aE^{(b+cE)}$, where the parameters a, b, and c are adjusted to fit existing $S(E)$ data. For example, the stopping power of Au for protons is

$$S(E) = 0.0651E^{(-0.4491-0.0168E)} \quad \text{MeV cm}^2/\text{mg}$$

for 0.8 MeV $\leqslant E \leqslant 6.0$ MeV, and for α particles it is

$$S(E) = 0.4733E^{(-0.42535-0.00318E)} \quad \text{MeV cm}^2/\text{mg}$$

for 3.2 MeV $\leqslant E \leqslant 20$ MeV. Stopping powers of other elements in the region of Au may be obtained on the assumption that $S(E)$ varies as $Z_2^{-1/2}$. This Z dependence can be deduced from the work of Lindhard and Scharff (1953).

It is generally more convenient to use a table of values for $S(E)$ when evaluating numerically the thick-target integrals for heavy ions. For this purpose, there is the extensive tabulation of Northcliffe and Schilling (1970) of $S(E)$ for ions with $1 \leqslant Z_1 \leqslant 103$ in 24 different material media. Since these tabulations by Northcliffe and Schilling (1970) are semi-empirical, it is difficult to estimate the accuracy of $S(E)$ for heavy ions. The uncertainty in $S(E)$ for heavy ions is the primary limitation to extracting accurate $B(E\lambda)$ from thick-target γ-ray yields. Accurate measurements of $S(E)$ for heavy ions in a variety of materials is clearly needed.

In the following sections, we discuss some of the more recent experimental approaches that are employed in the Coulomb excitation reaction and their interpretation in terms of a progressively refined theoretical description of the Coulomb excitation process. For the most part, these illustrative examples demonstrate the scope of nuclear structure information which results from γ-ray spectroscopy. In particular, the examples selected point up the fact that the Coulomb excitation reaction has provided much information on the knowledge of nuclear structure.

In Section IV, we give an interpretation of the level spectra and reduced electromagnetic transition probabilities from direct $E\lambda$ excitation of ^{232}Th in terms of the one-phonon octupole vibrational spectrum with associated rotational spectra of deformed nuclei. Deviations of the experimental results from this simple nuclear model can be reproduced by including the Coriolis interaction in this model. The $B(M1)$ reduced transition probabilities deduced from direct γ-ray spectroscopy resulting from Coulomb excitation of odd-A deformed nuclei with heavy ions are presented in Section V as an important test of the rotational model. In Section VI, the application of γ-ray coincidence spectroscopy to deduce nuclear structure information bearing on the collective vibrational model interpretation of spherical nuclei is discussed. Finally, in Section VII, the application of particle–γ coincidence spectroscopy to multiple Coulomb excitation of deformed nuclei together with the stringent tests of the rotational model provided by the experimental results is discussed. Also, a brief discussion of the Coulomb excitation mechanism which includes excitation of the projectile (multipole–monopole interaction) is given. Interest in E4 transition moments, which may result from the intrinsic shape of deformed nuclei, has stimulated new experiments. For these experiments, the cross sections must be measured to an accuracy of 1% or better. It is therefore necessary to assess the accuracy of the Coulomb excitation theory. Many of the theoretical models employed are discussed by Rasmussen, Chapter IX.B.

IV.   Direct E2 and E3 Coulomb Excitation with Light Ions

The Coulomb excitation reaction induced by light ions has a good built-in filter, i.e., it selectively excites 2^+ and 3^- states by direct E2 and E3 Coulomb excitation in even-A nuclei. From the analysis of γ-ray spectroscopy experiments, it is possible to extract fairly accurate values of $B(E\lambda)$ for excitation of the nuclear states. A compilation of the $B(E2, 0 \rightarrow 2)$ values for excitation of the first 2^+ state in even-A nuclei has been published by Stelson and Grodzins (1965). A reaction list (bibliographic information) for Coulomb excitation has been prepared from the journal literature for the period from 1956 through June 1971 by McGowan and Milner (1971).

The virtues of the γ-ray spectroscopy method are that, in addition to obtaining $B(E\lambda)$ for excitation of each state, one also obtains information on the spins of the excited states and the reduced transition probabilities for the different decay modes of these states. To illustrate these points, the results from a simple Coulomb excitation experiment, 18-MeV ^{4}He ions on a thick target of ^{232}Th, are presented and the nuclear structure information extracted from the results is discussed.

Twenty states in ^{232}Th have been observed from the Coulomb excitation reaction with 18-MeV ^{4}He ions and the 44 transitions from the decay of these states are displayed in Fig. 5. Eight 2^+ states and three 3^- states are

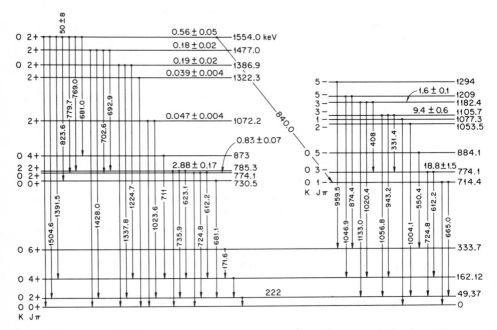

Fig. 5. Energy level diagram for Coulomb-excited states in ^{232}Th with ^{4}He ions. The $B(E2, 0 \to 2)$ values for the positive parity states and the $B(E3, 0 \to 3)$ values for the negative parity states are given in units of $B(E\lambda)_{sp}$ above the appropriate levels where $B(E\lambda, 0 \to J = \lambda)_{sp} = [(2\lambda + 1)/4\pi][3/\lambda + 3)]^2 (1.2)^{2\lambda} A^{2\lambda/3} e^2$ fm$^{2\lambda}$. For ^{232}Th, $B(E2, 0 \to 2)_{sp} = 4.23 \times 10^{-50}$ e^2 cm^4 and $B(E3, 0 \to 3)_{sp} = 2.24 \times 10^{-74}$ e^2 cm^6.

observed to be directly populated by either E2 or E3 Coulomb excitation. The remaining nine states either are excited by two-step Coulomb excitation or are populated by the γ-ray decay of the directly excited states.

For the conditions of this experiment, the strength parameter $\chi^{(2)}_{0 \to 2}$ which couples the ground state with the first 2^+ state is about 0.40 at the incident projectile energy. In the case of the higher $2'^+$ states, the parameter $\chi^{(2)}_{0 \to 2'}$ is 0.04 and smaller. Therefore, the use of first-order perturbation treatment of the excitation process for many of the states outside the ground-state rotational band is adequate. The main features of the results from such an analysis are correct. The computer program for multiple Coulomb excitation may be used to make estimates of effects higher than first order and to apply corrections to results deduced from the data analyzed with first-order perturbation theory.

Gamma-ray spectra were observed at $\theta_\gamma = 0°$, 55°, and 90° with respect to the beam direction with a 30-cm^3 Ge(Li) detector located 7 cm from the target. An example of a portion of a 4000-channel γ-ray spectrum is shown in Fig. 6. This spectrum required an accumulation time of 9 h of 100-nA

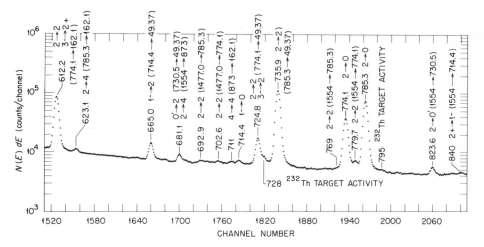

Fig. 6. Portion of a direct pulse-height spectrum of the γ rays at $\theta_\gamma = 0°$ for 18-MeV ^{4}He ions on a ^{232}Th target obtained with a 30-cm^3 Ge(Li) detector located 7 cm from target. Above each peak are given the energy of the γ ray in keV and the transition assignment.

^{4}He^{2+} ions on target. The assignment of J values to the states in Fig. 5 is based on γ-ray angular distribution measurements and on accurate γ-ray energies for the de-excitation γ rays. Since the positions of the 0$^+$, 2$^+$, 4$^+$ and 6$^+$ states in the ground-state rotational band are very accurately known (Schmorak et al., 1972), the energies of the γ rays from the decay of each state display a very definite signature in the pulse-height spectrum, e.g., note the 2$'$ → 0$^+$, 2$'$ → 2$^+$, and 2$'$ → 4$^+$ transitions from the decay of the 785.3-keV state.

Sample angular distributions of the γ rays expected for 18-MeV ^{4}He ions on ^{232}Th are given in Table 1. For $J_i = 2'$, the angular distributions of the

TABLE 1

SAMPLE ANGULAR DISTRIBUTIONS OF THE γ RAYS FOR
18-MeV ^{4}He IONS ON ^{232}Th

E (keV)	$J_i^{\pi} \rightarrow J_f^{\pi}$	$R = W(0°)/W(90°)$
785	$2'^+ \rightarrow 0^+$	1.364
	$2'^+ \rightarrow 4^+$	1.097
1106	$3^- \rightarrow 2^+$	0.777
	$3^- \rightarrow 4^+$	0.903

$2' \rightarrow 0$ and $2' \rightarrow 4$ transitions are unique. Similarly, for $J_i = 3^-$, the angular distribution of the $3^- \rightarrow 4^+$ transition is unique. Thus, definite J assignments for the Coulomb excited states are deduced from the angular distribution measurements of the γ rays. Also, if the anisotropy $R = W(\theta_\gamma = 0°)/W(\theta_\gamma = 90°)$ of a $J_i \rightarrow 2^+$ transition is greater than 0.9, the assignment $J_i = 3^-$ is excluded. Since the anisotropy R of a $2 \rightarrow 0$ γ-ray transition is unique, the accuracy of $\gamma(\theta)$ measurements is considerably improved by using $\gamma(\theta)$ of one of the $2 \rightarrow 0$ transitions for internal calibration of the effective solid angles of the detector during the collection of data.

The $3^- \rightarrow 2^+$ transition of 1056.3 keV from the decay of the 3^- state at 1105.7 keV is a relatively intense γ ray in the singles spectrum. As a result, the $\gamma(\theta)$ distribution measurement provides a sensitive test of the accuracy of the semiclassical particle parameter for interpretation of γ-ray angular distributions from direct E3 excitation. The accuracy of the thick-target particle parameter $a_2^{E3}(\eta_i \rightarrow \infty, \xi)$ is about $\pm 4\%$ from this experimental measurement with $\eta_i = 13.4$–18. This accuracy is certainly sufficient to obtain J assignments from direct E3 excitation.

The results from $\gamma(\theta)$ measurements for the γ rays from the decay of the 1554-keV state shown in Table 2 clearly demonstrate the assignment of $J_i = 2^+$. The interband branching ratio $B(E2, 2' \rightarrow 4)/B(E2, 2' \rightarrow 0)$ is indicative of a "β-vibrational-like" state with $K_i = 0$. The $B(E2)$ branching ratios are, within rather large errors, consistent with band mixing to first order, i.e., the coupling between vibrational and rotational motion (Nathan and Nilsson, 1965). The $B(E1, 2'^+ \rightarrow 1^-)$ value for the 840-keV transition is $4 \times 10^{-4} B(E1)_{sp}$.

The occurrence of numerous 2^+ states between 0.8 and 1.5 MeV is a striking feature of the Coulomb excitation reaction in ^{232}Th. For ^{232}Th, the $B(E2, 0 \rightarrow 2)$ in single-particle units range between 0.04 and 3 (see Fig. 5).

TABLE 2

THE 1554-keV STATE[a]

$R(823.6, J_i \rightarrow 0') = 1.51 \pm 0.08$	$J_i = 2'^+$
$R(840, 2' \rightarrow J_f) = 0.74 \pm 0.11$	$J_f = 1^-$
$R(1504, 2' \rightarrow 2) = 0.8 \pm 0.1$	$\delta = -7.9$
$B(E2, 2' \rightarrow 2)/B(E2, 2' \rightarrow 0) = 1.78 \pm 0.30$	$Z_0 = (17 \pm 15) \times 10^{-3}$
$B(E2, 2' \rightarrow 4)/B(E2, 2' \rightarrow 0) = 3.45 \pm 0.59$	$Z_0 = (7.6 \pm 4.5) \times 10^{-3}$
$R(779.7, 2 \rightarrow 2') = 1.22 \pm 0.04$	$\delta = 3.1$
$B(E2, 2 \rightarrow 0') = (50 \pm 8)B(E2)_{sp}$	
$B(E2, 2 \rightarrow 2')/B(E2, 2 \rightarrow 0') = 1.34 \pm 0.10$	
$B(E1, 2' \rightarrow 1^-) = 4 \times 10^{-4}B(E1)_{sp}$	

[a] $\delta = (E2/M1)^{1/2}$ and the phase convention corresponds to Biedenharn and Rose (1953). Z_0 is the first-order band-mixing parameter (Nathan and Nilsson, 1965). $B(E\lambda, 0 \rightarrow J = \lambda) = [(2\lambda + 1)/4\pi][3/(\lambda + 3)]^2(1.2)^{2\lambda} A^{2\lambda/3} e^2 \text{ fm}^{2\lambda}$.

These $B(E2)$ values include only excitations which decay by γ-ray transitions plus internal conversion. It is well known that an appreciable fraction of the excitations for several of these states decay by E0 radiation (Diamond and Stephens, 1967). The value of $B(E2, 2 \rightarrow 0') = 50B(E2)_{sp}$ between the states at 1554 and 730.5 keV is very large. The interpretation of a transition between two- and one-phonon vibrational states seems unlikely because the $B(E2, 0 \rightarrow 2)$ for excitation of the 1554- and 774-keV states are comparable (see Fig. 5). Another striking feature of these higher $2'^+$ states is that the values of $B(M1, 2' \rightarrow 2)$ cluster about 10^{-4} nm^2. The interband $B(E2)$ branching ratios for decay of the 785- and 774-keV states are not consistent with the interpretation of band mixing to first order.

The $\gamma(\theta)$ distributions for the γ rays from decay of the 774-keV state require the placement of a 3^- state as well as a 2^+ state at 774 keV. The assignment of 1^- to the state at 714.4 keV is based on $\gamma(\theta)$ distribution measurements. This level is populated primarily by the E2 transition cascading from the 3^- state at 774 keV. The $\gamma(\theta)$ of the 714- and 665-keV γ-ray transitions are consistent with the expected distributions for the 1–3 correlation (intermediate transition not observed) for the sequences $0(E3)3–(E2) J_z(E1)0$ and $0(E3)3–(E2) J_z(E1)2$ with $J_z = 1^-$. Direct E1 excitation of the 1^- state is not at all consistent with the γ-ray yields from Coulomb excitation with ^{4}He ions and protons. The assignment of 5^- to states shown in Fig. 5 are based primarily on the observance of γ-ray transitions to the 6^+ state of the ground-state rotational band. The excitation yield of the 5^- state at

884 keV is about 2% compared to the excitation yield of the 3$^-$ state at 774 keV. The yield of this 5$^-$ state is consistent with the excitation processes E3E2 via 3$^-$ state and E2E3 via 2$^+$ state. These excitation processes also account for approximately 20% of the population of the 1$^-$ state.

There has been considerable interest in the response of the nucleus to odd parity deformations, in particular the pear-shaped deformation $\beta_{30}Y_{30}$. The low-lying 1$^-$ states in the lighter actinide nuclei have very low energies (200–300 keV). These could possibly be interpreted as rotational states of a static octupole deformation. However, theoretical calculations (Vogel, 1968) of the sum of single-particle energies as a function of deformations β_{20}, β_{30}, and β_{40} indicate a stable equilibrium shape with the octupole deformation equal to zero. An alternative interpretation is that these states are members of the one-phonon octupole vibrational spectrum which contains four states $K = 0$, 1, 2, and 3 with associated rotational spectra.

This vibrational interpretation provides an interesting theoretical framework with which to compare the experimental information. In no case have all of the expected modes $K = 0$, 1, 2, and 3 of the octupole vibrational spectrum been observed for a given nucleus. Calculations of the $B(E3, 0 \rightarrow 3)$ values for the 3$^-$ members of the one-phonon octupole quadruplet by Neergard and Vogel (1970), which included the influence of Coriolis coupling between the negative parity bands, are in excellent agreement with the $B(E3, 0 \rightarrow 3)$ deduced from the experimental results for ^{232}Th shown in Fig. 5 in units of $B(E3)_{sp}$. The inclusion of the Coriolis interaction in the microscopic calculations is required to explain satisfactorily the distribution of the $B(E3)$ strength among the one-phonon octupole vibrational states. In fact, the fourth 3$^-$ state of the quadruplet, which was not observed in the Coulomb excitation experiment on ^{232}Th, is predicted to contain a very small amount of the $B(E3)$ strength, namely $0.6B(E3)_{sp}$.

A striking feature of the negative parity band with $K = 0$ is the especially large moment of inertia $(\hbar^2/2I = 6.0 \text{ keV})$ compared to that of the ground-state band $(\hbar^2/2I = 8.23 \text{ keV})$. This large difference in the moment of inertia would seem to imply the occurrence of a $K = 1^-$ band coupled to the $K = 0^-$ states by the Coriolis force. Thus, one should expect to observe the other members of the intrinsic octupole quadruplet with distorted rotational level spacings. The rotational level spacing for the negative parity states shown in Fig. 5 is distorted and implies that the Coriolis interaction has an important influence on the rotational spectrum.

The $B(E1)$ branching ratios for transitions from states of the octupole quadruplet to states of the ground-state rotational band should provide critical information concerning the Coriolis interaction. It is well known that the Coriolis interaction between states of K and $K \pm 1$ can have a profound influence on the moment of inertia and reduced transition probabilities (Kerman, 1956). Experimental evidence for the other members of the octupole quadruplet in a given nucleus would provide a good test of the interpretation of octupole vibrational states of deformed nuclei. The deviations of the $B(E1)$ branching ratios from the Alaga branching ratios (Alaga *et al.*, 1955) in Table 3 are very large. The relative $B(E1)$ values calculated from the Coriolis-coupled octupole state wave functions (Kocbach and Vogel, 1970; Neergard and Vogel, 1970) are also given in Table 3. The parameter Z is the ratio of the intrinsic E1 matrix elements between $K = 1^-$ and 0^+ and between $K = 0^-$ and 0^+ intrinsic wave functions (Kocbach and Vogel, 1970). The parameter Z has been adjusted to give the best fit to the $B(E1)$ branching ratios. This adjusted value for Z is in qualitative agreement

TABLE 3

$B(E1)$ Branching Ratios for Transitions between the Octupole Vibrational
States and the Ground-State Rotational Band of ^{232}Th

Transition[a] $J_i^{\pi}\alpha \to J_f^{\pi}K_f$	$B(E1, J_i \to J_i - 1)/B(E1, J_i \to J_i + 1)$		
	Vogel[b] $Z = 1.8$	Exp.	Alaga *et al.* (1955)
$1^-0 \to 0^+0$ $1^-0 \to 2^+0$	0.13	0.13 ± 0.01	0.50
$3^-0 \to 2^+0$ $3^-0 \to 4^+0$	0.025	0.045 ± 0.022	0.75
$5^-0 \to 4^+0$ $5^-0 \to 6^+0$	1×10^{-4}	$\leqslant 0.09$	0.83
$1^-1 \to 0^+0$ $1^-1 \to 2^+0$	2.9	$\leqslant 3$	2.0
$3^-1 \to 2^+0$ $3^-1 \to 4^+0$	3.5	4.0 ± 0.14	1.33
$5^-1 \to 4^+0$ $5^-1 \to 6^+0$	3.0	0.8 ± 0.3	1.2
$3^-2 \to 2^+0$ $3^-2 \to 4^+0$	2.6	—	Forbidden
$3^-3 \to 2^+0$ $3^-3 \to 4^+0$	2.1	3.7 ± 0.5	Forbidden

[a] The label α is chosen as the K corresponding to the largest component of the Coriolis-coupled octupole state wave function.

[b] Kocbach and Vogel, 1970; Neergard and Vogel, 1970.

with an estimate of the $|Z|$ obtained from the absolute $B(E1)$ extracted from the Coulomb excitation data for ^{232}Th. There the estimate of $|Z|$ is between 0.9 and 3.7.

In summary, the general features of the experimental $B(E1)$ branching ratios are reproduced by the calculations using the Coriolis-coupled wave functions of Neergard and Vogel (1970). This agreement between the calculated and experimental $B(E1)$ ratios and $B(E3, 0 \rightarrow 3)$ confirms the basic assumption about the vibrational structure of octupole states in ^{232}Th.

V. Direct Gamma-Ray Spectra from Coulomb Excitation with Heavy Ions

A. $B(M1)$ FOR ODD-A DEFORMED NUCLEI

The symmetric rotor model of the nucleus predicts that the relative $B(E2)$ and $B(M1)$ reduced transition probabilities within a rotational band in deformed odd-A nuclei are given in terms of angular momentum coupling coefficients. Extensive measurements of inelastically scattered projectiles following Coulomb excitation of the rotational states in deformed odd-A nuclei have verified these predictions for $B(E2, J_iK \rightarrow J_fK)$ (Olesen and Elbek, 1960; Hansen et al., 1961). More recently, the predictions of the relative $B(M1, J_iK \rightarrow J_fK)$ values have been verified by γ-ray spectroscopy measurements (Seaman et al., 1967; Boehm et al., 1966).

The reduced M1 transition probability within an odd-A rotational band is given by (Nathan and Nilsson, 1965)

$$B(M1, J_iK \rightarrow J_fK) = (3/4\pi) (g_K - g_R)^2 K^2 \langle J_i 1 K 0 \mid J_i 1 J_f K \rangle^2$$
$$\times [1 + \delta_{K, 1/2} b_0 (-1)^{J_> + 1/2}]^2 \quad \text{nm}^2 \qquad (28)$$

where g_K and g_R are gyromagnetic ratios, b_0 is an intrinsic parameter and only occurs for $K = \frac{1}{2}$, and $J_>$ denotes the larger of J_i and J_f. Values of $B(M1, J+2 \rightarrow J+1)$ can be obtained from an experimental measurement of the γ-ray branching ratio cascade/crossover, with the assumption that the E2 transition probabilities are given by the symmetric rotor model prediction, i.e., by the following expression:

$$B(M1, J+2 \rightarrow J+1) = 6.98 \times 10^{-1} \frac{(\Delta E_2)^5}{(\Delta E_1)^3} B(E2, J+2 \rightarrow J)$$
$$\times \left[\frac{\text{cascade}}{\text{crossover}} - \left(\frac{\Delta E_1}{\Delta E_2}\right)^5 \frac{B(E2, J+2 \rightarrow J+1)}{B(E2, J+2 \rightarrow J)} \right]$$
$$(29)$$

The $B(E2)$ in units of e^2b^2 are given by Eq. (18), where the intrinsic moment Q_{20} can be obtained from the measurements of the $B(E2)$ for excitation of the first excited state (Olesen and Elbek, 1960; Hansen et al., 1961). The energies ΔE_1 and ΔE_2 of the cascade and crossover transitions $J + 2 \rightarrow J + 1$ and $J + 2 \rightarrow J$ are in MeV. The M1 transition probability for the decay of the first excited state of the ground-state rotational band is determined from a knowledge of the E2/M1 transition probability ratio. Normally, this ratio can be obtained from a γ-ray angular distribution measurement following direct E2 excitation with light ions and low bombarding energy or from a γ–γ angular correlation following radioactive decay.

Levels in the ground-state rotational bands of odd-A rare earth nuclei have been populated by multiple Coulomb excitation with ^{16}O ions and the deexcitation γ rays were observed by direct γ-ray spectroscopy with calibrated high-resolution Ge(Li) detectors. Since only the branching ratio cascade/crossover is extracted from the γ-ray spectra, a detailed knowledge of the Coulomb excitation mechanism is not required. Therefore, the use of targets thick with respect to the range of the projectile is convenient for these measurements. The ground-state rotational band transitions are easily identified because of their characteristic energy sequence and strong intensities. Four to six excited states of the ground-state rotational band were observed in measurements by Boehm et al. (1966) and Seaman et al. (1967). The values of $B(M1, J_i \rightarrow J_f)$ and $|g_K - g_R|$ deduced from the measurements for each nucleus provide a test of the symmetric rotor model, i.e., the degree to which $|g_K - g_R|$ is constant within a rotational band. Within the error limits of 3–5%, the $|g_K - g_R|$ are constant for each nucleus as predicted by Eq. (28).

Boehm et al. (1966) combined the available experimental results on $B(M1, J_i \rightarrow J_f)$ with the ground-state magnetic moment to obtain the gyromagnetic ratios g_K and g_R associated with the intrinsic motion and with the collective motion, respectively. The general features of the experimental results are in reasonable agreement with theoretical calculations by Nilsson and Prior (1961).

B.   DOPPLER-BROADENED γ-RAY LINE SHAPE

We have emphasized the great advantage offered by the superior energy resolution of Ge(Li) detectors for measuring γ rays from Coulomb excitation. However, when heavy ions are used for Coulomb excitation, the increased linear momentum imparted to the excited nucleus can result in appreciable Doppler shifts of the energy of the emitted γ-ray and may produce a broadened peak: In fact, the broadening can be much larger than the resolution width of the Ge(Li) detector. Thus, as one proceeds to heavier projectiles, the Doppler broadening tends to offset the great gain in resolution offered by the Ge(Li) detector. Figure 7 shows four γ-ray peaks observed when a ^{93}Nb target was bombarded with 36-MeV ^{16}O ions. There is obvious Doppler broadening of the 744- and 979-keV peaks. Actually, all four peaks

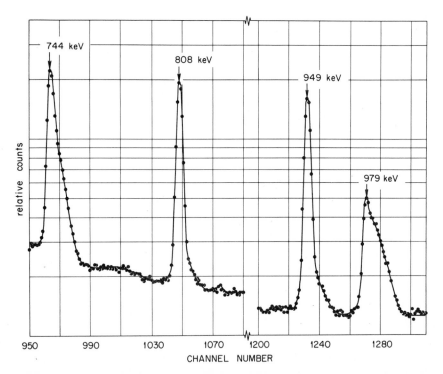

Fig. 7. Portions of the pulse-height spectrum of the γ rays resulting from 36-MeV ^{16}O ions on a thick ^{93}Nb target. The Ge(Li) detector was located at 0° to the beam. The Doppler broadening of the peaks is evident.

contain some Doppler broadening. It turns out that the shape of the Doppler-broadened peak can lead to valuable information on the total mean life of the γ-emitting state. The method is useful for mean lives in the range of a few psec (barely detectable broadening) down to about 10^{-2} psec (fully broadened peak).

Clearly, the γ-ray peak shape depends on many quantities besides the lifetime. It depends on the incident projectile mass and energy, the mass of the target nucleus, the gamma-ray energy, the angular distribution of the excited recoiling nuclei, the angular distribution of the γ rays emitted by the recoiling nuclei, the stopping power of the target material for the recoiling nuclei, and the size, energy resolution, and location of the Ge(Li) detector. Fortunately, the theory of Coulomb excitation tells us the angular distribution of the recoiling nuclei. We can also use a thick target because we can calculate the variation of the cross section with decreasing energy of the incident projectile.

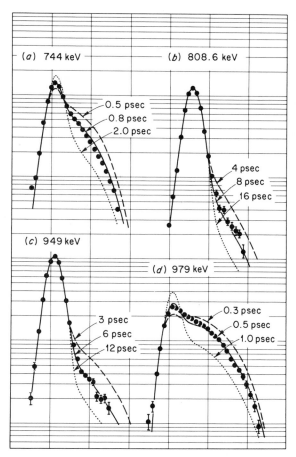

Fig. 8. Detailed comparison of the observed Doppler-broadened shapes of the peaks shown in Fig. 7 with calculated shapes for different assumed values for the total mean life of the Coulomb-excited states.

The central problem in calculating the Doppler-broadened peak shape is the lack of detailed knowledge about the slowing down process of the recoiling nucleus, especially near the end of its range. When 36-MeV ^{16}O ions strike ^{93}Nb, the maximum kinetic energy imparted to the ^{93}Nb nucleus is 18 MeV. The maximum energy shift for a 1-MeV γ ray is about 20 keV. Since the Doppler shift is proportional to velocity, it is important to know the stopping power at quite low ion energies such as 0.1 MeV.

Figure 8 shows in greater detail the observed shapes of the γ-ray peaks from ^{16}O Coulomb excitation of ^{93}Nb with the Ge(Li) detector placed at 0° to the beam direction. We also show calculated peak shapes for different assumed mean lives for the γ-emitting states.

VI. Gamma-Ray Coincidence Spectroscopy

A. DIRECT E2 AND E3 AND DOUBLE E2 EXCITATION OF EVEN-A
MEDIUM-WEIGHT NUCLEI

Nearly two decades ago, Scharff-Goldhaber and Weneser (1955) pointed out that certain regularities exhibited by the low-lying levels of even-A medium-weight nuclei, such as systematic trends in the ratio of level positions, spins of the second excited states, and the relative transition probabilities, could be described by a collective vibrational model of the nucleus. Within the model framework of quadrupole vibrations, the electromagnetic transitions are pure E2 transitions. This model has enjoyed some success by explaining qualitatively the large E2 transition probabilities of the γ rays from the first 2^+ states and by predicting that states with spins 0^+, 2^+, and 4^+ occur at approximately twice the energy of the first 2^+ states. However, more recent experimental evidence has revealed some disturbing flaws of this description in its simplest form. For instance, the ratios $B(E2, 2' \rightarrow 2)/B(E2, 2 \rightarrow 0)$ and $B(E2, 0' \rightarrow 2)/B(E2, 2 \rightarrow 0)$ for even-A ($A = 98$–116) nuclei are between 0.5 and 1.0 instead of 2.0 as predicted by the model (McGowan et al., 1965, 1968; Robinson et al., 1969b; Milner et al., 1969). (Here, 2' and 0' designate the second 2^+ and 0^+ states, respectively.) A more serious flaw in the model has been the observation of large static electric quadrupole moments for the first 2^+ state in many of these nuclei (see Häusser, Chapter VII.B).

Although considerable effort has been made to test the vibrational model to determine if the $0'^+$, $2'^+$, and 4^+ states have characteristics predicted for two-quadrupole phonon states, many of the E2 transition probabilities for decay of these states are still unknown. Direct E2 and double E2 excitation are excellent tools for investigation of these E2 transition probabilities. In general, the direct observation of the gamma rays from decay of these states in the singles spectrum is limited by the chemical purity of the targets. This was especially true before the era of Ge(Li) detectors. Nuclear reactions with these impurities produce γ rays whose Compton distributions in the pulse-height spectra mask the γ's from Coulomb excitation. This limitation in detecting weakly excited states can be overcome to a certain extent by a coincidence measurement on the cascade γ rays.

By the use of Ge(Li) γ-ray detectors with their inherent higher-energy resolution and electrodeposited targets of the isotopes of Cd and Pd, it is possible to observe in the direct singles spectra the weak excitation of the two-phonon-like states. These detectors make it possible to resolve close-lying states which occur frequently for the two-phonon-like states, to determine the branching ratio of cascade to crossover γ rays from the second 2^+ states, and to measure the angular distributions of the transitions from the decay of the 2^+ states (Robinson et al., 1969b; Milner et al., 1969).

A summary of the Coulomb-excited states and the reduced E2 transition probabilities for the even-A Cd nuclei is given in Fig. 9. Although the reduced transition probabilities for decay of the two-phonon-like states in the even-A nuclei of Ru, Pd, and Cd have the general behavior predicted by the vibrational model, they do not agree in detail. The ratios $B(\text{E2}, 4 \to 2)/B(\text{E2}, 2 \to 0)$ are between the asymmetric rotor model prediction of 1.37 and the phonon model prediction of 2.0. The $B(\text{E2}, 0 \to 2')$ values exhibit some uniformity, being approximately one single-particle value. The $B(\text{M1}, 2' \to 2)$ values scatter between 10^{-4} and 10^{-2} nm^2. For the even-A Cd nuclei the $B(\text{M1}, 2' \to 2)$ are not small, clustering around 10^{-2} nm^2 and represent a significant deviation from the vibrational model prediction.

The systematic occurrence of collective octupole states in spherical nuclei is now generally acknowledged; the evidence for collective 3^- states has mostly come from direct nuclear reaction experiments. A state at approximately 2 MeV was Coulomb-excited with ^{16}O ions for most of the even-A Ru, Pd, and Cd nuclei. These states decay predominantly by a cascade transition to the first 2^+ states. The states were so weakly excited that they could only be measured by taking advantage of the increased sensitivity offered by the γ–γ coincidence technique. If these states are interpreted to be the result of direct E3 Coulomb excitation, the $B(\text{E3}, 0 \to 3)$ values range between 13 and 25 times $B(\text{E3})_{sp}$. The rms octupole deformation parameters β_3 obtained from the $B(\text{E3}, 0 \to 3)$ values are between 0.12 and 0.17.

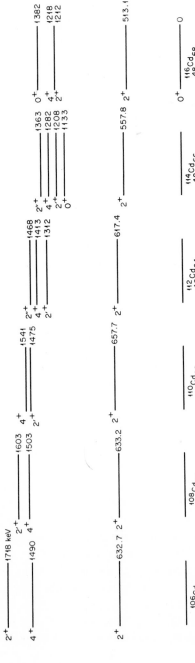

Fig. 9. Level diagram of the states observed by Coulomb excitation of the even-A Cd nuclei and a summary of the reduced E2 transition probabilities (from Milner *et al.*, 1969). Used with permission of North-Holland Publishing Co., Amsterdam.

B. POLARIZATION–DIRECTION CORRELATION AND γ–$\gamma(\theta)$ CORRELATIONS FOLLOWING COULOMB EXCITATION

Measurements of the angular distributions of γ rays following Coulomb excitation has provided a great deal of information on spins of the excited states and on the E2/M1 γ-ray mixing ratios. However, cases have been encountered in which the angular distributions are consistent with two rather different values of $\delta = (E2/M1)^{1/2}$ or with more than one spin assignment. Measurements of the polarization–direction correlation and of the triple correlation arising from the reaction induced by the incident beam followed by a cascade of two γ rays have resolved this ambiguity either in the value of δ or in the spin of the excited state (McGowan and Stelson, 1958; Robinson et al., 1969a).

Coulomb excitation of the first 2^+ state in even-A nuclei produces a convenient source of γ rays with known linear polarization with respect to the incident beam of ions. Biedenharn and Rose (1953) have expressed in a convenient form the polarization–direction correlation $W(\theta, \varphi)$, where φ is the angle between the direction of polarization and the normal to the plane defined by the directions of propagation of the two γ rays in cascade. For γ rays following Coulomb excitation, one replaces A_ν appearing in the correlation functions by $A_\nu a_\nu^{E2}$. A polarimeter based on the Compton scattering mechanism and its effectiveness for 200- to 800-keV γ rays has been described by McGowan and Stelson (1958). Since the beam defines one of the directions in the polarization–direction correlation, only a double coincidence is required in the measurements. The ratio of the polarization intensities $P(\theta) = W(\theta, \pi/2)/W(\theta, 0)$ is large for the $2 \rightarrow 0$ transitions in even-A nuclei, e.g., $P(\pi/2)$ is about 2.5. For mixed M1–E2 transitions, such as are generally observed in odd-A nuclei, measurements of the polarization–direction correlation have resolved ambiguities resulting from γ-ray angular distribution measurements. For example, if the angular distribution is isotropic, the polarization–direction correlation is not necessarily isotropic, i.e., $P(\theta) \neq 1.0$. Such a case was encountered in Coulomb excitation measurements of ^{113}Cd, where the angular distribution of the 300-keV γ ray was nearly isotropic but the polarization–direction correlation was quite large, $P(\theta = \pi/2) = 0.61 \pm 0.03$ (McGowan and Stelson, 1958).

More recently, Robinson et al. (1969a) have removed ambiguities resulting from Coulomb excitation of ^{107,109}Ag, ^{110}Pd, ^{117}Sn, and ^{190,192}Os by measuring the triple correlation. The angle functions for the general triple correlation are products of spherical harmonics (Ferguson, Chapter VII.G;

Ferguson, 1965). By choosing the beam direction along the positive direction of the z axis, the spherical harmonics can be written in terms of the associated Legendre polynomials. In the experiments by Robinson *et al.* (1969a), two "geometries" were chosen to further simplify the angle function in terms of a sum of Legendre polynomials. One of the γ-ray detectors was fixed at an angle of either $0°$ or $90°$ with respect to the incident beam direction and the other γ-ray detector, moving in a plane defined by the fixed detector and the incident beam, was located at an angle θ with respect to the beam. In addition, it is a trivial matter to interchange experimentally the roles of the two detectors with regard to which γ ray of the cascade is detected.

VII.   Particle–Gamma Coincidence Spectroscopy

A.   Multiple Coulomb Excitation of Deformed Nuclei

The level energies of the first few rotational states of even-A rare earth nuclei are reasonably well described by $E(J) = AJ(J + 1) - BJ^2(J + 1)^2$, where $B/A \cong 10^{-3}$ for nuclei in the middle of the rare earth region and $B/A \cong 10^{-2}$ near the edge of this region. More recently, the variable-moment-of-inertia (VMI) model has been proposed which gives an excellent fit to the level energies (Mariscotti *et al.*, 1969). An interesting question is how best to view these observed departures from the $J(J + 1)$ spacings in terms of changes in the intrinsic structure of the nucleus. Several possible mechanisms have been suggested, but at present the situation is unclear which of these are responsible: (1) At high angular momenta, one might expect a departure of the moment of inertia and of the $B(E2)$ values from the rigid-rotor prediction, due to centrifugal stretching. (2) The Coriolis force operating in a rotating nucleus tends to decouple the pairing correlations (reduce the gap) and this effect could account for the observed energy level spacings (Mottelson and Valatin, 1960).

The departures of the $B(E2)$ values within a rotational band from those for a rigid rotor, implied by the analysis of level positions in terms of a centrifugal stretching hypothesis, are not large. For example, a good rotor, such as ^{166}Er, might exhibit a 1% departure from the rigid-rotor prediction for the ratio $B(E2, 4 \rightarrow 2)/B(E2, 2 \rightarrow 0)$. As one proceeds up the rotational band, the possible departures become larger. For the ratio $B(E2, 10 \rightarrow 8)/B(E2, 2 \rightarrow 0)$, one might expect a 10% deviation from the rigid-rotor value, which is within the realm of experimental detection.

Considerable effort has been made to improve the experimental accuracy of $B(E2)$ values in the ground-state rotational band obtained from multiple

Coulomb excitation with ^{16}O and ^{32}S ions on several even-A rare earth nuclei (Fraser *et al.*, 1969b; Sayer *et al.*, 1970). The high-resolution Ge(Li) γ-ray detectors have played a significant role in obtaining the increased accuracy. Most of the experimental work involved coincidence measurements between the γ rays and the exciting projectiles backscattered into an annular silicon surface barrier detector. Normally, thick targets are used in these experiments and the actual target thickness for the Coulomb excitation measurements is determined by the energy cutoff in the annular detector. Therefore, a good knowledge of the energy response of the annular detector is required for ^{16}O and ^{32}S ions. This may be done either by variation of the beam energy incident on the target or by using targets of different atomic weight for a fixed incident beam energy. A typical target thickness for ^{16}O ions in the experiments by Sayer *et al.* (1970) was 10 MeV. The experimental Coulomb excitation probabilities $P(J)_{\text{exp}}$ for state J are compared to those expected for a rigid rotor. The latter excitation probabilities $P(J)_{\text{calc}} \equiv \langle d\sigma(J)\rangle/\langle d\sigma_R\rangle$ are calculated with the Winther–de Boer computer program using rigid-rotor model matrix elements given by Eq. (18). The brackets imply an integration of $d\sigma(J)$ and $d\sigma_R$, weighted by the reciprocal of the stopping power $S(E)$, over the projectile energy E and scattering angle θ.

For ^{152}Sm, the $B(\text{E2}, J \rightarrow J - 2)$ values for $J = 6$, 8, and 10 are larger than the rigid-rotor values (Fraser *et al.*, 1969b; Sayer *et al.*, 1970). It is instructive to interpret these deviations in terms of a band mixing model or centrifugal stretching. The effect of the band mixing on the $B(\text{E2}, J_i \rightarrow J_f)$ values is that the intrinsic quadrupole moment Q_{20} is no longer a constant but is slightly dependent on J_i and J_f. For this specific mechanism, the $B(\text{E2}, J_i \rightarrow J_f)$ are of the form (Symons and Douglas, 1967)

$$B(\text{E2}, J_i \rightarrow J_f) = B_0(\text{E2}, J_i \rightarrow J_f)\{1 + \tfrac{1}{2}\alpha[J_i(J_i + 1) + J_f(J_f + 1)]\}^2 \quad (30)$$

for $J_i \neq J_f$, where $B_0(\text{E2}, J_i \rightarrow J_f)$ is the rigid-rotor value. For ^{152}Sm, Fraser *et al.* (1969b) obtained $\alpha = (3.4 \pm 1.0) \times 10^{-3}$ and Sayer *et al.* (1970) obtained $\alpha = (2.7 \pm 1.7) \times 10^{-3}$. These results are in agreement with $\alpha = (2.2 \pm 0.7) \times 10^{-3}$ deduced from rather accurate lifetimes measured by the recoil distance method following Coulomb excitation of ^{152}Sm with ^{40}Ar ions (Diamond *et al.*, 1971). Within the framework of the classical centrifugal stretching model (Diamond *et al.*, 1964), the stretching parameter α is equal to B/A. The deviations in the ground-state band energy-level spacings would require $\alpha = (6.7 \pm 1.1) \times 10^{-3}$ for ^{152}Sm, which is about three times larger than the value deduced from the $B(\text{E2}, J_i \rightarrow J_f)$ measurements.

A similar analysis of the $B(E2, J_i \rightarrow J_f)$ values (Fraser *et al.*, 1969b) for the ground-state rotational band of [154]Sm indicates that α is not significantly different from zero. For [166]Er, [172]Yb, [174]Yb, and [176]Yb, Sayer *et al.* (1970) derived, from $B(E2)$ ratios in the ground-state rotational bands, values of α which were slightly negative. These $B(E2)$ ratios were based on a semi-classical treatment of multiple Coulomb excitation. Corrections to the semi-classical treatment were not available for the analysis of these data. Although extensive corrections are still not available for higher-order Coulomb excitation processes (higher than second order), it appears that such corrections for excitation of the 6^+ and 8^+ states could account for the departures observed by Sayer *et al.* (1970). Alder (1970) has given the corrections to the semiclassical Coulomb excitation process by an exact coupled channel calculation for the 0^+, 2^+, 4^+, and 6^+ rotational bands and for $\eta_i = 12$. The corrections are 4% for the 4^+ state and 25% for the 6^+ state. There is a distinct need for more work on the corrections to the semiclassical multiple Coulomb excitation process in order to make stringent tests of the rigid-rotor model, e.g., constancy of the intrinsic quadrupole moment Q_{20} within a rotational band.

B. PARTICLE–$\gamma(\theta)$

As already mentioned in Section VI.B, ambiguities are frequently encountered in the determination of $\delta = (E2/M1)^{1/2}$ from angular distribution measurements of γ rays following direct E2 Coulomb excitation of higher 2^+ states because the particle parameter $a_4^{E2}(\eta_i, \xi)$ is almost always small and does not provide any useful information. The particle–$\gamma(\theta)$ angular correlation can usually be decisive in removing this ambiguity. The particle parameter $a_4^{E2}(\eta_i, \xi, \theta_p)$ for projectiles inelastically scattered at an angle θ_p near 180° is large, being about -1.5. As a result, the angular correlation coefficients $A_\nu a_\nu^{E2}$ are relatively large from a particle–$\gamma(\theta)$ measurement for the γ rays from a $2'^+ \rightarrow 2^+$ transition.

Casten *et al.* (1969) have measured the particle–$\gamma(\theta)$ angular correlations for $2 \rightarrow 0$, $2' \rightarrow 2$, $4' \rightarrow 2$, and $4' \rightarrow 4$ transitions from Coulomb excitation of [186-192]Os with [16]O ions. Similar measurements for [152]Sm have been done by Fraser *et al.* (1969a) using [32]S ions. The $\delta = (E2/M1)^{1/2}$ deduced from the measurements for the $2' \rightarrow 2$ transitions in [190,192]Os are in accord with results deduced from triple angular correlation measurements by Robinson *et al.* (1969a) and from angular distribution measurements by Milner *et al.* (1971). The results for δ in [190,192]Os and in [182,184,186]W by Milner *et al.* (1971) have provided a rather sensitive test of the nuclear wave functions from the pairing-plus-quadrupole model of Kumar and Baranger (1968) and Kumar (1969). The sign and magnitude of the $(E2/M1)^{1/2}$ matrix element ratio are determined from the angular correlation and angular distribution measurements. A comparison between theory and experiment can be made concerning the sign of the ratio of matrix elements because each nuclear state appears an even number of times and therefore is independent of phase conventions used in defining the wave functions. The calculated values of δ by Kumar (1969) have the experimental sign in five of eight cases in Os and W nuclei. The disagreements occur in the W nuclei where two $2' \rightarrow 2$ transitions are observed in each even-A nucleus. It is not clear which experimental and calculated values should be compared (Milner et al., 1971).

C. COULOMB EXCITATION OF THE PROJECTILE

The results from Coulomb excitation experiments which have been discussed up to this point have been analyzed on the basis of the monopole–multipole interaction, i.e., the interaction of the projectile monopole with multipoles of the target. The use of projectiles such as the proton, ^{4}He, ^{12}C, and ^{16}O which are structureless or are not susceptible to Coulomb excitation fit this description. With projectiles such as ^{19}F, ^{20,22}Ne, ^{28}Si, ^{32}S, ^{35}Cl, or ^{36,40}Ar, the multipole–monopole interaction can give rise to excitation of the projectile. This description corresponds to the interaction of the projectile multipoles with the target monopole. The interaction strength $\chi^{(\lambda)}_{i \rightarrow f}$ for excitation of the projectile is obtained from Eq. (15) by replacing the projectile charge number Z_1 by the charge number of the target Z_2 and $\mathcal{M}(E\lambda)$ becomes the reduced matrix element for the projectile. The parameter ξ in Eq. (11) corresponds to an energy loss which is the excitation energy in the projectile. Calculations for Coulomb excitation of the projectile on the basis of the multipole–monopole interaction can be obtained from the existing computer code of Winther and de Boer (1966) by exchanging the roles of target and projectile.

A reformulation of the Coulomb excitation mechanism which includes excitations in both target and projectile demonstrates these modes more clearly (Eisenstein and Smilansky, 1970; Häusser and Cusson, 1970). Alder and Winther (1969) have given in a convenient form the interaction potential between two nonoverlapping charge distributions. An expansion for the interaction is obtained which is a sum of tensor products constructed from multipole operators for target and projectile. This sum contains interaction terms of the form: monopole–monopole + monopole–multipole + multipole–monopole + multipole–multipole. The monopole–monopole interaction is responsible for the relative motion (scattering) of the two centers of charge and does not contribute to the excitation. The next two terms describe the cases where either the target or projectile (but not both) is excited. The multipole–multipole interaction is responsible for simultaneous excitation of both target and projectile. All of these interactions contribute to the Coulomb excitation cross section in the usual first-order treatments.

In general, the multipole–multipole interaction contribution is small compared to first-order contributions from the monopole–multipole or multipole–monopole interaction and can be neglected. For instance, a calculation by Häusser and Cusson (1970) in first-order perturbation theory for the E2–E2 interaction (simultaneous E2 excitation of both target and projectile) shows that this process is typically 10^{-5} compared to the single excitation cross section Eq. (7). The smallness is due to the E4 orbital integrals which appear in the cross section for the E2–E2 process.

By second-order processes, the monopole–multipole and multipole–monopole interactions can give rise to excitation of both target and projectile by sequential excitation along the semiclassical orbit at different times. This is analogous to the usual multiple Coulomb excitation process within the target nucleus. Häusser and Cusson (1970) have modified the computer code of Winther and de Boer (1966) to solve the coupled differential equations for the composite system of target and projectile describing the sequential excitation. A sample calculation for 68.5-MeV ^{28}Si ions on ^{62}Ni scattered at $\theta_{CM} = 180°$ shows that the influence of target excitation on the cross section for projectile excitation can amount to several percent (Häusser and Cusson, 1970).

Coulomb excitation of the odd-A projectile has been observed by bombarding thick targets of ^{24}Mg and ^{12}C in the case of E2 excitation of ^{35}Cl and targets of ^{28}Si in the case of E2 and E3 excitation of ^{19}F (Häusser et al., 1969b; Alexander et al., 1969). The thick-target yields of γ rays observed with Ge(Li) detectors were analyzed by first-order perturbation theory for Coulomb excitation to deduce the $B(E\lambda)$. Higher-order excitations in projectile excitation can be relatively more important compared to those in target excitation. The perturbation approximation of the reorientation effect $d\sigma^{(1,2)}/d\sigma^{(1)}$ from Eq. (16) is proportional to A_1/Z_2 for target excitation and A_1/Z_1 for projectile excitation. With other factors being equal, the reorientation effect in the projectile is enhanced over that in the target by a factor Z_2/Z_1. For 57-MeV ^{35}Cl ions on ^{24}Mg, the influence of the reorientation effect in ^{35}Cl due to the static electric quadrupole moment of the $\frac{3}{2}$ ground state is $d\sigma^{(1,2)}/d\sigma^{(1)} = 4.4\%$ in the excitation of the $\frac{5}{2}$ state at 1.762 MeV.

Most of the experiments oriented to Coulomb excitation of the even-A projectiles have been designed to measure both the $B(E2, 0 \rightarrow 2)$ values and the static electric quadrupole moment of the 2^+ state. At Berkeley, the γ-ray yields from excited projectile nuclei (^{20,22}Ne, ^{28}Si, ^{32}S, and ^{36,40}Ar) and target nuclei (^{120}Sn, ^{130}Te, ^{148}Sm, and ^{206}Pb) at two projectile scattering angles $\theta_L = 160°$ and $90°$ were measured simultaneously as particle–γ coincidences between a NaI detector and surface barrier detectors (Nakai et al., 1970a, b, 1971). The $B(E2, 0 \rightarrow 2)$ and $Q(2^+)$ values for the projectile are extracted from the ratio of the excitation probability of the projectile to that of the target at each angle using known values of $B(E2, 0 \rightarrow 2)$ for the target nuclei. The $B(E2, 0 \rightarrow 2)$ values for the projectile have been determined to an accuracy of 12–15%.

Häusser et al. (1969a) have also exploited the dependence of the Coulomb excitation cross section on the $Q(2^+)$ value in the case of projectile excitation. By a suitable choice of target, it is possible to measure the inelastic scattering cross section at two widely separated θ_{CM} angles in a single surface barrier located at θ_L in which both the scattered projectiles and the target recoil ions are observed. The particle–γ coincidence technique is used to eliminate the elastic particle groups. The γ rays are detected with an array of several NaI detectors which simultaneously measure the γ-ray angular distribution. With this method, Coulomb excitation of the projectile ^{28}Si was observed by bombarding thin targets of ^{62}Ni (Häusser et al., 1969a). The analysis of the data was performed with the modified computer code (Häusser and Cusson, 1970) of Winther and de Boer. The virtual excitation of either target or projectile by the sequential excitation process gives rise to interference terms which also change the cross sections. The neglect of these terms would amount to about 10% of the effect due to the reorientation caused by the $Q(2^+)$ value of the projectile. Analyzing the data to extract $Q(2^+)$ is relatively insensitive to the $B(E2, 0 \rightarrow 2)$ values for ^{28}Si.

D. COULOMB EXCITATION WITH PROJECTILE ENERGIES EXCEEDING THE COULOMB BARRIER

Coulomb excitation of target nuclei has also been investigated by particle–γ spectroscopy where the scattered projectiles are detected at forward angles (Afonin et al., 1968). Since small scattering angles correspond to large impact parameters [see Eq. (5)], one can use projectile energies exceeding the Coulomb barrier energy. This makes possible the Coulomb excitation of

high-lying states in light nuclei. The relation between the selected forward
scattering angle and maximum permissible projectile energy can be established as follows by Eq. (5): For the condition that S should be at least
5–6 fm for a head-on collision, we have in the case of elastic scattering

$$E_{max} = (E_S/2) \left[1 + \csc(\theta_{CM}/2) \right] \qquad (31)$$

where E_S is the maximum safe bombarding energy for a head-on collision.

Afonin et al. (1968) have measured $B(E2)$ values from Coulomb excitation
of states in target nuclei between ^{19}F and ^{54}Fe with 37-MeV ^{12}C ions by
the method of particle–γ coincidence spectroscopy. The mean scattering angle
θ_L of the ^{12}C ions ranged from 20° for a ^{23}Na target to 40° for a ^{54}Fe target.
Within the accuracy of $\pm 20\%$ for the $B(E2)$ values, there is fair agreement
with values measured by more conventional methods.

E.   DIRECT E4 COULOMB EXCITATION

Probably one of the most interesting developments in the Coulomb
excitation reaction has been the observation of large E4 transition moments
in the excitation of 4^+ rotational states of deformed nuclei (Stephens et al.,
1970, 1971; McGowan et al., 1971). This E4 moment is of particular interest
because it may result from the intrinsic shape of the deformed nuclei.
Nilsson et al. (1969) have included a hexadecapole deformation in the nuclear
potential for the rare earth and actinide nuclei and have extended their calculations, based on a modified harmonic oscillator potential, into the region of
proposed "superheavy" nuclei. It is quite apparent from these theoretical
predictions that the presence of β_4 distortions exerts an appreciable influence on nuclear structure properties and in particular the inertial mass
parameter B in a fissioning nucleus (Sobiczewski et al., 1969). These should
be properly accounted for in calculations for the actinide nuclei before
extrapolation into the region of "superheavy" elements.

The experiments of Stephens et al. (1970, 1971) involved an accurate
measurement of the excitation probability for the 4^+ states in $^{152,154}Sm$
relative to the excitation probability of the 2^+ states following Coulomb
excitation with 10–12-MeV ^4He ions. The excitation probabilities were
extracted from singles γ-ray spectra and from particle–γ coincidence spectra.
The E2 excitation probability for the 2^+ state is obtained from known values
of $B(E2, 0 \rightarrow 2)$ and $B(E2, 2 \rightarrow 4)$. For actinide nuclei, the elastically and
inelastically scattered ^4He ions from 20 to 30 $\mu g/cm^2$ targets at an incident
energy of 17 MeV were observed in an Enge split-pole magnetic spectrometer equipped with a position-sensitive gas proportional counter. The

excitation probabilities for the 2^+ and 4^+ states are determined relative to the elastic scattering from the measured peak areas (McGowan et al., 1971). One principal advantage of this method over the detection of the γ ray is that a knowledge of the number of cascade transitions from higher-lying states to the states of interest is not required.

The experimental excitation probabilities for the 4^+ states are 10–25% larger than the values computed with the aid of the Winther–de Boer computer program using only E2 matrix elements. The excess excitation of the 4^+ state is attributed to the presence of E4 Coulomb excitation. The influence of the hexadecapole moments on the Coulomb excitation cross section for the 4^+ state in ^{230}Th is displayed in Fig. 10 as a function of the reduced E4 matrix element $\langle 4 \| \mathcal{M}(\text{E4}) \| 0 \rangle$. These calculations were done in the limit of the rigid-rotor model with states 0^+, 2^+, 4^+, and 6^+, where the six reduced E2 matrix elements and the seven reduced E4 matrix elements are given by Eq. (18). The intrinsic quadrupole moment Q_{20} for a prolate shape was obtained from the experimental value for $B(\text{E2}, 0 \rightarrow 2)$. The use of reduced E4 matrix elements based on the rigid-rotor model seems justified because the $B(\text{E4}, 0 \rightarrow 4)$ value for ^{230}Th is very large, being $100B(\text{E4})_{\text{sp}}$.

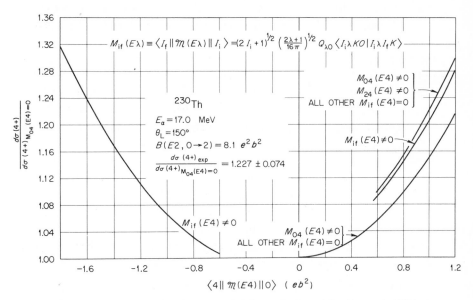

Fig. 10.   Ratio of the calculated cross section for excitation of the 4^+ state in ^{230}Th, including E4 excitation, to the cross section without any E4 contribution, as a function of $\langle 4 \| \mathcal{M}(\text{E4}) \| 0 \rangle$.

Several other effects which could influence the calculated excitation probabilities of the 2^+ and 4^+ states have been considered in the analyses by Stephens *et al.* (1971) and by McGowan *et al.* (1971): (1) quantum mechanical corrections to the excitation probabilities which are obtained from the semi-classical treatment of multiple Coulomb excitation; (2) deviation of the reduced E2 matrix elements from the rigid-rotor limit; and (3) influence of higher 2^+ states on the excitation probabilities of the 2^+ and 4^+ states in the ground-state rotational band. The quantal corrections to the E2 processes are significant for the determination of the reduced E4 matrix element $\langle 4\|\mathcal{M}(\mathrm{E}4)\|0\rangle$. To illustrate this point, the quantum mechanical corrections of the Coulomb excitation process in second-order perturbation theory by Alder *et al.* (1972) are presented for ^{230}Th. The use of quantum corrections in second-order perturbation theory for the E2 excitation processes is justified because 98% of the excitation of the 4^+ state would be pure double E2 excitation under the condition $\langle 4\|\mathcal{M}(\mathrm{E}4)\|0\rangle = 0$. The quantal correction reduces the excitation probability for the 4^+ state by the pure double E2 process by 3% for ^{230}Th. Omitting this quantal correction would decrease the measured value of $\langle 4\|\mathcal{M}(\mathrm{E}4)\|0\rangle = (1.05 \pm 0.21)$ eb^2 by 10% for ^{230}Th. Finally, the quantal correction for the case of a pure double E2 excitation and a simultaneous direct E4 excitation in first order is $+0.7\%$ for excitation of the 4^+ state in ^{230}Th. This correction has not been included in the analysis of the data for the actinide nuclei because higher-order effects for these cases are more important, e.g., compare the calculations for the case $M_{04}(\mathrm{E}4) \neq 0$ and all other $M_{if}(\mathrm{E}4) = 0$ with the case $M_{if}(\mathrm{E}4) \neq 0$ in Fig. 10. The quantal corrections for the higher-order excitations are not available. It is possible that the quantal corrections for these higher-order effects could have a significant influence on the extraction of the reduced E4 matrix elements from experimental data. There is a distinct need for theoretical studies of the quantal corrections for higher-order excitations.

In the limit of the rigid-rotor model the measured transition probabilities are related to the intrinsic multipole moments by $B(\mathrm{E}\lambda, 0 \to J = \lambda) = (2\lambda + 1)e^2 Q_{\lambda 0}^2/16\pi$. For a homogeneous charge distribution of shape $R = R_0(1 + \beta_{20}Y_{20} + \beta_{40}Y_{40})$, the intrinsic moments are related to the model-dependent deformation parameters β_{20} and β_{40} by the following expressions (Owen and Satchler, 1964) to second order in $\beta_{\lambda 0}$:

$$Q_{20} = [3ZR_0^2/(5\pi)^{1/2}]\, \beta_{20}[1 + 0.360\beta_{20} + 0.967\beta_{40} + 0.328(\beta_{40}^2/\beta_{20})] \quad (32)$$

$$Q_{40} = (ZR_0^4/\pi^{1/2})\, \beta_{40}[1 + 0.983\beta_{20} + 0.725(\beta_{20}^2/\beta_{40}) + 0.411\beta_{40}] \quad (33)$$

Using a charge radius parameter $R_0 = 1.2A^{1/3}$ fm, the values for β_{40} from the measurements on ^{152,154}Sm, ^{230}Th, and ^{238}U are somewhat larger than the theoretical calculations of Nilsson *et al.* (1969).

The excitation probabilities could also be analyzed on the basis of a negative value for $\langle 4\|\mathcal{M}(E4)\|0\rangle$ (see Fig. 10). In the case of ^{230}Th, the extracted value would be $-(1.57 \pm 0.21)\,eb^2$, which corresponds to $B(E4, 0 \rightarrow 4)$ $= 220B(E4)_{sp}$ and $\beta_{40} = -(0.26 \pm 0.03)$. This solution seems rather unlikely because the calculations of Nilsson *et al.* (1969) of the sum of single-particle energies as a function of β_{20} and β_{40} indicate that the equilibrium value of β_{40} is positive for ^{230}Th.

References

Afonin, O. F., Grinberg, A. P., Lemberg, I. Kh., and Chugunov, I. N. (1968). *Sov. J. Nucl. Phys.* **6**, 160.

Alaga, G., Alder, K., Bohr, A., and Mottelson, B. R. (1955). *Klg. Danske Videnskab. Selskab, Mat.-Fys. Medd.* **29**, No. 9.

Alder, K. (1970). Private communication.

Alder, K., and Pauli, H. (1969). *Nucl. Phys.* **A128**, 193.

Alder, K., and Winther, A. (1955). CERN Rep. T/KA-AW-4.

Alder, K., and Winther, A. (1956). *Klg. Danske Videnskab. Selskab, Mat.-Fys. Medd.* **31**, No. 1.

Alder, K., and Winther, A. (1969). *Nucl. Phys.* **A132**, 1.

Alder, K., Bohr, A., Huus, T., Mottelson, B. R., and Winther, A. (1956). *Rev. Mod. Phys.* **28**, 432.

Alder, K., Roesel, F., and Morf, R. (1972). *Nucl. Phys.* **A186**, 449.

Alexander, T. K., Häusser, O., Allen, K. W., and Litherland, A. E. (1969). *Can. J. Phys.* **47**, 2335.

Andersen, H. H., Hanke, C. C., Sorensen, H., and Vajda, P. (1967). *Phys. Rev.* **153**, 338.

Andersen, H. H., Hanke, C. C., Simonsen, H., Sorensen, H., and Vajda, P. (1968). *Phys. Rev.* **175**, 389.

Andersen, H. H., Simonsen, H., Sorensen, H., and Vajda, P. (1969). *Phys. Rev.* **186**, 372.

Bichsel, H. (1963). *In* "American Institute of Physics Handbook," 2nd ed. McGraw-Hill, New York.

Biedenharn, L. C., and Rose, M. E. (1953). *Rev. Mod. Phys.* **25**, 729.

Biedenharn, L. C., Goldstein, M., McHale, J. L., and Thaler, R. M. (1956). *Phys. Rev.* **101**, 662.

Boehm, F., Goldring, G., Hagemann, G. B., Symons, G. D., and Tvetar, A. (1966). *Phys. Lett.* **22**, 627.

Bohr, A., and Mottelson, B. R. (1969). "Nuclear Structure," Vol. I, p. 382. Benjamin, New York.

Camp, D. C., and Van Lehn, A. L. (1969). *Nucl. Instrum. Methods* **76**, 192.

Casten, R. F., Greenberg, J. S., Sie, S. H., Burginyon, G. A., and Bromley, D. A. (1969). *Phys. Rev.* **187**, 1532.

Chen, M. (1970). *Phys. Rev. C* **1**, 1176.

Cline, D., Gertzman, H. S., Gove, H. E., Lesser, P. M. S., and Schwartz, J. J. (1969). *Nucl. Phys.* **A133**, 445.

Cote, R. E., Prestwich, W. V., Gaigalas, A. K., Raboy, S., Trail, C. C., Carrigan, Jr., R. A., Gupta, P. D., Sutton, R. B., Suzuki, M. N., and Thompson, A. C., (1969). *Phys. Rev.* **179**, 1134.

Curtis, T. H., Eisenstein, R. A., Madsen, D. W., and Bockelman, C. K. (1969). *Phys. Rev.* **184**, 1162.

de Boer, J., and Eichler, J. (1968). *Advan. Nucl. Phys.* **1**, 1–65.

De Jager, J. L., and Boeker, E. (1972). *Nucl. Phys.* **A186**, 393.

De Wit, S. A., Backenstoss, G., Daum, C., Sens, J. C., and Acker, H. L. (1967). *Nucl. Phys.* **87**, 657.

Diamond, R. M., and Stephens, F. S. (1967). *Arkiv Fysik* **36**, 221.

Diamond, R. M., Stephens, F. S., and Swiatecki, W. J. (1964). *Phys. Lett.* **11**, 315.

Diamond, R. M., Poskanzer, A. M., Stephens, F. S., Swiatecki, W. J., and Ward, D. (1968). *Phys. Rev. Lett.* **20**, 802.

Diamond, R. M., Stephens, F. S., Nakai, K., and Nordhagen, R. (1971). *Phys. Rev. C* **3**, 344.

Douglas, A. C. (1963). *Nucl. Phys.* **42**, 428.

Eisenstein, R. A., and Smilansky, U. (1970). *Phys. Lett.* **31B**, 436.

Ferentz, M., and Rosenzweig, N. (1955). Argonne Nat. Lab. Rep. ANL-5324.

Ferguson, A. J. (1965). "Angular Correlation Methods in Gamma-Ray Spectroscopy." North-Holland Publ., Amsterdam.

Ford, J. L. C., Jr., Stelson, P. H., Bemis, C. E., Jr., McGowan, F. K., Robinson, R. L., and Milner, W. T. (1971). *Phys. Rev. Lett.* **27**, 1232.

Fraser, I. A., Greenberg, J. S., Sie, S. H., Stokstad, R. G., Burginyon, G. A., and Bromley, D. A. (1969a). *Phys. Rev. Lett.* **23**, 1047.

Fraser, I. A., Greenberg, J. S., Sie, S. H., Stokstad, R. G., and Bromley, D. A. (1969b). *Phys. Rev. Lett.* **23**, 1051.

Gauvin, H., Beyec, Y., Lefort, M., and Deprun, C. (1972). *Phys. Rev. Lett.* **28**, 697.

Greenwood, R. C., Helmer, R. G., and Gehrke, R. J. (1970). *Nucl. Instrum. Methods* **77**, 141.

Griffy, T. A., and Biedenharn, L. C. (1962). *Nucl. Phys.* **36**, 452.

Hansen, O., Olesen, M. C., Skilbreid, O., and Elbek, B. (1961). *Nucl. Phys.* **25**, 634.

Häusser, O., and Cusson, R. Y. (1970). *Can. J. Phys.* **48**, 240.

Häusser, O., Alexander, T. K., Pelte, D., and Hooton, B. W. (1969a). *Phys. Rev. Lett.* **23**, 320.

Häusser, O., Pelte, D., Alexander, T. K., and Evans, H. C. (1969b). *Can. J. Phys.* **47**, 1065.

Helmer, R. G., Greenwood, R. C., and Gehrke, R. J. (1971). *Nucl. Instrum. Methods* **96**, 173.

Holm, A. (1969). Private communication.

Horikawa, Y., Torizuka, Y., Nakada, A., Mitsunobu, S., Kojima, Y., and Kimura, M. (1971). *Phys. Lett.* **36B**, 9.

Itoh, K., Oyamada, M., and Torizuka, Y. (1970). *Phys. Rev. C* **2**, 2181.

Jardine, L. J. (1971). *Nucl. Instrum. Methods* **96**, 259.

Kerman, A. K. (1956). *Klg. Danske Videnskab. Selskab, Mat.-Fys. Medd.* **30**, No. 1.

Kocbach, L., and Vogel, P. (1970). *Phys. Lett.* **32B**, 434.

Kumar, K. (1969). *Phys. Lett.* **29B**, 25.

Kumar, K., and Baranger, M. (1968). *Nucl. Phys.* **A122**, 273.

Lindhard, J., and Scharff, M. (1953). *Kgl. Danske Videnskab. Selskab, Mat.-Fys. Medd.* **27**, No. 15.

Mariscotti, M. A. J., Scharff-Goldhaber, G., and Buck, B. (1969). *Phys. Rev.* **178**, 1864.

McGowan, F. K., and Milner, W. T. (1971). *Nucl. Data* A9, 469.

McGowan, F. K., and Stelson, P. H. (1958). *Phys. Rev.* **109**, 901.

McGowan, F. K., Robinson, R. L., Stelson, P. H., and Ford, Jr., J. L. C. (1965). *Nucl. Phys.* **66**, 97.

McGowan, F. K., Robinson, R. L., Stelson, P. H., and Milner, W. T. (1968). *Nucl. Phys.* **A113**, 529.

McGowan, F. K., Bemis, Jr., C. E., Ford, Jr., J. L. C., Milner, W. T., Robinson, R. L., and Stelson, P. H. (1971). *Phys. Rev. Lett.* **27**, 1741.

McKee, R. J. (1969). *Phys. Rev.* **180**, 1139.

Milner, W. T., McGowan, F. K., Stelson, P. H., Robinson, R. L., and Sayer, R. O. (1969). *Nucl. Phys.* **A129**, 687.

Milner, W. T., McGowan, F. K., Robinson, R. L., Stelson, P. H., and Sayer, R. O. (1971). *Nucl. Phys.* **A177**, 1.

Mottelson, B. R., and Valatin, J. G. (1960). *Phys. Rev. Lett.* **5**, 511.

Nakai, K., Quebert, J. L., Stephens, F. S., and Diamond, R. M. (1970a). *Phys. Rev. Lett.* **24**, 903.

Nakai, K., Stephens, F. S., and Diamond, R. M. (1970b). *Nucl. Phys.* **A150**, 114.

Nakai, K., Stephens, F. S., and Diamond, R. M. (1971). *Phys. Lett.* **34B**, 389.

Nathan, O., and Nilsson, S. G. (1965). *In* "Alpha-, Beta-, and Gamma-Ray Spectroscopy" (K. Siegbahn, ed.), Vol. I, pp. 601–700. North-Holland Publ., Amsterdam.

Neergard, K., and Vogel, P. (1970). *Nucl. Phys.* **A149**, 217.

Neiler, J. H., and Bell, P. R. (1965). *In* "Alpha-, Beta-, and Gamma-Ray Spectroscopy" (K. Siegbahn, ed.), Vol. I, pp. 245–302. North-Holland Publ., Amsterdam.

Nilsson, S. G., and Prior, O. (1961). *Klg. Danske Videnskab. Selskab, Mat.-Fys. Medd.* **32**, No. 16.

Nilsson, S. G., Tsang, C. F., Sobiczewski, A., Szymanski, Z., Wycech, S., Gustafson, C., Lamm, I. L., Moller, P., and Nilsson, B. (1969). *Nucl. Phys.* **A131**, 1.

Northcliffe, L. C., and Schilling, R. F. (1970). *Nucl. Data* A7, 233.

Olesen, M. C., and Elbek, B. (1960). *Nucl. Phys.* **15**, 134.

Owen, L. W., and Satchler, G. R. (1964). *Nucl. Phys.* **51**, 155.

Robinson, R. L., McGowen, F. K., Stelson, P. H., Milner, W. T., and Sayer, R. O. (1969a). *Nucl. Phys.* **A123**, 193.

Robinson, R. L., McGowan, F. K., Stelson, P. H., Milner, W. T., and Sayer, R. O. (1969b). *Nucl. Phys.* **A124**, 553.

Sayer, R. O. (1968). Oak Ridge Nat. Lab. Rep. ORNL-TM-2211.

Sayer, R. O., Stelson, P. H., McGowan, F. K., Milner, W. T., and Robinson, R. L. (1970). *Phys. Rev. C* **1**, 1525.

Scharff-Goldhaber, G., and Weneser, J. (1955). *Phys. Rev.* **98**, 212.

Schmorak, M. R., Bemis, Jr., C. E., Zender, M. J., Gove, N. B., and Dittner, P. F. (1972). *Nucl. Phys.* **A178**, 410.

Seaman, G. G., Bernstein, E. M., and Palms, J. M. (1967). *Phys. Rev.* **161**, 1223.

Smilansky, U. (1968). *Nucl. Phys.* **112**, 185.

Sobiczewski, A., Szymanski, Z., Wycech, S., Nilsson, S. G., Nix, J. R., Tsang, C. F., Gustafson, C., Moller, P., and Nilsson, B. (1969). *Nucl. Phys.* **A131**, 67.

Stelson, P. H., and Grodzins, L. (1965). *Nucl. Data* **A1**, 21.

Stelson, P. H., and McGowan, F. K. (1958). *Phys. Rev.* **110**, 489.

Stephens, F. S., Diamond, R. M., Glendenning, N. K., and de Boer, J. (1970). *Phys. Rev. Lett.* **24**, 1137.

Stephens, F. S., Diamond, R. M., and de Boer, J. (1971). *Phys. Rev. Lett.* **27**, 1151.

Symons, G. D., and Douglas, A. C. (1967). *Phys. Lett.* **24B**, 11.

Thaler, R. M., Goldstein, M., McHale, J. L., and Biedenharn, L. C. (1956). *Phys. Rev.* **102**, 1567.

Thomas, T. D. (1968). *Ann. Rev. Nucl. Sci.* **18**, 343.

Vitoux, D., Haight, R. C., and Saladin, J. X. (1971). *Phys. Rev. C* **3**, 718.

Vogel, P. (1968). *Nucl. Phys.* **A112**, 583.

Whaling, W. (1958). "Handbuch der Physik," Vol. 34, p. 193. Springer, Berlin.

Winther, A., and de Boer, J. (1966). *In* "Coulomb Excitation" (K. Alder and A. Winther, eds.), pp. 303–374. Academic Press, New York.

Wu, C. S. (1967). *Proc. Int. Conf. Nucl. Phys., Gatlinburg* (R. L. Becker, C. D. Goodman, P. H. Stelson, and A. Zucker, eds.), p. 409. Academic Press, New York.

Wu, C. S., and Wilets, L. (1969). *Ann. Rev. Nucl. Sci.* **19**, 527.

Yates, M. J. L. (1965). *In* "Alpha-, Beta-, and Gamma-Ray Spectroscopy" (K. Siegbahn, ed.), Vol. II, pp. 1691–1703. North-Holland Publ., Amsterdam.

VII.B COULOMB REORIENTATION

O. Häusser

CHALK RIVER NUCLEAR LABORATORIES
ATOMIC ENERGY OF CANADA LIMITED
CHALK RIVER, ONTARIO, CANADA

I. Introduction

Individual nuclear states are known to exhibit nonvanishing even electric and odd magnetic multipole moments. The study of these moments provides information on the distributions of electric charges and currents inside the nucleus and is complementary to the study of electromagnetic transitions between different nuclear states. The static electric quadrupole moments Q of excited states in even-even nuclei are of particular interest. They are directly related to the shape of the nucleus and may provide, together with other electromagnetic properties, a crucial test of nuclear structure theories. For example, the discovery of large quadrupole moments in nuclei exhibiting a vibrational energy spectrum was quite unexpected and has necessitated a revision of the theoretical concepts of these nuclei.

The Coulomb excitation of nuclei with heavy-ion beams provides a generally applicable and effective technique for the measurement of electric quadrupole moments of low-lying nuclear states. The use of heavy ions enhances higher-order terms occurring in a perturbation expansion of Coulomb excitation cross sections. The change of observables in Coulomb excitation introduced by the quadrupole moments of states is called the "reorientation effect." The reorientation method was extensively reviewed by de Boer and Eichler in 1968. By that time, the basic theoretical framework and some experimental techniques had been developed allowing the measurement of some 14 quadrupole moments of excited 2^+ states. In the last few years, this number has increased by more than a factor of four and significant progress has been made on the experimental and theoretical sides. Higher-order effects competing with the reorientation effect have been scrutinized and isolated in a few cases. A number of innovative experimental techniques have been developed which broaden the applicability of the method, particularly to the light nuclei. An important factor in the exploitation of the reorientation effect has been the availability of high-quality heavy-ion beams provided, for example, by tandem accelerators with terminal voltages in excess of 9 MV. The construction of accelerators capable of producing heavy-ion beams with $A > 40$ at energies near or above the Coulomb barrier of heavy nuclei will undoubtedly improve our understanding of higher-order effects in Coulomb excitation and increase our knowledge of static electric moments.

It is the main purpose of this chapter to provide an introduction to the reorientation method. After defining relevant quantities, a comparison is made of the merits and limitations of various methods that exist for the measurement of electric moments of excited states. It will be seen that the reorientation method is presently most suitable for determining nuclear quadrupole moments. We then introduce some theoretical considerations which are essential for the planning and the analysis of reorientation experiments. For a detailed development of the theory we refer to the articles of

Alder *et al.* (1956) and of de Boer and Eichler (1968) and to the original literature. In Section IV, some experimental techniques used in Coulomb reorientation are discussed, with the emphasis on recent developments. Finally, the data on some 60 quadrupole moments of 2^+ states in even-even nuclei are presented and their relevance to nuclear structure theory is briefly discussed.

A. STATIC AND INTRINSIC MOMENTS

The interaction energy of an external potential $V(r)$ with a nuclear charge distribution $\varrho(r)$ can be written as a tensor product

$$E_Q = \sum_{\lambda\mu} (-)^\mu T_\mu^{(\lambda)} V_{-\mu}^{(\lambda)} \tag{1}$$

In this expression, the $T_\mu^{(\lambda)}$ are the multipole tensors of $\varrho(r)$ and the $V_\mu^{(\lambda)}$ describe the tensor components of the external field. The odd multipole moments $\langle \Psi | T_\mu^{\lambda=\text{odd}} | \Psi \rangle$ are identically zero because the operator $T_\mu^{\lambda=\text{odd}}$ is odd under the time-reversal or the parity operation, whereas the nuclear wave functions Ψ have generally well-defined parity and time-reversal properties. Deviations of the nuclear charge distribution from sphericity are then described to lowest order by the quadrupole moment tensor

$$T_\mu^{(\lambda=2)} = \left(\tfrac{4}{5}\pi\right)^{1/2} \int \varrho(r)\, r^2 Y_{2\mu}(\theta, \phi)\, d\tau \tag{2}$$

where $Y_{2\mu}$ are spherical harmonics and (r, θ, ϕ) are the nuclear coordinates. The electrostatic field produced by external charges at coordinates (r_c, θ_c, ϕ_c) with respect to the nuclear center is (Ramsey, 1955; Steffen and Frauenfelder, 1964)

$$V_\mu^{(\lambda=2)} = \left(\tfrac{4}{5}\pi\right)^{1/2} \int \left[\varrho(r_c)/r_c^{2\lambda-1}\right] Y_{2\mu}(\theta_c, \phi_c)\, d\tau_c \tag{3}$$

or in terms of Cartesian gradients $V_{x'x'}, V_{x'y'}, \ldots,$

$$V_0^{(2)} = \frac{1}{2} V_{z'z'}, \qquad V_{\pm 1}^{(2)} = \mp \frac{1}{\sqrt{6}} (V_{x'z'} \pm iV_{y'z'})$$

$$V_{\pm 2}^{(2)} = \frac{1}{2\sqrt{6}} (V_{x'x'} - V_{y'y'} \pm 2iV_{x'y'}) \tag{4}$$

By a suitable rotation, the principal coordinate system xyz can always be found such that the mixed derivatives of V vanish, i.e.,

$$V_0^{(2)} = 1/2\, V_{zz}, \qquad V_{\pm 1}^{(2)} = 0, \qquad V_{\pm 2}^{(2)} = (1/2\sqrt{6})\,\eta V_{zz} \tag{4a}$$

where $\eta = (V_{xx} - V_{yy})/V_{zz}$. In the case of a field which is axially symmetric about the z axis, $\eta = 0$ and the interaction energy with a state of spin J and z component M is simply

$$E_Q = e\langle JM| \tfrac{1}{2}(\tfrac{4}{5}\pi)^{1/2} r^2 Y_{20}(\theta, \phi) |JM\rangle V_{zz} \tag{5}$$

We now introduce reduced matrix elements by applying the Wigner–Eckart theorem

$$\langle J_s M_s| i^\lambda r^\lambda Y_{\lambda\mu} |J_r M_r\rangle \equiv \langle J_s M_s| \mathcal{M}(E\lambda, \mu) |J_r M_r\rangle$$
$$= \langle J_r\lambda, M_r\mu | J_s M_s\rangle M_{rs}^{(\lambda)}/(2J_s + 1)^{1/2} \tag{6}$$

where the coupling coefficient is defined by Condon and Shortley (1935) and the reduced matrix element is $M_{rs}^{(\lambda)} \equiv \langle J_s \| i^\lambda r^\lambda Y_\lambda \| J_r\rangle$.[†] The $M_{rs}^{(\lambda)}$ can be made real by a suitable choice of phases of the nuclear wave functions, in which case they have the following symmetry:

$$M_{rs}^{(\lambda)} = (-)^{J_r - J_s - \lambda} M_{sr}^{(\lambda)} \tag{7}$$

The so-called static electric quadrupole moment Q is conventionally defined as

$$eQ \equiv e\langle JJ| (3z^2 - r^2) |JJ\rangle = e\langle JJ| (16\pi/5)^{1/2} r^2 Y_{20}(\theta, \phi) |JJ\rangle \tag{8}$$

This definition can be generalized to higher-order moments

$$eQ^{(\lambda)} = e\langle JJ| [16\pi/(2\lambda + 1)]^{1/2} r^\lambda Y_{\lambda 0}(\theta, \phi) |JJ\rangle \tag{8a}$$

After applying the Wigner–Eckart theorem to Eq. (8), we have

$$Q = -\left(\frac{16\pi}{5}\right)^{1/2} \left[\frac{J(2J - 1)}{(J + 1)(2J + 1)(2J + 3)}\right]^{1/2} M_{JJ}^{(2)} \tag{8b}$$

The simplified interaction energy [Eq. (5)] can now be written in terms of Q:

$$E_Q = eQV_{zz}[3M^2 - J(J + 1)]/4J(2J - 1) \tag{5a}$$

It is seen that states whose magnetic quantum numbers differ only in sign are degenerate. This degeneracy does not exist in the presence of an additional magnetic field. Furthermore, it follows from Eq. (8a) and the Wigner–

[†] The reduced matrix elements $M_{rs}^{(\lambda)}$ are used in this form by Winther and de Boer (1966) in their computer code for calculating multiple Coulomb excitation processes. For a comparison with definitions by other authors, see the review article by Rose and Brink (1967).

Eckart theorem that only states with $J \geqslant \lambda/2$ can have a nonvanishing moment $Q^{(\lambda)}$.

It is worth noting at this point that the quantity of theoretical interest is frequently not the observable static moment but the moment of the intrinsic body-fixed system, which is defined by the relevant nuclear model. Consider, for example, the adiabatic rotation of a highly deformed axially symmetric nucleus which is described by the wave function (Rogers, 1965)

$$\Psi^J_{MK} \equiv |JK, M\rangle$$

$$= [(2J + 1)/16\pi^2]^{1/2} [D^J_{MK}(\theta_i) X_K(q) + (-)^{J+K} D^J_{M-K}(\theta_i) X_{-K}(q)] \tag{9}$$

In this expression, the $D^J_{MK}(\theta_i)$ are the matrix elements of finite rotations (see, e.g., Brink and Satchler, 1968) and describe the rotation of the deformed nuclear potential, whereas the $X_K(q)$ describe the internal motion of the nucleons within the potential, K being the projection of the total spin J onto the symmetry axis. The *intrinsic* moment of order λ, $Q^{(\lambda)}_0$, is

$$Q^{(\lambda)}_0 = \langle X_K| \left(\frac{16\pi}{2\lambda + 1}\right)^{1/2} r'^\lambda Y'_{\lambda 0}|X_K\rangle$$

$$= \langle X_K| \left(\frac{16\pi}{2\lambda + 1}\right)^{1/2} \sum_\mu D^\lambda_{\mu 0}(\theta_i) r^\lambda Y_{\lambda 0}|X_K\rangle \tag{10}$$

where the (r', θ', ϕ') are defined in the body-fixed system. Substituting the wave function (9) into Eq. (8), one obtains for the static moments the expression

$$Q^{(\lambda)} = \langle J\lambda J0|JJ\rangle \langle J\lambda K0|JK\rangle Q^{(\lambda)}_0 + \text{small terms} \qquad \text{if} \quad K \geqslant 1 \tag{11}$$

This reduces for $\lambda = 2$ to the familiar relation between static and intrinsic quadrupole moments of a rotational nucleus

$$Q = \{[3K^2 - J(J + 1)]/(J + 1)(2J + 3)\} Q_0 \tag{11a}$$

This relation implies that in a prolate band (Q_0 positive) with $K = 0$, the measured static moments Q have a negative sign.

Similarly, one obtains for the reduced E2 transition probability $B(\text{E2}; J_i \rightarrow J_f)$, where J_i and J_f are different states of the same rotational band:

$$B(\text{E2}; J_i \rightarrow J_f) = (5/16\pi) \langle J_i 2K0|J_f K\rangle^2 Q_0^2 \tag{12}$$

The factorization of rotational and intrinsic motion [Eq. (9)] is only justified for strongly deformed, heavy nuclei and should more generally be replaced by a projection integral (Hill and Wheeler, 1953; Peierls and Yoccoz, 1957):

$$|JK, M\rangle = a_{JK} \int D^J_{MK}(\theta_i)\, X_K\, d\theta_i \tag{9a}$$

Wave functions of this form are used, for example, in the nuclear SU_3 model (Elliott, 1958) which takes the motion of individual nucleons into account.

II. Methods for Measuring Quadrupole Moments of Excited States

A. Mössbauer and Perturbed Angular Correlation Measurements

The hyperfine interaction [Eq. (6)] between the electric field gradients (EFG) in solids and the quadrupole moment of excited nuclear states can be observed by the Mössbauer effect or by the method of perturbed angular correlations (Steffen and Frauenfelder, 1964; Shirley, Chapter VIII.D). The experiments determine generally the product QV_{zz} and thus require the knowledge of the EFG. The EFG arises from electrons in incompletely filled shells of the decaying atomic nucleus, and from distant ions in the host crystal, provided the crystal symmetry is lower than cubic. Another important contribution comes from the distortion of the wave function of the electrons which surround the decaying nucleus by the EFG of the crystal (Sternheimer, 1951). Field gradients of up to a few 10^{18} V/cm^2 can be obtained, although the theoretical calculations are not yet very reliable. The resulting energy shifts [Eq. (5a)] are of the order of 10^{-7}–10^{-6} eV and can be easily detected by the Mössbauer effect (Kienle, 1968). The Mössbauer measurements provide at present accurate ratios of quadrupole moments for different isotopes. The Mössbauer technique is limited to $E_\gamma \lesssim 130$ keV, which implies that only strongly deformed 2^+ states in heavy nuclei can be studied.

The quadrupole moment interaction can also be detected from a perturbation of the angular distribution of deexcitation gamma rays. The perturbation is normally a function of the square of the angular frequency $\omega_Q = E_Q/\hbar$. A determination of the sign of the interaction E_Q requires either polarization of the initial state or detection of the polarization of the deexcitation gamma rays. The orientation of the initial state could be obtained by implanting polarized recoils from a nuclear reaction into a suitable environment (Grodzins and Klepper, 1970). With this method, states with mean lifetimes down to $\sim 10^{-10}$ sec can be studied provided time-integrated perturbations are observed.

B. REORIENTATION MEASUREMENTS

In Coulomb excitation experiments with heavy-ion beams, the bombarding energies are chosen such that the long-range Coulomb force is the only significant interaction and nuclear forces can be ignored. The projectile with charge $Z_1 e$, mass A, and lab energy E is well localized and moves in an approximate hyperbolic orbit past the target nucleus (Z_2, A_2) giving rise to a time-dependent hyperfine splitting of levels in both target and projectile

$$E_Q(t) = \frac{ZeQ}{r^3(t)} \frac{3M^2 - J(J+1)}{4J(2J-1)} \tag{5b}$$

The largest EFG is realized in a head-on collision at the distance of closest approach, $d_{min} = 2a$

$$d_{min} = 1.44 Z_1 Z_2 \frac{(1 + A_1/A_2)}{E} \tag{13}$$

where d is in fermis and E in MeV. In the projectile excitation of 90-MeV ^{24}Mg in the Coulomb field of ^{208}Pb, for example, the $M = 0$ substate of the 2^+ level $(Q \sim -0.2$ eb) is raised as much as 440 keV, which is comparable to ΔE, the excitation energy of the 2^+ level. Since the Coulomb excitation cross sections depend strongly on the level energy ΔE (Alder et al., 1956), the quadrupole moment interaction can in our example be detected as a strong decrease in the cross section for backscattered particles. It is also apparent that the sign of Q generally can be determined. From time-dependent perturbation theory, it can also be seen that the distribution of magnetic substates of the 2^+ state (density matrix) changes as a consequence of the quadrupole moment interaction, which can be detected by measuring the angular distribution of the deexcitation gamma rays. It is this latter feature which inspired Breit and collaborators (1955) to use the term reorientation effect.[†]

The reorientation method can be applied to all low-lying collective states that can be directly Coulomb-excited. This includes all the lowest 2^+ states

---

[†] The term "reorientation effect" is somewhat unfortunate. In a head-on collision, the initial magnetic substate is preserved during the scattering process, i.e., no reorientation occurs. On the other hand, we have seen that E_Q and thus the deviation of the cross section is maximized. For this reason, most quadrupole moments have been measured from deviations in the scattering cross sections rather than in the angular distributions of the deexcitation gamma rays.

in even-even nuclei that have a stable or long-lived ground state. The method is suited mainly to the measurements of quadrupole moments since the hyperfine interaction [Eq. (1)] for the hexadecapole and higher moments is at least an order of magnitude smaller.

C. MUONIC X RAYS

The lower Bohr orbits in muonic atoms are largely inside the nucleus and allow probing of the nuclear charge distribution in detail (see Anderson and Jenkins, Chapter V.A). The electric quadrupole "hyperfine" interaction, which is proportional to $\langle eQ/r_\mu{}^3 \rangle$, amounts to about 100 keV in the heavy, deformed nuclei and is comparable to the spin–orbit splitting and to the spacing of nuclear levels. The electric quadrupole interaction couples the nuclear and muonic levels and thus hyperfine structure can even be observed for nuclei with a spin-zero ground state. From the hyperfine spectra observed in Ge(Li) detectors, one may extract the nuclear quadrupole moment, although the value of Q depends somewhat on the model of the nuclear charge distribution used (Wu, 1968; Devons and Duerdoth, 1970). The quadrupole hyperfine splitting is proportional to the product QZ^3 and cannot be resolved with present Ge(Li) detectors in light nuclei. The measurement of electric moments from muonic X rays is thus limited to nuclei with $Z \geqslant 25$.

D. INELASTIC SCATTERING

The previous experimental methods made use of well-understood electrostatic interactions to determine electric moments of excited nuclear states. It is also possible to obtain a measure of nuclear deformation by measuring cross sections for inelastic scattering of protons, deuterons, and alpha particles at energies considerably above the Coulomb barrier. The analysis of these data makes use of deformed optical potentials and of the method of coupled channels (Tamura, 1965) which includes multiple excitation processes. In strongly deformed nuclei, for example, such analysis yields the parameters describing the nuclear surface

$$R = R_0 \left[1 + \sum_{\lambda = 2, 4, \ldots} \beta_\lambda Y_{\lambda 0}(\theta', \phi') \right] \tag{14}$$

where θ' and ϕ' are defined in the body-fixed system. Application of Eq. (8b) yields the intrinsic quadrupole and hexadecapole moments of the nuclear mass

$$Q_0^{(2)} = [3/(5\pi)^{1/2}] ZR_0{}^2(\beta_2 + 0.36\beta_2{}^2 + 0.33\beta_4{}^2 + 0.97\beta_2\beta_4)$$

$$Q_0^{(4)} = ZR_0{}^4(0.56\beta_4 + 0.41\beta_2{}^2 + 0.23\beta_4 + 0.56\beta_2\beta_4)$$

which in turn are related to the static moments by Eq. (11). The (α,α') experiments have the advantage that large momentum transfers are possible enabling the study of $\lambda = 4$ and $\lambda = 6$ deformations if the interference between single-stage and multiple-stage excitation probabilities is taken into account (Hendrie *et al.*, 1968; Rebel *et al.*, 1971).

Much the same information is provided by inelastic electron scattering experiments at intermediate energies ($E \gtrsim 100$ MeV). The form factors for inelastic scattering, extracted in first-order Born approximation, can be related in a model-dependent way to higher electric moments. The $\lambda = 4$ moments obtained from (e,e') experiments in sd-shell nuclei (Horikawa *et al.*, 1971) are in good agreement with those from (α,α') and (p,p') experiments.

III. Theoretical Considerations in Coulomb Reorientation

In Coulomb excitation experiments with heavy ions, a large number of higher-order effects come into play of which the reorientation effect is generally the dominant one. The extraction of accurate nuclear quadrupole moments requires an understanding of all processes which may influence the observable cross sections appreciably, say by more than 0.5%. In the experiments, it is desirable to enhance, as much as possible, the reorientation effect compared to other higher-order processes. In this section, we would like to give approximate estimates of the relative sizes of the reorientation effect, of other higher-order effects, and of significant corrections that originate from the approximations commonly made in the semiclassical treatment of the Coulomb excitation process.

A. Semiclassical Perturbation Theory

The calculation of Coulomb excitation cross sections is greatly simplified if one assumes that the projectile follows closely a classical hyperbolic trajectory. A necessary requirement for this is that the parameter $\eta = a/\lambda$, i.e., the ratio of half the distance of closest approach in a head-on collision a to the de Broglie wavelength λ of the incoming projectile, is large,

$$\eta_i = Z_1 Z_2 e^2 / \hbar v_i \gg 1 \qquad (15)$$

Furthermore, the energy of the excited state ΔE should be much smaller than the bombarding energy E, so that the effect of the excitation on the particle motion can be neglected. Both conditions are well fulfilled in Coulomb excitation with heavy ions, $Z_1 \gtrsim 8$.

One may then write the cross section for excitation of the initial state $|i\rangle$ to a final state $|f\rangle$ in the symmetrized form [†]

$$d\sigma^{\text{sym}}(\theta) = (v_f/v_i)\, P_{if}[(a_{if}^{\text{sym}})^2/4]\, \sin^{-4}(\theta/2)\, d\Omega \qquad (16)$$

where $a_{if}^{\text{sym}} = Z_1 Z_2 e^2 (1 + A_1/A_2)/(A_1 v_i v_f)$ and P_{if} is the excitation probability. If polarization of initial and final state is not observed, we have

$$P_{if} = (2I_i + 1)^{-1} \sum_{M_i M_f} |b_{if}|^2 \qquad (17)$$

The amplitudes b_{if} can be evaluated in first-order time-dependent perturbation theory denoted $b_{if}^{(1)}$, which yields (Dirac, 1947)

$$b_{if}^{(1)} = (i\hbar)^{-1} \int_{-\infty}^{\infty} \langle f|\, H_{\text{int}}(t)\, |i\rangle\, e^{i\omega_{fi} t}\, dt \qquad (18)$$

where $\omega_{fi} = (E_f - E_i)/\hbar$ is the nuclear frequency associated with the nuclear excitation energy ΔE. The time-dependent Coulomb interaction is

$$H_{\text{int}}(t) = \int \frac{\varrho(r_1)\, \varrho(r_2)\, d\tau_1\, d\tau_2}{|r_1 - r_2 - r_p(t)|} \qquad (19)$$

where r_1 and r_2 denote radius vectors with respect to the centers of mass of projectile and target, respectively, ϱ is the charge density operator, and $r_p(t)$ is the position vector connecting the centers of mass. Expanding $H_{\text{int}}(t)$ into multipoles and retaining only the quadrupole terms yields approximately (cf. Häusser and Cusson, 1970)

$$H(t) = -\frac{4\pi e}{5} \sum_m \frac{Y_{2m}(\theta_p \phi_p)}{r_p^3(t)} [Z_2 \mathcal{M}_{2m}^*(1) + Z_1 \mathcal{M}_{2m}^*(2)] \qquad (19a)$$

This expression describes excitation of target or projectile and allows the reduction of Eq. (17) into the product of a nuclear matrix element and an orbital integral, i.e.,

$$P_{if}(\theta) = [\chi_{if}^{(2)}]^2\, \tfrac{1}{5} \sum K_{2\mu}(\theta, \xi) \qquad (20)$$

[†] The symmetrization procedure is used to make the results of the semiclassical treatment correspond more closely to the correct quantal solution (Alder et al., 1956). It consists in replacing in the characteristic parameters for the excitation the velocity by the average $(v_i v_f)^{1/2}$. The factor v_f/v_i in the expression for the cross section is used to satisfy time-reversal invariance.

The strength of the nuclear $E\lambda$ transition is given by the dimensionless, symmetrized parameter $\chi_{if}^{(\lambda)}$, which is, in the case of target excitation,

$$\chi_{if}^{(\lambda)} = \frac{(16\pi)^{1/2}(\lambda - 1)!}{(2\lambda + 1)!!} \frac{Z_1 e}{\hbar(v_i v_f)^{1/2}} \frac{M_{fi}^{(\lambda)}}{a_{if}^{\lambda}(2J_i + 1)^{1/2}} \tag{21}$$

The orbital integrals $K_{2\mu}$ (de Boer and Eichler, 1968; Alder *et al.*, 1956) depend on the center-of-mass scattering angle and on the parameter ξ, representing the ratio of the collision time $\tau_{col} = a/v \approx 10^{-21}$ sec to the nuclear period $\tau_{nuc} = 1/\omega$, i.e., $\xi = a\,\Delta E/\hbar v$. In the symmetrized form,

$$\xi = (Z_1 Z_2 e^2/\hbar)\left[(1/v_f) - (1/v_i)\right] = \eta_f - \eta_i \tag{22}$$

In practical cases, ξ ranges between 0.05 and 1. The smaller ξ is, the larger are the Fourier components in $H_{int}(t)$ with the nuclear frequency ω and the larger are the orbital integrals.

In heavy-ion Coulomb excitation, the parameter $\chi_{if}^{(2)}$ can become of the order unity. Consequently, multiple excitation processes can occur and the probability P_{if} will depend on Q of the excited state. Useful estimates can often be made in second-order perturbation theory, which yields

$$b_{if}^{(2)} = b_{if}^{(1)} + b_{inf} = b_{if}^{(1)} + (i\hbar)^2 \int_{-\infty}^{\infty} dt \langle f| H_{int}(t) |n\rangle e^{i\omega_{fn}t}$$

$$\times \int_{-\infty}^{t} dt' \langle n| H_{int}(t') |i\rangle e^{i\omega_{ni}t'} \tag{23}$$

where $|n\rangle$ is an intermediate state. The second-order amplitude b_{inf} can be written as a sum of contributions on and off the energy shell (de Boer and Eichler, 1968)

$$b_{inf} = \tfrac{1}{2} b_{in}^{(1)}(\Delta E_{ni})\, b_{nf}^{(1)}(\Delta E_{fn})$$

$$+ (i/2\pi) P \int_{-\infty}^{\infty} b_{in}^{(1)}(\Delta E_{ni} + E)\, b_{nf}^{(1)}(\Delta E_{fn} - E)\,(dE/E) \tag{23a}$$

where P denotes the principal value of the integral. The off-energy shell term can be the only contribution in cases where the intermediate state is not significantly populated. Such virtual excitations are, for example, important if the states of the giant dipole resonance serve as the intermediate state $|n\rangle$.

Evaluation of Eq. (23) yields for the probability of a $0^+ \rightarrow 2^+$ transition, formally,

$$P_{02}(\theta) = A_1(\theta, \xi)|M_{02}|^2 + A_2(\theta, \xi)|M_{02}|^2 M_{22} + A_3(\theta, \xi) M_{0\lambda}M_{\lambda 2}M_{02}$$
$$(24)$$

where the second-order contributions from Q_{2^+} and from higher excited states have been written out separately. It is now formally evident that both sign and magnitude of Q_{2^+} can be determined from measurements of $P_{02}(\theta)$, and that the accuracy will depend on our knowledge of matrix elements connecting 0^+ and 2^+ states to higher excited states. For most cases, the size of the reorientation effect can be estimated from (de Boer and Eichler, 1968)

$$P_{02} = A(\theta, \xi) B(E2, 0 \to 2) (1 + 1.32 \frac{A_1}{Z} \frac{\Delta E}{1 + A_1/A_2} QK(\xi, \theta) + \cdots) \qquad (25)$$

where ΔE is in MeV, Q in eb, and Z refers to the nucleus being excited. The function $K(\xi, \theta)$ is shown in Fig. 1 as a function of ξ for several scattering angles θ.

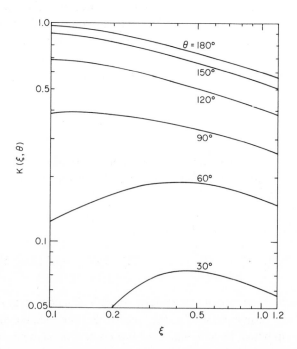

Fig. 1. The function $K(\xi, \theta)$ used in Eq. (25) to estimate the reorientation effect (after de Boer and Eichler, 1968, with permission of Plenum Publishing Corp.).

B. QUANTITATIVE CALCULATIONS

Although second-order perturbation theory provides a good grasp of the dependence of cross sections on nuclear matrix elements, the numerical accuracy is not sufficient in heavy-ion bombardments where multiple excitations are appreciable. For these cases, exact numerical methods have been developed by Winther and de Boer (1966) to solve the time-dependent Schrödinger equation

$$i\hbar\dot{\Psi}(t) = [H_0 + H_{int}(t)] \, \Psi(t) \tag{26}$$

where H_0 the is Hamiltonian of the free nucleus. Expanding $\Psi(t)$ in terms of eigenstates of H_0, $\Psi(t) = \sum_s a_s(t)\Psi_s \exp(-iE_s t/\hbar)$, one obtains a set of coupled differential equations for the amplitudes,

$$i\hbar\dot{a}_r(t) = \sum_s \langle r| H_{int}(t) |s \rangle \, e^{i\omega_{rs}t} a_s(t) \tag{27}$$

From the final amplitudes $a_s(t = \infty)$ obtained from integration of Eq. (27), one can calculate excitation probabilities and the statistical tensors $\alpha_{k\kappa}$, which determine the angular distribution of emitted radiation. If the ground-state spin is zero, one obtains for the Nth excited state

$$
\begin{aligned}
P(N) &= \sum_{M_N} |a_{J_N M_N}|^2 \\
\alpha_{k\kappa}(N) &= \sum_{M_N M_N'} (-)^k (J_N k, M_{N'}\kappa \, | \, J_N M_N) \, a_{J_N M_N'}^* a_{J_N M_N}
\end{aligned}
\tag{28}
$$

If the ground state has nonzero spin, appropriate averages over the initial polarizations have to be taken.

C. APPROXIMATIONS AND CORRECTIONS

From the previous sections, it has become apparent that for an accurate determination of Q_{2^+}, one has to know the sign and magnitude of all significant matrix elements connecting higher-lying states to both ground state and 2^+ state. In addition, corrections have to be applied which result from the approximations of the semiclassical treatment. In Figs. 2–4, we show the relative magnitude of all significant processes for two cases which are representative for light and heavy nuclei. The results are for projectile excitation of the 1.369-MeV, 2^+ state in ^{24}Mg and target excitation of the 0.558-MeV, 2^+ state in ^{114}Cd and were obtained chiefly with the Winther–de Boer computer code by including up to six excited states. Shown is the percentage change in the excitation probability of the 2^+ state, $\Delta P/P$, for a variety of experimental situations. The reorientation effect has been calculated for a prolate "rotational" intrinsic state using the known $B(E2, 0 \to 2)$ and Eqs. (11a) and (12). Figures 2–4 demonstrate the relative importance of the various processes in light and heavy nuclei and also give an indication of how the reorientation effect can be measured most favorably.

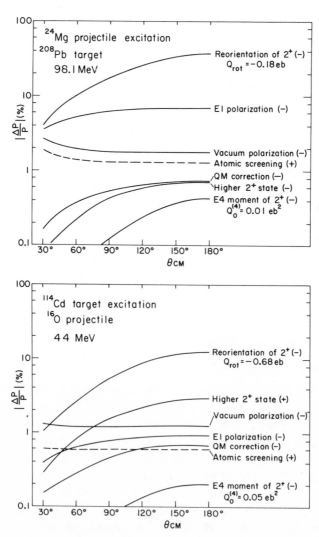

Fig. 2. Angular dependence of $\Delta P/P$ for two selected examples.

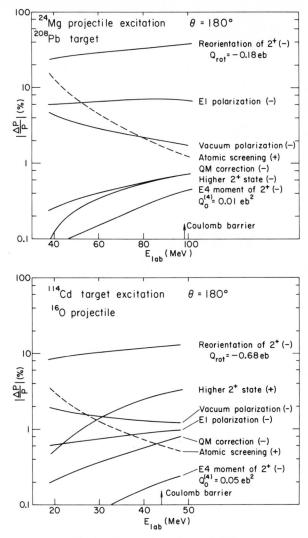

Fig. 3. Energy dependence of $\Delta P/P$.

D. CORRECTIONS DEPENDING ON NUCLEAR STRUCTURE

1. *Dipole Polarization*

One of the largest corrections $\Delta P/P$ comes from virtual E1 excitation of states of the giant dipole resonance. The underlying physical picture is that

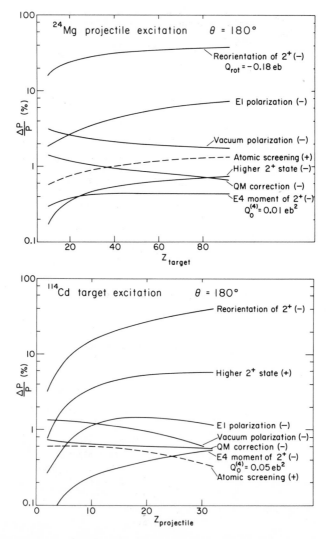

Fig. 4. $\Delta P/P$ as a function of Z. The bombarding energies of Eq. (35) were used.

the excited nucleus is polarized in the electric field of the partner. The potential energy arising from this polarization can be estimated (Winther, 1966) if one makes the reasonable assumption that a coordinate frame can be found in which the polarization tensor is diagonal, with principal polarizabilities proportional to the square of the nuclear radii in these directions. In the framework of the hydrodynamic model (Bohr and Mottelson, 1953), Winther (1966) obtains

$$V_{\text{pol}} = -\tfrac{1}{2}\alpha E^2[1 + 2\sum_{\mu} \alpha_{2\mu}^* Y_{2\mu}(\theta, \phi)] \tag{29}$$

where $E = Z'e/r^2$ is the electric field supplied by the partner and the $\alpha_{2\mu}^*$ are expansion coefficients of the nuclear surface, which are related to the operator for collective E2 transitions by

$$\mathcal{M}(E\lambda, \mu) = -(3/4\pi) ZeR_0{}^\lambda \alpha_{\lambda\mu}^* \tag{30}$$

The polarizability α is defined as by Levinger (1957) and can be related to the (-2) moment of the total photoabsorption cross section, $\sigma_{-2} = \int\sigma(E)(dE/E^2)$, by

$$\alpha \equiv 2e^2 \int \frac{|\langle n| z |i\rangle|^2}{E_n - E_i} = \frac{\hbar c}{2\pi^2}\sigma_{-2} \tag{31}$$

From photoabsorption measurements, one finds (Levinger, 1957) $\sigma_{-2} = 3.5kA^{5/3}$ µb/MeV, where $k \approx 1$ for most nuclei with $A \gtrsim 20$.

The first term in Eq. (29) gives rise only to a slight decrease in the elastic scattering cross section which is typically a small fraction of 1%. The second term is proportional to $\mathcal{M}^*(E2, \mu)$ and reduces the quadrupole interaction (19a) to

$$H(t) = -\frac{4\pi e}{5} Z' \sum_m \frac{Y_{2m}(\theta_p, \phi_p)}{r_p{}^3(t)} \mathcal{M}(E2, m)\left(1 - 0.0056k\frac{A}{Z^2} E_{\text{CM}}\frac{a}{r_p(t)}\right) \tag{19b}$$

In this expression, A and Z are the mass and charge numbers of the nucleus being excited, Z' refers to the partner supplying the electric field, and E_{CM} is the center of mass bombarding energy in MeV. The corrected interaction (19b) can be conveniently incorporated into the de Boer–Winther computer code.

The E1 polarization effect has been derived with equivalent assumptions in perturbation theory (de Boer and Eichler, 1968), yielding for the change in the excitation probability of $0^+ \to 2^+$ transitions [after correcting for trivial errors in (de Boer and Eichler, 1968)]

$$\Delta P/P = -0.069k(A/Z^2) E_{\text{CM}}\phi(\theta, \xi) \tag{32}$$

The function $\phi(\theta, \xi)$ is shown in Fig. 5 as a function of ξ for several scattering angles θ.

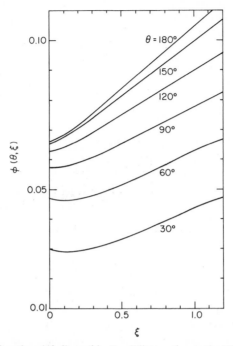

Fig. 5. The function $\phi(\theta, \xi)$ used in Eq. (32) to estimate the E1 polarization contribution (after de Boer and Eichler, 1968, with permission of Plenum Publishing Corp.).

Both Eqs. (19a) and (32) show that a large reduction of the excitation cross section is expected in very light nuclei, since the E1 polarization effect is proportional to $1/Z$. Furthermore, k can be much larger than unity in light nuclei (Levinger, 1957). Experimental evidence for the E1 polarization effect has been obtained in ^{7}Li (Häusser *et al.*, 1972; Smilansky *et al.*, 1972) and probably in ^{6}Li (Disdier *et al.*, 1971). In these nuclei, the E1 polarization effect exceeds the reorientation effect. The dependence of $\Delta P/P$ on the scattering angle from dipole polarization is much weaker than that from the reorientation effect, as is evident from Figs. 1 and 5. Inclusion of the E1 polarization effect in the analysis therefore increases mainly the extracted $B(E2)$, whereas the renormalization of Q_{2^+} is only slight.

2. *Interference from Higher* 2^+ *States*

Second-order contributions from $2'^+$ states above the first excited 2^+ state introduce a relatively large correction $\Delta P/P$. Even if the absolute magnitude of all relevant matrix elements in Eq. (24) is known, the extracted Q depends on the sign of the product

$$P_3 = M_{02}M_{02'}M_{2'2} \tag{33}$$

For each sign of P_3, one obtains a solution for Q_{2^+} and the two solutions may differ by as much as $\sim 0.4Q_{\text{rot}}$. The interference term $\Delta P/P$ from a $2'$ state is strongly energy dependent, in contrast to the reorientation effect (see Fig. 3), and can in principle be isolated by varying the bombarding energy; the required statistical accuracy, however, is difficult to obtain, particularly at the lower bombarding energies.

In cases where the two solutions for Q_{2^+} have the same sign, the interference term can be more conveniently earmarked by the product $P_4 = P_3 M_{22}$, which is independent of phase conventions as well as of the i^λ factor which we have included into the definition of reduced matrix elements. In many nuclei the sign of P_4 can be estimated fairly reliably from theoretical arguments, at least for the most important lowest $2'$ level. Kumar (1969) finds that P_4 is generally negative for nuclei in the transition region which can be described by an anharmonic vibrational model. On the other hand, in approximately "rotational" nuclei, the sign of P_4 is positive if the $2'$ level belongs to the gamma band ($K = 2$) and negative if the $2'$ level belongs to the beta band ($K = 0$). The negative (positive) sign of P_3 corresponds to the more negative (positive) value of Q_{2^+}, provided we use our definition of matrix elements [Eq. (6)].

3.   Hexadecapole Moments

The renormalization of Q_{2^+} resulting from the $\lambda = 4$ moment of the 2^+ state amounts generally to $\lesssim 0.1Q_{\text{rot}}$. The $\lambda = 4$ moment could, in principle, be determined from the strong dependence of its contribution $\Delta P/P$ on the bombarding energy. A more feasible approach at present appears to be to infer $Q_{2^+}^{(\lambda=4)}$ from the $0^+ \rightarrow 4^+$ E4 transition moment. The $B(E4, 0 \rightarrow 4)$ can be measured with light-ion beams taking the interference between single E4 and double E2 excitation via the 2^+ state into account (Stephens *et al.*, 1971).

4.   Simultaneous Excitation of Target and Projectile

The use of heavy projectiles, desirable to enhance the higher-order effects, may result in a comparable excitation of target and projectile [see Eq. (19a)]. The effect of this mutual excitation process on the measurement of Q_{2^+} has been discussed by Häusser and Cusson (1970) and by Eisenstein and Smilansky (1970). The renormalization is only significant if both individual excitation probabilities exceed 1% and if $\xi \gtrsim 0.3$ for both target and projectile.

E. INADEQUACY OF THE SEMICLASSICAL APPROXIMATION

1. *Quantal Corrections*

The semiclassical treatment is known to give accurate inelastic scattering cross sections provided the characteristic parameters of the excitation process are properly symmetrized and $\eta \gtrsim 50$ (Alder *et al.*, 1956). For smaller values of η, exact quantal calculations are necessary. Results on the deviations between classical and quantum mechanical treatments have been presented by Smilansky (1968) and Alder and Pauli (1969).

It is found that the cross sections agree in both approaches provided $Q_{2+} = 0$. Sizable deviations occur only for nonzero Q_{2+} and are proportional to $1/\eta$. Alder and Pauli write the excitation probability in analogy to Eq. (25) as

$$P_{02} = P_{02}(\eta = \infty, Q = 0)\,(1 - C_\eta \chi_{22}) \tag{34}$$

where χ_{22} is defined in Eq. (21) and

$$\frac{C_\eta}{C_\infty} = 1 + \frac{\eta_0}{\eta}\left(\frac{C_{\eta_0}}{C_\infty} - 1\right) \tag{34a}$$

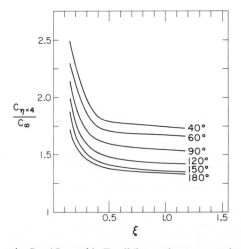

Fig. 6. The ratio $C_{\eta=4}/C_\infty$ used in Eq. (34) to estimate quantal corrections to Coulomb excitation cross sections. $\chi_{22} = 0.224$. (After Alder and Pauli, 1969, with permission of North-Holland Publishing Co., Amsterdam.)

The latter equation implies that it is only necessary to calculate C_η/C_∞ for one value of $\eta = \eta_0$. This ratio is given for $\eta_0 = 4$ in Fig. 6 as a function of ξ and the CM scattering angle θ for a fixed $\chi_{22} = +0.224$, i.e., a $Q_{2+} < 0$. The parameter C_η depends only slightly on χ_{22} and negligibly on χ_{02}. The result of the quantal correction is a decrease of $|Q_{2+}|$.

The angular distribution tensors and thus the gamma-ray angular distributions are strongly affected by the quantal correction except at the extreme backward angles ($\theta \gtrsim 160°$). The corrections are sizable even if $Q_{2+} = 0$ and are given in graphical form by Alder and Pauli (1969).

2.   Vacuum Polarization and Atomic Screening

Corrections $\Delta P/P$ arise from small deviations of the actual from the idealized Coulomb potential. In quantum electrodynamics, the Coulomb repulsion between two charges is increased at small distances through interaction with the electron–positron field (Foldy and Eriksen, 1954). This increase is $\sim 0.4\%$ at $d \simeq 10$ fm and causes an effective decrease of the bombarding energy and, consequently, of the excitation probability.

An effect of similar magnitude but opposite sign arises from screening of the nuclear Coulomb field by the atomic electrons of both target and projectile. An approximate expression for the resulting increase of the effective bombarding energy is quoted by Saladin et al. (1969).

It is seen from Fig. 2 that the effects of vacuum polarization and atomic screening are only weakly angle dependent; their combined contribution, if not canceling completely, would tend to renormalize the $B(E2; 0 \rightarrow 2)$ rather than Q_{2+}.

IV.   Experimental Methods in Coulomb Reorientation

From inspection of Figs. 2–4, it is evident that the reorientation effect can be observed most favorably by varying the particle scattering angle and/or the Z of the partner that is the source of the electric field. Measurements with ^{4}He beams are very useful since the reorientation effect and other high-order effects are small [see Eq. (25)]. The ^{4}He data chiefly determine the $B(E2, 0^+ \rightarrow 2^+)$. The quadrupole moment Q_{2+} is obtained mainly from measurements with heavy ions, such as ^{16}O, ^{32}S, etc., after inferring the $B(E2; 0^+ \rightarrow 2^+)$ from the ^{4}He data. In the actual analysis, one attempts, of course, to fit all data with a consistent set of E2 matrix elements. Unfortunately, this technique is not applicable to light nuclei because the excitation probabilities obtainable with ^{4}He beams are much too small. On the other hand, large reorientation effects can be obtained if these nuclei are used as projectiles on heavy targets [see Eq. (25) and Fig. 4]. Then Q_{2^+} is measured from the angular variation of the reorientation effect alone (see Fig. 2). In the following, some of the experimental techniques used in light and heavy nuclei for the measurement of precise cross sections with heavy-ion beams will be discussed.

A. BOMBARDING ENERGIES

The excitation probability $P_{02}(\theta)$ increases sharply with increasing bombarding energy, particularly for large values of the parameters ξ. It is therefore desirable to work at the highest possible energy at which the effects of nuclear interactions can still be neglected. This maximum "safe" energy may be reliably determined by observing the energy variation of the elastic scattering cross section near $\theta \sim 180°$, since at this angle, deviations from the Rutherford scattering cross section are expected to occur first. One conveniently measures, with a thin target, the ratio of the number of particles elastically scattered into an annular surface barrier detector at $\sim 180°$, to those observed in other detectors at forward angles (see e.g. Cline et al., 1969). High statistical accuracy is required since the deviation from the Rutherford ratios should be $\leqslant 0.5\%$. A conservative estimate for the highest "safe" bombarding energy is (Cline 1970)

$$E = \frac{1.44(1 + A_1/A_2)\, Z_1 Z_2}{1.25(A_1^{1/3} + A_2^{1/3}) + 5.1} \tag{35}$$

This equation should only be considered as a rough guide, although it is compatible with most published data obtained with projectiles heavier than ^{4}He.

B. PARTICLE SPECTROSCOPY

The most straightforward way of arriving at the inelastic scattering cross sections is by resolving, either with a solid-state detector or a high-resolution magnet, the elastically and inelastically scattered particles. The excitation probability $P_{02}(\theta)$ is simply equal to the ratio of peak intensities $I_{inel}/(I_{el} + I_{inel})$. The experimental difficulty lies in resolving the weak inelastic peak in the presence of a very strong elastic peak which often exhibits a low-energy tail. Measurements at forward angles are particularly sensitive to systematic errors in determining the background and scattering angle because $P(\theta)$ changes rapidly with θ, particularly if ξ is large. Several precautions must be taken to minimize these systematic errors. The incident beam should be well collimated to define the scattering angle to better than 0.1° and beam particles scattered from the defining collimator should be eliminated by a system of antiscattering apertures. Similar precautions should be taken for the collimators of the detector system. Very thin targets of a few micrograms on equally thin backings must be used to keep the contributions of energy loss and energy straggling to the overall resolution to a minimum. Heavy projectiles with $A_1 \gtrsim 40$ cannot be used in most cases because of the increase in energy loss, energy straggling, and kinematic dispersion, although the use of such beams would be desirable to increase the reorientation effect. A further limitation may arise from small target impurities of lower mass than A_2 whose elastic peak overlaps the inelastic peak of interest. If P_{02} is of the order of 1%, the amount of such harmful impurities should be below 10^{-4}.

Excellent particle energy resolution is obtained with the new generation of magnetic spectrometers of the split-pole and the QD^3 type (Hendrie, Chapter III.C; Spencer and Enge, 1967; Milton et al., 1970). These instruments have a large solid angle and the energy resolution for heavy ions is limited entirely by energy loss and straggling. Relatively large solid angles can be used if kinematic compensation fields are employed. Corrections may have to be applied to I_{el} and I_{inel} resulting from the energy dependence of the charge-state distribution. This energy dependence should then be measured experimentally because frequently the targets are thinner than required for the observation of equilibrium charge-state distributions (Grodzins et al., 1967; Moak et al., 1968). The correction turns out to be negligible if measurements at most of the abundant charge states are combined.

The energy resolution of the surface barrier detectors is sometimes sufficient to separate the inelastic peak. This has the advantage that several detectors can be used simultaneously to measure $P_{02}(\theta)$. The energy resolution can be improved slightly by cooling the detectors to $\lesssim -30°C$. If at the same time the detector bias is increased, the charge collection time may be shortened and the number of pulses in the low-energy tails may decrease. Figure 7 shows as an example spectra observed after bombarding thin ^{208}Pb targets with 65-MeV ^{18}O and ^{16}O beams (Disdier et al., 1971). The spectra show very weak, yet resolved inelastic peaks from Coulomb excitation of the 1.98-MeV 2^+ state in ^{18}O and the 2.61-MeV 3^- state in ^{208}Pb.

C. Particle–Gamma Coincidences

Unresolved elastic and inelastic peaks may be separated if coincidences between inelastically scattered particles and deexcitation gamma rays are observed. At any scattering angle $\theta \neq 180°$, the statistical tensors $\alpha_{k\kappa}$ and thus

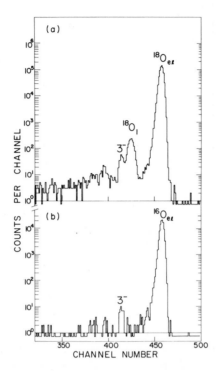

Fig. 7. Particle spectra obtained with a surface barrier detector using 65-MeV(a)^{18}O and (b) ^{16}O beams and a thin ^{208}Pb target. The inelastic ^{18}O line is wider than the elastic ^{18}O line because of the Doppler shift of the unobserved $2^+ \rightarrow 0^+$ gamma ray; $\theta = 134°$.

the gamma-ray angular distributions depend on Q_{2^+}. However, the effects are generally not as large as those obtainable from the variation of the particle scattering angle or the projectile Z, and, furthermore, the gamma-ray angular distributions are frequently distorted through hyperfine interactions in vacuum (deorientation effects). In the great majority of experiments, the coincidence data are used primarily to extract the transition probabilities $P_{02}(\theta)$. For this purpose, the intensity $(I_{el} + I_{inel})$ has to be measured simultaneously with the coincidence data, the coincidence efficiency must be determined, and corrections for random coincidences have to be applied to the data.

1. Coincidence Yields

If we denote the laboratory angles of particle and gamma-ray detectors by $(\theta_p, 0)$ and $(\theta_\gamma, \phi_\gamma)$, respectively, and ignore the finite size of the detectors for the sake of simplicity, the coincidence yield $Y(\theta_p, \theta_\gamma, \phi_\gamma)$, can be written

$$\frac{Y(\theta_p, \theta_\gamma, \phi_\gamma)}{I_{el} + I_{inel}} = \varepsilon(E_\gamma) \frac{d\sigma_{inel}(\theta_p)}{d\sigma_{Ruth}(\theta_p)} \sum_{\substack{k = 0, 2, 4 \\ -k \leqslant \kappa \leqslant k}} G_k(\theta_p) \frac{dW_{k\kappa}}{d\Omega_{TN}^\gamma} \frac{d\Omega_{TN}^\gamma}{d\Omega_L^\gamma} \tag{36}$$

where $\varepsilon(E_\gamma)$ is the full energy peak efficiency of the gamma detector, and, for the case of a $2^+ \rightarrow 0^+$ transition with a negligible internal conversion rate,

$$\frac{dW_{k\kappa}}{d\Omega_{TN}^\gamma} = (4\pi)^{1/2} \alpha_{k\kappa} F_k(2202) Y_{k\kappa}(\theta_\gamma \phi_\gamma) \tag{37}$$

The angular distribution tensor $\alpha_{k\kappa}$ is as defined in Eq. (28) and the F coefficients are tabulated, for example, by Rose and Brink (1967). The kinematic correction factor, $d\Omega_{TN}^\gamma / d\Omega_L^\gamma$ is required because the $\alpha_{k\kappa}$ are defined in the rest frame of the recoil nucleus, i.e.,

$$d\Omega_{TN}^\gamma = d\Omega_L^\gamma (1 - \beta^2)/(1 - \beta \cos \theta')^2 \tag{38}$$

where $\beta = v/c$ is the recoil velocity and θ' is the angle between the directions of the recoil nucleus and the gamma ray. The coefficients G_k describe the attenuation of the gamma-ray angular distributions by randomly oriented hyperfine fields acting on the highly ionized nuclei recoiling in vacuum (discussed later). The G_k depend on the charge-state distribution of the recoils and thus on the recoil velocity and scattering angle. They have to be measured carefully by observing gamma rays at several angles $(\theta_\gamma, \phi_\gamma)$ since inclusion of the attenuation in the analysis may drastically change the value of Q_{2^+} (see, e.g., Kleinfeld et al., 1970).

In practice, several gamma-ray detectors are employed in coincidence with one or a few particle detectors. In cases where the type of projectile can be varied, a single annular particle detector near $\theta_p \sim 180°$ is sufficient; the sum in Eq. (36) then contains only terms with $\kappa = 0$. For the measurement of the angular variation of $P_{02}(\theta_p)$ in light nuclei, multiple-particle, multiple Na I(Tl) detector systems have been developed at Berkeley, Chalk River, and Rochester (see, e.g., Nakai et al., 1970; Häusser et al., 1971; Towsley et al., 1972). In some cases, both recoil target nuclei and scattered projectiles can be observed in the same detector, providing $P_{02}(\theta_p)$ at two separate scattering angles θ_p (see, e.g., Häusser et al., 1969).

2. Deorientation

The perturbation of the gamma-ray angular distributions caused by hyperfine fields acting on the recoils in vacuum seriously affects the values of Q_{2^+} extracted from particle–gamma coincidence data, and are therefore briefly commented upon in the following section. It is now generally accepted that the attenuation is caused by magnetic fields and that two fundamentally different deorientation processes may be distinguished.

In Coulomb excitation of heavy nuclei with ^{16}O and ^{32}S beams, recoil velocities $\beta \sim 0.02$ occur and the charge-state distributions are approximately Gaussian with a full-width of about four charge units and an average charge state of about ten (Betz et al., 1966). The outer electrons are distributed over many excited states and the unpaired electrons produce hyperfine fields at the nucleus which fluctuate at a time scale $\tau_c \approx 3 \cdot 10^{-12}$ sec (Ben Zvi et al., 1968) determined mainly by the atomic transition rates. For long nuclear lifetimes $\tau \gg \tau_c$ and low nuclear spins, the theory of time-dependent interactions (Abragam and Pound, 1953) is applicable, which yields for the time-integrated attenuation coefficients

$$G_k = (1/\tau) \int_0^\infty G_k(t)\, e^{-t/\tau}\, dt = (1 + P_k \omega^2 \tau \tau_c)^{-1} \qquad (39)$$

The numerical constants P_k for a 2^+ state are $P_2 = 2$ and $P_4 = 20/3$ and the precession frequency $\omega = \mu_n g H/\hbar$, H being typically of the order of tens of megagauss (Ben Zvi et al., 1968).

In Coulomb excitation of light nuclei, large recoil velocities are encountered ($\beta \sim 0.06$) and an appreciable number of recoil atoms may be produced in hydrogenlike electron configurations. The hydrogenlike atoms may end up in the $1s_{1/2}$ configuration in a time short compared to the mean nuclear lifetime τ and then generate a randomly oriented static field at the nucleus with $H(1s_{1/2}) = 0.167 Z^3$ MG, which exceeds the fields from other electron configurations by at least an order of magnitude. The time-integrated attenuation coefficients are, in the static case (Alder, 1952),

$$G_k = \sum_{FF'} \frac{(2F + 1)(2F' + 1)}{2J + 1} \begin{Bmatrix} F & F' & k \\ I & I & J \end{Bmatrix}^2 [1 + (\omega_{FF'} \tau)^2]^{-1} \qquad (40)$$

where $\mathbf{F} = \mathbf{I} + \mathbf{J}$ is the total spin of the ion, the term in curly brackets is a 6-j symbol (Brink and Satchler, 1968), and the hyperfine frequencies are $\omega_{FF'} = (E_F - E_{F'})/\hbar$. If one retains only terms arising from the most important $1s_{1/2}$ electron configuration, Eq. (40) simplifies to (Faessler et al., 1971)

$$G_k = 1 - p(s_{1/2}) \frac{k(k + 1)}{(2I + 1)^2} \frac{(\omega\tau)^2}{1 + (\omega\tau)^2} - \cdots \qquad (40a)$$

In this expression, $p(s_{1/2})$ is the fraction of atoms reaching the single-electron $1s_{1/2}$ configuration and the hyperfine splitting is characterized by $\omega = \mu_n g(2I + 1)H(1s_{1/2})/\hbar$.

Whenever one of the two limiting situations arises, the relation between G_2 and G_4 is given by Eqs. (39) and (40a) and essentially only one deorientation parameter needs to be determined experimentally.

D. DIRECT GAMMA-RAY SPECTRA

1. *Gamma-Ray Line Shapes*

The observation of detailed gamma-ray line shapes in Ge(Li) detectors after Coulomb excitation provides an elegant way of measuring simultaneously particle–gamma correlations for all particle scattering angles θ. The Doppler shift is, in first order,

$$\Delta E_\gamma = E_\gamma^0 \beta(\cos \Psi \cos \theta_\gamma + \sin \Psi \sin \theta_\gamma \cos \phi_\gamma) \qquad (41)$$

where E_γ^0 is the unshifted gamma-ray energy and Ψ is the angle of the recoil to the beam axis and is related by the reaction kinematics to the CM scattering angle θ. If the gamma detector is at $\theta_\gamma = 0°$ to the beam axis, there is a unique correlation between the energy shift ΔE_γ and θ, thus allowing a measurement of Q_{2+} to be made through the θ dependence of the reorientation effect. The gamma-ray line shape can be formally written (see, e.g., Warburton *et al.*, 1967; Häusser *et al.*, 1968)

$$\frac{dN(E_\gamma)}{dE_\gamma} \sim \int \frac{dE}{dE/d\varrho x} \int d\Omega \frac{d\sigma(\theta, v_i)}{d\Omega_p} \int d\Omega_L{}^\gamma \sum_{k\kappa} \frac{dW_{k\kappa}(\theta, \Omega_{TN}^\gamma)}{d\Omega_L{}^\gamma} g(vv_i\tau) \frac{dv}{dE_\gamma} \qquad (42)$$

where $dE/d\varrho x$ is the energy loss of the incident beam in the target, the slowing-down function $g(vv_i\tau)$ describes the distribution of excited recoils starting with initial velocity v_i and decaying in the interval dv with velocity v, and other quantities have been defined before. At $\theta_\gamma = 0°$, the sum in Eq. (42) contains only $\kappa = 0$ terms.

In most applications, the slowing down of the recoils is avoided by using very thin, solid targets or gas targets (see, e.g., Schwalm and Povh, 1969). The deorientation correction affects mainly the high-energy part of the $0°$ line shape corresponding to high initial velocities, as can be seen from Fig. 8.

Thick, solid targets can nevertheless be used provided the mean slowing-down time is longer than the mean nuclear lifetime τ (Pelte *et al.*, 1969). The smearing out of the correlation between ΔE_γ and θ is partially com-

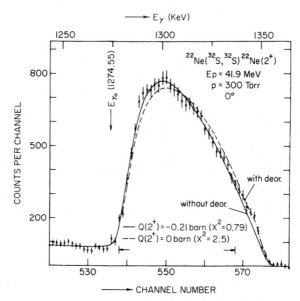

Fig. 8. Gamma-ray line shape observed after Coulomb excitation of the 1.275-MeV level in ^{22}Ne (after Schwalm and Povh 1969, with permission of North-Holland Publishing Co., Amsterdam).

pensated by the improved statistical accuracy and by the absence of de-orientation effects. The $\theta_\gamma = 90°$ line shape does not depend on Q_{2^+} and allows τ to be fitted.

2. Thick-Target Yields

Gamma-ray yields observed with Ge(Li) detectors after Coulomb excitation of thick targets have so far been used to deduce relative quadrupole moments. The thick-target yield in a detector at an angle θ_γ to the beam axis is

$$Y(E_0, \theta_\gamma) = \varepsilon(E_\gamma)\, nN \int \frac{\sigma(E, \theta_\gamma)}{dE/dx}\, dE \qquad (43)$$

where $\varepsilon(E_\gamma)$ is the detection efficiency, n is the number of projectiles, N is the density of atoms of a certain isotope in the target, and dE/dx is the energy loss of the incident beam. The quantity $\sigma(E, \theta_\gamma)$, the integrated cross section for E2 excitation followed by gamma emission in the direction θ_γ, depends on Q_{2^+} with a sensitivity that is typically one-third of the effect obtainable with the backscatter geometry ($\theta_p \sim 180°$). If the thick target contains several isotopes, the relative yields depend very slightly on dE/dx (if ξ differs for the isotopes). The ratios of yields for each isotopic component and several different beams can be measured very accurately since they are independent of n and $\varepsilon(E_\gamma)$. Using beams of ^{4}He, ^{16}O, and ^{32}S, several relative quadrupole moments have been measured from the relative yields from thick, multiisotopic targets (see, e.g., Steadman *et al.*, 1970).

In principle, Q_{2+} can be measured absolutely from thick-target yields. The beam current can be integrated with good accuracy provided secondary electrons from the target are suppressed and the pressure in the beam transport system is low enough ($\lesssim 10^{-7}$ Torr) to avoid charge exchange of the incident beam. Recently, convenient methods have been developed to measure the energy loss of heavy ions in solids (Ward *et al.*, 1972). It is necessary, however, that the reorientation effect with the heaviest available projectile considerably exceeds the accuracy of *relative dE/dx* measurements.

E. FUTURE DEVELOPMENTS

A large number of precise heavy-ion techniques have so far been developed to study the reorientation effect. Several of these have first been applied to measuring Q_{2+} of ^{114}Cd. Initially, discrepancies appeared which have brought the reorientation method into some ill repute. However, more recently, the importance of the deorientation effect in particle–gamma coincidence measurements has been realized and at present most of the ^{114}Cd data are in reasonable agreement (see, e.g., Kleinfeld *et al.*, 1970). One should keep in mind that in the great majority of experiments on ^{114}Cd, ^{16}O was used as the heaviest projectile, limiting the size of the reorientation effect to at most 10%. Obviously, small systematic errors in the experiment and in the application of corrections may then have serious consequences on the extracted Q_{2+}. The most significant improvement in the reliability and accuracy of reorientation measurements in heavy nuclei can be expected from the use of heavier projectiles such as ^{32}S, ^{40}Ar, and ^{58}Ni, in addition to the traditional ^{4}He and ^{16}O beams, to increase the size of the reorientation effect [see Fig. 4 and Eq. (25)]. The particle–gamma coincidence technique with an annular particle detector at 180° and an array of gamma detectors to determine accurately the attenuation coefficients G_k would appear most suitable.

In light nuclei ($A \lesssim 40$), large reorientation effects are obtained in projectile excitation from the angular dependence of $P_{02}(\theta)$. With very thin ^{208}Pb targets, one is able to resolve the inelastic peak in surface barrier detectors or magnetic spectrographs over a wide angular range. The high-quality beams could be provided by tandem accelerators with terminal voltages of ~14 MV. The only corrections to be applied in the analysis of these data would be those arising from the nuclear structure of the projectile.

V. Results and Discussion

The information on static quadrupole moments of the lowest 2^+ states in even-even nuclei available before February 1972 is shown in Figs. 9 and

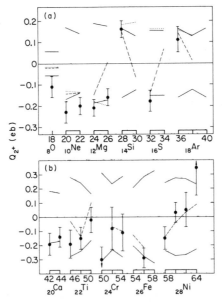

Fig. 9. Static quadrupole moments of 2^+ states in nuclei of the (a) (sd) and (b) (fp) shells. The solid lines are the rigid rotor predictions; – – –, shell model; ······ Hartree-Fock.

10. The experimental moments are taken from the compilation by Christy and Häusser (1972). Data which did not include appropriate corrections for deorientation in vacuum or which were taken more than 10% above the "safe" bombarding energy [Eq. (35)] were rejected. Frequently, the agreement between different experiments was poor, in which case average values with rather generous errors have been adopted. In cases where the interference term from the 2′ levels was important, the larger value of $|Q_{2^+}|$

was chosen as is indicated by the anharmonic model. The solid lines in Figs. 9 and 10 show the rotational values $\pm|Q_{rot}|$ derived from the measured $B(E2, 0^+ \rightarrow 2^+)$ (see Stelson and Grodzins, 1965; Christy and Häusser, 1972) using Eqs. (11) and (12).

The relatively poor accuracy of the data encourages only a rather qualitative comparison with theory, which will be undertaken separately for light

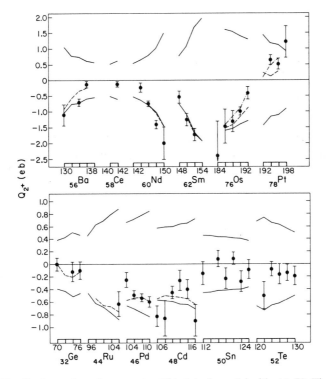

Fig. 10.   Static quadrupole moments of 2^+ states in nuclei with $A \geqslant 70$. The solid lines are the rigid rotor predictions; $-\cdot-\cdot-$, Kumar–Baranger; $-\,-\,-$, Davydov.

and heavy nuclei. This task has benefitted from previous reviews by Cline (1970). For a detailed introduction to the theoretical models mentioned in the following, the reader is referred to the articles of Ripka (1968) on the nuclear Hartree–Fock method, of French et al. (1969) on large-shell-model calculations, and of Bes and Sorensen (1969) on the pairing-plus-quadrupole model. An overview, which emphasizes the connections between these various approaches, can be found in Chapter IX.A by Harvey and Khanna and Chapter IX.B by Rasmussen.

A. LIGHT NUCLEI

The results in light nuclei ($A < 70$) are of particular interest because for these nuclei one may attempt a full microscopic description of nuclear wave functions proceeding from basic properties of the nucleon–nucleon system and of nuclear matter. Such a program has been partially realized for (sd) shell nuclei using the unrestricted Hartree–Fock method (Ford *et al.*, 1971; Lee and Cusson, 1972). This approach approximates the intrinsic state by a single Slater determinant, which is obtained by a self-consistent variational procedure starting from a large-shell-model expansion basis (4–5 major shells) and a "realistic" two-body reaction matrix. The results for (sd) shell nuclei are particularly useful because the gaps between occupied and unoccupied orbits turn out to be several times the excitation energy of the first excited 2^+ states. One may then expect to obtain fairly realistic wave functions for the lowest states by projecting states of good angular momentum from the intrinsic state according to Eq. (9a). The Q_{2^+} values calculated by Lee and Cusson from these projected states are shown in Fig. 9 as dotted lines. The agreement with experiment for Q_{2^+} [and also $B(\text{E2}; 0^+ \rightarrow 2^+)$] is remarkable considering that free nucleon charges have been assumed in the calculation. The frequent change of deformation is reproduced except in ^{32}S, which exhibits a vibrational energy spectrum. It might be conjectured that vibrational modes of excitation cannot be suitably related to a single Slater determinant. The quadrupole moment and other properties of ^{28}Si are of interest because the Hartree–Fock calculations yield prolate and oblate solutions whose energy difference depends sensitively on the two-body reaction matrix used (Wong *et al.*, 1969).

The measured moments at the beginning of the (sd) shell appear to be larger than $|Q_{rot}|$, yet there is no evidence from either Hartree–Fock or large-shell-model calculations (Halbert *et al.*, 1971) that this is likely. More accurate experiments in ^{18}O, ^{20}Ne, and ^{24}Mg would appear worthwhile. The results of shell model calculations by Wildenthal *et al.* (1971) are shown as dashed lines. A surface delta interaction and effective charges of ~ 0.5 and 1.5 for neutron and proton, respectively, have been used in this work. For nuclei in the middle of the shell, different truncation schemes have been used which give comparable results. The moments for the mass $(4n + 2)$, $T = 1$ nuclei in the upper half of the (sd) shell come out considerably different from those of the neighboring $(4n)$, $T = 0$ nuclei, a prediction which is at variance with the Hartree–Fock results and with experiment in ^{26}Mg (Schwalm, 1972). Further experimental and theoretical work on quadrupole moments of $T = 1$ nuclei is desirable.

The Hartree–Fock method does not provide useful predictions of Q_{2^+} in (fp) shell nuclei because the gap between occupied and unoccupied orbits is about a factor of three smaller than in the (sd) shell (Parikh and Svenne, 1968). One therefore expects that particle–hole excitations and pair correlation effects are important in a description of the low-lying states, in contrast to findings in the (sd) shell (Satpathy et al., 1969). The dashed lines in the lower half of Fig. 9 are results of shell model calculations by the Rochester group in Ti, Cr, and Fe isotopes (see, e.g., Lesser et al., 1972) and by Auerbach (1969) in the Ni isotopes. The shell model basis is restricted to (fp)^n configurations and large effective charges are required to reproduce the magnitude of E2 matrix elements. The observed sizable prolate deformation in the $N = 26$ and $N = 30$ nuclei is roughly reproduced by the calculations and is supported in ^{50}Cr and ^{56}Fe by the more nearly "rotational" energy spectrum. The prolate deformations of 2^+ states in ^{42,44}Ca and ^{46}Ti (and similarly in ^{18}O) are to a large extent caused by admixtures of strongly deformed core excited components into the (fp)^n [or (sd)^2] part of the wave function (see, e.g., Engeland, 1965; Flowers and Skouras, 1968).

B.   HEAVY NUCLEI

In heavy nuclei, meaningful shell model calculations become prohibitively large and one has to settle for an intermediate approach which combines collective and single-particle features. Impressive results have been obtained with the pairing-plus-quadrupole model, which recognizes the fact that the low-lying levels of even-even nuclei are mainly determined by quadrupole deformation $(J = 2, T = 0)$ and pairing gap $(J = 0, T = 1)$. Baranger and Kumar (1968) used a "renormalized" force and variational procedures to obtain the equilibrium shape of the intrinsic state. The equidensity contours of the nuclear shape (potential) are given by

$$R(\theta, \phi) = R_0\{1 + \beta(\cos \gamma)\, Y_{20}(\theta\phi) + (1/\sqrt{2})\, \beta(\sin \gamma)\, [Y_{22}(\theta\phi) + Y_{2-2}(\theta\phi)]\}$$

$$(44)$$

where β measures the deformation, and γ, the angle with the prolate axis, denotes the deviation from axial symmetry. Spectra of excited states are obtained from the intrinsic state by a numerical solution of Bohr's generalized collective Hamiltonian (Bohr, 1952; Kumar and Baranger, 1967). The calculations, which have only been performed for a few nuclei, predict oblate deformations in strongly neutron-deficient Ba isotopes and in Pt isotopes. The Os–Pt isotopes provide a beautiful example of a nuclear shape transition which has been convincingly demonstrated by the Pittsburgh group (Glenn et al., 1969; Pryor and Saladin, 1970). The change of sign of the quadrupole moment is accompanied by a crossing of the 2' ($K = 2$) level and the first 4^+ level. It has been pointed out by Kumar (1970) that both features are strongly correlated with the energy difference of the prolate and oblate minima in the potential energy surface. The correspondence between collective energy surface and nuclear energy spectra has been further investigated by Gneuss and Greiner (1971).

The relative sizes of $|Q_{2+}|$ and other E2 matrix elements can be estimated from the observed experimental level scheme using purely collective models. The predictions by Davydov et al. (1960) of the rotation–vibration coupling model whose parameters were obtained from the energies of the lowest 2^+, 2', and 4^+ levels are shown in Fig. 10. The model takes only β vibrations and a fixed nonaxial deformation γ into account and is therefore unsuitable for nuclei that are soft with respect to gamma vibrations. However, the model is reasonably successful in the stable isotopes of Pt, Os, Sm, Nd, and Ba. The nuclei near the $N = 82$ closed neutron shell provide an example of a spherical-to-prolate phase transition, reflected by Q_{2+} and the level spacings. A quadrupole moment equal to that of a prolate rotor is already observed in isotopes that are four neutrons away from the closed shell. The moments in ^{148}Nd and ^{150}Nd apparently exceed $|Q_{\text{rot}}|$, a feature which can be reproduced by the variable-moment-of-inertia model of Mariscotti et al. (1969).

The stable isotopes of Pd, Cd, Sn, and Te exhibit energy spectra which deviate little from those of a harmonic vibrator. It was the original measurement by de Boer et al. (1965) of a large Q_{2+} in ^{114}Cd that created a great deal of interest in the theoretical interpretation of vibrational nuclei. It is now recognized that large quadrupole moments can arise from relatively small anharmonicities that cause admixtures of two- into one-phonon states (see, e.g., Tamura and Udagawa, 1966). The magnitude of $|Q_{2+}|$ can be estimated from the matrix element connecting the mainly two-phonon 2' state and the ground state. This matrix element vanishes for a harmonic vibrator. Reasonable agreement with experiment has been obtained by Sips (1971) for the Cd and Te isotopes using the particle–vibration coupling model to estimate the anharmonicities.

It is apparent from the preceding survey that at present a reasonably large fraction ($\sim 40\%$) of all Q_{2+} values in stable, even-even nuclei that are accessible to the reorientation method have been studied at least once. The measured quadrupole moments have contributed to the identification of (sometimes unexpected) shape transitions in various regions of the periodic table and have helped in broadening our understanding of the structure of nuclear states. It appears almost a certainty that in the near future substantially more, and more precise, E2 and higher moments will become available from Coulomb excitation experiments.

References

Abragam, A., and Pound, R. V. (1953). *Phys. Rev.* **92**, 943.

Alder, K. (1952). *Helv. Phys. Acta* **25**, 235.

Alder, K., and Pauli, H. K. A. (1969). *Nucl. Phys.* **A128**, 193.

Alder, K., Bohr, A., Huus, T., Mottelson, B., and Winther, A. (1956). *Rev. Mod. Phys.* **28**, 432.

Auerbach, N. (1969). *Phys. Rev.* **163**, 1203.

Baranger, M., and Kumar, K. (1968). *Nucl. Phys.* **A110**, 490; **A122**, 241.

Ben Zvi, I., Gilad, P., Goldberg, M., Goldring, G., Schwarzschild, A., Sprinzak, A., and Vager, Z. (1968). *Nucl. Phys.* **A121**, 592.

Bes, D. R., and Sorensen, R. A. (1969). *Advan. Nucl. Phys.* **2**, 129–222.

Betz, H. D., Hortig, G., Leischner, E., Schmelzer, C., Stadler, B., and Weihrauch, J. (1966). *Phys. Lett.* **22**, 643.

Bohr, A. (1952). *Mat. Fys. Medd. Dan. Vid. Selsk.* **26**, no. 14.

Bohr, A., and Mottelson, B. R. (1953). *Mat. Fys. Medd. Dan. Vid. Selsk.* **27**, no. 16.

Breit, G., and Lazarus, J. P. (1955). *Phys. Rev.* **100**, 942; and Breit, G., Glückstern, R. L., and Russell, J. E. (1956). *ibid.* **103**, 727.

Brink, D. M., and Satchler, G. R. (1968). "Angular Momentum," Oxford Univ. Press, London and New York.

Christy, A., and Häusser, O. (1972). *Nucl. Data Tables* **11**, 281.

Cline, D. (1967). *Bull. Amer. Phys. Soc.* **14**, 726, Univ. of Rochester Rep. UR-NSRL-22 (1969).

Cline, D. (1970). Univ. of Rochester Rep. UR-NSRL-37 and UR-NSRL-40.

Cline, D., Gertzman, H. S., Gove, H. E., Lesser, P. M. S., and Schwartz, J. J. (1969). *Nucl. Phys.* **A133**, 445.

Condon, E. U. and Shortley, G. H. (1935). "Theory of Atomic Spectra," Cambridge Univ. Press, London and New York.

Davydov, A. S., and Chaban, A. A. (1960). *Nucl. Phys.* **20**, 444; Davydov, A. S., and Ovcharenko, V. I. (1966). *Sov. J. Nucl. Phys.* **3**, 770; (1968). *Ibid.* **7**, 41.

de Boer, J., and Eichler, J. (1968). *Advan. Nucl. Phys.* **1**, 1–66.

de Boer, J., Stokstad, R. G., Symons, G. D., and Winther, A. (1965). *Phys. Rev. Lett.* **14**, 546.

Devons, S., and Duerdoth, I. (1970). *Advan. Nucl. Phys.* **2**, 295–424.

Dirac, P. A. M. (1947). "The Principles of Quantum Mechanics." Oxford Univ. Press, London and New York.

Disdier, D. L., Ball, G. C., Häusser, O., and Warner, R. E. (1971). *Phys. Rev. Lett.* **27**, 1391.

Eisenstein, R. A., and Smilansky, U. (1970). *Phys. Lett.* **31B**, 436.

Elliott, J. P. (1958). *Proc Roy. Soc.* **A245**, 128, 562.

Engeland, T. (1965). *Nucl. Phys.* **72**, 68.

Faessler, M. A., Povh, B., and Schwalm, D. (1971). *Ann. Phys.* **63**, 577.

Flowers, B. H., and Skouras, L. D. (1968). *Nucl. Phys.* **A116**, 529.

Foldy, L. L. and Eriksen, E. (1954). *Phys. Rev.* **95**, 1048.

Ford, W. F., Braley, R. C. and Bar-Touv, J. (1971). *Phys. Rev. C* **4**, 2099.

French, J. B., Halbert, E. C., McGrory, J. B., and Wong, S. S. M. (1969). *Advan. Nucl. Phys.* **3**, 193–258.

Glenn, J. E., Pryor, R. J., and Saladin, J. X. (1969). *Phys. Rev.* **188**, 1905.

Gneuss, G., and Greiner, W. (1971). *Nucl. Phys.* **A171**, 449.

Grodzins, L., and Klepper, O. (1970). *Phys. Rev. C* **3**, 1019.

Grodzins, L., Kalish, R., Murnick, D., Van de Graaff, R. J., Chmara, F., and Rose, P. H. (1967). *Phys. Lett.* **24B**, 282.

Halbert, E. C., McGrory, J. B., Wildenthal, B. H., and Pandya, S. P. (1971). *Advan. Nucl. Phys.* **4**.

Häusser, O., and Cusson, R. Y. (1970). *Can. J. Phys.* **48**, 240.

Häusser, O., Pelte, D., Alexander, T. K., and Evans, H. C. (1968). *Can. J. Phys.* **47**, 1065.

Häusser, O., Hooton, B. W., Pelte, D., Alexander, T. K., and Evans, H. C. (1969). *Phys. Rev. Lett.* **22**, 359; (1970). *Can. J. Phys.* **48**, 35.

Häusser, O., Alexander, T. K., McDonald, A. B., and Diamond, W. T. (1971). *Nucl. Phys.* **A175**, 593.

Häusser, O., McDonald, A. B., Alexander, T. K., Ferguson, A. J., and Warner, R. E. (1972). *Phys. Lett.* **38B**, 75.

Hendrie, D. L., Glendenning, N. K., Harvey, B. G., Jarvis, O. N., Duhm, H. M., Saudinos, J. and Mahoney, J. (1968). *Phys. Lett.* **26B**, 127.

Hill, D. L., and Wheeler, J. A. (1953). *Phys. Rev.* **89**, 1106; and Griffin, J. J., and Wheeler, J. A. (1957). *Ibid.* **108**, 311.

Horikawa, Y., Torizuka, Y., Nakada, A., Mitsunobu, S., Kojima, Y., and Kimura, M. (1971). *Phys. Lett.* **36B**, 9.

Kienle, P. (1968). *In* "Hyperfine Structure and Nuclear Radiations" (E. Matthias and D. A. Shirley, eds.), pp. 27–45. North-Holland Publ., Amsterdam.

Kleinfeld, A. M., Rogers, J. D., Gastebois, J., Steadman, S. G., and de Boer, J. (1970). *Nucl. Phys.* **A158**, 81.

Kumar, K. (1969). *Phys. Lett.* **29B**, 25.

Kumar, K. (1970). *Phys. Rev. C* **1**, 369.

Kumar, K., and Baranger, M. (1967). *Nucl. Phys.* **A92**, 608; see also (1963). *Phys. Rev. Lett.* **12**, 72; (1966). *Ibid.* **17**, 1146; (1968) *Nucl. Phys.* **A110**, 529; (1969). *Ibid.* **A122**, 273.

Lee, H. C., and Cusson, R. Y. (1972). *Ann. Phys.* **72**, 353; private communication.

Lesser, P. M. S., Cline, D., Goode, P., and Horoshko, R. N. (1972). Univ. of Rochester preprint UR-NSRL-43.

Levinger, J. S. (1957). *Phys. Rev.* **107**, 554.

Mariscotti, M. A. J., Scharff-Goldhaber, G., and Buck, B. (1969). *Phys. Rev.* **178**, 1864.

Milton, J. C. D., Ball, G. C., Davies, W. G., Ferguson, A. J. and Fraser, J. S. (1970). Chalk River Rep. AECL-3563.

Moak, C. D., Lutz, H. O., Bridwell, L. B., Northcliffe, L. C. and Datz, S. (1968). *Phys. Rev.* **176**, 427.

Nakai, K., Stephens, F. S., and Diamond, R. M. (1970). *Nucl. Phys.* **A150**, 114; (1971). *Phys. Lett.* **34B**, 389.

Parikh, J. C., and Svenne, J. P. (1968). *Phys. Rev.* **174**, 1343.

Peierls, R. E., and Yoccoz, J. (1957). *Proc. Phys. Soc.* **A70**, 381.

Pelte, D., Häusser, O., Alexander, T. K., and Evans, H. C. (1969). *Phys. Lett.* **29B**, 660.

Pryor, R. J., and Saladin, J. X. (1970). *Phys. Rev. C* **1**, 1573.

Ramsey, N. F. (1955). "Molecular Beams," Oxford Univ. Press, London and New York.

Rebel, H., Schweimer, G. W., Specht, J., Schatz, G., Löhken, R., Hauser, G., Habs, D., and Klewe-Nebenius, H. (1971). *Phys. Rev. Lett.* **26**, 1190.

Ripka, G. (1968). *Advan. Nucl. Phys.* **1**, 183–260.

Rogers, J. D. (1965). *Annu. Rev. Nucl. Sci.* **15**, 241–290.

Rose, H. J., and Brink, D. M. (1967). *Rev. Mod. Phys.* **39**, 306.

Saladin, J. X., Glenn, J. E., and Pryor, R. J. (1969). *Phys. Rev.* **186**, 1241.

Satpathy, L., Goss, D., and Banerjee, M. K. (1969). *Phys. Rev.* **183**, 887.

Schwalm, D. (1972). Private communication from BNL.

Schwalm, D., and Povh, B. (1969). *Phys. Lett.* **29B**, 103.

Sips, L. (1971). *Phys. Lett.* **36B**, 193.

Smilansky, U. (1968). *Nucl. Phys.* **A112**, 185.

Smilansky, U., Povh, B., and Traxel, K. (1972). Weizmann Inst. preprint WIS-71/52-Ph.

Spencer, J. E., and Enge, H. A. (1967). *Nucl. Instrum. Methods* **49**, 181.

Steadman, S. G., Kleinfeld, A. M., Seaman, G. G., de Boer, J. and Ward, D. (1970). *Nucl. Phys.* **A155**, 1.

Steffen, R. M. and Frauenfelder, H. (1964). *In* "Perturbed Angular Correlations" (E. Karlsson, E. Matthias and K. Siegbahn, eds.), pp. 1–89. North-Holland Publ., Amsterdam.

Stelson, P. H., and Grodzins, L. (1965). *Nucl. Data* **A1**, 21.

Stephens, F. S., Diamond, R. M., and de Boer, J. (1971). *Phys. Rev. Lett.* **27**, 1151.

Sternheimer, R. M. (1951). *Phys. Rev.* **84**, 244; (1952). *Ibid.* **86**, 316; (1963). *Ibid.* **130**, 1423.

Tamura, T. (1965). *Rev. Mod. Phys.* **37**, 679.

Tamura, T., and Udagawa, T. (1966). *Phys. Rev.* **150**, 783.

Towsley, C. W., Cline, D., and Horoshko, R. N. (1972). *Phys. Rev. Lett.* **28**, 368.

Warburton, E. K., Olness, J. W., and Poletti, A. R. (1967). *Phys. Rev.* **160**, 938.

Ward, D., Graham, R. L., and Geiger, J. S. (1972). *Can. J. Phys.* **50**, 2302; see also Häusser, O., Khanna, F. C., and Ward, D. (1972). *Nucl. Phys.* **A194**, 113.

Wildenthal, B. H., McGrory, J. B., and Glaudemans, P. W. M. (1971). *Phys. Rev. Lett.* **26**, 96.

Winther, A. (1966). *In Proc. Int. Symp. Nucl. Phys. Tandems, Heidelberg*; Private communication.

Winther, A., and de Boer, J. (1966). *In* "Coulomb Excitation" (K. Alder and A. Winther, eds.), pp. 303–374. Academic Press, New York.

Wong, S. K. M., Le-Tourneaux, J., Quang-Hoc, N., and Saunier, G. (1969). *In Proc. ICPNS* (M. Harvey, R. Y. Cusson, J. S. Geiger, J. M. Pearson, eds.), Contribution 7.18.

Wu, C. S. (1968). *In* "Hyperfine Structure and Nuclear Radiations" (E. Matthias and D. A. Shirley, eds.), pp. 46–64. North-Holland Publ., Amsterdam.

VII.C MAGNETIC MOMENTS OF EXCITED STATES

E. Recknagel

PHYSICS DEPARTMENT, FREE UNIVERSITY OF BERLIN
AND
NUCLEAR AND RADIATION PHYSICS DEPARTMENT
HAHN-MEITNER INSTITUTE FOR NUCLEAR RESEARCH
BERLIN, GERMANY

I. Introduction

Since the postulation of nuclear magnetic moments by Pauli (1924) and the first discoveries of such moments in the 1930's, experimental and theoretical research on electromagnetic moments of nuclear states has had a continuous impact on our knowledge of nuclei and nuclear structure. The discussion of all the material on electromagnetic moments and of the broad scope of experimental techniques for measuring these moments is beyond the limited space of this chapter, so here the main emphasis will be on the magnetic dipole moments of excited nuclear states.

This is well justified, since the determination of nuclear ground-state moments of stable isotopes using conventional methods was already quite complete years ago and no essential development has occurred during the last decade. These techniques—optical hyperfine structure, atomic and molecular beam resonance, nuclear magnetic resonance—were summarized by Frauenfelder and Steffen (1960), in several chapters of the book "Hyperfine Interactions," edited by Freeman and Frankel (1967), and in the books of Kopfermann (1958) and Ramsey (1953). Considerable refinements of these techniques led to more accurate measurements. The combination of conventional techniques with nuclear radiation techniques extended the information to ground-state moments of unstable isotopes as well as long-lived isomeric states (Nierenberg and Lindgren, 1965; de Groot et al., 1965; Otten, 1970).

The rapid increase of information about magnetic dipole moments of excited states is best illustrated by a series of conferences. The Uppsala Conference on Extranuclear Perturbations in Angular Correlations (Karlsson et al., 1964) summarized the development of the first decade of research. About 40 magnetic moments had been measured at that time, most of them by the technique of perturbed angular correlation (PAC), which can be traced back to an idea of Brady and Deutsch (1950), who suggested that the precession of a magnetic moment of a short-lived excited state in a magnetic field should be reflected in the rotation of the angular distribution of the deexciting γ radiation. Besides this most successful way of investigating hyperfine interactions via the Larmor frequency, Mössbauer's discovery of recoilless resonance absorption of γ radiation introduced the other fundamental method of determining the hyperfine interaction by the direct measurement of the energy splitting of the magnetic sublevels of a short-lived excited state. Though the majority of experiments were carried out by PAC using radioactive isotopes and external magnetic fields, future developments had already been pointed out at the time of the Uppsala Conference, namely investigating magnetic moments by applying nuclear reactions, internal magnetic fields, NMR in excited states, etc.

The following International Conference on Hyperfine Interactions Detected by Nuclear Radiations (Matthias and Shirley, 1968) held in 1967 in Asilomar demonstrated the expansion and activity in this field. The extensive use of implantation techniques—first applied to study internal fields with known magnetic moments as probes, and afterward, with the knowledge of the nuclear environment, making it possible to determine new magnetic moments—contributed to a large extent to the 200 known moments of excited states in 1967. In 1970, the Conference on Hyperfine Interactions Detected by Nuclear Radiations (Goldring and Kalish, 1971) held in Rehovot and Jerusalem and the more specialized Conference on Angular Correlations in Nuclear Disintegration (Krugten and Nooijen, 1971) held in Delft represented the state of the art at that time. They showed especially the advances made in the understanding of the action of transient and fluctuating magnetic fields on magnetic moments and in precision experiments on longer-lived excited states. The proceedings of these conferences are a comprehensive source of information on experimental details and new theoretical approaches, which cannot all be covered in this chapter. [†]

In Section II, only a short outline on hyperfine interactions and the theory of PAC will be given, since both are treated in detail by Shirley in Chapter VIII.D and by Frauenfelder and Steffen (1965). In Section III, experimental methods used to measure magnetic moments of excited states will be introduced. The emphasis will be on methods using nuclear reactions, since the PAC methods involving radioactive sources are described in the article by Frauenfelder and Steffen (1965). Section IV is devoted to the discussion of theoretical models which have proved to be very successful in interpreting the experimental results. Limited space does not permit the representation of all experimental results; if not otherwise mentioned, they are taken from the table of Shirley (1971).

II. Hyperfine Interaction and Perturbed Angular Correlations

The determination of magnetic dipole moments of nuclear states is based on the observation of the magnetic hyperfine interaction. The magnetic degeneracy of a level with spin I and energy E is removed under the influence of a magnetic field and $(2I + 1)$ sublevels with magnetic quantum number m_I, $I \geqslant m_I \geqslant -I$, will occur, separated by the difference of potential energy of the magnetic moment μ with respect to the magnetic field $\mathbf{H}$:

$$\Delta E = \langle Im| -\mu\mathbf{H}|Im\rangle - \langle Im - 1| -\mu\mathbf{H}|Im - 1\rangle$$

$$= -g\mu_N H = \omega_L \hbar \tag{1}$$

[†] In 1972, there was a conference devoted to the topic of Nuclear Moments and Nuclear Structure held in Osaka (Horie and Sugimoto, 1973).

with g being the nuclear g factor, μ_N the nuclear magneton, and ω_L the Larmor frequency. Information about this energy difference can be extracted from the deexciting γ radiation in two ways:

(a)   The energy splitting can be directly observed via the Mössbauer effect (see discussion by Shirley, Chapter VIII.D) as long as the separation energy is at least comparable with the natural linewidth of the γ radiation:

$$\Delta E \gtrsim h/\tau \tag{2}$$

where τ is the lifetime of the excited state.

(b)   The splitting can be deduced from the perturbed angular distribution of the γ radiation provided an unequal population of the magnetic substates exists.

The general formula for perturbed angular distributions between two correlated radiations $\mathbf{k}_1$ and $\mathbf{k}_2$ is given by Frauenfelder and Steffen (1965):

$$W(\mathbf{k}_1, \mathbf{k}_2, t) = \sum_{\substack{k_1 k_2 \\ N_1 N_2}} A_{k_1}(1)\, A_{k_2}(2)\, G_{k_1 k_2}^{N_1 N_2}(t) \left[(2k_1 + 1)(2k_2 + 1) \right]^{-1/2}$$
$$\times\, Y_{k_1}^{N_1}(\theta_1, \varphi_1)\, Y_{k_2}^{N_2}(\theta_2, \varphi_2) \tag{3}$$

The influence of an extranuclear perturbation is completely described by the perturbation factor $G_{k_1 k_2}^{N_1 N_2}(t)$. For vanishing perturbation, Eq. (3) reduces to the unperturbed directional correlation function, expressed in terms of Legendre polynomials,

$$W(\theta) = \sum_k A_k(1)\, A_k(2)\, P_k(\cos\theta) \tag{4}$$

where θ is the angle between the two radiation directions $\mathbf{k}_1$ and $\mathbf{k}_2$. The coefficients $A_k(1)$ for the first radiation are orientation parameters, describing to what extent the intermediate state is oriented (aligned or polarized) with respect to the quantization axis $\mathbf{k}_1$. They depend on the nuclear spins I_i and I and on the angular momentum L_1 transferred by the population process (see Fig. 1). In the case of a nuclear reaction, additional factors involving the reaction mechanism and the particle have to be considered. [In the literature they are usually labeled $B_k(1)$.] The angular distribution of the second radiation is described by $A_k(2)$, depending on I, I_f, and L_2. The maximum value of k is determined by the selection rules $k_{max} \leqslant (2I, 2L_1,$ or $2L_2)$. For parity-conserving radiations—α decay, γ radiation—only even values of k contribute, unless the circular polarization of the γ radiation is also observed.

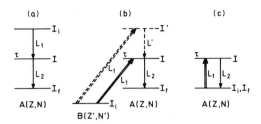

Fig. 1. Principal population schemes of isomeric states and the angular momenta involved. (a) Angular correlation; (b) nuclear reaction; (c) Coulomb excitation and Mössbauer effect.

Sometimes it is convenient to use the equivalent form of Eq. (4):

$$W(\theta) = \sum_k b_k \cos k\theta \tag{5}$$

where, for $k_{max} = 4$ and with the abbreviation $A_{kk} = A_k(1)A_k(2)$, the coefficients b_k are given by

$$b_0 = 1 + \tfrac{1}{4}A_{22} + \tfrac{9}{64}A_{44}, \quad b_2 = \tfrac{3}{4}A_{22} + \tfrac{5}{16}A_{44}, \quad b_4 = \tfrac{35}{64}A_{44}$$

The A_{kk} are tabulated, for example, by Ferentz and Rosenzweig (1965). The corresponding coefficients $B_k(1)$ for reaction processes are tabulated by Yamazaki (1967).

If a perturbation by extranuclear magnetic fields is present—and only those should be considered in this context—the perturbation factor $G_{k_1k_2}^{N_1N_2}(t)$ can be written as

$$G_k^N(t) = \exp(-\lambda_k^N t) \tag{6}$$

with t the time spent in the intermediate state and the λ_k^N being in general complex perturbation constants.

If only a static magnetic field H acts upon the aligned nuclei, λ_k^N takes a simple, pure imaginary form:

$$\lambda_N = iN(E_{m+1} - E_m)/\hbar = i\omega_L N \tag{7}$$

i.e., the nuclei in the intermediate state precess and consequently the connected angular distribution rotates with the Larmor frequency ω_L. Equation (3) becomes in the case of H applied perpendicular to the correlation plane

$$W(\theta, t, H) = \sum_k A_k(1) A_k(2) P_k[\cos(\theta - \omega_L t)] \tag{8}$$

If the lifetime of the isomeric state is short compared to the resolving time of the detection system, the precession angle integrated over the lifetime of the state is measured:

$$W(\theta, t, H) = \sum_k \int_0^\infty (1/\tau)\, e^{-t/\tau} A_k(1)\, A_k(2)\, P_k[\cos(\theta - \omega_L t)]\, dt \qquad (9)$$

or, in the equivalent Fourier form,

$$W(\theta, t, H) = \sum_k \{b_k/[1 + (k\omega_L \tau)^2]^{1/2}\} \cos[k(\theta - \Delta\theta)] \qquad (10)$$

with the rotation angle $\tan(k\,\Delta\theta) = k\omega_L \tau$.

The time-integrated perturbed angular correlation results in a rotation $k\,\Delta\theta$ as well as an attenuation $[1 + (k\omega_L \tau)^2]^{-1}$ compared to the unperturbed correlation. Both values depend on the Larmor frequency ω_L and the lifetime τ of the excited state. To extract the g factor from time-integrated measurements, the lifetime has to be known. The sign of the g factor can be determined from the direction of the rotation.

In case only fluctuating magnetic fields are present, the perturbation constant λ_k^N is, according to Abragam and Pound (1953),

$$\lambda_k = \tfrac{3}{2}\tau_c I(I+1)\,(1/\hbar)\, g^2 \mu_N^2 H_{int}^2 [1 + (2I+1)\, W(I1kI, II)] \qquad (11)$$

(W is a Racah coefficient) where τ_c is the correlation time of the fluctuating magnetic field H_{int}. Since λ_k is real, the time-integrated angular correlation will be attenuated:

$$W(\theta, \infty, H_{int}) = \sum_k A_k(1)\, A_k(2)\, \frac{1/\tau}{(1/\tau) + \lambda_k}\, P_k(\cos\theta) \qquad (12)$$

III.   Methods for Measuring Magnetic Moments of Excited States

A.   GENERAL REMARKS

As already mentioned in the introduction, an important advancement in the determination of nuclear magnetic moments of excited states during the last ten years resulted from perturbed angular distribution experiments following nuclear reactions or Coulomb excitation, hereafter referred to as PAD. This progress is mainly due to the following reasons:

(1)   Many isomeric states not accessible by conventional perturbed angular correlation experiments with radioactive isotopes (PAC) can be populated by nuclear reactions.

(2)   The investigation of isomeric states is only weakly limited by the

lifetime of the states, i.e., states with lifetimes ranging from picoseconds up to hours can in principle be explored. New methods as well as decisive improvements of experimental techniques opened up new areas of investigation.

(3) Increasing knowledge of internal hyperfine fields made it possible to extract nuclear moments even from rather complex hyperfine interactions.

1. *Nuclear Reactions*

The nuclear reaction process as well as inelastic scattering and Coulomb excitation are discussed in detail in other chapters of this book, so that only the relevant phenomena for the measurement of hyperfine interactions will be summarized. The proper choice of target material and beam particles and energies enables one to populate nearly any isomeric state, i.e., one is not restricted to the very selective population of suitable γ cascades in the radioactive decay. Systematic studies of corresponding isomeric states in a series of isotopes or isotones are thus much easier and sometimes the only possibility, particularly if nuclei far from the valley of stable isotopes are involved, where the lifetimes of parent nuclei are too short to prepare radioactive sources.

In all reaction processes, the energy of the incoming particles has to exceed the threshold energy and the Coulomb barrier, if one disregards sub-Coulomb stripping. The choice of the proper energy, especially in heavy-ion reactions, or use of heavy targets allows one to produce final nuclei very selectively.

Besides the energetic conditions which have to be fulfilled for nuclear reactions, the difference in angular momentum I between the ground state of the initial nucleus and the isomeric state of the final nucleus—or a higher excited level which cascades down to that state—has to be provided by the incoming particle. In most cases, an orbital angular momentum transfer is connected with this process, which usually results in an anisotropic population of the magnetic sublevels of the final state and, because of the axial symmetry, nuclei are aligned either perpendicular or (seldom) parallel with respect to the beam axis, depending on the different moments involved in this process. Polarized nuclei can be obtained either by observing recoiling final nuclei at a certain reaction angle or employing polarized beam particles or targets. Subsequently, the electromagnetic radiation deexciting the isomeric state will show an anisotropic angular distribution, which is given by Eq. (4), where the angle θ is taken with respect to the beam axis. In heavy-ion reactions, usually high angular momenta are transferred—up to

$30\hbar$. Since the evaporated neutrons (or other particles) carry away none or only little orbital angular momentum, final states with very high spins are excited (also see Newton, Chapter VII.E).

Due to the nuclear reaction, the final nucleus receives a recoil, which may be of the order of millions of electron volts. This frequently will complicate investigations in solid materials, because the final position of the nucleus in the lattice is poorly known, so that internal fields in addition to the externally applied fields may act on the nuclear moments. On the other hand, one can take advantage of this recoil by implanting the nuclei into vacuum, gas, or condensed material, especially in ferromagnetic substances. Thus any nucleus can in principle serve as a probe to study the properties of a given host lattice. Extensive use of this possibility is made in the experiments discussed in Section III.

2. Survey of the Methods

Many techniques for the determination of magnetic dipole moments of excited states are currently used and have reached a high degree of perfection, so that almost all states living longer than a few picoseconds can be investigated. In Fig. 2, a summary of different methods for studying hyperfine interactions and their applicability to different nuclear lifetime ranges is given (adapted from Matthias, 1967). The accuracy of a measurement for favorable cases is determined by the natural line width

$$\Delta\omega_{1/2} = 1/\tau \tag{13}$$

which can be related for magnetic hyperfine interactions to a line width $\Delta H_{1/2}$, using Eq. (1):

$$\Delta H_{1/2} = \hbar/\tau g\mu_N \tag{14}$$

This line width $\Delta H_{1/2}$ is drawn as an additional scale in Fig. 2, assuming the g factor to be $g = 1$. The arrows indicate the range of application where experiments are presently being carried out.

Time-differential methods may be applied for $\tau \approx 10^{-9}$–10^{-2} sec, limited at shorter lifetimes by the necessity of observing approximately one revolution within the lifetime, and at longer lifetimes by counting rates and relaxation processes. Here the application of PAD proves to be superior to PAC, since the coincidence condition restricts the γ–γ correlation experiments to $\tau \lesssim 1$ μsec. The accuracy of these measurements, depending on the lifetime, ranges from 1% to 0.1%. Time-integral methods are limited by the rotation angle which a magnetic field causes within the lifetime of the state. Thus angular resolution and the highest internal fields so far known ($\sim 10^6$ G) restrict these investigations to $\tau \gtrsim 10^{-11}$ sec. The accuracy of these measurements is on the order of 10%, mainly due to inaccurate lifetime determinations and uncertainties in the knowledge of inner fields. NMR experiments on excited states can only be carried out for lifetimes longer than a few

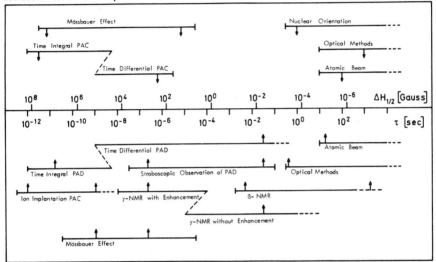

Fig. 2. Summary of experimental techniques for measuring magnetic dipole moments of exited states. Adapted from a figure by Matthias (1967). The arrows indicate the range of application where experiments have been carried out.

microseconds, due to the radio frequency power condition. In ferromagnetic substances, this can be extended to shorter lifetimes, $\tau \gtrsim 10^{-8}$ sec, by taking advantage of the enhancement factor.

The limiting conditions for the other techniques, not extensively discussed in what follows, are summarized in Table 1.

B. TIME-DIFFERENTIAL PERTURBED ANGULAR DISTRIBUTION METHODS

For isomeric states with lifetimes longer than about 1 nsec, the time-differential observation of the perturbed angular distribution is the most appropriate method for determining their magnetic moments.

TABLE 1

LIMITING CONDITIONS FOR THE APPLICATION OF METHODS FOR MEASURING NUCLEAR MAGNETIC DIPOLE MOMENTS OF EXCITED STATES OR UNSTABLE GROUND STATES

Method	Shorter lifetimes		Longer lifetimes	
	τ (sec)	Limiting condition	τ (sec)	Limiting condition
Time-integral PAC	10^{-11}	Rotation angle resolution $\omega\tau > 1$ mrad; internal hyperfine field $H \approx 10^6$ G	$\sim 10^{-8}$	Time-differential PAC
Time-differential PAC	10^{-9}	Time resolution $t_1 \lesssim \tau$; Larmor period $1/\omega_L \gtrsim \tau$	10^{-5}	Coincidence condition
Time-integral PAD	10^{-11}	Rotation angle resolution $\omega\tau > 1$ mrad; internal hyperfine field $H \approx 10^6$ G	$\sim 10^{-8}$	Time-differential PAD
Ion implantation PAC	10^{-12}	Rotation angle resolution $\omega\tau > 1$ mrad; internal fluctuating hyperfine field $H \approx 10^7$ G	$\sim 10^{-8}$	Time-differential PAD
Time-differential PAD	10^{-9}	Time resolution $t_1 \gtrsim \tau$; Larmor period $1/\omega_L \gtrsim \tau$	10^{-2}	Counting rates; relaxation (T_1, T_2)
Stroboscopic observation of PAD	10^{-7}	Time condition: repetition time $\gg$ time resolution	10^{-2}	Line width; relaxation (T_1, T_2)
γ-NMR/RD with enhancement	10^{-8}	Rf power	—	γ-NMR/RD without enhancement
γ-NMR/RD without enhancement	10^{-5}	Rf power	10^2	Relaxation (T_1)
β-NMR/RD	10^{-3}	Lifetime of β-unstable nuclei	10^2	Relaxation (T_1)
Mössbauer effect	10^{-11}	Energy resolution; internal fields	10^{-4}	Line width, velocity curve
Nuclear orientation	10^3	Source handling of parent nuclei	—	None
Atomic beam resonance	10	Source handling	—	None
Optical methods	10^{-1}	Source handling	—	None

1. *Differential Perturbed Angular Correlation (DPAC)*

The various techniques applied in γ–γ angular correlation experiments using radioactive sources are summarized by Frauenfelder and Steffen (1965), so that only some recent developments should be mentioned, which are aimed at extending the range of application of DPAC. Usually the following conditions have to be fulfilled: To observe at least one rotation of the PAC within the lifetime τ of the excited state, the Larmor frequency ω_L has to be larger than or comparable to $1/\tau$; $\omega_L \gtrsim 1/\tau$. This requires the use of high magnetic fields for short-lived states. Since external fields are limited to about 10^5 G, internal hyperfine fields, which range up to more than 10^6 G are increasingly used. (Tables of hyperfine fields are given by Koster and Shirley, 1971.) The short Larmor periods thus obtained become comparable with the resolving time Δt of the detection device, which has to be smaller than $1/\omega_L$; $\Delta t < 1/\omega_L$. Resolving times of about 1 nsec or better for γ energies $E_\gamma \gtrsim 100$ keV are feasible with sophisticated electronic systems.

A typical example is an experiment by Rhaghavan *et al.* (1971) shown in Fig. 3, where the perturbed γ–γ correlation of the 173–247-keV cascade

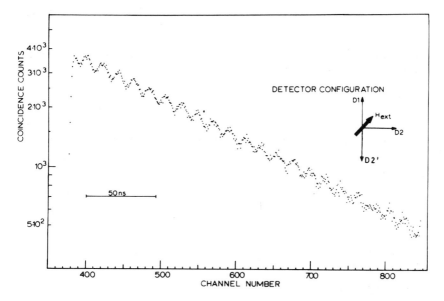

Fig. 3.   DPAC results for the 247-keV state of ^{111}Cd in iron, polarized in the detector plane (Rhagavan *et al.*, 1971).

of ^{111}Cd ($\tau = 84$ nsec) is observed in the polarized internal field of iron, $|H_{int}| = (366 \pm 7)$ kG. Beyond that, this experiment demonstrates a new technique, in which the magnetic field is applied in the detector plane instead of perpendicular to the plane. (See inset of Fig. 3.) The first detector is fixed at an angle of 45° with respect to the magnetic field, while the second one is placed successively at 90° and 180° to the first. The difference in coincidence counting rates $C(t)$ at the two positions is

$$C(t) = W(180°, t, H) - W(90°, t, H) = \tfrac{3}{4}A_2 \cos \omega_L t \qquad (15)$$

assuming $A_4 = 0$. The modulation frequency is only ω_L instead of the lowest frequency component $2\omega_L$ in the usual case according to Eq. (8), thus gaining a factor of two for a given resolving time.

Even in unpolarized magnetic substances, the hyperfine interaction can be observed. According to Matthias *et al.* (1965), the perturbation factor $G_k(t)$ for such randomly oriented static magnetic fields amounts to

$$G_k(t) = (1/2k + 1) \sum_{N=0}^{k} \cos(N\omega_L t) \qquad (16)$$

Again in this case, the predominant frequency component is given by ω_L.

2. Differential Perturbed Angular Distribution (DPAD) or Spin Rotation Method

In contradistinction to the techniques just discussed, where radioactive sources are involved, DPAD refers to experiments following nuclear reactions. The theoretical formulation of the spin rotation method is identical to the theory of DPAC. In pulsed-beam spin-precession experiments, the Larmor precession of nuclei excited and aligned by a nuclear reaction can be determined by time-differential measurements of the decay γ-ray distribution, i.e., one has to carry out a delayed coincidence experiment between the exciting pulse and the subsequent γ-ray decay. Therefore, this method can only be applied as long as the lifetime of the excited state is shorter than the pulse repetition time T_0. The application of a pulsed beam offers two advantages:

(a) More than one γ-ray can be observed with respect to one beam pulse ($t = 0$). Thus, for longer lifetimes and consequently longer time intervals between exciting pulses, the counting rates are not as much affected as for γ–γ correlations.

(b) Since only γ rays in between the pulses are observed, the background

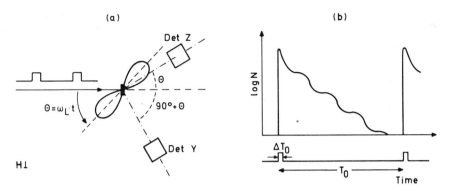

Fig. 4. Schematic illustration of the spin rotation method (DPAD). (a) Experimental arrangement. The magnetic field is perpendicular to the detector plane. (b) Modulated γ-decay spectrum as a function of time. ΔT_0 is the width of exciting beam pulse; T_0 is the repetition time of beam pulses.

radiation can easily be suppressed by electronic means. In Fig. 4a an experimental setup is schematically shown. An external magnetic field H at the target position is directed perpendicular to the beam–detector plane. Usually two detectors are mounted at angular positions θ_1 and θ_2 with respect to the beam direction. Then the observed γ-ray intensity of the decaying state at a time t with respect to the beam pulse ($t = 0$) is given by

$$I(\theta, t, H) = I_0 e^{-t/\tau} W(\theta, t, H) \tag{17}$$

where $W(\theta, t, H)$ is given by Eq. (8). Figure 4b shows the principal intensity distribution as a function of time in relation to the pulsing process. Generally the intensity ratio of two counters separated by 90° is formed:

$$R(\theta, t, H) = \frac{I(\theta, t, H) - I(\theta + 90°, t, H)}{I(\theta, t, H) + I(\theta + 90°, t, H)} \tag{18}$$

which, by substituting the explicit expression of Eq. (17), directly allows one to obtain the Larmor frequency. For $k_{max} = 4$,

$$
\begin{aligned}
&R(\theta, t, H) \\
&= \frac{[(3/2)\, B_2 A_2 + (5/8)\, B_4 A_4]\cos[2(\theta - \omega_L t)]}{2 + (1/2)\, B_2 A_2 + (9/32)\, B_4 A_4 + (35/32)\, B_4 A_4 \cos[4(\theta - \omega_L t)]}
\end{aligned} \tag{19}
$$

If $B_4 A_4 = 0$, or $|B_4 A_4/B_2 A_2| \ll 1$, this expression reduces to

$$R(\theta, t, H) = [3 B_2 A_2/(4 + B_2 A_2)]\cos[2(\theta - \omega_L t)] \tag{20}$$

By choosing an asymmetric geometry for the two detectors with respect to the beam direction, the sign of the g factor can be easily determined from the phase angle θ at $t = 0$.

To summarize, the following conditions have to be fulfilled in conventional spin rotation experiments:

(a) $\tau < T_0$;

(b) $\pi/\omega_L < \tau$;

(c) $\Delta T < \pi/\omega_L$ (ΔT is the pulse width).

As an example of DPAD, the determination of the g factor of the first excited state in $^{19}\text{Ne}(I = \frac{5}{2}^+; E_\gamma = 239 \text{ keV}; \tau = 25.5 \text{ nsec})$, which was populated by the reaction $^{19}\text{F}(p, n)^{19}\text{Ne}$, shall be discussed (Bleck et al., 1969a). A thick CaF_2 target was bombarded with 5.5-MeV protons using a pulsed beam of width $\Delta T = 10$ nsec and repetition time $T = 1000$ nsec. Figure 5 shows the spin precession spectra of two counters α and β. The higher counting rates near $t = 0$ result from prompt radiation during the beam pulse. In the sum spectrum $\gamma = \alpha + \beta$, a small modulation with the double frequency $4\omega_L$ can be seen, which is due to the contribution of the $k = 4$ term. The external magnetic field was about 34 kG. The inset of Fig. 5a indicates the geometric position of the counters. Figure 5b shows the ratio $R(\theta, t, H)$ according to Eq. (18). To calibrate this spectrum in time and magnetic field, the spin precession spectrum of the second excited state of $^{19}\text{F}(I = \frac{5}{2}^+; E_\gamma = 197 \text{ keV}; \tau = 80 \text{ nsec})$ was taken at the same time (Fig. 5c). Since the g factor of this level is known with high accuracy (Bleck et al., 1969b), the determination of $g(^{19}\text{Ne})$ reduces to the evaluation of the frequency ratio $\omega_L(^{19}\text{Ne})/\omega_L(^{19}\text{F})$. From this ratio, the g factor is $g(^{19}\text{Ne}) = -0.296(3)$ [the error in the least significant digit(s) is also given].

3. Stroboscopic Observation of Perturbed Angular Distribution (SOPAD)

For excited states with lifetimes longer than about 1 μsec, the DPAD technique suffers from intensity difficulties. This is overcome by the stroboscopic observation of the nuclear Larmor precession (Christiansen et al., 1968, 1970). In principle, this method measures directly the Fourier transform of the spin precession spectrum, which results in a resonance curve. From the position of the resonance, the magnetic moment can be determined to higher accuracy within shorter measuring times, while the amplitude, width, and position of the curve are sensitive quantities to other perturbations present during the lifetime of the state (e.g. Bertschat et al., 1970). The unambiguous correlation between exciting pulse and subsequent γ decay

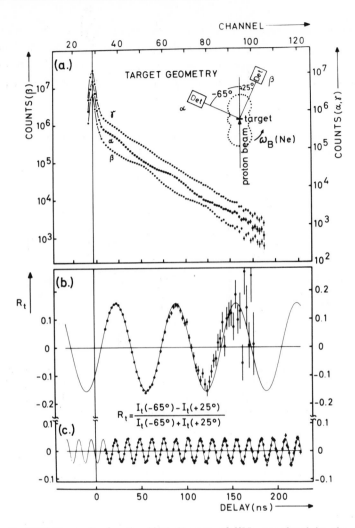

Fig. 5. DPAD results for the 239-keV state of ^{19}Ne populated by the reaction ^{19}F(p, n) ^{19}Ne. External magnetic field $H \approx 34$ kG. (a) Original modulated decay spectra taken with detectors α and β separated by $\Delta\theta = 90°$ [$\alpha : I_t(-65°)$; $\beta : I_t(+25°)$; $\gamma : I_t(-65°) + I_t(+25°)$]. The spectrum γ is the sum of α and β. (b) Intensity ratio evaluated by Eq.(18). (c) Calibration: intensity ratio from 197-keV state of ^{19}F. (Bleck *et al.*, 1969a, with permission of North-Holland Publishing Co., Amsterdam.)

is no longer necessary in SOPAD; instead, many beam pulses may occur within the lifetime of the excited state ($\tau > T_0$). If the Larmor frequency of the excited nuclei is equal to a multiple of half the pulse frequency which characterizes the resonance condition $\omega_L = m\pi/T_0$ ($m = \pm 1, 2, \ldots$ harmonics), all nuclei precess with a constant phase with respect to the beam pulses, and one gets a coherent superposition of the intensity modulation produced by the perturbed angular distribution. Figure 6 explains the principle of the method.

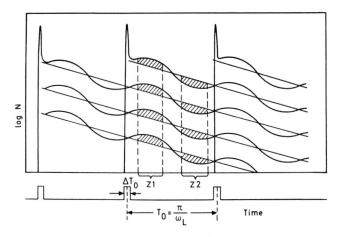

Fig. 6. Schematic illustration of the stroboscopic observation of the Larmor frequency (SOPAD). In the resonance case, $T_0 = \pi/\omega_L$, the counting rates Z_1 and Z_2 are sensitive to the anisotropy of the perturbed angular distribution. ΔT_0 is the width of exciting beam pulse; T_0 is the repetition time of beam pulses (Christiansen *et al.*, 1968).

Summation of the γ-ray intensities according to Eq. (17) produced by different beam pulses results in a γ-intensity distribution within the pulse interval

$$I(\theta, t, H, T_0, \tau) = I_0' \sum_{n=0}^{\infty} \{\exp[-(t + nT_0)/\tau]\} \, W[\theta - \omega_L(t + nT_0)] \qquad (21)$$

Inserting an expression equivalent to Eq. (8), one gets

$$I(\theta, t, H, T_0, \tau) = I_0' e^{-t/\tau} \sum_{k} B_k' \cos[k(\theta - \omega_L t - \delta_k)] \qquad (22)$$

with

$$B_k' = b_k \{1 - 2[\cos(k\omega_L T_0)] \, e^{-T_0/\tau} + e^{-2T_0/\tau}\}^{-1/2} \qquad (23)$$

and a corresponding expression for the phase constant δ_k. The modulation amplitudes B_k' exhibit a typical resonance behavior depending on the relation of the two frequencies ω_L and $2\pi/T_0$, a fundamental of any stroboscopic process. If one varies one of the two characteristic frequencies (e.g., the Larmor frequency by changing the magnetic field H), the intensity near the maximum of the modulation, measured at a fixed time t_0 and an appropriate detector position, is given by ($k_{max} = 2$):

$$I(t_0, H, T_0, \tau) \sim 1 \pm b_2 \frac{(1/\tau)^2}{(1/\tau)^2 + 4(\omega - \omega_L)^2} \qquad (24)$$

The line width of this Lorentzian is the natural line width $\Delta\omega_{1/2} = 1/\tau$.

As indicated in Fig. 6, one usually measures the counting ratio Z_1/Z_2 in two time intervals, $t_0 \pm \Delta t$ and $t_0 + (T_0/2m) \pm \Delta t$, taken in the maximum and minimum of modulation:

$$\frac{Z_1}{Z_2} = \frac{I[t_0, \Delta t, \tau, T_0, H, \theta]}{I[(t_0 \pm T_0/2m), \Delta t, \tau, T_0, H, \theta]} \qquad (25)$$

For a two-counter experiment, a double ratio $Z_1 Y_2/Z_2 Y_1$ is formed, where the counters are separated by $\Delta\theta = 90°$, corresponding to the detector geometry in spin precession experiments. This procedure has the advantage of eliminating, to first order, changes in energy- and time-window settings of the electronic detection device during a measurement.

As in DPAD experiments, the sign of the g factor can be easily determined by a proper choice of the detector positions. A detailed discussion, especially about corrections which lead to shifts and broadenings of the resonance curve, is given by Christiansen et al. (1970).

The first SOPAD experiment was carried out by Christiansen et al. (1968) on the $\frac{9}{2}^+$ excited state of $^{69}Ge(E_\gamma = 398$ keV; $\tau = 4.2$ μsec) using the reaction $^{69}Ga(p, n)^{69}Ge$ with liquid gallium as a target. In Fig. 7 the double ratio $(Z_1/Z_2)(Y_2/Y_1)$ is plotted as a function of the external magnetic field. The parameters according to Eq. (25) in this experiment—and they are typical for most SOPAD experiments—are: $T_0 = 1$ μsec; $t_0 = \frac{1}{4}T_0$ and $\frac{3}{4}T_0$; $\Delta t = 100$ nsec; and $\theta_{1,2} = \pm 45°$ with respect to the beam axis. From the resonance field $H_0 = 2.950(15)$ kG, the resulting g factor is $g = -0.2224$ (7). The measured line width $\Delta H_{1/2} = 230(10)$ G agreed with the natural one determined from the lifetime of the state $\Delta H_{1/2}(\tau) = 230(5)$ G. The solid line in Fig. 7 is calculated from Eq. (24) using the given parameters and $b_2 = 0.093$.

C. TIME-INTEGRAL PERTURBED ANGULAR DISTRIBUTION METHODS

Magnetic moments of excited states with lifetimes shorter than a few

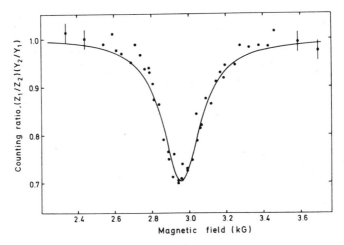

Fig. 7. SOPAD results for the 398-keV state of ^{69}Ge populated by the reaction ^{69}Ga(p,n)^{69}Ge. The magnetic field and hence the Larmor frequency is changed. The solid line is calculated from Eq. (24) with $\tau = 4.2\ \mu$sec (Christiansen *et al.*, 1968).

nanoseconds are determined by measuring the Larmor precession angle integrated over the lifetime of the state and/or by the integral attenuation of the angular distribution [Eqs. (9) and (10)]. The two values are in some sense complementary. While for a large angle, $\tan(k\ \Delta\theta) = k\omega_L\tau$ saturates, the attenuation $|1 + (k\omega_L\tau)^2|^{-1}$ is more pronounced. However, other time-dependent processes may cause the same effect, and therefore the origin of the attenuation has to be carefully studied in each case. For small rotations, $\omega_L\tau \ll 1$, the attenuation can usually be neglected. With present techniques, rotation angles as small as a few milliradians can be measured, which enables one to investigate states with $\tau = 10^{-11}$ sec assuming internal fields of 10^6 G.

1. Integral Perturbed Angular Correlation (IPAC)

A general outline of IPAC techniques is given by Frauenfelder and Steffen (1965), so that only one example need be mentioned, which exhibits some features of recent advances. The low-lying vibrational states of the even Pt isotopes were investigated by taking advantage of the high internal field of dilute Pt in iron. Figure 8a shows the angular correlation for the 1151–329-keV and 939–329-keV 0^+–2^+–0^+ cascades of ^{194}Pt for the two field directions perpendicular to the correlation plane (Keszthelyi *et al.*, 1965). The rotation angle for the 2^+ state [$E = 329$ keV; $\tau = 45(5)$ psec] is

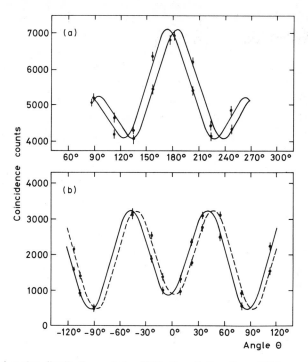

Fig. 8. Angular distributions of the 329-keV radiation from [194]Pt precessed by the hyperfine field of Pt in iron for the two polarized field directions. (a) IPAC results from Keszthelyi *et al.* (1965), with permission of North-Holland Publishing Co., Amsterdam. Radioactive [194]Ir melted in iron was used as a source. (b) IMPAC results from Kalish *et al.* (1967). Coulomb-excited [194]Pt was implanted into iron. (The original figure of Keszthelyi *et al.* was changed in scale for easier comparison with Fig. 8b.)

determined to be $\omega\tau = 0.0916(67)$, in agreement with the results of other authors (Agarwal *et al.*, 1966; Buyrn and Grodzins, 1964). Taking for the internal magnetic field the value of [195]Pt in iron of $|H_{\text{int}}| = 1320(80)$ kG, independently measured by Mössbauer and NMR experiments, the g factor is calculated to be $g = 0.23(3)$. No $\omega\tau$ value could be extracted from the attenuation since contributions from other, disturbing cascades diminished the angular correlation coefficients b_2 and b_4 in Eq. (10).

2. *Ion Implantation Perturbed Angular Correlation (IMPAC)*

As in the case of time-differential experiments, the population of excited states by radioactive sources is rather limited. The application of nuclear reactions and particularly Coulomb excitation extends the number of accessible states considerably and generally results in large anisotropies in the angular distribution. Basically, all IMPAC experiments are very similar, though different approaches have been made in studying the resultant hyperfine interaction. The excited and aligned nuclei recoil from a thin target layer into a suitable stopping environment. This could be (a) a perturbation-free material (i.e., a diamagnetic metal) with an applied external magnetic field (IPAD); or (b) a ferromagnet, where one takes advantage of the high static internal field (this, strictly speaking, is IMPAC); or (c) a gas or vacuum in which the fluctuating magnetic fields of the highly ionized nuclei are exploited (CEAD: Coulomb excitation angular distribution). The general layout of most of these experiments is shown in Fig. 9. A particle beam,

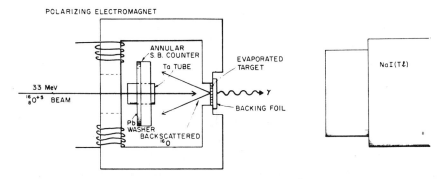

Fig. 9. Schematic drawing of experimental setup for implantation studies (Murnick *et al.*, 1967). For recoil-into-gas experiments, the backing foil is replaced by a gas cell.

^{16}O, passes through the hole of a surface barrier ring counter and Coulomb-excites the target nuclei. The deexciting γ radiation is detected in coincidence with the backscattered particles. By this means, only the $m = 0$ sublevel of the excited state is populated, resulting in a large anisotropy of the angular distribution (if one ignores the finite solid angle of the detector), and γ rays are counted only from those nuclei that recoil with maximum energy in the direction of the beam ($v/c \approx$ const).

By means of two examples, some principal features of IMPAC experiments can be discussed. Figure 8b shows the angular correlation of the same 2^+ state of ^{194}Pt, this time Coulomb-excited and implanted into an iron foil magnetized perpendicularly to the beam–detector plane (Kalish *et al.*, 1967). By comparing the rotation angle $\omega\tau = 0.070(4)$ with the result of the IPAC measurement shown in Fig. 8a, one has to conclude that a much lower average magnetic field acts upon the implanted nuclei. This discrepancy is observed in various similar experiments and has been explained as due to a transient magnetic field H_1 present on the nuclei during the slowing down process, which is short compared to the lifetime of the excited state ($\tau_1 \ll \tau$). In Fig. 10, $(\omega\tau)_{\text{IMPAC}}$ is plotted versus $(\omega\tau)_{\text{IPAC}}$ for several excited states in iron. All data are in agreement if one assumes an additional rotation

$$\omega_1\tau_1 = (\omega\tau)_{\text{IMPAC}} - (\omega\tau)_{\text{IPAC}} \tag{26}$$

which can be attributed to an aligned magnetic field caused by an additional

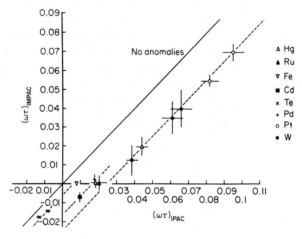

Fig. 10. Precession angles obtained from implantation recoil for short-lived 2^+ states versus radioactivity results for the same levels (Murnick, 1971).

density of polarized electrons at the nucleus which comes from Coulomb scattering of the nearly free polarized electrons of the magnetic host material with the recoiling nucleus.

Lindhard's theory (Lindhard, 1969; Lindhard and Winther, 1971) postulates an amplification factor $2\pi Z_1 e^2/\hbar v_{\text{rel}}$ of the Fermi contact term $|\psi_s^2(0)\uparrow| - |\psi_s^2(0)\downarrow|$, where Z_1 is the atomic number of the recoil and v_{rel} is the relative velocity of electron and ion. From experiment, an average positive H_1 of the order of 10^7 G, acting during a time $\tau_1 \approx 1$ psec, can be deduced. Several experiments using different layers of stopping material were carried out to establish this effect (Murnick, 1971). Once this additional perturbation is understood, other excited states, at least of isotopes of the same element, can be determined. The IMPAC technique has been extensively applied to 2^+ vibrational states with lifetimes $\tau \leqslant 0.5$ nsec and many of the data to be shown later were measured this way.

The second example deals with a recoil-into-gas implantation experiment, a technique usually referred to as CEAD. Highly ionized atoms recoiling into gas undergo collisions with the molecules of the gas, in which the time between successive collisions is $\tau_c = l/v$, l being the mean free path of the ion, given by the gas kinetic theory, and v being the velocity of the ion. The collisions affect the ion in two ways:

(a)   They establish an equilibrium ionization state if the velocity does not change too much during the lifetime of the excited state, i.e., $\tau \ll$ stopping time t. Thus usually a large hyperfine field will be present.

(b)   They cause a random change of the atomic angular momentum J of the ion, due to the large momentum transfer in the ion–molecule collisions. This involves a random fluctuation of the hyperfine field at the nucleus. To apply the Abragam and Pound theory (1953) of perturbations by randomly fluctuating fields, the collision time τ_c has to be short compared to the lifetime τ of the excited state, $\tau_c \ll \tau$. Under these assumptions, the integral perturbation factor is given by

$$G_k(\infty) = [1 + p_k \omega^2 \tau \tau_c(p)]^{-1} \tag{27}$$

where the p_k are coefficients which depend upon whether a magnetic or quadrupole interaction is involved, and $\tau_c(p)$ is the correlation time, which is a function of the gas pressure.

Figure 11 shows the perturbed angular correlation of the 2^+ excited state of ^{172}Yb, recoiled into argon (Ben-Zvi et al., 1970). The least-squares fitted curve coincides with a pure magnetic perturbation. This behavior is established for many examples of recoil-into-gas experiments (Ben-Zvi et al., 1968, 1970). The g factor has to be extracted from the directly observable quantity $\omega^2 \tau \tau_c(p)$, which is a product of an atomic factor $\lambda^2 = H^2 \tau_c(p)$ and a nuclear factor $g^2 \tau$. Though the atomic factor is already understood quite well by systematic experiments (Ben-Zvi et al., 1970), the easiest way to determine an unknown g factor is to relate it to a known g factor which is investigated under the same experimental conditions. The result in the case of the 2^+ state of ^{172}Yb is $g = 0.558(28)$, where the average field was determined to be $\bar{H} \approx 2 \cdot 10^7$ G. Experiments in which the conditions $\tau_c \ll \tau$ and $\tau \ll t$ as well as $J \gg I$ are not fulfilled are discussed by Ben-Zvi et al. (1970).

The case of recoils implanted into vacuum can be treated in a similar way. There, the time-dependent hyperfine interaction is caused by the optical

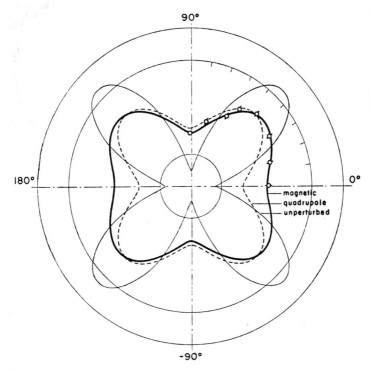

Fig. 11. Polar plot for the angular distribution of ^{172}Yb recoiling into argon. The least-squares fitted curve coincides with a pure magnetic perturbation prediction (Ben-Zvi *et al.*, 1970, with permission of North-Holland Publishing Co., Amsterdam).

transitions occurring during the lifetime of the state (Nordhagen *et al.*, 1970; Nordhagen, 1971).

D. RADIATIVE DETECTION OF NUCLEAR MAGNETIC RESONANCE

1. *Gamma-Nuclear Magnetic Resonance Perturbed Angular Distribution (γ-NMR/RD)*

This method was developed in analogy to classical nuclear magnetic resonance. While in classical NMR the difference in population of the magnetic sublevels m is realized by a deviation from the Boltzmann distribution by means of strong external magnetic fields and low temperature, in NMR/RD the anisotropic population of the magnetic sublevels of the isomeric state is caused by the selective population process. If an external magnetic field H is applied parallel to the symmetry axis of the angular distribution, i.e., the beam axis or the direction of the first γ ray, the sublevels will split up by the energy difference $\Delta E = \hbar\omega_L$, but the angular distribution will not be perturbed by the magnetic field, since the precession will be around the symmetry axis. If an alternating magnetic field H_1 with frequency ω polarized perpendicularly to H is applied, the differences in population of the magnetic substates will be removed in the resonance case $\omega = \omega_L$ ($\Delta m = \pm 1$ transitions), thus destroying the original alignment. Consequently, the subsequent γ radiation will lose its anisotropic angular distribution. The perturbation is reflected in a change of counting rates in- and off-resonance in suitably placed detectors, usually at $0°$ and $90°$ with respect to the beam direction. According to Matthias et al. (1971), the time-dependent perturbation factor $G_{k_1k_2}^{N_1N_2}(t)$ can be calculated by using the time-dependent interaction Hamiltonian for a static magnetic field H and a circularly polarized rf field H_1

$$\mathbf{H}(t) = -g\mu_N\{HI_z + H_1[I_x\cos(\omega t + \Delta) + I_y\sin(\omega t + \Delta)]\} \tag{28}$$

Without going into further details of the theory, which is described by Matthias et al. (1971), the final expression for the perturbation factor can be greatly simplified by a suitable geometry and detection method:

(a)   If the resonance is observed in a time-integrated manner, which is the case in most experiments, $G(t)$ has to be integrated over the lifetime τ of the isomeric state:

$$\hat{G}_{k_1k_2}^{N_1N_2} = (1/\tau)\int_0^\infty G_{k_1k_2}^{N_1N_2}(t)\, e^{-t/\tau}\, dt \tag{29}$$

(b)   If the measurement is not phase sensitive, one has to integrate over phase constants Δ from 0 to 2π (random phase experiments). From this it follows that only terms with $N_1 = N_2$ occur.

(c)   For certain observation angles, for example, $\theta = 0°$, only $\hat{G}_{k_1k_2}^{00}$ exists. In Fig. 12, the perturbation factor $\hat{G}_{22}^{00}$ is plotted for the most probable case $k = 2$ as a function of ω/ω_L. The resonance is actually split k fold, i.e., for $k = 2$, twofold. The splitting is proportional to the rf field strength H_1, and can be used to determine the effective rf amplitude at the nucleus.

Some other requirements have to be fulfilled for γ-NMR/RD experiments:

(a)   The splitting of the substates should be large compared to the natural line width, i.e., $\omega_L\tau \gg 1$ (see Fig. 12).

(b)   The amplitude of the rf field H_1 must be big enough to induce tran-

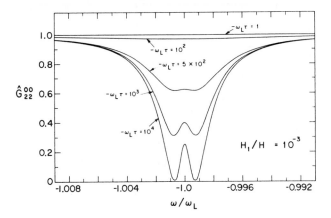

Fig. 12.   Form of angular correlation NMR for $k_{max} = 2$ and random phase. The curves for different values of $-\omega_L\tau$ reflect the saturation behavior (Matthias, 1968, with permission of North-Holland Publishing Co., Amsterdam).

sitions between the substates in times comparable to the lifetime of the isomeric state, i.e., $\omega\tau > 1$. For a lifetime of 200 nsec and a g factor $g = 1$, the rf amplitude has to be about 1 kG, which corresponds to a power of 10^6 W. This usually restricts measurements to lifetimes $\tau > 10^{-6}$ sec. In special cases, namely in NMR in ferromagnets, one can take advantage of the enhancement of the rf field (Portis and Lindquist, 1965). The rf field H_1 is amplified according to the relation

$$2H_1^{\text{eff}} = [1 + (H_{\text{hf}}/H)]\, H_1 \tag{30}$$

where H_{hf} is the magnetic hyperfine field and H is the external polarizing field. In this way, a factor of 10^3 easily can be gained (Matthias *et al.*, 1966).

(c)   Nuclear relaxation times should be longer than the nuclear lifetime, i.e., $\tau < \tau_{\text{rel}}$, so as not to broaden or wipe out the resonance line.

As an example, a measurement by Bräuer *et al.* (1971) is given. They investigated the 37 μsec excited state in ^{81}Br$(I = \frac{9}{2}^+; E = 550$ keV), which was populated by the reaction ^{80}Se(d, n)^{81}Br. A molten Se–Tl alloy was used as a target. Since the γ-NMR/RD is essentially a time-integral method, a pulsed beam $(\Delta T \approx 10\ \mu\text{sec},\ T_0 \approx 400\ \mu\text{sec})$ was used and the integral counting rates between the pulses were measured. In Fig. 13, the NMR resonance of the isomeric state can be seen. The experiment was carried out with a fixed radio frequency of $\omega = 330$ kHz and a variable external magnetic field. The

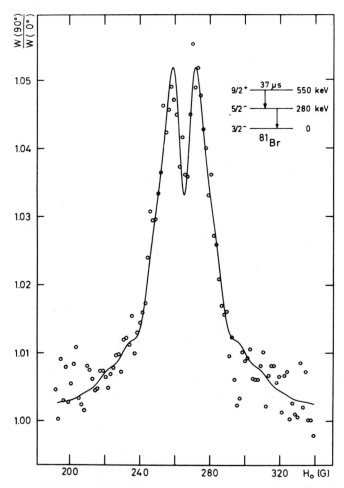

Fig. 13. γ-NMR/RD result for the 550-keV state of ^{81}Br populated by the reaction ^{80}Se(d, n)^{81}Br. From Bräuer, N., Focke, B., Lehmann, K., Nishiyama, K., Riegel, D.: Magnetic Moments of Isomeric Levels in the Br-Isotopes Measured with In-Beam NMR-PAC. *Z. Physik* **244**, 375–382 (1971). Berlin-Heidelberg-New York: Springer.

resonance occurred at $H \approx 260$ G, from which the g factor was calculated to be $|g| = 1.297(15)$.

2. β-*Nuclear Magnetic Resonance Perturbed Angular Distribution (β-NMR/ RD)*

Essentially this method is the same as the γ-NMR, so that only the

distinguishing features will be discussed. The idea is to use the parity viola-
tion of the β decay as a detection method for the degree of polarization.
Therefore, the nuclei under investigation have to be polarized (and not only
aligned) before applying any kind of perturbation.

For allowed transitions, the angular distribution for outgoing β particles
is given by Shapiro (1958) as

$$W(\theta) = 1 + (v/c)\, PA \cos\theta \tag{31}$$

where P determines the degree of nuclear polarization $\langle I_z \rangle / I$; v/c is the
electron velocity versus the velocity of light; θ is the angle between the direction of electron emission and polarization axis; and the coefficient A is a function of the relevant matrix elements for the β decay. For maximum anisotropy, θ should be $0°$ or $180°$, i.e., two β counters should be placed along the
polarization axis above and below the probe. Then the asymmetry R is given by

$$R = (W_U / W_L) - 1 = \alpha/(1 - \alpha/2) \tag{32}$$

where W_U and W_L are the upper and lower counting rates, respectively, and
$\alpha = (\langle I_z \rangle / I)(v/c)A$.

This asymmetry is largest if no perturbation is present and will be wiped
out, if, for example, the proper rf field induces transitions between the
unequally populated m substates. An experiment of this type was carried out
by Sugimoto et al. (1966) on the β-unstable ground state of $^{17}F(\tau = 95\ \text{sec})$,

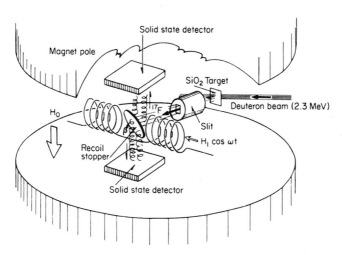

Fig. 14. Schematic drawing of experimental arrangement for the β-NMR/RD measurement on ^{17}F (Sugimoto et al., 1966).

which was populated by the reaction $^{16}O(d,n)^{17}F$. Figure 14 shows the experimental arrangement. A deuteron beam of 2.3 MeV hits a thin SiO_2 target. The ^{17}F nuclei, having an energy of 300–400 keV, recoiled into vacuum and were collimated within a certain angle ($\sim 20°$) with respect to the beam direction to select only those that are polarized parallel to the direction of a strong magnetic field. This field has to maintain the polarization during the flight of the nuclei through the vacuum (I and J are decoupled; Paschen–Back effect) and has to provide the static magnetic field to split the magnetic substates. The recoils were stopped in a pure CaF_2 crystal (fcc lattice). Perpendicular to the magnetic field, a linearly polarized rf field was applied.

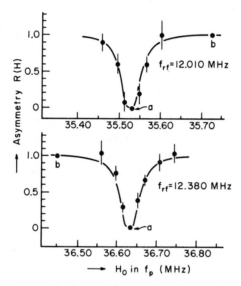

Fig. 15. Experimental results for β-NMR on ^{17}F. The relative asymmetries $R(H)$ normalized by the values at the points a and b are plotted with statistical errors (Sugimoto *et al.*, 1966).

In Fig. 15, the normalized asymmetry is plotted as a function of the magnetic field. The g factor of the $\frac{5}{2}^+$ ground state of ^{17}F was determined to be $g = 1.2889(5)$. In a further experiment with crossed rf fields, where the direction of the circular component of the rf field could be chosen, the sign of the moment was determined to be positive.

Similar experiments were carried out with polarized slow neutrons which are captured by the target nuclei. The polarization of the neutrons is accomplished by a magnetized neutron mirror. To discriminate against unwanted background radiation, the neutron beam is pulsed by a mechanical chopper. Since the binding energy of the captured neutron appears as excitation of the compound nucleus, typically 7 MeV, the ground state is populated by the emission of one or more capture γ rays. Thus the ground state will have a polarization different from that of the excited state, usually a somewhat smaller one. Tsang and Connor (1963) determined the g factor of the unstable ground state of ^{20}F($\tau = 16.3$ sec, $I = 2^+$), using the reaction ^{19}F(n$\uparrow$, $x\gamma$)^{20}F, to be $g = +1.0463(2)$.

E. OTHER METHODS

In recent years, some well-known techniques developed to measure nuclear moments of ground states have been successfully applied to long-lived radioactive states.

The atomic beam resonance method has been extensively used to investigate unstable ground states with lifetimes of the order of hours (Nierenberg and Lindgren, 1965). The radioactive atoms, evaporated from an oven in the usual way, pass the polarizing, resonance, and analyzing magnetic fields and are accumulated on a suitable collector. The detection takes place by counting the activity of the interchangeable collectors for each frequency point. Besides these off-line experiments, some on-line measurements have been carried out in which the oven is replaced by a target and the production of radioactive atoms is accomplished by a nuclear reaction, e.g., ^{24}Mg (p,α)^{21}Na($\tau = 32$ sec), (Ames et al., 1965). A further variation has been developed by Calaprice et al. (1965) for short-lived noble gases (^{19}Ne, ^{35}Ar) where the polarized radioactive atoms were stored in a bulb and the change in β-decay asymmetry was exploited for resonance detection.

A similar development has occurred with optical pumping techniques. First, double resonance experiments were carried out with small amounts of radioactive material ($\sim 10^{12}$ atoms), taking advantage of the spin exchange by a buffer gas and leaving the optical detection system unchanged. Recently, on-line experiments have been successfully applied to short-lived β emitters, again making use of the resonance destruction of the β asymmetry, e.g., ^{21}Na($\tau = 32$ sec), ^{37}K($\tau = 1.7$ sec), ^{20}Na($\tau = 0.45$ sec) (Otten, 1971).

Nuclear orientation experiments have also been extended to radioactive isotopes and in a few cases to longer-lived excited states (Stone, 1971).

Finally, inbeam Mössbauer effect experiments should be mentioned, in which low-lying states are investigated that could not be populated by suitable radioactive parents (Wheeler et al., 1969).

IV.   Results and Discussion

The interpretation of magnetic dipole moments of nuclei still lacks a consistent theoretical description, and one is forced to compare different classes of moments with theories which are based on certain model assumptions most appropriate to the kind of nuclei involved. This discussion will concentrate on the most successful and widely applicable models, since limited space does not allow our going into details of the theory. Our knowledge of magnetic moments of excited states has influenced the understanding of nuclear structure in an important way. The extensive material allows further checks of the single-particle and shell model, especially for cases of comparable single-particle states and odd-odd nuclei. The same holds for the core polarization model, where other configurations are taken into account as admixtures to the single-particle value. Collective behavior of spherical and deformed nuclei is decisively studied by the investigation of excited states of vibrational or rotational character. The combination of collective excitation and single-particle wave functions is described in the rotational model for deformed odd nuclei and the core excitation model for spherical ones. Some remarks will be devoted to mirror nuclei, where the measurement of magnetic moments of corresponding excited or unstable ground states has considerably extended the material available for a test of theory. (For additional discussion, see Harvey and Khanna, Chapter IX.A and Rasmussen, Chapter IX.B.)

A.   SINGLE-PARTICLE AND SHELL MODEL

In the nuclear shell model, the nucleons—protons and neutrons—are assumed to move in a common attractive central nuclear potential $V(r)$ which is only modified by an additional spin–orbit force. According to the Pauli principle, the nucleons occupy successive shells with discrete single-particle energies. All residual forces between the nucleons are neglected. From this it follows that for an odd number of nucleons, only the last unpaired nucleon determines the moments of the nucleus. The angular momentum j of the odd particle is the sum of the orbital angular momentum l and spin s

$$\mathbf{j} = \mathbf{l} + \mathbf{s} \tag{33}$$

with j having the two values $j = l \pm \frac{1}{2}$. The single-particle angular momentum j is identical with the spin I of the nucleus. The g factor of an odd nucleus is thus only connected with the angular momentum of the odd particle and can be calculated from the shell model magnetic moment operator

$$\mathbf{\mu} = (g_s \mathbf{s} + g_l \mathbf{l}) \, \mu_N \qquad (34)$$

to be

$$\mu/\mu_N = I g_I = I\{g_l \pm [1/(2l + 1)] (g_s - g_l)\} \quad \text{for} \quad I = l \pm \tfrac{1}{2} \qquad (35)$$

In the independent-particle model the free nucleon moments are assumed to remain unchanged and the orbital contribution is assumed to be the classical electromagnetic value. Thus, by introducing the g_s and g_l values for protons and neutrons

$$g_s(\mathrm{p}) = + 5.587 \qquad g_s(\mathrm{n}) = - 3.826$$

$$g_l(\mathrm{p}) = 1.00 \qquad g_l(\mathrm{n}) = 0$$

Eq. (35) gives the so-called Schmidt values. In Figs. 16 and 17 these values are plotted versus j for odd-proton and odd-neutron nuclei, respectively (Schmidt lines), together with the experimental moments of unstable ground and excited states of odd nuclei. The dotted lines (Dirac lines) represent magnetic moment values under the assumption of normal g_s factors for protons and neutrons:

$$g_s(\mathrm{p}) = 2, \qquad g_s(\mathrm{n}) = 0$$

As in the case of nuclear ground-state moments, most of the experimental values taken from the compilation of Shirley (1971) are well between the Schmidt lines and, except for a few cases which will be discussed later, can be attributed to one of the lines, if one also considers the parity of the state, which is determined by the orbital angular momentum l of the nucleon. Experimental values lying outside the Schmidt lines are well explained by collective contributions (Sections IV.C.2 and IV.D.2).

The extreme single-particle model implies a very strong pairing force. Taking into account realistic forces of about 1 MeV, which can be deduced from nuclear binding energies, it can happen that more than one nucleon of an unclosed shell may contribute to the spin and magnetic moment of the nucleus.

A configuration $(j)_I^q$ of q nucleons in the state j coupled to the total spin I gives rise to a magnetic moment

$$\mu = \mu_j I/j \qquad (36)$$

where μ_j is the Schmidt value of the single particle. The g factor of such a

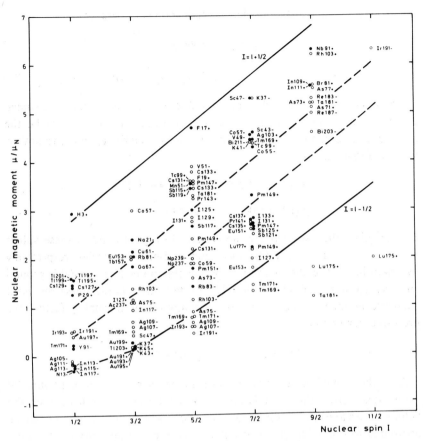

Fig. 16. Schmidt diagram of nuclear magnetic dipole moments of excited states (open circles) and unstable ground states (full circles) for odd-proton nuclei.

configuration equals the single-particle value. For three or more nucleons coupled in this way, the total spin I of the nucleus is sometimes smaller than the single-particle value j (usually $I = j - 1$). Thus some evident discrepancies in Fig. 16 and 17 can be explained where one would expect the magnetic moment to be nearer to the other Schmidt line. For example, the magnetic moment $\mu_{exp} = 4.22(72)$ of the 320-keV, $I = \frac{5}{2}$ level of ^{51}V would be attributed to a $d_{5/2}$ configuration with positive parity. The observed negative parity suggests a $(f_{7/2})^3_{5/2}$ configuration, which also results in an expected magnetic moment of $\mu_{th} = 4.14$. The same may hold for the $I = \frac{9}{2}^-$ states of ^{181}Ta $(E_\gamma = 6.2\,\text{keV})$ and ^{187}Re $(E_\gamma = 206\,\text{keV})$, though the agreement of the experi-

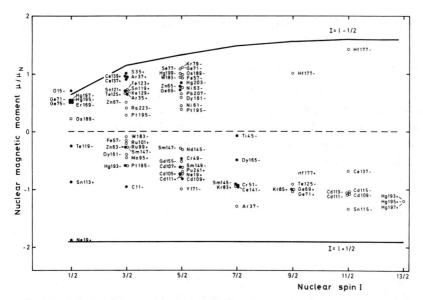

Fig. 17. Schmidt diagram of nuclear magnetic moments of excited states (open circles) and unstable ground states (full circles) for odd-neutron nuclei.

mental magnetic moments with the theoretical ones is not as good and can be explained to some extent by the discussion in Section IV.B.

The magnetic moments of many odd-odd and some even-even excited states can be calculated by coupling the resulting spin of the unfilled proton shell I_p of configuration $(j_p)_{I_p}^{q_p}$ and the unfilled neutron shell I_n of configuration $(j_n)_{I_n}^{q_n}$ to the total spin $\mathbf{I} = \mathbf{I}_p + \mathbf{I}_n$. According to Eq. (36), the g factors of these configurations are g_p and g_n, respectively, and the magnetic moment of the configuration $(j_p)_{I_p}^{q_p}(j_n)_{I_n}^{q_n}$ is calculated to be

$$\mu/\mu_N = \tfrac{1}{2}(g_p + g_n) + \tfrac{1}{2}(g_p - g_n)\left[_p(I_p + 1) - I_n(I_n + 1)\right]/(I + 1) \qquad (37)$$

where the most probable spin I of the lowest odd-odd state is given by the empirical rules of Nordheim (1951) as

$$I = I_p - I_n \quad \text{for} \quad j_p = l_p \pm \tfrac{1}{2} \quad \text{and} \quad j_n = l_n \mp \tfrac{1}{2}$$
$$|I_p - I_n| \leqslant I \leqslant I_p + I_n \quad \text{for} \quad j_p = l_p \pm \tfrac{1}{2} \quad \text{and} \quad j_n = l_n \pm \tfrac{1}{2} \qquad (38)$$

A somewhat better agreement between experimental magnetic moments of odd-odd states and the theoretical values calculated by Eq. (37) can be obtained if one replaces the free-particle g factors g_p and g_n by experimental ones g_p' and g_n' of corresponding configurations in neighboring odd-proton and odd-neutron nuclei.

B. CORE POLARIZATION MODEL

In the single-particle model, all nucleons are assumed to move in an average central potential $V(r)$ and residual interactions between individual nuclei, especially between the odd nucleon and the core, are neglected. These residual interactions are taken into account in the configuration mixing model of Noya et al. (1958) and Blin-Stoyle and Perks (1954). In many cases, the deviation of the experimental magnetic moments from the Schmidt values are well interpreted by this model. The interaction of the odd nucleon with the core will give rise to a total wave function

$$|I\rangle = \alpha_0 |j, I_c = 0; I = j\rangle + \sum_n \alpha_n |j, (j_1^{-1} j_2^1) I_c; I = j\rangle \qquad (39)$$

where the first term denotes the wave function of the pure single-particle state, while the other terms are first-order admixtures of polarized core states which contribute to the total wave function with amplitudes α_n. The magnetic moment of the nucleus is then given by the expectation value of the magnetic moment operator with respect to $|I\rangle$. Since the operator has non-vanishing matrix elements only between the core ground state, $I_c = 0$, and excited states, $I_c = 1^+$, corrections to the magnetic moment arise from excitations in which a nucleon (hole) is transferred from a state $j = l + \frac{1}{2} (j = l - \frac{1}{2})$ to its partner orbit $j = l - \frac{1}{2} (j = l + \frac{1}{2})$.

The polarizing influence of the odd nucleon on the core particles will always result in a quenching of the free-nucleon g_s factor. Naively speaking, the odd nucleon prefers seeing like nucleons with their spin antiparallel— only the singlet interaction is present—while it prefers seeing unlike nucleons with their spin parallel, due to the stronger triplet interaction of the n–p system. Since the g_s factors of protons and neutrons are opposed, the net effect of the polarization will be a decrease of the free g_s values, $|g_s^{\text{eff}}|$ $< |g_s^{\text{free}}|$. Due to the assumptions of short-range interaction forces and linear superposition of contributing moments, the magnitude of the deviation will depend on the number of polarizable spins in the core, i.e., the number of particles (holes) in the $l + \frac{1}{2} (l - \frac{1}{2})$ orbits that may be transfered by the spin–spin force into empty $l - \frac{1}{2} (l + \frac{1}{2})$ orbits. This accounts for the negligible effects near the doubly closed shells of ^{16}O and ^{40}Ca, where the one-particle or one-hole states agree with the Schmidt values, whereas near the doubly magic ^{208}Pb, the core polarization effect can transfer

TABLE 2

COMPARISON OF SOME RECENTLY MEASURED MAGNETIC DIPOLE MOMENTS μ_{exp} OF EXCITED STATES WITH SINGLE-PARTICLE MOMENTS μ_{sp} AND VALUES CALCULATED BY THE CORE POLARIZATION MODEL μ_{NAH}[a]

Isotope	I^π	Proton configuration[b]	Neutron configuration[c]	μ_{sp}	μ_{NAH}[c]	μ_{exp}[d]
61Ni	$\frac{5}{2}^-$	$(f_{7/2})^8$	$(f_{7/2})^8 (p_{3/2})^4 f_{5/2}$	+1.37	+0.48	+0.43
63Ni	$\frac{5}{2}^-$	$(f_{7/2})^8$	$(f_{7/2})^8 (p_{3/2})^4 (f_{5/2})^3$	+1.37	+0.75	+0.75
69Ge	$\frac{9}{2}^+$	$(f_{7/2})^8 (p_{3/2})^4$	$(f_{7/2})^8 (p_{3/2})^4 (f_{5/2})^4 g_{9/2}$	−1.91	−0.93	−1.00
71Ge	$\frac{9}{2}^+$	$(f_{7/2})^8 (p_{3/2})^4$	$(p_{3/2})^4 g_{9/2}$	−1.91	−1.21	−1.04
71As	$\frac{9}{2}^+$	$(f_{7/2})^8 (p_{3/2})^2 (g_{9/2})^3$	$(p_{3/2})^4$	+6.79	+5.19	+5.12
73As	$\frac{9}{2}^+$	$(f_{7/2})^8 (p_{3/2})^2 (g_{9/2})^3$	—	+6.79	+5.27	+5.23
81Br	$\frac{9}{2}^+$	$(f_{7/2})^8 (p_{3/2})^2 (f_{5/2})^4 g_{9/2}$	$(g_{9/2})^8$	+6.79	+6.03	+5.84
109Cd	$\frac{11}{2}^-$	$(g_{9/2})^8$	$(d_{5/2})^6 (g_{7/2})^4 h_{11/2}$	−1.91	+1.01	−1.09
111Cd	$\frac{5}{2}^+$	$(g_{9/2})^8$	$(d_{5/2})^5$	−1.91	−0.74	−0.79
115Sn	$\frac{11}{2}^-$	$(g_{9/2})^{10}$	$(d_{5/2})^6 h_{11/2}$	−1.91	−1.36	−1.36
121Sb	$\frac{7}{2}^+$	$(g_{9/2})^{10} g_{7/2}$	$(d_{5/2})^6 (h_{11/2})^6$	+1.72	+2.61	+2.51
125Te	$\frac{3}{2}^+$	$(g_{9/2})^{10} (g_{7/2})^2$	$(h_{11/2})^8 d_{3/2}$	+1.15	+0.51	+0.60
133Cs	$\frac{5}{2}^+$	$(g_{9/2})^{10} (g_{7/2})^2 (d_{5/2})^3$	$(d_{5/2})^6 (h_{11/2})^{12} (d_{3/2})^2$	+4.79	+3.34	+3.44

[a] All values are in nuclear magnetons μ_N.

[b] Only configurations which contribute to the configuration mixing are listed.

[c] Calculated with harmonic oscillator wave function parameter $c = 40$ MeV of Noya et al. (1958).

[d] Experimental values taken from Shirley (1971).

particles to the $h_{11/2}$ proton and $i_{13/2}$ neutron orbits. Some recently measured magnetic moments are compared with the results of configuration mixing calculations and Schmidt values in Table 2. In a corresponding way, the core polarization model can also be applied to odd-odd nuclei by calculating the effect independently for the odd proton and the odd neutron. (For additional corrections see Noya *et al.*, 1958.)

Several theoretical approaches exist to correct the magnetic moment values given by the independent-particle model. One correction includes the momentum-dependent force into the internucleon potential. A numerical evaluation of this effect is given by Chemtob (1969). An important contribution can be attributed to the mesonic effect. In addition to early explanations for the anomalous g_s factors of the free nucleons, which are quenched in nuclei, recent investigations on high-spin states revealed experimental evidence for the anomalous orbital magnetism, caused by mesonic exchange currents (Yamazaki, 1973). Average corrections to the free nucleon g_s factors amount to $\delta g_s(p) = 0.10(3)$ and $\delta g_s(n) = -0.04(3)$. The influence of configuration mixing and magnetic core polarization is discussed in the following section. A comprehensive description of all corrections is given by Nagamiya (1972).

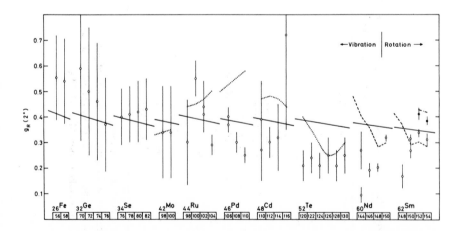

Fig. 18a. Values of g_R factors for first 2^+ excited states of even-even nuclei. Values for spherical and deformed nuclei are shown in one figure to demonstrate the overall continuity and similarity of g_R values. $\blacklozenge$ Mössbauer effect; ϕ IMPAC, CEAD, IPAC; —— Greiner; – – – Kumar and Baranger, –·–·– Prior, Boehm, and Nilsson; ··· Lombard and Campi-Benet.

C. Collective States in the Deformed Region

1. Even-Even Nuclei

A first approach in explaining the g factors of collective states, denoted g_R, arises through looking at the nucleus as a classic rotor of charge Ze and mass A. If protons and neutrons contribute to the rotation in the same way, g_R is given by

$$g_R = J_p/(J_p + J_n) = Z/A \simeq 0.4 \qquad (40)$$

where J_p and J_n are the moments of inertia of the proton group and neutron group, respectively. From Figs. 18a and 18b, one can see that these values lie well above the average of the experimental results, and, of course, the scattering of the individual values cannot be explained by such a classical model. Considerable improvement has been obtained by Nilsson and Prior (1961), who considered, on the basis of a microscopic theory, the influence of the different pairing forces for protons and neutrons on J_p and J_n. Since the pairing energy is higher for protons than for neutrons, the contribution of the protons to the total moment of inertia J is less than that of the neutrons and one gets from Eq. (40)

$$g_R < Z/A \qquad (41)$$

Prior *et al.* (1968) calculated g_R with different empirical values of the pairing force, taken from the odd-even mass difference in the mass region $150 < A < 190$. The best agreement was found for proton and neutron pairing interaction constants of $G_p = (23.5/A)$ MeV and $G_n = (18/A)$ MeV, respectively. Though the moments of inertia are very sensitive to the degree of deforma-

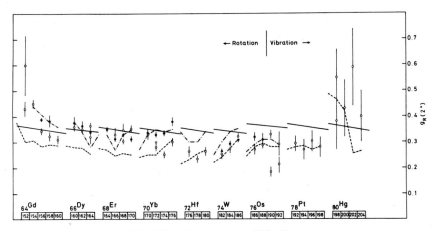

Fig. 18b. Continuation of Fig. 18a.

tion, the dependence of g_R on this is small, since J_p and J_n vary similarly with deformation. The theoretical g_R values of Prior *et al.* (1968) are shown in Fig. 18 together with results of Baranger and Kumar (1968), who calculated g_R factors for 2^+ states of all even-even nuclei in a unified model based on a microscopic theory of pairing-plus-quadrupole forces, thus covering nuclei in the spherical, deformed, and transition regions. In their model, they demonstrate the competition between the quadrupole force in deforming the nucleus and the pairing force in returning it to sphericity.

Another phenomenological approach was undertaken by Greiner (1966) in explaining the g_R factors of collective states. His basic idea was the assumption that the proton distribution in nuclei is less deformed than the neutron distribution because of the greater pairing force of the protons. Therefore the neutrons contribute more to the collective motion. In the framework of the rotation–vibration model of Faessler and Greiner (1962, 1964), he calculated the magnetic properties of the 2^+ rotational states in deformed nuclei to be given by

$$g_R = (Z/A)(1 - 2f) \tag{42}$$

with

$$f = (N/A)\left[(G_n/G_p)^{1/2} - 1\right] \tag{43}$$

Assuming the pairing strength parameters G_p and G_n to be the same as given by Prior *et al.* (1968), the g_R factors for rotational states are approximately $g_R \simeq 0.33$ (Figs. 18a and 18b).

2. *Odd Nuclei*

In odd-A nuclei, the extra nucleon makes a substantial contribution to the moment of inertia J of the nucleus and therefore to the magnetic moment. From Eq. (40), it immediately follows that

$$g_R \text{ (even-even + proton)} > g_R \text{ (even-even)}$$

$$g_R \text{ (even-even + neutron)} < g_R \text{ (even-even)}$$

which qualitatively can be seen in Fig. 19, where on the average the odd-proton g_R factors are larger than the odd-neutron g_R factors.

A more detailed discussion has to consider the coupling of the odd particle to the deformed core. On the basis of Nilsson wave functions, the g factor of an odd-A rotational state (I, K) may be split up into an internal g factor g_K, which accounts for the extra particle and a rotational one g_R due to the collective motion. This g_R value might not be the same as for the even-even core, but is modified by the collective contribution of the odd particle. The magnetic dipole moments of members of a rotational band with projection K on the nuclear symmetry axis are then given by

$$\mu(I, K)/\mu_N = g_R I + (g_K - g_R) [K^2/(I + 1)]$$
$$\times [1 + (2I + 1)(-1)^{I+(1/2)} b_0 \, \delta_{K, 1/2}] \qquad (44)$$

where b_0 is the $K = \frac{1}{2}$ band decoupling parameter.

Since one would like to know both contributions to the magnetic moment separately—to compare them with theoretical predictions—a further, independent measurement has to be carried out which also depends on g_R and g_K. This is given by the magnetic dipole transition probabilities $B(M1)$ of an interband transition (Rogers, 1965):

$$B[M1; (I + 1, K) \to (I, K)] = \tfrac{3}{4}\mu_N^2 (g_K - g_R)^2 \frac{K^2(I + 1 + K)(I + 1 - K)}{(I + 1)(2I + 3)}$$
$$\times [1 + (-1)^{I-(1/2)} b_0 \, \delta_{K, 1/2}] \qquad (45)$$

If one knows g_R and g_K, the magnetic moments of all states of a rotational band can be calculated from Eq. (44), a typical example of a model-dependent determination of magnetic moments.

The experimental data derived from magnetic dipole moment and transition probability measurements can be compared with the following theoretical expression for g_R (Grin and Pavlichenkov, 1962):

$$g_R = g_R^0 + \delta g_R + \delta' g_R \qquad (46)$$

where g_R^0 is the rotational g_R factor of the neighboring even-even nucleus $(A - 1)$, δg_R is the contribution of the variation in the moment of inertia J, and $\delta' g_R$ accounts for the spin polarization of the core nucleons. In first approximation, the g_R factor for an odd nucleus is given by

$$g_R = 1 - (1 \mp g_R^0)(J_{even}/J_{odd}) + \delta' g_R \qquad (47)$$

where the negative (positive) sign has to be taken for an odd number of protons (neutrons). The g_R factors calculated by this method are compared with experimental results in Fig. 19. Also shown in this figure are the results of Prior et al. (1968) for odd, deformed nuclei and, for comparison, their g_R values for neighboring even-even nuclei.

The internal g_K factor may be calculated by means of Nilsson wave functions (Nilsson, 1955):

$$g_K K = g_l K + (g_s - g_l)\langle v | s_3 | v \rangle \qquad (48)$$

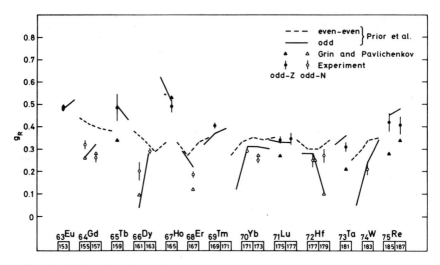

Fig. 19. Values of g_R derived from measured magnetic moments of rotational bands in deformed, odd-A nuclei according to Eq. (44). For comparison, g_R values predicted by Prior *et al.* (1968) for neighboring even-even nuclei are shown by the dashed line.

where the mean value of the operator s_3 is the projection of the spin on the symmetry axis of the nucleus in the Nilsson state $|v\rangle$. De Boer and Rogers (1963) analyzed the experimental values of g_K and deduced that Eq. (48) would give almost correct results if one introduces an effective g_s factor g_s^{eff} $\approx 0.6 g_s^{free}$. These effective g_s factors in terms of the free-nucleon g_s factor are plotted in Fig. 20 for deformed odd nuclei in the mass region $150 < A < 190$. Bochnacki and Ogaza (1965, 1968) have calculated the contribution to the renormalization of g_s assuming a residual spin–spin interaction. Even better agreement with experimental results was obtained by Bochnacki and Ogaza (1968) by discriminating between a longitudinal and a transverse spin polarization of the even nuclear core by the unpaired nucleon.

In Table 3, g_R and g_K values of rotational bands of odd-A deformed nuclei are given and experimental magnetic moments of excited states are compared with predictions from Eq. (44) (for references, see Grodzins, 1968).

D. Collective States in the Spherical and Transition Regions

1. *Even-Even Nuclei*

A first approach toward calculating g factors of 2^+ states in the spherical region was done by Kisslinger and Sorensen (1963), in which the particles

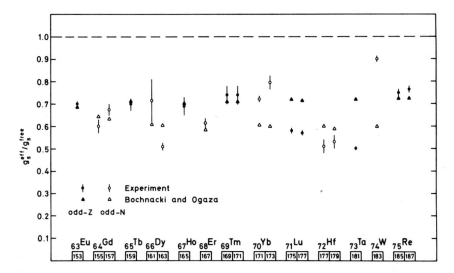

Fig. 20.   Values of g_s^{eff}/g_s^{free} calculated from g_K factors, which are derived from measured magnetic moments of rotational bands in deformed, odd-A nuclei according to Eq. (44).

outside the closed shells were assumed to interact through a pairing-plus-quadrupole force. The low-lying states are described in terms of quasi-particles or phonons. The 2^+ states can than be interpreted as a consequence of harmonic quadrupole vibrations about a spherical equilibrium shape. Evident discrepancies of this model could be overcome by refined calculations (Lombard and Campi-Benet, 1972) which additionally take into account core excitations of the particle–hole type. These results are shown in Figs. 18a and 18b.

The unified model of Baranger and Kumar (1968) has already been mentioned in Section IV.C.1 and their results are also compared with the experimental g_R factors in Figs. 18a and 18b. The same holds for Greiner's phenomenological model (1966), in which the g_R factors of 2^+ vibrational states in nearly spherical nuclei are calculated to be

$$g_R = (Z/A)(a - \tfrac{4}{3}f) \tag{49}$$

where f is given by Eq. (43). These values are somewhat larger than those in the deformed region.

TABLE 3   MAGNETIC MOMENTS OF EXCITED STATES OF ODD-A DEFORMED NUCLEI[a]

Isotope	$I = K$	$\mu_{g.s.}$[b]	g_R	g_K	I	E (keV)	μ_{pred}	μ_{exp}[c]
153Eu	5/2+	+1.529	0.484(10)	+0.664(3)	7/2+	83.4	1.95	1.80(8)
155Gd	3/2-	-0.242(2)	0.318(17)	-0.482(12)	5/2-	60.0	0.280	—
157Gd	3/2-	-0.3225(10)	0.262(20)	-0.533(14)	5/2-	54.5	0.144	—
159Tb	3/2+	+1.994	0.51(6)	+1.88(5)	5/2+	58	2.30(20)	2.30(13)
161Dy	5/2+	-0.472(13)	0.20(4)	-0.346(18)	7/2+	43.8	0.049	—
			0.083(25)					
163Dy	5/2-	+0.635(14)	0.23(2)	+0.25(2)	7/2-	75	0.884	—
165Ho	7/2-	+4.113	0.48(2)	+1.37(1)	9/2-	94.7	4.12	—
167Er	7/2+	-0.5647	0.184(10)	-0.259(3)	9/2+	79.3	-0.159	—
169Tm	1/2+	-0.2308	0.406(10)	-1.57	3/2+	8.42	—	+0.533(8)
					5/2+	118.2	0.74	0.73(5)
					7/2+	138.9	1.45	1.32(7)
171Tm[d]	1/2+	-0.277(5)	0.432(28)	-1.42(4)	3/2+	5.06	—	—
					5/2+	116.7	0.79	0.81(37)
					7/2+	129.1	1.45	1.44(14)
171Yb	1/2-	+0.4918	0.288(4)	+1.405(20)	3/2-	66.7	—	+0.347(2)
					5/2-	75.9	1.01(5)	1.01(1)
173Yb	5/2-	-0.6775	0.250(10)	-0.480(10)	7/2-	78.7	0.133	< 0.175
175Lu	7/2+	+2.230	0.326(10)	+0.726(3)	9/2+	113.8	2.36	1.81(20)
					11/2+	251.5	2.55	2.0(6)
177Lu	7/2+	+2.24	0.345(25)	+0.72(1)	9/2+	121.6	2.38	—
177Hf	7/2-	+0.61(3)	0.244(30)	+0.155(12)	9/2-	113	0.82(15)	1.04(6)
					11/2-	249.6	1.1(2)	1.43(50)
179Hf	9/2+	-0.47(3)	0.282(25)	-0.190(15)	11/2+	122.7	0.157	—
181Ta	7/2+	+2.35(1)	0.30(2)	+0.78(1)	9/2+	136.25	2.42	1.22(9)
183W	1/2-	+0.1172	0.212(25)	+0.77(10)	3/2-	46.5	-0.1(1)	-0.03(03)
					5/2-	99.1	1.3(2)	-0.930(25)

[a] Table taken from Grodzins (1968). References given there.
[b] Experimental values of ground state magnetic moments.
[c] Experimental data revised for new results (Shirley, 1971).
[d] Newly adopted values (Kaufmann, et al., 1968).

2. Odd Nuclei

Many magnetic moments are satisfactorily explained by the shell model and the core polarization model, already discussed in preceding sections.

Some low-lying excited states of odd nuclei in the spherical region can be accounted for by the phenomenological core excitation model of de-Shalit (1961), which assumes coupling between the unpaired particle, remaining in its lowest energy state, and core excitations, i.e., quadrupole surface vibrations. The total angular momentum I of the excited state is then given by

$$\mathbf{I} = \mathbf{j} + \mathbf{J}_c \tag{50}$$

where $\mathbf{j}$ and $\mathbf{J}_c$ are the spin operators of the single particle and core, respectively. Since the model implies that the interaction of the odd particle is weak compared to the interparticle interactions of the core, the energies of the (jJ_c) coupling multiplets are nearly the same as the excitation energy of the core alone, which can be taken from the energies of corresponding states of neighboring even-even nuclei. The magnetic moment is given by the additivity law

$$\boldsymbol{\mu} = (g_j \mathbf{j} + g_c \mathbf{J}_c)\, \mu_N \tag{51}$$

and the g factors of the excited states can be calculated by a corresponding expression to Eq. (37).

Thus for a specific single-particle state of $j = \tfrac{1}{2}^+$ and a core state $J_c = 2^+$, the following states $|J_c\, j,\, I\rangle$ and g factors will result:

$$
\begin{aligned}
|\tfrac{1}{2}^+\rangle_{gs} &= |0\ \tfrac{1}{2}^+,\, \tfrac{1}{2}^+\rangle & g_{1/2} &= g_j \\
|\tfrac{3}{2}^+\rangle &= |2^+\tfrac{1}{2}^+,\, \tfrac{3}{2}^+\rangle & g_{3/2} &= \tfrac{6}{5}g_c - \tfrac{1}{5}g_j \\
|\tfrac{5}{2}^+\rangle &= |2^+\tfrac{1}{2}^+,\, \tfrac{5}{2}^+\rangle & g_{5/2} &= \tfrac{4}{5}g_c + \tfrac{1}{5}g_j
\end{aligned}
\tag{52}
$$

Additional information can be gained from the transition probabilities between the states, which would obey the following selection rules:

(a) The E2 transitions between the excited states and the ground state will have equal strength, after correction for energy dependence.

(b) The M1 transitions to the ground state are forbidden, while the M1 transition between the excited states is given by the transition probability

$$B(\text{M1};\, I_i \rightarrow I_f) = (3/4\pi)\, \mu_N^{\,2}(2I_f + 1)j(j + 1)(2j + 1)$$

$$\times \begin{Bmatrix} I_i & I_f & 1 \\ j & j & J_c \end{Bmatrix}^2 (g_c - g_j)^2 \tag{53}$$

TABLE 4

g Factors of Core-Excited States of Nuclei with Ground-State Spins $\frac{1}{2}$ and Comparison with Core-Excitation Model Predictions[a]

Isotope	E (keV)	I^π	τ (psec)	g_J	g_I^{exp}	g_I^{th}	g_c^{exp}	g_c^{th}
^{77}Se	440	$\frac{5}{2}-$	34.5(3.5)	+1.0688	0.44(11)	0.53(10)	0.40(11)	0.28(11)
^{103}Rh	295.1	$\frac{3}{2}-$	11(1)	−0.1766	0.60(15)	0.55(5)	0.41(4)	0.55(12)
	377.4	$\frac{5}{2}-$	110(9)	−0.1766	0.45(10)	0.30(4)	0.41(4)	0.61(12)
^{107}Ag	325	$\frac{3}{2}-$	8.5(8)	−0.227	0.50(15)	0.43(5)	0.29(4)	0.35(12)
	423	$\frac{5}{2}-$	43(5)	−0.227	0.40(10)	0.19(4)	0.29(4)	0.55(10)
^{109}Ag	309	$\frac{3}{2}-$	10(2)	−0.261	0.55(15)	0.45(5)	0.30(4)	0.38(12)
	414	$\frac{5}{2}-$	50(7)	−0.261	0.40(10)	0.19(4)	0.30(4)	0.56(10)
^{195}Pt	98.8	$\frac{3}{2}-$	224(22)	+1.210	−0.41(4)	− 0.065(10)	−	−
	129.6	$\frac{5}{2}-$	895(43)	+1.210	−	0.41(6)	−	−
	211.2	$\frac{3}{2}-$	97(7)	+1.210	0.19(4)	−0.04(6)	−	−
	239.8	$\frac{5}{2}-$	180(26)	+1.210	0.15(8)	0.42(6)	−	−
^{197}Hg	134	$\frac{5}{2}-$	10500(300)	+1.048	0.380(25)	0.58(12)	0.55(11)	0.21(2)
^{199}Hg	158	$\frac{5}{2}-$	3350(120)	+1.005	0.404	0.57(12)	0.55(11)	0.25(2)
^{203}Tl	279	$\frac{3}{2}+$	400(30)	+3.223	0.11(3)	0.26(19)	0.59(15)	0.63(3)

[a] Table adopted from Bhattacherjee (1971). Experimental data are revised for new results (Shirley, 1971).

which for the specific case mentioned is

$$B(M1, \tfrac{5}{2} \to \tfrac{3}{2}) = (3/10\pi) \, \mu_N^{\,2} (g_c - g_j)^2 \tag{54}$$

In Table 4, experimental and predicted g_I factors as well as experimental g_c factors of the even-even core and predicted g_c factors are listed for several nuclei with ground-state spin $j = \tfrac{1}{2}$.

E.   MIRROR NUCLEI

The behavior of magnetic dipole moments of mirror nuclei was studied by Sachs (1946), who found the sum of the moments of a mirror pair to be equal to the sum of their Schmidt values. His theorem implied that the quenching contributions due to mesonic, core polarization, or other effects are, in first order, the same for the odd proton and neutron and enter with opposite sign.

Recent evaluations of the experimental results yielded some other regularities (Sugimoto, 1969; Leonardi and Rosa-Clot, 1968, 1970). The data were evaluated on the basis of the isospin formalism, where the magnetic moment can be separated into an isoscalar and an isovector part:

$$\mu = \langle \sum_i \mu_0^{(i)} \rangle_I + \langle \sum_i \mu_3^{(i)} \rangle_I \tag{55}$$

with

$$\mu_0^{(i)} = \tfrac{1}{2}[l_z^{(i)} + (\mu_p + \mu_n) \, \sigma_z^{(i)}]$$
$$\mu_3^{(i)} = \tfrac{1}{2}[\tau_3^{(i)} l_z^{(i)} + (\mu_p - \mu_n) \, \tau_3^{(i)} \sigma_z^{(i)}] \tag{56}$$

Assuming charge symmetry of the nuclear force and a negligible Coulomb interaction, the expectation values $\langle \sum \mu_0 \rangle_I$ for pairs of mirror states are independent of T_3, while the isovector parts differ only in sign for $T_3 = \pm\tfrac{1}{2}$.

Hence the sum of the moments of an isospin doublet is given by

$$\mu(T_3 = +\tfrac{1}{2}) + \mu(T_3 = -\tfrac{1}{2}) = \langle \sum_i l_z^{(i)} \rangle_I + (\mu_p + \mu_n) \langle \sum_i \sigma_z^{(i)} \rangle_I \tag{57}$$

If both magnetic moments are known, one can calculate the values of $\langle \sum l_z \rangle_I$ and $\langle \sum \sigma_z \rangle_I$.

Similar expressions can be deduced for the difference of the moments, where only the isovector part remains:

$$\mu(T_3 = +\tfrac{1}{2}) - \mu(T_3 = -\tfrac{1}{2}) = \langle \sum_i \tau_3^{(i)} l_z^{(i)} \rangle_{I, \, T_3 = +1/2}$$

$$+ (\mu_p - \mu_n) \langle \sum_i \tau_3^{(i)} \sigma_z^{(i)} \rangle_{I, \, T_3 = +1/2} \tag{58}$$

The expectation value $\langle \sum \tau_3^{(i)} \sigma_z^{(i)} \rangle_I$ can also be related to the Gamow–Teller matrix element of the mirror β decay. Thus both expectation values in Eq. (58) can be determined.

In Fig. 21a, the spin parts $\langle \sum \sigma_z \rangle_I$ and $\langle \sum \tau_3 \sigma_z \rangle_I$ are plotted versus mass number for mirror nuclei up to $A = 40$ (Sugimoto, 1969). A pronounced regularity can be seen for the members of each subshell. The successive

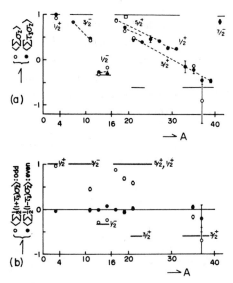

Fig. 21. Spin parts of the magnetic moment distribution of mirror nuclei. The solid lines represent the single-particle values. The dashed lines connect the values of the same I^{π}. The square represents the metastable $\frac{1}{2}^{+}$ pair of $A = 19$. Parts (a) and (b) are discussed in the text (Sugimoto, 1969).

decrease from the Schmidt value with increasing mass number has been explained by Leonardi and Rosa-Clot (1968) as the core polarization contribution of an additional proton or neutron pair, respectively, which approximately is the same in each subshell as indicated by the equal decline of the dashed lines.

From the sum and the difference of the two mentioned expectation values, the contributions of the odd group and even group of nucleons in an isospin doublet can be deduced. The results are plotted in Fig. 21b, which clearly demonstrates that the contribution of the even-group nucleons to the deviation from the Schmidt values is negligible compared to those of the odd group.

A similar evaluation was carried out by Sugimoto for the orbital contribution to the magnetic moment, but no clear results could be deduced.

References

Abragam, A., and Pound, R. V. (1953). *Phys. Rev.* **92**, 943.

Agarwal, Y. K., Baba, C. V., and Bhattacherjee, S. K. (1966). *Nucl. Phys.* **79**, 437.

Ames, O., Phillips, E. A., and Glickstein, S. S. (1965). *Phys. Rev.* **137B**, 1157.

Baranger, M., and Kumar, K. (1968). *Nucl. Phys.* **A110**, 490, 529.

Ben-Zvi, I., Gilad, P., Goldberg, M., Goldring, G., Schwarzschild, A., Sprinzak, A., and Vager, Z. (1968). *Nucl. Phys.* **A121**, 592.

Ben-Zvi, I., Gilad, P., Goldberg, M. B., Goldring, G., Speidel, K.-H., and Sprinzak, A. (1970). *Nucl. Phys.* **A151**, 401.

Bertschat, H., Christiansen, J., Mahnke, H.-E., Recknagel, E., Schatz, G., Sielemann, R., and Witthuhn, W. (1970). *Phys. Rev. Lett.* **25**, 102.

Bhattacherjee, S. K. (1971). *Ann. Phys.* **63**, 613.

Bleck, J., Haag, D. W., Leitz, W., Michaelsen, R., Ribbe, W., and Sichelschmidt, F. (1969a). *Nucl. Phys.* **A123**, 65.

Bleck, J., Haag, D. W., and Ribbe, W. (1969b). *Nucl. Instrum. Methods* **67**, 169.

Blin-Stoyle, R. J., and Perks, M. A. (1954). *Proc. Phys. Soc. (London)* **A67**, 885.

Bochnacki, Z., and Ogaza, S. (1965). *Nucl. Phys.* **69**, 186.

Bochnacki, Z., and Ogaza, S. (1968). *In* "Hyperfine Structure and Nuclear Radiations" (E. Matthias and D. A. Shirley, eds.), p. 106. North-Holland Publ., Amsterdam.

Brady, E. L., and Deutsch, M. (1950). *Phys. Rev.* **78**, 558.

Bräuer, N., Focke, B., Lehmann, B., Nishiyama, K., and Riegel, D. (1971). *Z. Phys.* **244**, 375.

Buyrn, A., and Grodzins, L. (1964). MIT Progr. Rep. 2098, No. 251.

Calaprice, F. B., Commins, E. D., and Dobson, D. A. (1965). *Phys. Rev.* **137B**, 1453.

Chemtob, M. (1969). *Nucl. Phys.* **A123**, 449.

Christiansen, J., Mahnke, H.-E., Recknagel, E., Riegel, D., Weyer, G., and Witthuhn, W. (1968). *Phys. Rev. Lett.* **21**, 554.

Christiansen, J., Mahnke, H.-E., Recknagel, E., Riegel, D., Schatz, G., Weyer, G., and Witthuhn, W. (1970). *Phys. Rev. C* **1**, 613.

de Boer, J., and Rogers, J. D. (1963). *Phys. Lett.* **3**, 304.

de Groot, S. R., Tolhoek, H. A., and Huiskamp, W. J. (1965). *In* "Alpha-, Beta-, Gamma-Ray Spectroscopy" (K. Siegbahn, ed.), Vol. 2, Chapter XIX B. North-Holland Publ., Amsterdam.

de-Shalit, A. (1961). *Phys. Rev.* **122**, 1530.

Faessler, A., and Greiner, W. (1962). *Z. Phys.* **168**, 425.

Faessler, A., and Greiner, W. (1964). *Z. Phys.* **177**, 190.

Ferentz, M., and Rosenzweig, N. (1965). *In* "Alpha-, Beta-, and Gamma-Ray Spectroscopy" (K. Siegbahn, ed.), Vol. 2, Appendix 8, North-Holland Publ., Amsterdam.

Frauenfelder, H., and Steffen, R. M. (1960). *In* "Nuclear Spectroscopy" (F. Ajzenberg-Selove, ed.), Vol. 2, Chapter IV.C. Academic Press, New York.

Frauenfelder, H., and Steffen, R. M. (1965). *In* "Alpha-, Beta-, and Gamma-Ray Spectroscopy" (K. Siegbahn, ed.), Vol. 2, Chapter XIX A. North-Holland Publ., Amsterdam.

Freeman, A. J., and Frankel, R. B. (eds.) (1967). "Hyperfine Interactions" Academic Press, New York.

Goldring, G., and Kalish, R. (eds.) (1971). "Hyperfine Interactions in Excited Nuclei," Gordon and Breach, New York.

Greiner, W. (1966). *Nucl. Phys.* **80**, 417.

Grin, Y. T., and Pavlichenkov, I. M. (1962). *Sov. Phys. JETP* **14**, 679.

Grodzins, L. (1968). *Ann. Rev. Nucl. Sci.* **18**, 291.

Horie, H., and Sugimoto, K. (eds.) (1973). Nuclear Moments and Nuclear Structure, *J. Phys. Soc. Japan, Suppl.* **34**.

Kalish, R., Grodzins, L., Borchers, R. R., Bronson, J. D., and Herskind, B. (1967). *Phys. Rev.* **161**, 1196.

Karlsson, E., Matthias, E., and Siegbahn, K. (eds.) (1964). "Perturbed Angular Correlations," North-Holland Publ., Amsterdam.

Kaufmann, E. N., Bowman, J. D., and Bhattacherjee, S. K. (1968). *Nucl. Phys.* **A119**, 417.

Keszthelyi, L., Berkes, I., Dézsi, I., and Pócs, L. (1965). *Nucl. Phys.* **71**, 662.

Kisslinger, L. S., and Sorensen, R. (1963). *Rev. Mod. Phys.* **25**, 853.

Kopfermann, H. (1958). "Nuclear Moments," Academic Press, New York.

Koster, T. A., and Shirley, D. A. (1971). *In* "Hyperfine Interactions in Excited Nuclei" (G. Goldring and R. Kalish, eds.), Vol 4, p. 1239. Gordon and Breach, New York.

Leonardi, R., and Rosa-Clot, M. (1968). *Phys. Rev. Lett.* **21**, 377.

Leonardi, R., and Rosa-Clot, M. (1970). *Phys. Rev. Lett.* **24**, 407.

Lindhard, J. (1969). *Proc. Roy. Soc.* **A311**, 11.

Lindhard, J., and Winther, A. (1971). *Nucl. Phys.* **A166**, 413.

Lombard, R. J., and Campi-Benet, Y. (1972). *Nucl. Phys.* (to be published).

Matthias, E., Rosenblum, S. S., and Shirley, D. A. (1965). *Phys. Rev. Lett.* **14**, 46.

Matthias, E., Shirley, D. A., Klein, M. P., and Edelstein, N. (1966). *Phys. Rev. Lett.* **16**, 974.

Matthias, E. (1967). *In* "Hyperfine Interactions" (A. J. Freeman and R. B. Frankel, eds.), p. 598. Academic Press, New York.

Matthias, E. (1968). *In* "Hyperfine Structure and Nuclear Radiations" (E. Matthias and D. A. Shirley, eds.) p. 822. North-Holland Publ., Amsterdam.

Matthias, E., and Shirley, D. A. (eds.) (1968). "Hyperfine Structure and Nuclear Radiations," North-Holland Publ., Amsterdam.

Matthias, E., Olsen, B., Shirley, D. A., Templeton, J. E., and Steffen, R. M. (1971). *Phys. Rev. A* **4**, 1626.

Murnick, D. E., Grodzins, L., Bronson, J. D., Herskind, B., and Borchers, R. R. (1967). *Phys. Rev.* **163**, 254.

Murnick, D. E. (1971). *In* "Angular Correlations in Nuclear Disintegration" (H. van Krugten and B. van Nooijen, eds.), p. 455. Rotterdam Univ. Press, Wolters-Noordhoff Publ., Groningen.

Nagamiya, S. (1972). Scientific Papers of the Inst. of Phys. and Chem. Research 66, 39. Wako-shi, Saitama, Japan.

Nierenberg, W. A., and Lindgren, I. (1965). *In* "Alpha-, Beta-, Gamma-Ray Spectroscopy" (K. Siegbahn, ed.), Vol. 2, Chapter XX. North-Holland Publ., Amsterdam.

Nilsson, S. G. (1955). *Mat. Fys. Medd. Dan. Vid. Selsk.* **29**, No. 16.

Nilsson, S. G., and Prior, O. (1961). *Mat. Fys. Medd. Dan. Vid. Selsk.* **32**, No. 16.

Nordhagen, R., Goldring, G., Diamond, R. M., Nakai, K., and Stephens, F. S. (1970). *Nucl. Phys.* **A142**, 577.

Nordhagen, R. (1971). *In* "Hyperfine Interactions in Excited Nuclei" (G. Goldring and R. Kalish, eds.) Vol. 3, p. 893. Gordon and Breach, New York.

Nordheim, L. W. (1951). *Rev. Mod. Phys.* **23**, 322.

Noya, H., Arima, A., and Horie, H. (1958). *Progr. Theor. Phys. Suppl.* **8**, 33.

Otten, E. W. (1970). CERN Rep. 70-30, Vol. 1, p. 361. CERN, Geneva.

Otten, E. W. (1971). *In* "Hyperfine Interactions in Excited Nuclei" (G. Goldring and R. Kalish, eds.) Vol. 2, p. 363. Gordon and Breach, New York.

Pauli, W. (1924). *Naturwissenschaften* **12**, 741.

Portis, A. M., and Lindquist, R. H. (1965). *In* "Magnetism" (G. T. Rado and H. Suhl, eds.), Vol. II A, p. 359. Academic Press, New York.

Prior, O., Boehm, F., and Nilsson, S. G. (1968). *Nucl. Phys.* **A110**, 257.

Ramsey, N. F. (1953). "Nuclear Moments," Wiley, New York.

Rhaghavan, R. S., Rhaghavan, P., and Sperr, P. (1971). *In* "Hyperfine Interactions in Excited Nuclei" (G. Goldring and R. Kalish, eds.), Vol. 3, p. 462. Gordon and Breach, New York.

Rogers, J. D. (1965). *Ann. Rev. Nucl. Sci.* **15**, 241.

Sachs, R. G. (1946). *Phys. Rev.* **69**, 611.

Shapiro, F. D. (1958). *Usp. Fiz. Nauk.* **65**, 133.

Shirley, V. S. (1971). *In* "Hyperfine Interactions in Excited Nuclei" (G. Goldring and R. Kalish, eds.), Vol. 4, p. 1255. Gordon and Breach, New York.

Stone, N. J. (1971). *In* "Hyperfine Interactions in Excited Nuclei" (G. Goldring and R. Kalish, eds.), Vol. 1, p. 237. Gordon and Breach, New York.

Sugimoto, K., Mizobuchi, A., Nakai, K., and Matuda, K. (1966). *J. Phys. Soc. Japan* **21**, 213.

Sugimoto, K. (1969). *Phys. Rev.* **182**, 1051.

Tsang, T., and Connor, D. (1963). *Phys. Rev.* **132**, 1141.

van Krugten, H., and van Nooijen, B. (eds.) (1971). "Angular Correlations in Nuclear Disintegration," Rotterdam Univ. Press, Wolters-Noordhoff Publ., Groningen.

Wheeler, R. M., Atzmony, U., and Walker, J. C. (1969). *Phys. Rev.* **186**, 1280.

Yamazaki, T. (1967). *Nucl. Data* **A3**, 1.

Yamazaki, T. (1973). *In* Nuclear Moments and Nuclear Structure (H. Horie and K. Sugimoto, eds.) *J. Phys. Soc. Japan, Suppl.* **34**, 17.

VII.D GAMMA RAYS FROM
CAPTURE REACTIONS

C. Rolfs and *A. E. Litherland*

UNIVERSITY OF TORONTO, TORONTO, CANADA

I. Introduction

Most nuclear reactions produce a variety of fragments, such as protons, neutrons, and alpha particles, which leave the residual nuclei in a variety of

143

excited states. If these excited states are bound, they can then decay by the emission of gamma rays. There is, however, a particular class of nuclear reactions in which the incident projectile is absorbed by the target nucleus followed by the emission of gamma radiation. These nuclear reactions are known as radiative capture reactions and form an important class of nuclear reactions for two main reasons. First, the radiative capture reaction is one of the most important reactions for the formation of various species of elements in the universe. Second, it is an important class of nuclear reactions because of its great value in facilitating the determination of the properties of nuclear states. In this chapter, we will be concerned solely with the use of radiative capture reactions of charged particles in the study of the properties of nuclear states and we will refer the reader to the literature for the astrophysical implications (Fowler *et al.*, 1968). The radiative capture of neutrons has been described in detail recently by Harvey (1970) and so will not be discussed.

The study of the radiative capture of charged particles is important in the determination of the properties of nuclear states for the following reasons:

(1)   The Q values for the radiative capture of protons and alpha particles are positive and vary from about 1 to 17 MeV in light nuclei and 5 to 8 MeV in heavier nuclei. This implies that the compound nucleus is usually highly excited and especially in the heavier nuclei the density of resonant nuclear states (Richter, Chapter IV.D.1 and Robson, Chapter IX.C) can be very high (Section V.B). If the apparatus used to excite the resonances is of a high enough quality (Section II), then the radiative capture process can be used to study a large variety of resonant nuclear states.

(2)   The capture by target nuclei of charged particles from a well-collimated beam frequently produces isolated resonant states which are often strongly aligned. This is because the relative orbital angular momentum of the particle in an ion beam and the target nucleus is at right angles to the direction of the ion beam. Consequently, the projection of the orbital angular momentum along the direction of motion is zero. In the simplest case of the capture of alpha particles by even-even nuclei at a resonance of angular momentum J, only the $M_J = 0$ magnetic substates are populated. This strong alignment often produces marked angular distributions of the capture gamma radiation which are of great value in determining the quantum numbers and other properties of the nuclear states (Ferguson, Chapter VII.G). The situation, when either target or projectile has spin, is more complicated but strong alignment can still be obtained.

(3)   The energy E and energy spread ΔE of the ion beams from electrostatic accelerators (Allen, Chapter I.A) can have values of $E/\Delta E$ of greater than 10^4. This permits more precise studies to be made of many unbound nuclear states than, for instance, by the use of magnetic spectrometry of charged particles following nuclear reactions (Hendrie, Chapter III.C). Sharp resonances of width Γ at bombarding energy E_R with values of $E_R/\Gamma > 10^4$ are not uncommon and frequently they are of great theoretical interest (Temmer, Chapter IV.A.2 and von Brentano and Cramer, Chapter IV.A.3). As discussed later (Section V.B), nuclear states with high angular momentum or isospin frequently have large values of E_R/Γ.

(4)   In the case of overlapping resonances, the study of interference effects can be a fruitful source of additional information concerning the properties of the states involved (e.g., relative phases, widths, parities) which is not accessible otherwise (Devons and Goldfarb, 1957, and Section V.C).

(5)   In addition to the study of resonant nuclear states of special interest, with large values of E_R/Γ, the large number of isolated resonant states available permits a wide variety of bound nuclear states to be studied conveniently. This is in contrast with the study of gamma radiation following the beta decay of nuclei. The bound and sometimes unbound nuclear states excited by beta decay (Hardy, Chapter VIII.B) are often very limited in number by angular momentum and parity selection rules, whereas it is possible to use the radiative capture phenomenon to choose resonances which decay to bound nuclear states of particular interest.

The principal difficulty in the use of radiative capture for the study of the properties of nuclear states lies in the low cross section for the process. This in turn is due to the weakness of the electromagnetic forces compared with the nuclear forces (Wilkinson, 1960). Consequently, the decay of unbound nuclear states by the emission of a particle of the same type as that captured, or by the emission of some other type of particle, is often 10^3–10^6 times more probable than the decay by gamma-ray emission. This fact implies that for similar reaction counting rates in an experiment, higher ion beam currents must be used in the study of radiative capture than in the study of a nuclear disintegration with nucleon or alpha-particle emission. Fortunately, high currents are available so that the problem at low energies is usually one of heat dissipation in the targets. At higher beam energies, however, the emission of nucleons and alpha particles to excited nuclear states is in strong competition with the capture radiation and produces a gamma-ray background which makes the study of the much weaker capture radiation very difficult and usually prohibits the use of high beam currents.

It is the object of this chapter to elaborate upon the various problems involved in the study of radiative capture reactions. Section II discusses briefly the necessary properties of the accelerators used as well as their energy calibration. The description of gamma-ray detectors suitable for radiative capture studies follows in Section III. Section IV is devoted to the problems involved in selecting and preparing targets. The theoretical and experimental aspects of yield curves for capture reactions are outlined in Section V. Finally, in Section VI, miscellaneous topics associated with capture reaction studies are considered.

II. Accelerator Properties and Energy Calibration

Almost all investigations of the gamma rays from charged-particle capture reactions have been carried out using the ions from direct current electro-static accelerators to initiate the reactions (Allen, Chapter I.A; Herb, 1959). Accelerators of the Van de Graaff type (Van de Graaff, 1931; Van de Graaff *et al.*, 1933) have been used most frequently and accelerators of the Cock-croft–Walton type (Cockcroft and Walton, 1932a, b) and the Dynamitron type (Cleland, 1968) have also been used. It is expected that in the future all of these types will continue to be used. In addition, it is expected that accelerators of the Pelletron type (Herb, 1972) will become more widely used.

The principal properties of an accelerator and ion beam transport system needed for the study of charged-particle radiative capture reactions are as follows:

(1) Ion beam currents of several tens of microamperes of protons and alpha particles are desirable because of the low cross sections for radiative capture. Above about 5 MeV ion energy, smaller ion currents are sometimes useful because counting rates are then limited by the intense gamma rays from competing reactions.

(2) An ion beam energy spread less than or equal to 1 keV at several MeV ion energy is usually necessary because of the complexity (see Fig. 8) of the cross section as a function of energy. An ion beam energy spread much less than 100 eV is difficult to use because of the broadening due to the thermal motion of the target atoms (Staub, 1967).

(3) The energy of the ion beam should be easily variable in steps as small as 100 eV at several MeV ion energy. Steps smaller than 100 eV require very

special target techniques due to the thermal motion of the target atoms (Staub, 1967).

(4) The ion beam should be well collimated with a well-defined beam envelope. Ion beam cross-sectional areas as small as about 4 mm^2 should be available. This feature is of great value when differentially pumped gas targets are used (Section IV). If a larger area of beam is required, then it can be defocused.

(5) Pulsed ion beams for the study of radiative capture (Alter and Garbuny, 1949; Shoupp et al., 1949) are advantageous when capture gamma rays in competition with fast neutrons are studied. They are also useful for the study of the longer-lived (> 0.1 nsec) nuclear states.

In general, a well-engineered electrostatic accelerator will meet these requirements and a detailed discussion of such devices can be found in the literature (Parks et al., 1958; Armstrong, 1967; Allen, Chapter I.A).

The energy calibration of electrostatic accelerators is an important activity because many resonances are very narrow (Section V.B) and it is necessary to make precise ion beam energy adjustments in order to locate them.

Usually a 90° magnetic analyzer, with input and output slits and with a proton resonance magnetic field measuring device, is used to define the ion beam energy and to provide a signal for the energy stabilization feedback loop (Chapter I.A). Such a system must be calibrated using the energies of certain known resonances and lists of these energies are readily available (Marion, 1966).

The energy of an ion beam can be measured absolutely with the help of an electrostatic analyzer. If a sufficiently large radius of curvature of the cylindrical electrodes is used, it is possible to measure the voltage across the electrodes accurately. This voltage can then be related to the equivalent energy of the ion beam if the geometry of the electrostatic analyzer is known. The techniques required for such measurements have been described by Herb et al. (1949).

The velocity of the ion beams can also be measured directly if the ion beam is intensity modulated. The device used for this type of measurement has been described in the literature and is known as the absolute speed gauge (Alter and Garbuny, 1949; Shoupp et al., 1949).

Since the advent of the Ge(Li) gamma-ray detector, it has been possible to measure gamma-ray energies with a precision of considerably better than one part in 10^3. Consequently, a measurement of the energy of a primary gamma ray following resonant capture gives the energy of the captured particle if the Q value of the reaction is known. This procedure is usually not convenient because of the high energy of most primary capture gamma rays. However, the $^{16}O(p, \gamma)^{17}F$ reaction shows, over a wide range of proton energies, a primary gamma-ray transition to the first excited state of ^{17}F and this gamma ray is very nearly equal in energy to the energy of the proton beam. The measurement of the energy of this gamma ray with respect to some accurately known standard gamma-ray energy, such as the energy of the gamma rays following the beta decay of ^{24}Na (Marion, 1968), would permit the ion beam energy to be measured to better than one part in 10^3.

III. Gamma-Ray Spectrometers

A. Relative Merits of NaI(Tl) and Ge(Li) Gamma-Ray Spectrometers

The choice of a spectrometer for studying the gamma radiation from charged-particle capture depends upon the yield of gamma radiation and the complexity of the gamma-ray spectrum. The yield of gamma radiation integrated over all angles can easily be lower than 100 per second (Gove and Litherland, 1960) even though many tens of microamperes of particles are used. Efficient gamma-ray spectrometers are therefore highly desirable.

The two main types of gamma-ray spectrometers at present in use are the NaI(Tl) crystal scintillation counter (Paul, Chapter III.B) and the Ge(Li) crystal conduction pulse spectrometer (Goulding and Pehl, Chapter III.A). Typical efficiency curves for three of these devices are shown in Fig. 1. Large NaI(Tl) crystals can be grown and the efficiency of a 12.8-cm-diameter 15.4-cm-long cylindrical crystal with the front face 8 cm from the target is shown as a function of energy. The efficiency of the large NaI(Tl) crystal is clearly superior to that of the smaller NaI(Tl) crystal and a typical Ge(Li) crystal detector and so, for the weakest capture gamma radiation, the large NaI(Tl) crystal detector seems to be preferable. However, the choice is not a simple one to make because the much higher resolution of the Ge(Li) detector can compensate for the disadvantage of lower efficiency and in some cases the signal-to-noise ratio of a Ge(Li) detector for very weak capture gamma radiation is superior to that of the NaI(Tl) detector. There is no simple recipe for deciding which detector to use in a given situation and usually it is best to have both available. Sometimes two NaI(Tl) detectors used in coincidence are better than a single Ge(Li) detector. Diffraction, magnetic pair, Compton, or pair (crystal) spectrometers (Alburger, 1960) are not in frequent use for the study of capture gamma rays because of their low efficiency.

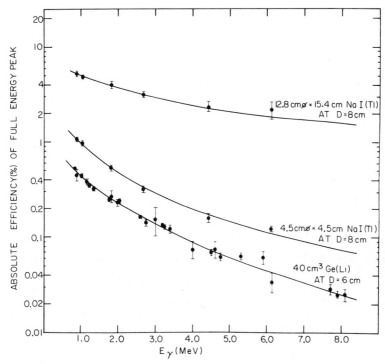

Fig. 1. The absolute efficiencies of the full energy peak for NaI(Tl) and Ge(Li) spectrometers are compared as a function of gamma-ray energy. The distances of the source to the front face of the detectors are given. The solid lines are drawn to guide the eye through the data points.

To illustrate the differences between the pulse-height spectrum from a large NaI(Tl) gamma-ray detector and a typical Ge(Li) detector, two spectra from the ^{24}Mg(p, γ)^{25}Al reaction (Dworkin, 1972) are shown in Fig. 2. These spectra were taken concurrently and illustrate the much higher resolution of the Ge(Li) detector. The resolution is over 50 times better for the 2.9-MeV primary transition $R \to 2$. The details of this decay scheme will be discussed in Section VI.A, but Fig. 2 emphasizes the great value of the high resolution of the Ge(Li) detector. It is, however, worth noting that the higher efficiency of the NaI(Tl) detector is often very valuable for finding the resonance in the first place and checking quickly the thickness and quality of the target (Section IV).

The higher efficiency of the NaI(Tl) detector is of great value if it is

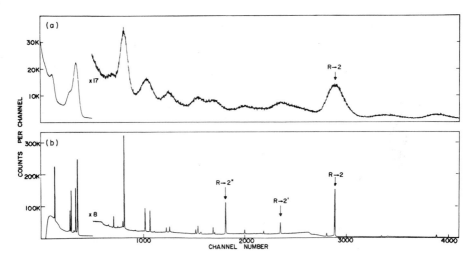

Fig. 2. Pulse-height spectrum from (b) a 40-cm³ Ge(Li) spectrometer is compared with the spectrum from (a) a 1840-cm³ (12.8 cm diameter by 15.4 cm long) NaI(Tl) spectrometer. In both cases, the gamma rays from the $E_p = 1660$ keV resonance of $^{24}Mg(p, \gamma)$ ^{25}Al are shown (Dworkin, 1972). The label $R \rightarrow 2$ refers to the gamma-ray transition from the resonance state to the second excited state of ^{25}Al. The unprimed, single-, and double-primed peaks denote full energy, single-, and double-escape peaks, respectively.

necessary to study gamma rays in coincidence. For example, the gamma rays from the $J^\pi = 0^+$, $T = 1$ resonance at $E_p = 561$ keV in the reaction $^{17}O(p, \gamma)^{18}F$ are emitted isotropically. However, the gamma–gamma angular correlations, which can be studied by observing the cascade gamma rays in coincidence, provide important information about spins, parities, and gamma-ray multipole mixing (Rolfs *et al.*, 1973b). Coincidence angular correlation measurements between NaI(Tl) and Ge(Li) counters combine the efficiency of the NaI(Tl) counter with the resolution of the Ge(Li) counter and are of value in particular cases such as the case just mentioned.

The higher efficiency of the NaI(Tl) detector is also useful (a) for the study of capture reactions with very low cross sections, (b) if ion beams of high intensity are not available, and (c) when capture gamma rays of energy higher than about 12 MeV are studied (see Paul, Chapter III.B). An example of an experiment which exploits these three advantages has been given by Feldman and Heikkinen (1969). Their experiments included a study of the resonant-like behavior of the radiative capture of heavy ions such as ^{12}C and ^{16}O. These studies would have been very difficult with Ge(Li) detectors.

A large NaI(Tl) detector has been used recently to investigate the high-energy gamma rays from the $^{11}\text{B}(p, \gamma)^{12}\text{C}$ reaction in the giant dipole resonance region (Glavish et al., 1972). Low currents (~ 2–5 nA) of polarized protons were used. These studies give valuable additional information when compared with studies with unpolarized protons. It is quite likely that as higher currents ($> 1\,\mu\text{A}$) of polarized protons become available, much new work in the lower-energy region of radiative capture resonances will be started.

Discussions of the electronic amplification of the signals from NaI(Tl) and Ge(Li) counters and the analysis of the multichannel pulse-height spectra can be found in Chapters IIIA and IIID. For energy and efficiency calibrations of gamma-ray detectors, the reader is referred to the literature (Marion, 1968; Van der Leun and de Wit, 1969; Jarczyk et al., 1962; Aubin et al., 1969).

B. GAMMA-RAY LINEAR POLARIMETERS

The gamma rays from radiative capture reactions are frequently strongly linearly polarized. At 90° to the incident ion beam, the important ratio P, to be defined, is given by

$$P = (a_0 + a_2)/(a_0 - 2a_2)$$

for magnetic dipole radiation and

$$P = (a_0 - 2a_2)/(a_0 + a_2)$$

for electric dipole radiation. In these relations a_0 and a_2 are the coefficients in the Legendre polynomial expansion of the gamma-ray angular distribution using a detector insensitive to the polarization of the gamma rays. P can be considered to be the number of gamma rays emitted with the electric vector parallel to the reaction plane divided by the number of gamma rays emitted with the electric vector perpendicular to the reaction plane. The reaction plane is defined by the direction of the ion beam and the direction of the emitted gamma ray. More complicated expressions for P can be obtained (Ferguson, 1965; Taras, 1971) for dipole–quadrupole mixtures and other related cases. The expressions given for P show that the polarization of capture gamma rays can be quite large and that the magnitude of the polarization is different for electric and magnetic dipole radiation. In general, the linear polarization ratio P, which can be measured (Gove and Litherland, 1960; Ferguson, 1965), gives valuable information on the multipolarities of the gamma rays and on the spins and parities of the nuclear states involved.

The Compton scattering process is sensitive to the linear polarization of gamma rays and this sensitivity is illustrated in Fig. 3. If monochromatic gamma rays, which are completely linearly polarized in some plane, are incident upon a scatterer, then there is a maximum asymmetry R in the Compton scattering at a certain scattering angle θ_{max}. The asymmetry R

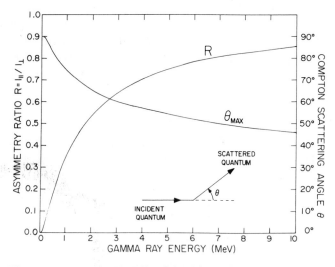

Fig. 3. The asymmetry ratio $R = I_{\parallel}/I_{\perp}$, defined in the text, is shown as a function of gamma-ray energy, as is the Compton scattering angle, where the asymmetry ratio R is a maximum.

is equal to the intensity $I_{\parallel}$ of the scattering parallel to the plane of polarization (the direction of the electric vector) divided by the intensity $I_{\perp}$ of the scattering perpendicular to the plane of polarization. The maximum asymmetry R occurs at an angle θ_{max} which is near 90° at low gamma-ray energies. Both R and θ_{max} are shown in Fig. 3. Clearly, the Compton scattering process is most sensitive at very low energies, the region where Compton scattering becomes Thomson scattering in the limit of zero photon energy, and the sensitivity decreases with increasing energy. If $N_{\parallel}$ and $N_{\perp}$ are the experimentally observed counting rates of gamma rays scattered into the reaction plane and perpendicular to the reaction plane, respectively, then P and R are related to these numbers as follows (Fagg and Hanna, 1959;

Gove and Litherland, 1960):

$$N_{\parallel}/N_{\perp} = (P + R)/(1 + PR) \qquad (1)$$

Successful linear polarimeters based on groups of NaI(Tl) crystals have been constructed (Metzger and Deutsch, 1950; Taras, 1971) but they have the disadvantage of low energy resolution. Ge(Li) crystals can be used in pairs to measure the linear polarization of gamma rays with high energy resolution (Broude et al., 1969). It is also possible to use a single, suitably oriented planar Ge(Li) counter (Litherland et al., 1970) to study the linear polarization of gamma rays from capture reactions (Lam et al., 1971). The sensitivity of the photodisintegration of the deuteron to the linear polarization of gamma rays (Hughes and Sinclair, 1956) is not frequently used because no suitable device has been developed that can compete with the Compton scattering device.

IV. Targets

A. TARGET CHAMBERS AND TARGET BACKINGS

The study of gamma rays produced by charged-particle capture is facilitated by the small attenuation of gamma rays in passing through matter. For example, a sheet of tantalum 0.5 mm thick, which is commonly used as a target backing, has approximately 90% transmission for 500-keV gamma rays and 96% transmission for 5-MeV gamma rays. The target materials themselves rarely play a significant role in the absorption of the gamma rays. This is because target thicknesses in the range from 10 to 100 $\mu g/cm^2$ are used very frequently. The choice and measurement of target thickness are discussed in the next section.

The high transmission of capture gamma rays over 500 keV by materials makes it possible to locate the gamma-ray counters outside the vacuum chamber containing the target. This is fortunate and convenient because of the complexity of the Ge(Li) and NaI(Tl) gamma-ray counters which are used to detect the capture gamma rays. Special low-Z materials must, however, be used for target chambers if gamma rays below about 500 keV are to be studied. For gamma rays over about 500 keV, thin stainless steel and brass target chamber components can be used quite satisfactorily.

A typical target chamber assembly for the study of capture gamma rays is shown in Fig. 4. This figure illustrates several of the important features necessary in a successful target chamber design.

(1) A minimum amount of material in between the target and the gamma-ray detector should be used in order to minimize the absorption of gamma rays. Cylindrical symmetry for the vacuum envelope is usually chosen in order to reduce the variation of gamma-ray absorption with angle of observation.

(2) The low cross sections for many interesting capture reactions (Section V.B) and the low efficiency of gamma-ray counters used (Section III)

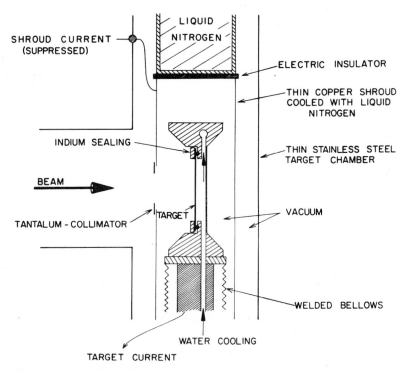

Fig. 4. A schematic diagram of a typical target chamber used for radiative capture studies with high beam power.

frequently requires the use of ion beam currents over 10 μA and as high as 100–200 μA in order to improve the signal-to-noise ratio. The power dissipation in the target backing is frequently several hundred watts and so cooling is essential for most target materials. A successful form of cooling is the direct cooling of the target backing shown in Fig. 4.

(3) The buildup of carbon deposits on targets bombarded with charged

particles is a well-known problem. In the study of proton (p, γ) or alpha-particle (α, γ) capture reactions, such a buildup causes a shift in the apparent position of the resonance due to the energy loss of the protons in traversing the carbon film. The buildup is more serious for (α, γ) reactions because the $^{13}C(\alpha, n)^{16}O$ reaction is a prolific source of neutrons $(Q_0 = 2.2$ MeV). These neutrons interact with the surrounding materials or the materials of the gamma-ray counters and produce consequently a background of gamma-ray-induced pulses (Chasman et al., 1965).

The buildup of carbon can be inhibited by surrounding the target by a thin copper shroud cooled with liquid nitrogen (Fig. 4). Care should also be taken to minimize the use of materials containing volatile hydrocarbons in the vacuum system near the target chamber.

(4) The copper shroud is usually insulated electrically so that the procedure of beam alignment can be facilitated by minimizing the current on the shroud. The shroud is shown in Fig. 4 together with a removable tantalum collimator which ensures that protons or alpha particles do not strike the lower-Z copper. The size of the tantalum collimator is chosen to intercept very little beam and the collimator should be easily removable for cleaning.

(5) An accurate measurement of the beam current on the target, and hence the accumulated charge, requires the suppression of the secondary electrons emitted from the target under beam bombardment conditions. This is accomplished by operating the insulated copper shroud at a sufficiently high negative voltage.

(6) The welded bellows, shown in the figure, permits the experimenter to change the region of the target being bombarded without opening the vacuum chamber and without using sliding and rolling rubber O rings. These O rings are a source of hydrocarbons and other contaminants. The target is deposited on a flat strip of a backing which is longer than it is wide so that the motion of the welded bellows can be used to expose different parts of the target by a simple linear motion along the length of the target.

(7) The target backing is usually either tantalum or tungsten because these materials have a high melting point and can be obtained relatively free of low-atomic-number contaminants. Nuclear reactions with tantalum and tungsten, other than Coulomb excitation, are essentially zero for proton and alpha-particle energies below 4 MeV. Indium wire can be used quite effectively as a water and vacuum seal. Care must be taken, however, to ensure that there is an adequate flow of water behind the target during an experiment; otherwise the consequences are as spectacular as they are inconvenient.

At proton and alpha particle energies higher than about 4 MeV, the target chamber shown in Fig. 4 becomes less useful. In the case of protons, nuclear reactions with the heavy element backings become significant and rapidly make it quite difficult to study proton capture reactions with backed targets. In the case of alpha particles, the nuclear reactions with ^{13}C and ^{18}O, which are hard to eliminate from the backing, become more troublesome because of the high cross sections of the reactions ^{13}C$(\alpha, n\gamma)^{16}$O and ^{18}O$(\alpha, n\gamma)^{21}$Ne. Capture reactions with alpha particles become increasingly difficult to study above about 5 MeV beam energy but the difficulty depends upon the Q value of the reaction. Low-Q-value capture reactions such as ^{16}O$(\alpha, \gamma)^{20}$Ne $(Q = 4.76$ MeV$)$ are harder to study than higher-Q-value reactions such as ^{24}Mg$(\alpha, \gamma)^{28}$Si $(Q = 10.00$ MeV$)$ because in the latter case, resonant gamma rays of energies $E_\gamma \gtrsim 10$ MeV can be readily observed well above the background radiation due to ^{13}C or ^{18}O $(E_\gamma \leqslant 10$ MeV$)$. The difficulties are independent of the Q value for resonant gamma rays of $E_\gamma \leqslant 10$ MeV.

The solution to these problems is, of course, the elimination of the target backing altogether. This implies that thin, 5–50 μg/cm^2, self-supporting targets must be used and in some cases this is possible. Such targets require a different target chamber from the one shown in Fig. 4. In this case, the beam must be allowed to pass through the target and be transported away to a shielded beam catcher, otherwise the problems of a target backing will remain.

Thin targets are unfortunately difficult to handle and the high beam currents readily destroy them. An alternative to a fragile, thin solid target is an almost indestructable, thin gaseous target which can be used quite successfully although the target chamber becomes more complex. These target chambers are known as differentially pumped target chambers (Bloch *et al.*, 1967; Litherland *et al.*, 1967; Bussière and Robson, 1971; Diamond *et al.*, 1971). A differentially pumped gas target chamber for radiative capture reactions is quite complex because many stages of pumping are required to lower the gas pressure from about 1 Torr (1 mm Hg) to 1 μTorr (10^{-6}mm Hg). It is of course necessary to ensure that the ion beam does not hit any of the differential pumping apertures; otherwise the target backing problem has not been solved. For this reason, the differential pumping restrictions must be larger than the ion beam diameter and this implies large gas flows (Diamond *et al.*, 1971; Alexander *et al.*, 1972). Consequently, the targets are not well localized in space and this creates problems when angular distributions are studied (see also Section VI.C). The engineering details of such systems are described in the references quoted.

Gas targets have been used for a variety of studies. It was found possible, for example, to study the $^{16}O(\alpha, \gamma)^{20}Ne$ reaction at an alpha-particle energy of about 9 MeV and to observe the weak capture at the $J^{\pi} = 8^{+}$ resonance in that reaction (Alexander et al., 1972). The experiment actually required the use of ^{16}O ions to bombard a ^{4}He gas target in order to eliminate the problem of nuclear reactions of alpha particles with the ^{17}O and ^{18}O components in the oxygen gas target. The success of this experiment illustrates that, with considerable expense and effort, gas targets can be used to observe capture reactions under conditions which would be impossible with solid targets.

For further information on target chamber designs, the reader is referred to the literature (Dahl et al., 1960; Walters et al., 1962; Costello et al., 1964; Donhowe et al., 1967).

B. CHOICE OF TARGET THICKNESS AND TARGET PREPARATION TECHNIQUES

The gamma-ray yield of a capture reaction is influenced by the rate at which an incident charged particle loses energy in traversing the target. If the target thickness is much greater than the width of the resonance (both given in energy units), then the gamma-ray yield, called the thick-target yield (Section V), is proportional to the number of *effective* target nuclei per cm^2 and not to the total number of target nuclei per cm^2. For example, a 2-MeV beam of alpha particles loses about four times more energy than a 2-MeV beam of protons in passing a given target thickness (Northcliffe and Schilling, 1970). Consequently, the number of effective target nuclei which can contribute to the resonant yield is four times smaller for alpha-particle capture than for proton capture. In addition, the de Broglie wavelength squared, $\lambdabar^2$, of a 2-MeV alpha particle is one-quarter of that for a 2-MeV proton, so that the resonant yield is reduced by a further factor of four (Section V). The lower gamma yield in (α, γ) capture reactions due to these effects therefore requires the use of beam currents 16 times higher for the same gamma-ray counting rates as in (p, γ) capture reactions.

When the target consists of a chemical compound containing active (resonant) and inert atoms, then the resonant thick-target gamma-ray yield is reduced by the amount by which the inert atoms participate in the slowing down process of the projectiles. In this situation, it is desirable to choose a chemical compound in which the number of inert atoms is as small as possible compared to the number of active atoms. The inert atoms should also be of low Z because the rate of energy loss increases with Z (Gove, 1959). Unfortunately, very light atoms can cause difficulties because they can also capture the incident charged particles and can therefore contribute to the background gamma rays. In order to avoid this difficulty, it is necessary to use a target compound containing a heavy inert atom at the expense of lower gamma-ray yield for the resonant capture reaction of interest.

The choice of target thickness for a particular experiment is often difficult because the capture reactions are usually very strongly resonant, with total resonance widths ranging from below microelectron volts up to megaelectron volts. In addition, resonances are often close together or they overlap. An optimum choice of target thickness requires some prior knowledge of the details of the resonance structure of a particular capture reaction. Consequently, exploratory experiments with targets of different thicknesses are often necessary before starting an experiment at a particular resonance or group of resonances.

It is possible to specify a minimum target thickness by determining the energy spread of the ion beams from the accelerator being used. The energy spread can be found quite readily with a thin target by observing the broadening of a narrow resonance of known width or by observing the slope of the leading edge of a narrow resonance with a thick target. Once the energy spread of the ion beam from a particular accelerator is known, there is clearly no advantage in using targets which are much thinner, in energy units, than the energy spread of the ion beam.

It is possible to decide upon a maximum target thickness if a preliminary experiment on the details of the resonant structure of the capture reaction has been made. The maximum thickness, in energy units, can then be chosen so that the thickness is several times the energy spread of the ion beam, if the energy spread is known to be greater than the natural width of the resonance. If the natural width of the resonance is known to be greater than the beam energy spread, then about 90% of the maximum gamma-ray yield can be obtained by using a target five times thicker, in energy units, than the natural width of the resonance (Fowler et al., 1948). If one is dealing with interfering resonances (Section V.C), the coefficients of the terms in the angular distribution expression arising from interference change sign as one passes through the resonance (for interfering resonances of quite different total widths) and therefore a target several times thicker than the natural width of the narrower resonance will cause such terms to be reduced significantly.

These simple criteria can be modified in obvious ways according to the requirements of the particular experiment. For example, the rate of energy loss varies with particle energy and type in a known way (Northcliffe and

Schilling, 1970) and consequently a target suitable for the study of a low-energy proton resonance will not necessarily be suitable for studying a higher-energy proton resonance. Alpha particles lose more energy in traversing a given target thickness, in $\mu g/cm^2$, than protons, so that targets suitable for (α, γ) reactions are frequently not suitable for (p, γ) reactions and vice versa. Techniques for measuring a target thickness have been described in detail in the literature (Gove and Litherland, 1960; Richards, 1960; Kato, 1969). Target thicknesses commonly used in radiative capture reactions range from 10 to 100 $\mu g/cm^2$.

Target preparation is an extensive subject which cannot be described in detail here. Solid targets are most commonly prepared by evaporation, and techniques for this are discussed by several authors (Holland, 1956; Richards, 1960; Muggleton and Howe, 1961; Maxman, 1967). It should be noted, however, that when chemical compounds are evaporated, one should not assume that the composition of the target will be the same as that of the original compound. In the case of noble gases, targets have been produced by implanting the gases into metal backings (Almén and Bruce, 1961).

C. Contaminants

Studies of gamma rays from the radiative capture of protons and alpha particles are frequently complicated by the presence of contaminants in the target material or in the backing (Donhowe et al., 1967). In the case of proton-induced reactions, the most troublesome contaminant is ^{19}F, which gives rise to 6- and 7-MeV gamma rays from the $^{19}F(p, \alpha\gamma)^{16}O$ reaction (Gove and Litherland, 1960). The cross section for this reaction is generally several orders of magnitude higher than for (p, γ) reactions and so traces of ^{19}F are troublesome. Another troublesome contaminant is nitrogen. This is because of the high yield of the exothermic $^{15}N(p, \alpha\gamma)^{12}C$ reaction, which produces a 4.43-MeV gamma ray. All reactions of the type $(p, \alpha\gamma)$, $(p, n\gamma)$, $(p, p'\gamma)$, etc., are sources of background gamma rays in the appropriate energy range.

The study of (α, γ) reactions is hindered by the neutrons from the reactions $^{13}C(\alpha, n)^{16}O$, $^{17}O(\alpha, n)^{20}Ne$, and $^{18}O(\alpha, n)^{21}Ne$ and the gamma rays of the residual nuclei at the appropriate alpha-particle energies. Sometimes the reactions $^{10}B(\alpha, p\gamma)^{13}C$ and $^{19}F(\alpha, p\gamma)^{22}Ne$ are also troublesome.

In order to minimize contaminants on the target backing, it is necessary, in all radiative capture work, to clean and handle the backings carefully and then to test the backings (obtained from several sources) under conditions similar to the conditions of the proposed experiment. The elimination of contaminants from target materials is usually more difficult, since each target material has to be treated as a special case. If, for example, it is necessary to avoid target oxidation, then the target should not be exposed to the air. Calcium is a typical example of such a case. If oxidation is unavoidable, then consideration should be given to deliberate oxidation by oxygen depleted in ^{17}O and ^{18}O in order to reduce the background from the reactions $^{17}O(\alpha, n\gamma)^{20}Ne$ and $^{18}O(\alpha, n\gamma)^{21}Ne$. The reactions $^{16}O(\alpha, n\gamma)^{19}Ne$ and $^{16}O(\alpha, p\gamma)^{19}F$ have high, negative Q values and usually cause no background problems. The buildup of carbon on targets during bombardment has already been mentioned in Section IV.A. The use of liquid-nitrogen-cooled shrouds has been found to be very effective in this case.

It is a general rule in capture gamma-ray studies that one spends nearly as much time studying contaminants as studying the nuclear reaction of interest and in many cases a contaminant will be unidentified for a long time because of the lack of detailed study of all the principal contaminants. For example, a particularly annoying contaminant was encountered in an experiment discussed later (Fig. 12). The important region near $R \rightarrow 5$ was observed to be sometimes obscured by a gamma ray which is probably from the $^{56}Fe(p, \gamma)^{57}Co$ reaction. This contaminant was difficult to identify because of the fragmentary nature of the published information on the radiative capture of protons by medium-weight nuclei.

V. Yield Curves

A. ISOLATED RESONANCES—THEORETICAL EXPRESSIONS

Let us first of all consider the simple case in which particles of zero intrinsic angular momentum (spin) are incident on target nuclei of spin zero. This is the case for the radiative capture of alpha particles by even-even nuclei. If the incident energy is such that the compound nucleus is formed in the vicinity of a resonance characterized by total angular momentum J, and if the yield of gamma radiation of multipolarity L emitted in a transition to the final state of spin I is measured, then the total yield as a function of particle energy, averaged over all angles with respect to the beam, can be written (Gove and Litherland, 1960) as

$$\sigma_{\alpha\beta} = \pi\lambda_\alpha^2(2J + 1)\, \Gamma_{\alpha l}\Gamma_{\beta L}/[(E - E_0)^2 + (\tfrac{1}{2}\Gamma)^2] \tag{2}$$

The subscript α is a channel index specifying the incoming particle and target and the subscript β is a channel index specifying the emitted gamma ray and residual nucleus; l is the orbital angular momentum of the incident particle (in this case equal to J); and L is the gamma-ray multipolarity, where $\mathbf{J} = \mathbf{I} + \mathbf{L}$. The wavelength of the incoming particle in the center-of-mass (CM) system can be written

$$\lambda_\alpha / 2\pi = \lambdabar_\alpha = [(m + M)/M] \, \hbar/(2mE_{\mathrm{L}})^{1/2} \tag{3}$$

where E_{L} is the laboratory energy of the incident particle of mass m and M is the mass of the target nucleus; E is the CM energy of the incident particle, $E = ME_{\mathrm{L}}/(M + m)$; E_0 is the resonance energy in the CM system; Γ is the total width of the resonance, again in the center-of-mass system, and is equal to the sum of the partial widths of all the possible modes or channels of decay of the particular resonance. If, in this case, we define for brevity

$$\omega\gamma = (2J + 1) \, \Gamma_{\alpha l} \Gamma_{\beta L} / \Gamma \tag{4}$$

then

$$\sigma_{\alpha\beta} = \pi\lambdabar_\alpha{}^2 \omega\gamma\Gamma / [(E - E_0)^2 + (\tfrac{1}{2}\Gamma)^2] \tag{5}$$

The equation connecting the thick-target yield Y_∞ (reactions per incident particle), $\omega\gamma$, the CM wavelength λ, and the rate of energy loss in the target (dE/dx in units of [laboratory energy/length]), can then be written in the form

$$\omega\gamma = \frac{2}{\lambda^2} \frac{M}{m + M} \left(\frac{dE}{dx}\right) Y_\infty \tag{6}$$

This equation allows one to readily determine the quantity $\omega\gamma$ in terms of the known quantities m, M, $\lambdabar$, and dE/dx and the measurable quantity Y_∞ (Richards, 1960). This result differs from results found in some textbooks because it is often forgotten that the expression for the rate of energy loss dE/dx assumes that the energy is measured in the laboratory system (Gove and Litherland, 1960; Gove, 1959). In many cases, the thick-target yield is the quantity determined, but in cases where the target can no longer be considered thick, corrections to these expressions are necessary (Gove and Litherland, 1960).

If the target and incident particle have spin and if more than one gamma-ray multipolarity is involved in the decay of the resonance, then the expression given earlier must be modified in the following manner. If s and j are the spins of the incident particle and target nucleus, respectively, and S is the incident channel spin, then the following vector equations interrelate the various quantities:

$$\mathbf{S} = \mathbf{s} + \mathbf{j}, \qquad \mathbf{J} = \mathbf{S} + \mathbf{l} = \mathbf{I} + \mathbf{L}$$

The quantity $\omega\gamma$ is now modified (Gove and Litherland, 1960) as follows:

$$\omega\gamma = \frac{(2J+1)}{(2s+1)(2j+1)} \sum_{SlL} \frac{\Gamma_{\alpha Sl}\Gamma_{\beta L}}{\Gamma} \tag{7}$$

In practice, the yield of the capture reaction as a function of energy (a yield curve) is usually measured with a gamma-ray detector at an angle θ to the incident particle beam direction, so that the differential cross section, rather than the total cross section, is required. If the angular distribution is expressed in terms of a sum of Legendre polynomials, $W(\theta) = \sum_k a_k P_k(\cos\theta)$, then the coefficient a_0 is proportional to the total cross section $\sigma_{\alpha\beta}$ for the capture reaction (Gove, 1959). The coefficients a_k for $k \neq 0$ can be calculated as discussed by Ferguson in Chapter VII.G.

In some cases a measurement of Γ can be made by observing the shape of the yield curve (Gove and Litherland, 1960; Richards, 1960; Mourad et al., 1967) provided that the ion-beam energy spread ΔE_s is such that $\Delta E_s \ll \Gamma$ and that the target thickness measured in energy units ΔE_t is such that $\Delta E_t \ll \Gamma$. However, the situation is usually more complicated than this; for example, $\Delta E_s \approx \Gamma$, and the measurement of Γ is often not straightforward.

The quantity Γ is usually very nearly equal to the sum of the partial widths $\sum_{Sl}\Gamma_{\alpha Sl}$ for formation of the level because usually $\Gamma_{\beta L} \ll \Gamma_{\alpha Sl}$. The individual partial widths $\Gamma_{\alpha Sl}$ are strongly influenced by the penetrability of the Coulomb and centrifugal barriers because the penetrability varies rapidly with energy and orbital angular momentum l. It is possible to remove a large part of the dependence of $\Gamma_{\alpha Sl}$ on incident particle energy and angular momentum by expressing $\Gamma_{\alpha Sl}$ in a different way (Gove, 1959; Gove and Litherland, 1960):

$$\Gamma_{\alpha Sl} = (2R_\alpha/A_{\alpha l}^2 \lambda_\alpha)\, \gamma_{\alpha Sl}^2 = (2\hbar^2/A_{\alpha l}^2 \lambda_\alpha \mu_\alpha R_\alpha)\, \theta_{\alpha Sl}^2 \tag{8}$$

where

$$A_{\alpha l}^2 = F_{\alpha l}^2 + G_{\alpha l}^2$$

R_α is known as the channel radius and is usually equal to or slightly larger than the radius of the nucleus, λ_α is the CM wavelength, and μ_α is the reduced mass given by $\mu_\alpha = mM/(m+M)$. Here, $F_{\alpha l}$ and $G_{\alpha l}$ are the conventional solutions to the radial wave equation outside the nuclear surface (Coulomb waves) evaluated at the channel radius R_α, and $\theta_{\alpha Sl}^2$ is called the (dimensionless) reduced width and is a quantity that can be predicted by nuclear models. For a single-particle state in a potential well, $\theta_{\alpha Sl}^2 \sim 1$. In cases where the resonance is very broad, a more accurate form of Eq. (8) should be employed (Gove, 1959).

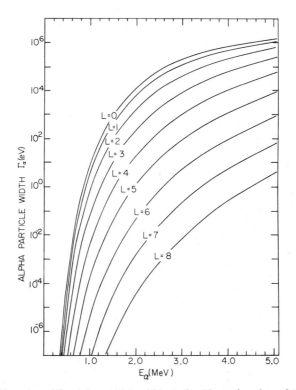

Fig. 5. The value of the alpha-particle width Γ_α (in eV) as a function of E_α, the energy of the alpha particles, is shown for the reaction $^{14}N(\alpha, \gamma)^{18}F$. A reduced alpha-particle width of the Wigner limit, i.e., $\theta^2_{\alpha Sl} = 1$, is assumed for an interaction radius of $R_\alpha = 4.8$ fm.

The factor in front of $\theta^2_{\alpha Sl}$ in Eq. (8) can readily be evaluated (Gove, 1959) and the graphs shown in Figs. 5 and 6 represent the value of $\Gamma_{\alpha Sl}$ for $\theta^2_{\alpha Sl} = 1$ and for the two capture reactions $^{14}N(\alpha, \gamma)^{18}F$ and $^{17}O(p, \gamma)^{18}F$. These reactions will be discussed in the next section. The graphs show clearly the strong dependence of $\Gamma_{\alpha Sl}$ on incident energy and orbital angular momentum.

B. Isolated Resonances—Experiment

The $^{14}N(\alpha, \gamma)^{18}F$ and $^{17}O(p, \gamma)^{18}F$ capture reactions have been chosen as examples because many of the features of capture reactions at isolated resonances can be demonstrated using these two examples (Rolfs *et al.*, 1973a) shown in Fig. 7.

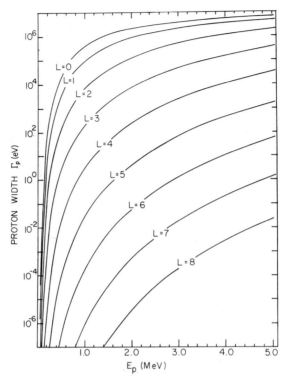

Fig. 6. The value of the proton width Γ_p (in eV) is shown as a function of proton energy E_p for the reaction $^{17}O(p, \gamma)^{18}F$. A reduced proton width of the Wigner limit is assumed for an interaction radius of $R_p = 4.3$ fm.

(1)   The Q values of the $^{14}N(\alpha, \gamma)^{18}F$ and the $^{17}O(p, \gamma)^{18}F$ capture reactions are 4.416 MeV and 5.609 MeV, respectively. This implies that above 5.61 MeV excitation energy in ^{18}F, both alpha-particle and proton channels are open and, under the appropriate conditions, the same resonant state can be observed in both reactions. Figure 7 shows several cases of this possibility. From the ratio of the thick target yields for (α, γ) and (p, γ) capture and Eq. (6) we can deduce the ratio

$$\omega\gamma(\alpha, \gamma)/\omega\gamma(p, \gamma) = \Gamma_\alpha/\Gamma_p$$

This ratio is obviously a valuable one to obtain experimentally and examples are discussed later in this section.

(2)   If isospin T is a good quantum number (see discussion by Temmer, Chapter IV.A.2), then the $^{14}N(\alpha, \gamma)^{18}F$ reaction should form only resonant

states with $T = 0$. This is because the isospins of both the ^{14}N and the alpha-particle ground states are zero. However, the ^{17}O(p, γ)^{18}F reaction can form both $T = 0$ and $T = 1$ states because the isospins of both the ^{17}O ground state and the proton are $T = \frac{1}{2}$. Consequently Γ_α/Γ_p, or more precisely $\theta_\alpha^2/\theta_p^2$, should be zero for isospin $T = 1$ states and this is very nearly the case for the resonant states at 6138 (0^+), 6164 (3^+), and 6284 (2^+) keV (Fig. 7) which are expected to be $T = 1$ states from a comparison of their excitation energies with the energies of the excited states in ^{18}O and from their gamma-ray decay schemes (Rolfs et al., 1973b, e). For the 6164-keV state, for example, one finds $\theta_\alpha^2 = 8 \times 10^{-7}$ and $\theta_p^2 = 0.3$, which indicates how significant the isospin selection rule is in this case. The formation of the $J^\pi = 0^+$ state at 6138 keV via the ^{14}N(α, γ)^{18}F reaction is also forbidden by a parity selection rule since a state with $J = 0$ can only be formed by $l = 1$ alpha particles in this example. The parity of such a state must clearly be negative if parity is a good quantum number. The formation of the negative-parity states with $T = 1$ in the (α, γ) reaction (Fig. 7), however, indicates that these states have mixed isospin. This complicated situation, which is due to the effect of the Coulomb interaction, is discussed in more detail in the literature (Charlesworth et al., 1967; Lindgren et al., 1971; Rolfs et al., 1973d, e).

The study of unbound states of mixed isospin has been pursued extensively in recent years (Warburton and Weneser, 1969; Hanna, 1969). Much effort has also been put into the location of resonances of high isospin in order to test, for example, the mass formula for isobaric spin multiplets (Elliott, 1969). States of high excitation are usually broad due to the increased penetrability of the Coulomb and centrifugal barriers by the emitted particles. However, the emission of particles from states of high isospin is often reduced considerably by isospin selection rules and hence such states of high isospin are often unusually sharp (see Temmer, Chapter IV.A.2). For example, many $T = \frac{3}{2}$ resonances in nuclei with ground-state isospin $T = \frac{1}{2}$ have been located (Morrison et al., 1968; Aitken et al., 1969; Bearse et al., 1970) as well as many $T = 1$ and $T = 2$ resonances in nuclei with $T = 0$ ground states (Hanna, 1969). Many of these resonances are characterized by low particle formation widths and large gamma-ray decay widths, usually M1 (Hanna, 1969). The $T = 1$, 6164-keV level in ^{18}F with a reduced alpha-particle width of $\theta_\alpha^2 = 8 \times 10^{-7}$ is but one example of this extensive class. Broad $T = \frac{3}{2}$ resonances are also known when the particle decay is allowed by the isospin selection rules and an example is given in Section V.C. The value of the radiative capture reaction for the study of these interesting nuclear states with high isospin and low reduced particle widths is based upon the fact

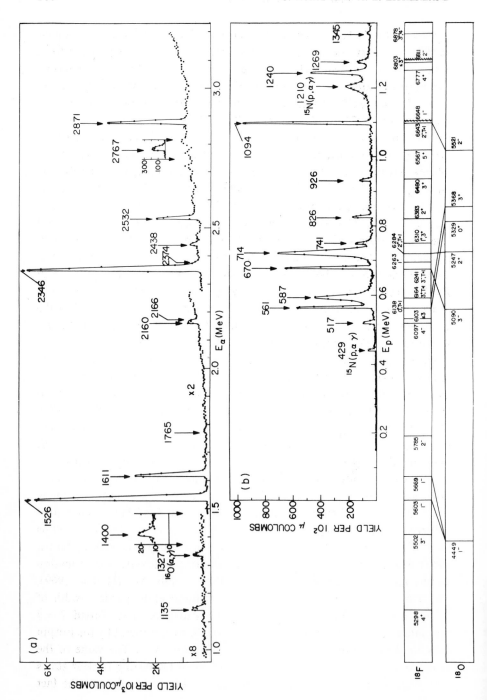

that usually Γ_p or $\Gamma_\alpha \gg \Gamma_\gamma$ due to the small intrinsic values of Γ_γ (see Section V.A). As a result of this, the thick-target yield of the capture reaction is often proportional to Γ_γ and in the case of high isospin resonances, this gamma-ray width is usually larger than the average. Consequently, even though the formation is strongly inhibited, the high-isospin resonances often dominate the yield curve.

(3) Sometimes Γ_α/Γ_p is much greater than unity. This is the case for the $J^\pi = 5^+$ state at an excitation energy of 6567 keV in ^{18}F which is observed only in the ^{14}N$(\alpha, \gamma)^{18}$F reaction (Rolfs *et al.*, 1973c). This state is of special theoretical interest and it was crucial to the understanding of the $K^\pi = 1^+$ rotational band in ^{18}F to find this state by the ^{14}N$(\alpha, \gamma)^{18}$F reaction. The resonance corresponding to this state, which has been interpreted as the fifth member of the $K^\pi = 1^+$ rotational band based upon the $J^\pi = 1^+$ state at 1701 keV, could not be located in the yield curve shown in Fig. 7 using a NaI(Tl) gamma-ray detector. However, the inset shown in Fig. 7 illustrates that with the help of the superior signal-to-noise ratio of the Ge(Li) crystal, the 2767-keV resonance can readily be located. The 6567-keV resonant state is not observed in the ^{17}O$(p, \gamma)^{18}$F reaction even though the angular momentum of the captured proton can be as small as $l = 2$. The value of Γ_α/Γ_p can be estimated to be greater than 96 from the experimental data $\omega\gamma = (2J + 1)\Gamma_\alpha\Gamma_\gamma/\Gamma = 280 \pm 60$ meV and $\omega\gamma = (2J + 1)\Gamma_p\Gamma_\gamma/\Gamma < 2.3$ meV. These results, together with the elastic alpha-particle scattering data, yield $\Gamma_\alpha \approx \Gamma \approx 560$ eV and hence $\Gamma_\gamma = 26 \pm 5$ meV and $\Gamma_p < 4.5$ eV (Rolfs *et al.*, 1973c). The reduced particle widths can be obtained with the help of the Figs. 5 and 6 and they are $\theta_\alpha^2(l = 4) \approx 1.1$ and $\theta_p^2(l = 2) < 10^{-3}$. Both reduced widths are consistent with the theoretical predictions that the $K^\pi = 1^+$ rotational band in ^{18}F should have a four particle–two hole configuration (Bassichis *et al.*, 1965). The reduced proton width θ_p^2 should be small because ^{17}O has

---

Fig. 7. The yields of capture gamma rays from the nuclear reactions (a) ^{14}N$(\alpha, \gamma)^{18}$F and (b) ^{17}O$(p, \gamma)^{18}$F are compared (Rolfs *et al.*, 1973a). The excited states of ^{18}F above 5.61 MeV are accessible to both reactions and the alpha-particle and proton energy scales are chosen so that the same excitation energies in ^{18}F lie above one another. For example, the 2346-keV (α, γ) resonance corresponds to the same resonant state as the 670-keV (p, γ) resonance. By analogy with the $T = 1$ level scheme of ^{18}O, one expects six $T = 1$ states in ^{18}F at the displayed region of excitation energy. The identification of these analog states is also indicated. The difference in excitation energies of some of these analog states is due to the Thomas–Ehrman level shift as a consequence of the nearby proton particle threshold in ^{18}F at 5609-keV (Rolfs *et al.*, 1973 b, e). Used with permission of North-Holland Publishing Co., Amsterdam.

presumably a dominant one particle–no hole ground-state configuration. This expectation is supported by the observation of $\theta_p^2 < 10^{-3}$.

(4) The $J^\pi = 4^+$ resonance at $E_\alpha = 1135$ keV corresponds to a state in ^{18}F at an excitation energy of 5298 keV and is the fourth member of the $K^\pi = 1^+$ rotational band (Rolfs et al., 1973c). The low alpha-particle energy combined with the high angular momentum $l = 4$ implies that the alpha-particle width Γ_α cannot be more than about 18 meV (Fig. 5). This is a very small alpha-particle width and so one cannot make the usual assumption that $\Gamma_\alpha \gg \Gamma_\gamma$. From the observed thick-target yield $\omega\gamma = 59 \pm 21$ meV and the total width $\Gamma = 22 \pm 3$ meV, deduced from an attenuated Doppler shift experiment of the resonant gamma rays, one finds $\Gamma_\gamma = 12 \pm 4$ meV and $\Gamma_\alpha = 10 \pm 4$ meV. This resonance is one of the rare cases where a Doppler shift attenuation measurement of the primary resonant gamma rays is possible (Section VI.B). The origin of the large reduced width $\theta_\alpha^2(l = 4)$ ≈ 0.55 is presumably the same as that for the large reduced width for the 2767-keV resonance.

The radiative capture of protons or alpha particles is useful in the study of resonances of high angular momentum (spin) for a similar reason to that given for the study of resonances of high isospin. Even if Γ_p or Γ_α is very small due to the high centrifugal barrier, the thick-target yield is independent of Γ_p or Γ_α as long as they are larger than Γ_γ. Consequently, the resonances of high spin can be observed readily if $\Gamma_\gamma < \Gamma_\alpha$ or Γ_p and this is frequently the case. The main problem in the study of high-spin resonances compared with the study of high-isospin resonances is that the gamma-ray decay widths for the former are usually smaller than for the latter. This is because the high-spin states are usually members of a rotational band and the energy of the gamma-ray transition, with the largest matrix element, is smaller than the usual energy for an isospin-allowed transition. Consequently Γ_γ is usually smaller for a high-spin resonance than for a high-isospin resonance and so the thick-target yield is smaller. The case of the $l = 8$ resonance in the ^{16}O$(\alpha, \gamma)^{20}$Ne capture reaction was discussed in Section IV.A. The difficulty in studying this resonance was mainly due to the low value of $\Gamma_\gamma = 6 \pm 2$ meV and not due to the low value of $\Gamma_\alpha = 35 \pm 10$ eV (Alexander et al., 1972).

(5) The yield curve in Fig. 7 shows how complex the situation can be at high excitation energies. There is strong evidence for doublets at 2160, 2166 keV and 2346, 2374 keV alpha-particle bombarding energies. These doublets have not yet been resolved as separated resonances but the gamma-ray spectra from a Ge(Li) detector show that, without doubt, there are two pairs of unresolved resonances (Rolfs et al., 1973a).

(6)   The rising background in the $^{14}N(\alpha, \gamma)^{18}F$ reaction is partly due to the $^{13}C(\alpha, n)^{16}O$ reaction (see Section IV.C). However, an unusual contaminant can be located at 1327 keV bombarding energy. The weak $^{16}O(\alpha, \gamma)^{20}Ne$ resonance (Van der Leun *et al.*, 1965) can be seen at $E_\alpha = 1327$ keV and it is approximately equal in intensity to the 1135 keV resonance in $^{14}N(\alpha, \gamma)^{18}F$. The strengths $\omega\gamma$ for these two resonances are similar; therefore, the target used for the data shown in Fig. 7 contained much oxygen. In the $^{17}O(p, \gamma)^{18}F$ reaction, the familiar $^{15}N(p, \alpha\gamma)^{12}C$ reaction makes an appearance at 429 and 1210 keV bombarding energies (Ajzenberg-Selove, 1971).

It has already been mentioned that the level density near 6 MeV in ^{18}F is high enough for unresolved doublets to create difficulties. The situation in

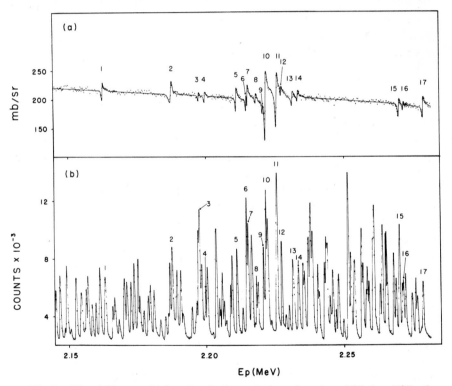

Fig. 8.   The yield curve (a) for the elastic scattering of protons [$^{58}Fe(p, p)^{58}Fe$, θ_L = 160°] is shown together with the yield curve (b) for the capture gamma rays from the reaction $^{58}Fe(p, \gamma)^{59}Co$ (1.0 MeV $\leqslant E_\gamma \leqslant$ 5.0 MeV). (Lindstrom *et al.*, 1968; Bilpuch, 1972.) The numbers refer to the same resonances in the two reaction channels. Used with permission of North-Holland Publishing Co., Amsterdam.

the reaction $^{58}\text{Fe}(\text{p}, \gamma)^{59}\text{Co}$ is much more complex because the level density at $E_x = 9.6$ MeV in ^{59}Co is much higher. This point is illustrated in the lower part of Fig. 8, which shows the large number of resonances in just over 100 keV in proton energy (Lindstrom et al., 1968; Bilpuch, 1972). The upper part of the figure also illustrates the effect of ion beam energy spread on the results for the elastic scattering of protons by ^{58}Fe. Many of the resonances in the lower part of the figure, shown unnumbered, are not observed in the elastic scattering because their proton widths are much smaller than a few tens of eV. The background in the elastic scattering is due to Rutherford scattering and this makes the observation of weak resonances very difficult. In contrast, the (p, γ) cross section has no continuous background provided all the usual precautions are taken to avoid contaminants. It is clear from the data shown in Fig. 8 that even the very high-resolution equipment used is not adequate for the study of the detail that is evidently present. The techniques for an automatic yield curve measurement have been described recently by Lam and Ferguson (1972).

C. INTERFERENCE BETWEEN RESONANCES

When the spacing between resonances becomes comparable with their natural widths, interference effects occur. These interference effects can be observed readily if the resonance widths are larger than the instrumental spreads resulting from the ion beam energy spread and target thickness. The detailed theoretical shape of a yield curve for a pair of interfering resonances can be obtained by summing two Breit–Wigner amplitudes and multiplying the result by the complex conjugate (Gove, 1959; Ferguson and Gove, 1959). The effects of angular correlations of the radiations must also be included as described by Devons and Goldfarb (1957) and Ferguson (1965). However, some general comments are in order here. First, if two resonances have the same parities but different spin, interference will contribute only terms with even k $(k > 0)$ in the Legendre polynomial expansion $P_k(\cos\theta)$ of the angular distribution. Second, if the interfering resonances have opposite parities, the interference terms give rise to odd terms in the polynomial expansion. Finally, if the interfering resonances have identical spins and parities, then the interference terms include also the $k = 0$ term in the Legendre polynomial expansion. The upper limit of k is determined by the values of the quantum numbers of the interfering resonances.

The experimental data for three resonances which interfere in pairs are shown in Fig. 9. This interesting example (Bartholomew et al., 1955) illustrates some of the comments on angular distributions. All three res-

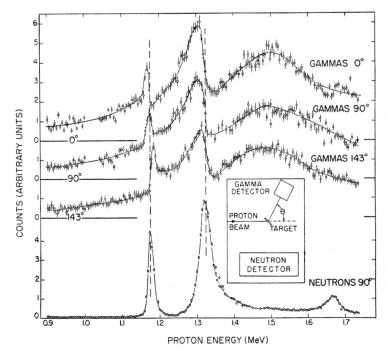

Fig. 9. The relative yield of the capture gamma rays from the $^{14}C(p, \gamma)^{15}N$ reaction is shown as a function of proton energy for three angles of observation. In addition, the relative neutron yield of the $^{14}C(p, n)^{14}N$ reaction is also illustrated. For details, see Bartholomew *et al.* (1955). Used with permission of the *Canadian Journal of Physics*.

onances are known to have spin $\frac{1}{2}$ with the 1.17-MeV resonance negative parity, the 1.31-MeV resonance positive parity, and the broad 1.50-MeV resonance also positive parity. The interference effects between the narrow 1.17-MeV and the broad 1.50-MeV resonance show up clearly in the three yield curves taken for the $^{14}C(p, \gamma)^{15}N$ reaction. The angular distribution of the ground-state gamma rays at particular energies in the vicinity of the 1.17-MeV resonance shows only terms with $k = 0$ and 1, which is to be expected for two interfering resonances of spin $\frac{1}{2}$ and opposite parity. It may come as a surprise to some readers to note that spin $\frac{1}{2}$ does not always imply isotropic angular distributions.

The shape of the yield curve in the vicinity of the 1.31-MeV resonance is very instructive. It illustrates in a simple manner the effect of the interference of two levels of spin $\frac{1}{2}$ with the same parities. The interference term associated with $k = 0$ can be most readily appreciated by noting the position

of the 1.31-MeV resonance in the lowest yield curve, which is for the
$^{14}C(p, n)^{14}N$ reaction. In the case of the (p, n) reaction, the relative intensity of
the 1.31-MeV and the 1.50-MeV resonances is much lower than in the (p, γ)
reaction, so that the interference effects are much smaller.

A detailed theoretical study (Ferguson and Gove, 1959) of these interfering resonances has been made and excellent agreement between theory
and experiment obtained. It is interesting to note that the broad 1.50-MeV
resonance can be identified as the $T = \frac{3}{2}$ ($T_z = \frac{1}{2}$) analog of the $J^\pi = \frac{1}{2}^+$
ground state of ^{15}C (Bartholomew et al., 1955). The inhibition of the decay
of this state by neutron emission can be understood simply as an example
of inhibition due to isospin selection rules. The formation of the 1.50-MeV
resonance is allowed, however, by the isospin selection rules.

D. Lewis Effect in Resonance Yield Curves

A thick-target yield curve taken at a narrow resonance with an ion beam
of high energy resolution shows a maximum just above the resonance energy
before it assumes a constant value (Fig. 10). This effect is often called the
Lewis effect (Lewis, 1962; Costello et al., 1964). In passing through the
target material, a charged particle loses energy in discrete steps, rather than
continuously, due to Coulombic collisions with electrons in the target
material. If some of these steps are larger than the natural width of a narrow
resonance, some of the particles incident on a target at an energy well above

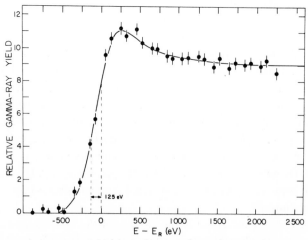

Fig. 10. Observed and calculated thick-target yield curves for the $E_p = 1424$ keV
resonance in the $^{58}Ni(p, \gamma)^{59}Cu$ capture reaction. For details, see Costello et al. (1964).
Used with permission of North-Holland Publishing Co., Amsterdam.

the resonance energy E_R will jump over the resonance and thus will not contribute to the resonant capture yield. If particles are incident at E_R, then all particles will have, for a finite time, the correct energy to interact. The yield curve should exhibit therefore a peak near E_R. This effect has been observed in several resonances and can be up to 23% above the thick-target yield (Fig. 10). Calculated yield curves based on a collision spectrum which takes into account the electron shell structure of the atoms agree very well with the experimental results (Costello et al., 1964).

This effect is very sensitive to surface conditions of the target (Costello et al., 1964) and therefore may be of interest for surface studies. It should be noted, furthermore, that the point at the half-plateau yield, which is commonly assumed to be E_R, does not fall at E_R due to this Lewis effect. In the case of $^{58}Ni(p, \gamma)^{59}Cu$, shown in Fig. 10, it is about 125 eV below the actual value of E_R.

The thick-target yield curve for a particular (p, γ) resonance is also different in shape when $H_2{}^+$ or $H_3{}^+$ rather than $H_1{}^+$ ion beams are used (Dahl et al., 1960; Walters et al., 1962), due to atomic and molecular effects.

The cross section in a charged-particle capture reaction has been observed to vary with the orientations of a monocrystalline target (Anderson et al., 1965). The cross section often shows a pronounced dip at certain orientations and this phenomenon can be used, for example, to align quickly a single crystal of aluminum.

E. DIRECT CAPTURE AND COULOMB EXCITATION

In many radiative capture reactions of protons by light nuclei, the resonances observed are superimposed on a small $(\sigma \leqslant 10\,\mu b)$, nonresonant yield which varies slowly with bombarding energy (Woodbury et al., 1954; Thomas and Tanner, 1963). The angular distribution of the gamma rays from protons captured by even-even nuclei to nuclear states of spin and parity $J^\pi = \frac{1}{2}^+$ show a very striking $\sin^2 \theta$ angular distribution which changes very little with bombarding energy. Figure 11 shows a good example of this phenomenon which was recently studied with a Ge(Li) counter (Rolfs et al., 1972).

Christy and Duck (1961) have shown that a simple extranuclear explanation is in good agreement with the experimental data. The phenomenon is known as direct capture and the matrix elements for the process can be obtained by considering only the contributions of the proton Coulomb wave functions well outside the nuclear potential of the capturing nucleus. The absence of the nuclear interaction in the initial state accounts for the smoothly varying

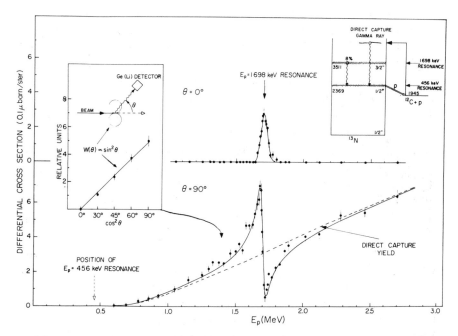

Fig. 11. The cross section for the gamma-ray transition to the 2369-keV state in ^{13}N, as observed in the ^{12}C(p, γp$'$)^{12}C reaction, is shown as a function of beam energy for two angles of observation (Rolfs *et al.*, 1972). The solid lines through the data points represent a theoretical fit (Woodbury *et al.*, 1954; Rolfs *et al.*, 1972) based on the interference between a Breit–Wigner amplitude for the 1698-keV resonance and a direct capture amplitude slowly varying with energy (dashed line). The insets show the gamma-ray angular distribution obtained at $E_p = 1.4$ MeV and the gamma-ray decay scheme.

character of the cross section. The predominance of electric dipole gamma transitions together with the absence of the nuclear spin–orbit interaction in the initial state accounts for the $\sin^2 \theta$ angular distribution in the case of p-wave capture leading to a nuclear s state.

Figure 11 illustrates the results obtained from an experiment on the direct capture of protons by ^{12}C from about 0.5 to 3.0 MeV. The smooth yield of the direct capture process is interrupted by a prominent feature near 1.7 MeV. Away from this feature, the smooth yield is associated with a $\sin^2 \theta$ angular distribution as shown in the inset and is also evident from a comparison of the yield of gamma rays at $\theta = 0°$ and $\theta = 90°$. The direct capture in this case shows predominantly an E1 transition to the unbound state at 2369 keV in ^{13}N which decays in turn mainly by proton emission ($\Gamma \approx 40$ keV). The

1698-keV resonance in the ^{12}C(p, γ)^{13}N reaction has an 8% branch to the 2369-keV unbound state and the resonant yield clearly interferes with the nonresonant direct capture yield.

It has been demonstrated recently that the study of the direct capture process can provide nuclear structure information on final states, i.e., spectroscopic factors (Rolfs, 1972a, 1973; Domingo, 1965).

Resonant capture is also observed sometimes to be superimposed on a slowly varying background due to the excitation of the nucleus by the Coulomb field of the projectile. For example, Temmer and Heydenberg (1954) observed the Coulomb excitation of the first excited state of ^{23}Na by alpha particles below 3 MeV. Above 2 MeV, resonances from the reaction ^{23}Na(α, pγ)^{26}Mg appeared, although the yield of the Coulomb excitation was still observed to be varying smoothly. Resonances from the reaction ^{23}Na(α, γ)^{27}Al are also prominent (Röpke and Anyas-Weiss, 1969; de Voigt et al., 1971) in the region where the Coulomb excitation of ^{23}Na is important.

VI. Miscellaneous Topics

A. GAMMA-RAY SPECTRA FROM CAPTURE REACTIONS—ILLUSTRATIONS OF SOME OF THE PROBLEMS

The comparison between the gamma-ray pulse-height spectra from a large NaI(Tl) detector and from a high-resolution but smaller Ge(Li) detector, shown in Fig. 2, illustrates the great importance of the development of the Ge(Li) detector for the study of gamma-ray spectra following the radiative capture of charged particles. We have chosen the gamma-ray pulse-height spectrum from the 1.66-MeV resonance in the ^{24}Mg(p, γ)^{25}Al reaction (Figs. 12 and 13) to illustrate the detailed information that can be obtained with current gamma-ray detection techniques. The example chosen is an interesting one partly because it was discussed in an article similar to this one (Gove and Litherland, 1960) before the invention of the solid-state gamma-ray detector. It is quite possible that the developments in new solid-state gamma-ray detectors (Miller, 1972) during the next 12 years will produce further improvements in detector efficiency and resolution.

The details of the decay scheme of the 1.66-MeV, $\Gamma = 100$ eV resonance in the ^{24}Mg(p, γ)^{25}Al reaction were first elucidated (Gove et al., 1958) by studying coincidences between a pair of large NaI(Tl) gamma-ray detectors. This procedure simplifies the gamma-ray spectra in an obvious way and thereby overcomes, to some extent, the poor resolution of the NaI(Tl)

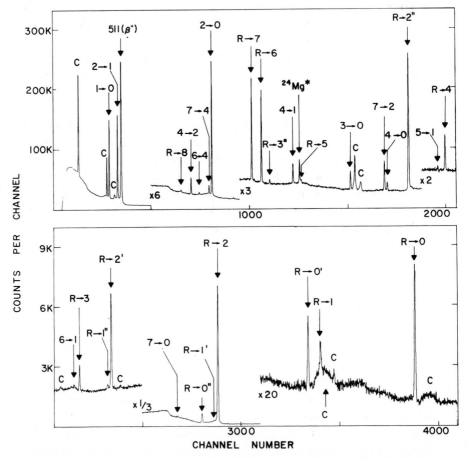

Fig. 12. Details of the gamma-ray pulse-height spectrum (shown in Fig. 2) obtained for the 1660-keV resonance in ^{24}Mg(p, γ)^{25}Al are illustrated (Dworkin, 1972) together with the assignment of the individual transitions in the decay scheme shown in Fig. 13. The label C refers to gamma rays from contaminant reactions. Unprimed, single-, and double-primed peaks denote full energy, single-, and double-escape peaks, respectively.

detector. However, this procedure is only applicable to gamma rays which are members of a coincident cascade. For example, the gamma rays marked $R \rightarrow 8$, $R \rightarrow 6$, and ^{24}Mg* were not observed in the earlier work but are important transitions (Dworkin, 1972). The existence of the transition $R \rightarrow 6$ was inferred previously (Gove *et al.*, 1958) from the intensities of unresolved gamma rays. The levels 8 and 6 in ^{25}Al are unbound and, since the values of Γ_γ/Γ

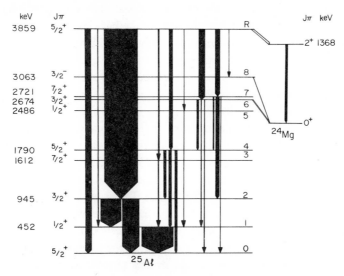

Fig. 13. The gamma-ray decay scheme of the 1660-keV resonance in ^{24}Mg(p, γ)^{25}Al is shown (Dworkin, 1972). The width of the solid arrows is proportional to the gamma-ray branching ratios. The open arrows indicate proton decay. The proton decay of the 1660-keV resonance to the ground state of ^{24}Mg is omitted.

are much less than one, the levels emit low-energy protons in strong competition with gamma rays. The gamma ray from the first excited state of ^{24}Mg, labeled ^{24}Mg* in Fig. 12, is from the competing nuclear reaction ^{24}Mg(p, p'γ)^{24}Mg. The low intensity of this gamma-ray line is due to the very low energy of the inelastic protons, ~ 200 keV, compared with the height of the Coulomb barrier.

Although the value of Γ_γ/Γ for the level marked 6 in Fig. 13 is much less than one, it is still possible to observe the gamma rays from the level 6 which are in unfavorable competition with proton emission. These gamma rays are represented by the peaks $6 \rightarrow 4$ and $6 \rightarrow 1$ in Fig. 12. If the efficiency curve for the Ge(Li) counter is known, as shown in Fig. 1, it is actually possible to deduce the value for Γ_γ/Γ from the data in spectra like the one shown in Fig. 12. The value obtained (Dworkin, 1972) was $\Gamma_\gamma/\Gamma = 0.12 \pm 0.02$.

The point made earlier about the superior signal-to-noise ratio of Ge(Li) counter measurements is illustrated by the observation of new transitions from the resonance and in the subsequent cascading. For example, the previously reported transition strength (Litherland et al., 1956) to the first excited state of ^{25}Al, $R \rightarrow 1$, was in error for reasons which are obvious in Fig. 12.

The transition $R \to 1$ sits on top of a broad peak due to the ^{15}N component of the nitrogen contamination of the target backing. The $^{15}N(p, \alpha\gamma)^{12}C$ reaction has a large cross section compared with the cross section for radiative capture and it is also resonant, so that background subtraction is difficult. There is also an additional contaminant line nearby. It is obvious that the high resolution of the Ge(Li) detector is playing the most important role in this case.

The most interesting gamma rays in the spectrum are those marked $R \to 5$ and $5 \to 1$. These are only just at the threshold of observation and are of great importance because the levels 5, 6, and R of Fig. 13 are strongly suspected to be the first three members of a rotational band, just as are the levels 0, 3 and 1, 2, 4, 7 (Dworkin, 1972). The strength of the E2 primary labeled $R \to 5$ is an important piece of evidence to support this assignment. The 40-cm^3 Ge(Li) detector used for this experiment is clearly marginally efficient to observe this weak transition. Higher-efficiency solid-state detectors will make this problem an easier one to solve in the future.

B. DOPPLER SHIFTS AND LINE SHAPE EFFECTS IN CAPTURE REACTIONS

Doppler shifts of gamma radiation emitted in a capture reaction can readily be observed (Carlson and Azuma, 1967). The attenuated Doppler shift method (Fossan and Warburton, Chapter VII.H) has been used in many cases to determine the lifetimes of the states to which the resonance decays. In some cases, the primary resonant gamma rays themselves show an attenuated Doppler shift (Anyas-Weiss et al., 1968; Rolfs et al., 1973c, d) and hence allow the full-width Γ of the resonance to be determined (see also Section V.B.4). These are the cases when $\Gamma_\gamma \approx \Gamma \lesssim 100$ meV. In most cases, however, the total width of the resonance is $\Gamma \geqslant 1$ eV and hence all primary gamma-ray transitions exhibit the full Doppler shift. This fact, however, can be quite useful as a consistency test in the Doppler shift measurements for the secondary transitions from the lower states. In addition, the analyses of such measurements is simplified by the uniquely defined recoil velocity of the residual nucleus.

Due to the $\cos\theta$ dependence of the Doppler shift, a particular resonant gamma ray can sometimes be observed as a sharp peak at $0°$ and a broadened peak at $90°$. This situation is common in (α, γ) reactions where the Doppler shift of the gamma rays, for the angle subtended at the target by the Ge(Li) detector, is significant. Since most cylindrical Ge(Li) detectors used are coaxially drifted with an inert center, the gamma-ray peak at $90°$ often appears to be a partly resolved doublet.

C. Angular Distribution Measurements

Gamma-ray angular distribution measurements at isolated resonances constitute a powerful technique for the determination of the spins involved in the gamma-ray transition as well as the multipole mixing ratio of the transition itself (Litherland, 1965). This determination is usually facilitated by the strong alignment of the resonant state which in turn produces marked angular distributions (see Section I). Details of such measurements as well as their analysis are described by Ferguson in Chapter VII.G of this book, as well as by Litherland (1965) and Ferguson (1965).

Two special features of such measurements in radiative capture work should be pointed out however.

(1)   The parity of the resonant state can be deduced from the angular distribution measurements for alpha-particle capture on even-even or odd-odd target nuclei. In the former case, the target nucleus has spin $j = 0$ and the spin of the resonance state J is therefore given by the orbital angular momentum l of the captured alpha particle, hence $\pi = (-)^l = (-)^J$. One can therefore form only resonance states of natural parity. In the case of odd-odd target nuclei (for example, ^{14}N), the population of the $m = 0$ magnetic substate $P(0)$ is proportional to the square of a Clebsch–Gordan coefficient, $P(0) \propto \langle j0, l0 | J0 \rangle^2$, and this coefficient vanishes for the case $j + l + J = $ odd. For example, if in the ^{14}N$(\alpha, \gamma)^{18}$F capture reaction ($j = 1$) a resonant state with $J = 2$ is formed, then this condition results in $3 + l = $ odd for $P(0) = 0$. If the experiment requires a population $P(0) > 0$, then the sum must be even and hence l must be odd. In this case, the parity of the resonance state is opposite to the parity of the target nucleus. This feature has been used in many cases of the ^{14}N$(\alpha, \gamma)^{18}$F reaction to determine the parity of resonant states (Rolfs *et al.*, 1973c–e).

(2)   In the case when two channel spins (S_1 and S_2) are possible for the formation of the resonance, then one can obtain the channel spin intensity ratio $t = I(S_2)/I(S_1)$ experimentally by expressing the angular distributions as $W(\theta) = [1/1 + t] [W_{S_1}(\theta) + tW_{S_2}(\theta)]$. This t value can then be compared with theoretical predictions based on the L–S or j–j coupling scheme and hence valuable information concerning the structure of the resonance state can be deduced. For example, in the ^{17}O$(p, \gamma)^{18}$F reaction, the 1240-keV resonance ($J^{\pi} = 4^+, T = 0$) is formed by pure d-wave capture in channels with spins $S_1 = 3$ and $S_2 = 2$ and $t = 2.0 \pm 0.2$ (Rolfs *et al.*, 1973e). In the j–j coupling scheme, one wishes to know the relative fractions of $d_{5/2}$ and

$d_{3/2}$ contributions in the formation and hence the dominant shell model configuration of the resonance. For pure $d_{5/2}$ and $d_{3/2}$ capture, one expects $t = 0.5$ and 2.0, respectively, favoring therefore a $d_{3/2}$ captured proton for the 6777-keV resonant state. This is in agreement with the predicted 100% pure $(d_{5/2}d_{3/2})^2$ configuration for this state (Rolfs et al., 1973e). Further examples are given in the literature (Gove, 1959).

When accurate angular distribution measurements are analyzed, it is necessary to remember the aberration correction due to the finite velocity v of the nucleus emitting the gamma ray. This correction is given by

$$\sigma(\theta)_{CM} = \sigma(\theta)_{L}[1 - 2(v/c)\cos\theta]$$

This correction is very important when the roles of the target and projectile are reversed. Corrections due to the hyperfine interaction must also be considered if the excited nuclei recoil into a vacuum (Berant et al., 1971).

D. RESONANT ABSORPTION OF CAPTURE GAMMA RADIATION

A capture reaction frequently produces a resonant state at an excitation energy E_0 which emits a gamma ray leading to the ground state of a stable residual nucleus. In this case, under the appropriate conditions, the ground-state gamma radiation can be used in turn to excite identical nuclei to the same resonant state E_0. This effect is known as nuclear resonant fluorescence (Metzger, 1956; Schopper, 1956; Knapp, 1957; Hanna and Meyer-Schützmeister, 1957).

The energy of the emitted gamma rays is reduced by an amount of $E_0^2/2Mc^2$ due to the recoil of the emitting nucleus, and because the nucleus upon absorbing the γ ray also recoils, a total extra energy of E_0^2/Mc^2 must therefore be supplied externally in order that resonant fluorescence can take place. This extra energy can be obtained from the Doppler shift by observing the capture gamma radiation at some forward angle so that the motion imparted to the residual nucleus by the captured particle increases the energy of the gamma radiation.

This type of resonant fluorescence was first demonstrated by Smith and Endt (1958) for the 12.329-MeV ground-state radiation of the 774-keV resonance in ^{27}Al$(p, \gamma)^{28}$Si. The transmission of this capture radiation through a sample of silicon powder was observed with a Ge(Li) counter (Rolfs, 1972b) as a function of detector angle θ. In Fig. 14, the resonant fluorescence of the 12.329-MeV gamma radiation is shown. At the predicted angle, $\theta = 71°$, a minimum in transmission was found. The absorption cross

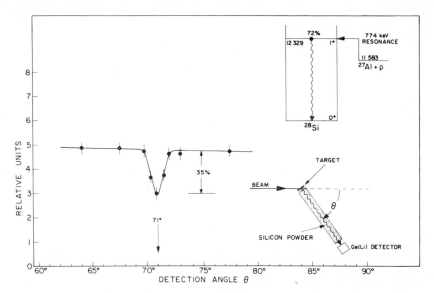

Fig. 14.   Resonant absorption of the 12.329-MeV gamma rays from the $E_p = 774$ keV resonance in ^{27}Al(p, γ)^{28}Si is shown as a function of the counter angle θ (Rolfs, 1972b). The predicted angle $\theta = 71°$, at which resonant absorption should occur, is in excellent agreement with observation. This example of resonant absorption was first demonstrated by Smith and Endt (1958) with the use of a NaI(Tl) crystal to detect the gamma rays. Lead collimator slits were used to define well the detection angle θ.

section for such experiments can be written as (Smith and Endt, 1958)

$$\sigma(E) \propto \lambda^2 \Gamma_\gamma \Gamma / [(E - E_0)^2 + (\tfrac{1}{2}\Gamma)^2] \qquad (9)$$

where Γ_γ and Γ are the partial gamma-ray width and total width of the resonant state, respectively, and E is the energy of the gamma rays incident on the absorbing sample. In some cases, the width (in degrees) of the transmission minimum is directly related to Γ and the area of the minimum to Γ_γ (Smith and Endt, 1958; Mouton and Smith, 1960). Such resonant absorption measurements have been used as absolute calibration points for the determination of relative resonance strengths $\omega\gamma$ of a given capture reaction (Engelbertink and Endt, 1966).

References

Aitken, J. H., Litherland, A. E., Dixon, W. R., and Storey, R. S. (1969). *Phys. Lett.* **30B**, 473.
Ajzenberg-Selove, F. (1971). *Nucl. Phys.* **A166**, 1.

Alburger, D. E. (1960). *In* "Nuclear Spectroscopy" (F. Ajzenberg-Selove, ed.), Part A, pp. 228–244. Academic Press, New York.

Alexander, T. K., Häusser, O., McDonald, A. B., Ferguson, A. J., Diamond, W. T., and Litherland, A. E. (1972). *Nucl. Phys.* **A179**, 477.

Almén, O., and Bruce, G. (1961). *Nucl. Instrum. Methods* **11**, 257.

Alter, W., and Garbuny, M. (1949). *Phys. Rev.* **76**, 496.

Anderson, J. U., Davis, J. A., Nielsen, K. O., and Anderson, S. L. (1965). *Nucl. Instrum. Methods* **38**, 210.

Anyas-Weiss, N., Litherland, A. E., and Röpke, H. (1968). *Phys. Lett.* **27B**, 161.

Armstrong, J. C. (1967). *In* "Nuclear Research with Low Energy Accelerators" (J. B. Marion and D. M. Van Patter, eds.), pp. 247–273. Academic Press, New York.

Aubin, G., Barette, J., Lamoureux, G., and Monaro, S. (1969). *Nucl. Instrum. Methods* **76**, 85.

Bartholomew, G. A., Brown, F., Gove, H. E., Litherland, A. E., and Paul, E. B. (1955). *Can. J. Phys.* **33**, 441.

Bassichis, W. H., Giraud, B., and Ripka, G. (1965). *Phys. Rev. Lett.* **15**, 980.

Bearse, R. C., Legg, J. C., Morrison, G. C., and Segel, R. E. (1970). *Phys. Rev. C* **1**, 608.

Berant, Z., Goldberg, M. B., Goldring, G., Hanna, S. S., Loebenstein, H. M., Plesser, I., Popp, M., Sokolowski, J. S., Tandon, P. N., and Wolfson, Y. (1971). *Nucl. Phys.* **A178**, 155.

Bilpuch, E. G. (1972). Duke Univ., private communication.

Bloch, R., Pixley, R. E., and Winkler, H. (1967). *Helv. Phys. Acta* **40**, 832.

Broude, C., Häusser, O., Malm, H., Sharpey-Schafer, J. F., and Alexander, T. K. (1969). *Nucl. Instrum. Methods* **69**, 29.

Bussière, J., and Robson, J. M. (1971). *Nucl. Instrum. Methods* **91**, 103.

Carlson, L. E., and Azuma, R. E. (1967). *Phys. Lett.* **24B**, 462.

Charlesworth, A. M., Azuma, R. E., and Kuehner, J. A. (1967). *Bull. Amer. Phys. Soc.* **12**, 53, 663.

Chasman, C., Jones, K. W., and Ristinen, R. A. (1965). *Nucl. Instrum. Methods* **37**, 1.

Christy, R. F., and Duck, I. (1961). *Nucl. Phys.* **24**, 89.

Cleland, M. R. (1968). *In* "Third Symposium on the Structure of Low-Medium Mass Nuclei" (J. P. Davidson, ed.), pp. 230–261. University Press of Kansas, Lawrence and London.

Cockcroft, J. D., and Walton, E. T. C. (1932a). *Nature* **129**, 242.

Cockcroft, J. D., and Walton, E. T. C. (1932b). *Proc. Phys. Soc.* **A136**, 619.

Costello, D. G., Skofronick, J. G., Morsell, A. L., Palmer, D. W., and Herb, R. G. (1964). *Nucl. Phys.* **51**, 113.

Dahl, P. F., Costello, D. G., and Walters, W. L. (1960). *Nucl. Phys.* **21**, 106.

de Voigt, M. J. A., Maas, J. W., Veenhof, D., and Van der Leun, C. (1971). *Nucl. Phys.* **A170**, 449.

Devons, S., and Goldfarb, L. J. B. (1957). *In* "Handbuch der Physik" (S. Flügge, ed.), Vol. 42, pp. 362–554. Springer, Berlin.

Diamond, W. T., Alexander, T. K., and Häusser, O. (1971). *Can. J. Phys.* **49**, 1589.

Domingo, J. J. (1965). *Nucl. Phys.* **61**, 39.

Donhowe, J. M., Ferry, J. A., Mourad, W. G., and Herb, R. G. (1967). *Nucl. Phys.* **A102**, 383.

Dworkin, P. B. (1972). Univ. of Toronto, private communication.

Elliott, J. P. (1969). *In* "Isospin in Nuclear Physics" (D. H. Wilkinson, ed.), pp. 73–114. North-Holland Publ., Amsterdam.

Engelbertink, G. A. P., and Endt, P. M. (1966). *Nucl. Phys.* **88**, 12.

Fagg, L. W., and Hanna, S. S. (1959). *Rev. Mod. Phys.* **31**, 711.

Feldman, W., and Heikkinen, D. W. (1969). *Nucl. Phys.* **A133**, 177.

Ferguson, A. J. (1965). "Angular Correlation Methods in Gamma-Ray Spectroscopy," North-Holland Publ., Amsterdam.

Ferguson, A. J., and Gove, H. E. (1959). *Can. J. Phys.* **37**, 660.

Fowler, W. A., Lauritsen, C. C., and Lauritsen, T. (1948). *Rev. Mod. Phys* **20**, 236.

Fowler, W. A., Bashkin, S., Bodansky, D., Brown, W. L., Clayton, D. D., Davies, J. A., Fossan, D., Mayer, J. W., Parker, P. D., Stephens, W. E., Whaling, W., and Wolicki, E. A. (1968). "New Uses for Low-Energy Accelerators." Nat. Acad. of Sci., Washington, D. C.

Glavish, H. F., Hanna, S. S., Avida, R., Boyd, R. N., Chang, C. C., and Diener, E. (1972). *Phys. Rev. Lett.* **28**, 766.

Gove, H. E. (1959). *In* "Nuclear Reactions" (P. M. Endt and P. B. Smith, eds.), Vol. I, pp. 259–317. North-Holland Publ., Amsterdam.

Gove, H..E., and Litherland, A. E. (1960). *In* "Nuclear Spectroscopy" (F. Ajzenberg-Selove, ed.), Part A, pp. 260–304. Academic Press, New York.

Gove, H. E., Litherland, A. E., Almquist, E., and Bromley, D. A. (1958). *Phys. Rev.* **111**, 608.

Hanna, S. S. (1969). *In* "Isospin in Nuclear Physics" (D. H. Wilkinson, ed.), pp. 591–664. North-Holland Publ., Amsterdam.

Hanna, S. S., and Meyer-Schützmeister, L. (1957). *Phys. Rev.* **108**, 1644.

Harvey, J. A. (1970). "Experimental Neutron Resonance Spectroscopy." Academic Press, New York.

Herb, R. G. (1959). *In* "Handbuch der Physik" (S. Flügge, ed.), Vol. 44, pp. 65–104. Springer, Berlin.

Herb, R. G. (1972). *Rev. Brasil. Fis.* **2**, 17.

Herb, R. G., Snowdon, S. C., and Sala, O. (1949). *Phys. Rev.* **75**, 246.

Holland, L. (1956). "Vacuum Deposition of Thin Films." Chapman and Hall, London.

Hughes, I. S., and Sinclair, D. (1956). *Proc. Phys. Soc.* **A69**, 125.

Jarczyk, L., Knöpfel, H., Lang, J., Müller, R., and Wölfli, W. (1962). *Nucl. Instrum. Methods* **17**, 310.

Kato, S. (1969). *Nucl. Instrum. Methods* **75**, 293.

Knapp, V. (1957). *Proc. Phys. Soc.* **A70**, 142.

Lam, S. T., Azuma, R. E., and Litherland, A. E. (1971). *Can. J. Phys.* **49**, 685.

Lam, S. T., and Ferguson, A. J. (1972). *Nucl. Instrum. Methods* **99**, 151.

Lewis, H. W. (1962). *Phys. Rev.* **125**, 937.

Lindgren, R. A., Young, F. C., and Cotton, B. (1971). *Phys. Lett.* **37B**, 358.

Lindstrom, D. P., Newson, H. W., Bilpuch, E. G., and Mitchell, G. E. (1968). *Nucl. Phys.* **A168**, 37.

Litherland, A. E. (1965). *In* "Nuclear Structure and Electromagnetic Interactions" (N. McDonald, ed.), p. 136. Oliver and Boyd, Edinburgh and London.

Litherland, A. E., Paul, E. B., Bartholomew, G. A., and Gove, H. E. (1956). *Phys. Rev.* **102**, 208.

Litherland, A. E., Ollerhead, R. W., Smulders, P. J. M., Alexander, T. K., Broude, C., Ferguson, A. J., and Kuehner, J. A. (1967). *Can. J. Phys.* **45**, 1901.

Litherland, A. E., Ewan, G. T., and Lam, S. T. (1970). *Can. J. Phys.* **48**, 2320.

Marion, J. B. (1966). *Phys. Lett.* **21**, 61.

Marion, J. B. (1968). *Nuc. Data* **A4**, 301.

Maxman, S. H. (1967). *Nucl. Instrum. Methods* **50**, 53.

Metzger, F. (1956). *Phys. Rev.* **103**, 983.

Metzger, F., and Deutsch, M. (1950). *Phys. Rev.* **78**, 551.

Miller, G. L. (1972). *IEEE Trans. Nucl. Sci.* **NS-19**, 251.

Morrison, G. C., Youngblood, D. H., Bearse, R. C., and Segel, R. E. (1968). *Phys. Rev.* **174**, 1366.

Mourad, W. G., Nielsen, K. E., and Petrilak, M. Jr. (1967). *Nucl. Phys.* **A102**, 406.

Mouton, W. L., and Smith, P. B. (1960). *Nucl. Phys.* **16**, 206.

Muggleton, A. H., and Howe, T. A. (1961). *Nucl. Instrum. Methods* **13**, 211.

Northcliffe, L. C., and Schilling, R. F. (1970). *Nuc. Data* **A7**, 233.

Parks, P. B., Newson, H. W., and Williamson, R. M. (1958). *Rev. Sci. Instrum.* **29**, 834.

Richards, H. T. (1960). *In* "Nuclear Spectroscopy" (F. Ajzenberg-Selove, ed.), Part A, pp. 99–138. Academic Press, New York.

Rolfs, C. (1972a). Proc. Panel Meeting on Charged Particle Induced Radiative Capture, IAEA, Vienna.

Rolfs, C. (1972b). Univ. of Toronto, internal report.

Rolfs, C. (1973). *Nucl. Phys.* (in press).

Rolfs, C., Beukens, R., and Litherland, A. E. (1972). Univ. of Toronto, internal report.

Rolfs, C., Charlesworth, A. M., and Azuma, R. E. (1973a). *Nucl. Phys.* **A199**, 257.

Rolfs, C., Kieser, W. E., Azuma, R. E., and Litherland, A. E. (1973b). *Nucl. Phys.* **A199**, 274.

Rolfs, C., Trautvetter, H. P., Azuma, R. E., and Litherland, A. E. (1973c). *Nucl. Phys.* **A199**, 289.

Rolfs, C., Berka, I., and Azuma, R. E. (1973d). *Nucl. Phys.* **A199**, 306.

Rolfs, C., Berka, I., Trautvetter, H. P., and Azuma, R. E. (1973e). *Nucl. Phys.* **A199**, 328.

Röpke, H., and Anyas-Weiss, N. (1969). *Can. J. Phys.* **47**, 1545.

Schopper, H. (1956). *Z. Phys.* **144**, 476.

Shoupp, W. E., Jennings, B., and Jones, W. (1949). *Phys. Rev.* **76**, 502.

Smith, P. B., and Endt, P. M. (1958). *Phys. Rev.* **110**, 397.

Staub, H. H. (1967). *Proc. Int. Conf. At. Masses, 3rd Winnipeg, Canada* (R. C. Barber, ed.), pp. 495–507. Univ. of Manitoba Press.

Taras, P. (1971). *Can. J. Phys.* **49**, 328.

Temmer, G. M., and Heydenberg, N. P. (1954). *Phys. Rev.* **96**, 426.

Thomas, G. C., and Tanner, N. W. (1963). *Nucl. Phys.* **44**, 647.

Van de Graaff, R. J. (1931). *Phys. Rev.* **38**, 1919.

Van de Graaff, R. J., Compton, K. T., and van Atta, L. C. (1933). *Phys. Rev.* **43**, 149.

Van der Leun, C., and de Wit, P. (1969). *Phys. Lett.* **30B**, 406.

Van der Leun, C., Sheppard, D. M., and Smulders, P. J. M. (1965). *Phys. Lett.* **18**, 134.

Walters, W. L., Costello, D. G., Skofronick, J. G., Palmer, D. W., Kane, W. E., and Herb, R. G. (1962). *Phys. Rev.* **125**, 2012.

Warburton, E. K., and Weneser, J. (1969). *In* "Isospin in Nuclear Physics" (D. H. Wilkinson, ed.), pp. 173–228. North-Holland Publ., Amsterdam.

Wilkinson, D. H. (1960). *In* "Nuclear Spectroscopy" (F. Ajzenberg-Selove, ed.), Part B, pp. 852–889. Academic Press, New York.

Woodbury, H. H., Tollestrup, A. V., and Day, R. B. (1954). *Phys. Rev.* **93**, 1311.

VII.E GAMMA RAYS FROM HEAVY-ION REACTIONS

J. O. Newton

DEPARTMENT OF NUCLEAR PHYSICS, INSTITUTE OF ADVANCED STUDIES
AUSTRALIAN NATIONAL UNIVERSITY, CANBERRA, AUSTRALIA

I. Introduction

Heavy-ion reactions provide a unique way of producing and studying nuclei in states of very high angular momentum. Until beams of heavy ions (HI) with sufficient energy and quality were available around 1958, there was little information on nuclear states with spin $J > 4$. Coulomb excitation

185

with heavy ions has enabled excitation of states with J up to 12 in ^{238}U, but this process is limited to stable or very long-lived nuclei. Although Coulomb excitation furnishes very valuable and specific information about nuclear electromagnetic matrix elements (see McGowan and Stelson, Chapter VII.A), one can normally only excite states which are connected to the ground state through a series of enhanced transitions, usually E2. So far, only the (HI, $xn\gamma$) reaction, first investigated by Morinaga and Gugelot (1963), has enabled study of states with J up to about 18.

The mode of population of the investigated levels is usually different in heavy-ion-induced reactions to those mostly used in traditional γ-ray spectroscopy. In the latter, the levels are usually populated directly, as for example, in the decay of a radioactive nucleus or in proton capture. Often, as in (d, pγ) reactions, a number of levels are populated and a number of different reactions such as (d, nγ) and (d, $\alpha\gamma$) occur. However, one can study the decay of a given level independently of the others by coincidence measurements between the γ rays and outgoing particles. In heavy-ion-induced reactions, apart from a few cases with light target nuclei, the major part of the cross section results in the formation of a compound system which is so highly excited that several particles, usually neutrons, are evaporated from it. Nevertheless, correct choice of bombarding energy can result in successive evaporations leading almost uniquely to a single product nucleus. This is formed in a great variety of excitation energies and excited states, from which γ rays cascade down via many different pathways. The multitude of γ rays originating from the highly excited states cannot be resolved with present detectors and therefore form a continuum in the observed spectrum. However, the pathways are eventually forced to pass through the limited number of well-separated levels at low excitation energies. Consequently, some of these levels are strongly populated and the γ rays from their decay produce peaks in the spectrum which stand above the continuum.

A spectrum showing the γ-ray peaks arising from the decay of the rotational states of a deformed even-even nucleus populated in the ^{160}Gd(^{4}He, 4n)^{160}Dy reaction is given in Fig. 1. States, usually of high angular momentum, with excitation energies up to a few MeV can be studied by this method. Since they are populated through many routes, their relative degrees of population depend very little on the details of their structure. This is in marked contrast to stripping reactions, inelastic scattering, and β decay, all of which preferentially populate states with particular structure. The lack of dependence on structure has the advantage that the decay of complex

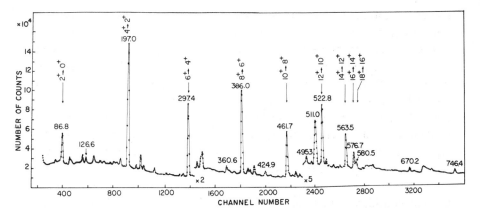

Fig. 1. Spectrum of γ rays from the ^{160}Gd(^{4}He, 4n)^{160}Dy reaction taken in a Ge(Li) detector; $E = 43$ MeV, $\theta = 125°$. The rotational transitions are indicated (redrawn from Johnson *et al.*, 1972, with permission of North-Holland Publishing Co., Amsterdam).

states, which might be difficult to form in other reactions, can be studied. These states are frequently isomeric and can be studied, without the disadvantage of the γ-ray continuum from prompt decay, if a pulsed beam is used and measurements taken outside of the beam pulses. The main virtue of heavy-ion-induced reactions of this type, of which the (HI, $xn\gamma$) reaction is the most important, is that they enable one to study states with very high angular momentum in a wide range of slightly to highly neutron-deficient nuclei. Of course, one can also study the decay of the radioactive decay products but that is outside of the scope of this section.

A few heavy-ion-induced reactions on light nuclei involve sufficiently low excitation energies that emission of a single particle, rather than many, is probable. In these cases, the conventional methods of particle–γ coincidence can be applied. Such reactions have produced valuable information on the high-spin states of a number of sd shell nuclei.

A number of direct reactions such as particle transfer and nuclear inelastic scattering (see Siemssen, Chapter IV.C.I and von Oertzen, Chapter IV.C.2) also occur with heavy ions. No experimental studies of the γ rays from these processes have so far been reported. Particle–γ coincidence methods would be necessary because of the much greater yield from compound nucleus reactions. Particle-identification techniques might also be required, since a variety of particles with nearby mass and charge numbers could be present. Such experiments are likely to be complex and difficult. However, they may prove valuable in the future because of the excellent energy resolution attainable in γ-ray measurements as compared to the very poor resolution in particle measurements.

The mechanism of heavy-ion reactions involving a compound system followed by evaporation is different from that of conventional γ-ray-producing reactions. Thus, considerable emphasis is placed on providing a thorough, though qualitative, description of this mechanism to give an understanding of the merits and limitations of these studies. Experimental methods are only given insofar as they are peculiar to measurements of this type; otherwise standard techniques are assumed. Measurements of lifetimes and magnetic moments, for which heavy-ion reactions are particularly well suited because of the large recoil velocities and strong nuclear alignments, are not considered since they are dealt with by Fossan and Warburton, Chapter VII.H and Recknagel, Chapter VII.C. The results from a few experiments are discussed in order to illustrate the type of nuclear structure information which can be obtained. They are chosen, somewhat arbitrarily, on the basis of their interest but no attempt is made to give a complete coverage of the now very numerous experiments in this field.

II. Qualitative Outline of Theory

Heavy ions usually have wavelengths short compared to nuclear dimensions, and to a fairly good approximation their motion can be considered classical, e.g., 64-MeV ^{16}O has $\lambda = 0.15$ F. In Fig. 2, we illustrate the various reactions which can occur with heavy ions. The three classical orbits correspond to cases where the two nuclei never come within the range of nuclear forces, where only the tails of the nuclear wave functions overlap, and where a major collision occurs. In the first two, complete fusion of the two nuclei to form a compound system is highly improbable, although in the second case, one or more nucleons may be transferred or nuclear inelastic scattering may take place. In the third case, complete fusion is most likely, though some grazing orbits can induce incomplete fusion reactions. These are not well understood but are likely to involve the transfer of groups of nucleons between target and projectile. Of course, when all possible orbits are considered, all of the indicated reactions occur, provided that the bombarding energy exceeds that of the Coulomb barrier E_B. Above this energy, the nuclear reaction cross sections are roughly of the order of 1 b for compound nucleus formation and about 5 mb for single-neutron transfer. Other transfer processes usually have still lower cross sections. Incomplete fusion reactions have not been much studied, but it

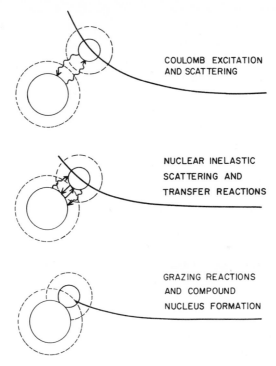

COULOMB EXCITATION
AND SCATTERING

NUCLEAR INELASTIC
SCATTERING AND
TRANSFER REACTIONS

GRAZING REACTIONS
AND COMPOUND
NUCLEUS FORMATION

Fig. 2. Schematic illustration of heavy-ion reactions. The large and small circles represent target and projectile, respectively. Dashed circles represent the "limits" of the tails of the nuclear wave functions. The straight arrows indicate particle transfers, and the others indicate photon transfers (redrawn with permission from Newton, "Progress in Nuclear Physics," 1969, Pergamon Press Ltd.).

seems that their cross sections increase with the energy and mass of the projectile. For projectiles with mass number less than 20 and bombarding energies not too much above E_B, the incomplete fusion cross section is usually considerably less than that for complete fusion. Coulomb excitation and nuclear inelastic scattering cross sections can sometimes be of the order of 1 b, but only for a few low-lying collective states.

Apart from these, the dominant cross section is that for compound nucleus formation. In heavy-ion reactions, the highly excited compound nucleus can usually evaporate several nucleons but, with suitable choice of bombarding energy, it is often possible to populate one final nucleus almost uniquely after the particle emission. Hence the γ rays following the compound nucleus reaction originate mainly from this nucleus. Incomplete fusion is likely to involve a great variety of reactions, none of which is dominant. Even if the cross section for this process were comparable with that for compound nucleus formation, it would seem most unlikely that the cross section for producing a particular γ ray would be large. Thus the radiation spectrum is dominated by the γ rays from the decay of excited states of the final nucleus originating from compound nucleus decay. (A few γ lines resulting from Coulomb excitation and nuclear inelastic scattering may be present also.) It is this special, and perhaps surprising, feature of the γ radiation following heavy-ion reactions which has enabled much new and exciting nuclear structure information to be obtained through rather simple experiments.

A. The Decay of the Compound Nucleus

The minimum excitation energy at which a compound nucleus can be formed in a heavy-ion reaction is the sum of the minimum energy which can be brought into the center-of-mass system by the incoming projectile, which is of the order of E_B, and the Q value for compound nucleus formation. Estimates of minimum excitation energies for a number of cases are shown in Fig. 3; they are usually in the range 30–50 MeV, except for the lightest projectiles. The Q values were deduced from the mass tables of Myers and Swiatecki (1965) and E_B from the simple estimate of $Z_1 Z_2/(A_1^{1/3} + A_2^{1/3})$ MeV, corresponding to $r_0 = 1.44$ F. Here Z and A are the charge and mass numbers, 1 and 2 referring to the projectile and target nuclei, respectively.

The decay of such a highly excited nucleus is treated by statistical methods, since the number of possible final states and emitted particles is very large. For details of such calculations, see Thomas (1968), Ericson (1960), and Blatt and Weisskopf (1952). Roughly speaking, the probability for the decay of a compound nucleus with excitation energy E into another nucleus at excitation energy E' plus an outgoing particle with energy $(E - E' - B)$ is proportional to the product $T(E - E' - B)\varrho(E')\,dE'$. Here B is the binding energy of the particle in the compound nucleus, $T(E - E' - B)$ is a transmission coefficient for the outgoing particle, and $\varrho(E')$ is the density of states in the final nucleus. We have ignored the spins of the final states and the orbital angular momenta of the outgoing particles. The shape of the spectrum of emitted particles can be obtained from this expression. It is the familiar evaporation spectrum, an example of which is shown for neutrons in the inset of Fig. 5. The spectrum falls off at high particle energies because the level density declines with decreasing excitation energy and at low energies because the transmission coefficient tends to zero at zero

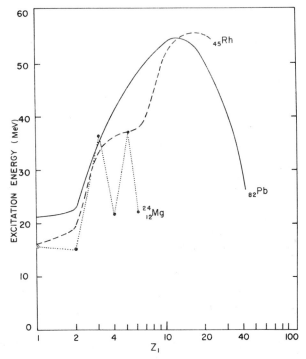

Fig. 3. Minimum excitation energies of the compound systems initiated by projectiles of charge number Z_1. The continuous and dashed lines refer to the production of Rh and Pb nuclei from the appropriate target and projectile of greatest isotopic abundance. The dotted line represents the production of a particular nucleus $^{24}_{12}$Mg. In this case, not all of the indicated reactions are possible in practice due to instability of target or projectile.

particle energy. The spectra for charged particles are roughly of similar shape but displaced up in energy by an amount of the order of E_B since below this energy, the transmission coefficient becomes small.

The highly excited compound nucleus can decay in a variety of ways, but which of these is most likely? In Fig. 4, we show two possible modes of decay for an excited rare earth nucleus. The relative probabilities for decay can be roughly assessed by taking the ratio of the level densities in the two final nuclei at excitation energies corresponding to average energies for the outgoing particles. (We have ignored the difference in the two transmission coefficients.) If we assume identical shapes for the two spectra, then from Fig. 4 we see that the excitation energy in the nucleus to which α decay occurs is 7 MeV lower than that fed by neutron decay. Although the energy

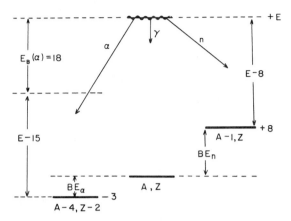

Fig. 4. Scheme for decay of a highly excited rare earth nucleus, showing the energies available for neutron and alpha emission (redrawn with permission from Newton, "Progress in Nuclear Physics," 1969, Pergamon Press Ltd.).

available for α emission is 11 MeV greater than that for neutron emission, a Coulomb barrier of 18 MeV has to be surmounted. Neglecting angular momentum effects, the relative level densities can be roughly estimated from the expression $\varrho(E) \propto E^{-2} \exp(2aE)^{1/2}$, where a is given empirically as $A/8\,\mathrm{MeV}^{-1}$. Thus there is a large difference in the level densities for the two excitation energies and α decay is quite negligible compared to neutron decay. Similarly, proton emission is relatively improbable since, if the compound nucleus is not too far from the β-stability line, the neutron and proton binding energies are similar but the protons have to overcome a Coulomb barrier of about 10 MeV. Hence a moderately heavy nucleus would be expected to decay almost entirely by neutron emission. Several neutrons can be evaporated before reaching particle-stable states in the final nucleus, which then decay by γ-ray cascades to the ground state. This is known as a (HI, xnγ) reaction, x referring to the number of neutrons emitted. At present, it is the most important and generally useful reaction for γ-ray spectroscopy with heavy ions.

B. THE HEAVY-ION–xn REACTION

One of the major advantages of the (HI, xnγ) reaction is that one final nucleus can be made almost uniquely if the bombarding energy is chosen appropriately. How this comes about is illustrated in Fig. 5. The compound nuclei (A, Z) at 25 MeV excitation energy evaporate neutrons, with the

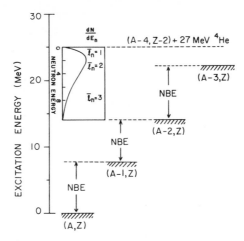

Fig. 5. Scheme of (^{4}He, xn) reaction (redrawn from Newton *et al.* 1967, with permission of North-Holland Publishing Co., Amsterdam).

energy spectrum shown, forming mostly neutron-unstable states in $(A - 1, Z)$ nuclei. These likewise emit further neutrons which lead mainly to bound states in $(A - 2, Z)$, which then decay by γ emission. Hence at this bombarding energy, most of the decay leads to the nucleus $(A - 2, Z)$ and very little to the nuclei on either side. This occurs because most of the evaporated neutrons have energies considerably less than the neutron binding energies. If the bombarding energy were raised by a suitable amount, it would be possible to make mainly the nucleus $(A - 3, Z)$ and so on. Because of the finite width of the neutron spectrum, relatively more of the nuclei on either side of the desired final nucleus would be formed as x increased. According to this simple picture, the excitation function for a (HI, xnγ) reaction consists of a series of peaks, one for each product nucleus, separated by about the sum of the appropriate neutron binding energy and average neutron kinetic energy. This behavior is illustrated in the calculated excitation functions in Fig. 6.

The (HI, xn) reactions produce neutron-deficient nuclei, i.e., nuclei which have fewer neutrons than do the β-stable nuclei with the same mass number. The reason is that the ratio N/Z, which is approximately unity for very light nuclei, increases steadily to about 1.6 for the heaviest β-stable nuclei observed in nature. Hence the compound system formed in heavy-ion bombardment is neutron deficient and after the evaporation of several neutrons, the final nucleus is even more so. Very neutron-deficient nuclei are not easily and cleanly produced by other methods. We can expect to do useful measurements in γ-ray spectroscopy only if a major part of the reaction cross section ($\gtrsim 10\%$) goes into production of the nucleus of interest. How neutron-deficient a nucleus can we make while still satisfying this criterion? The decay probability for, say, neutrons and protons is equal when $B_n - B_p = E_{Bp}^*$, where E_{Bp}^* is an effective Coulomb barrier energy for the outgoing proton. The value of E_{Bp}^* is less than that of E_{Bp} because the barrier is not

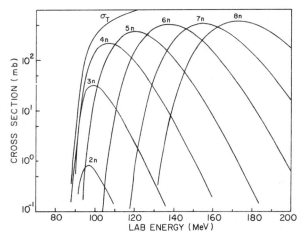

Fig. 6. Excitation functions for the ^{181}Ta(^{16}O, xn)$^{197-x}$Tl reactions calculated with the program of Sikkeland (1967) (redrawn with permission from Newton, "Progress in Nuclear Physics," 1969, Pergamon Press Ltd.).

sharp for light particles and appreciable penetration occurs well below it. As nuclei become more neutron deficient, neutron binding energies increase and both proton- and alpha-binding energies decrease.

An example for the cerium isotopes is shown in Fig. 7. Once one reaches the point where the neutron and proton decay probabilities are equal, the (HI, xnγ) cross section drops off very rapidly with further neutron emission. The relative probability for proton emission continues to increase but, even if it did not, the cross section would drop by a factor of two for each extra neutron emitted. For example, if there were four such steps with equal probabilities, the cross section would decrease to one-sixteenth of the total. From Fig. 7 and the level density formula previously given, one might expect that the relative decay probabilities would alternately be larger and smaller after each neutron emission. In fact this is not so because the alternation in $(B_n - B_p)$ arises from pairing effects, which should also be taken into account in the level density formula [Eq. (1)]. The criterion for deciding when the probabilities are equal is that both final nuclei are of the same type, i.e., odd-mass nuclei (the upper set of points in Fig. 7).

The result of a simple calculation for the cerium isotopes on the variation with A of the relative probabilities for proton, neutron, and α-particle emission is shown in Fig. 8. It is entirely in accordance with the qualitative

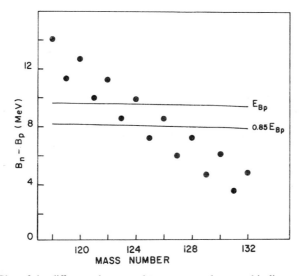

Fig. 7. Plot of the difference between the neutron and proton binding energies against mass number for the cerium isotopes.

argument given. The probability for α emission is expected to be less than that for proton emission, except perhaps for states of high angular momenta. Thus equality of neutron and proton emission probabilities gives a good estimate for the most neutron-deficient nuclei which we can study by (HI, $xn\gamma$) reactions. For heavier nuclei with $Z \gtrsim 78$, fission, rather than proton decay, determines this limit. The limiting line shown in Fig. 9 is always slightly on the neutron-rich side of the line $Z = A/2$, which can be simply understood in terms of the semiempirical mass formula (Newton, 1969). Hence the (HI, $xn\gamma$) reaction will not be very useful for studying neutron-deficient light nuclei since stable light nuclei have $Z \simeq A/2$. However, between the limiting line and the region of stable nuclei, there are about 1000 nuclei which could be studied. They can all be reached with reactions

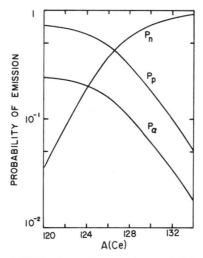

Fig. 8. Relative probabilities for neutron, proton, and alpha emission from Ce nuclei calculated from $P_n/P_p = \exp - [B_n - B_p^*]/T$; $P_p/P_\alpha = 3$. The nuclear temperature is denoted by T and $B_p^* = B_p + E_{Bp}^*$ (redrawn from Stephens *et al.*, 1971, with permission of North-Holland Publishing Co., Amsterdam).

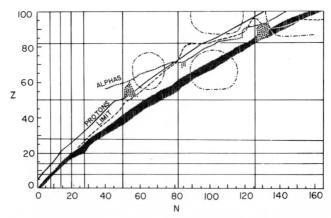

Fig. 9. Regions of nuclei which can be reached with (HI, $xn\gamma$) reactions. The shaded area indicates the β-stable nuclei, the vertical and horizontal lines the closed shells, and the dot-dashed contours the approximate boundaries of rotational regions. The region which can be studied lies between the dashed "limit" line and the right-hand edge of the shaded area. Lifetimes for proton and alpha emission are estimated to be less than 1 sec to the left of the lines labeled protons and alphas (redrawn with permission from Newton, "Progress in Nuclear Physics," 1969, Pergamon Press Ltd.).

involving not more than four emitted neutrons and projectiles with $A \leqslant 40$ (Newton, 1969; Stephens *et al.*, 1971).

C. The Effect of High Angular Momentum

One of the principal features of heavy-ion-induced reactions is the very high angular momentum brought into the compound system. This aspect is of crucial importance to consider. The first effect, still not well understood, concerns the fraction of the total reaction cross section which goes into compound nucleus formation. It is now well known that there are no levels of a given angular momentum in a nucleus below some minimum energy, called the yrast energy. If a heavy ion brings in angular momentum higher than the yrast angular momentum for the appropriate excitation energy of the compound system, then a compound nucleus cannot be formed. However, it is not easy to calculate the yrast energies at very high angular momenta since the nuclei get deformed. Further, there may be some critical angular momentum for the nucleus, which, if exceeded, will result in fission. Such a critical angular momentum is quite familiar in classical physics; any rotating body will disintegrate if spun too fast. It may be independent of the excitation energy, so that even if the yrast condition for forming a compound nucleus is satisfied, the nucleus may still fly apart almost immediately. Theoretical estimates of some of these phenomena have been attempted by Beringer and Knox (1961), Cohen *et al.* (1963), and Kalinkin and Petrov (1964). An empirical approach by Natowitz (1970a, b) assumes that a compound nucleus can be formed when the angular momentum $J < J_{\mathrm{crit}}$ and cannot when $J > J_{\mathrm{crit}}$. His estimate for J_{crit} as a function of the mass of the compound system A_{CF}, shown in Fig. 10, was derived from measured ratios of the cross sections for complete fusion σ_{CF} to those for the total reaction σ_{R}. From this curve, he calculated σ_{CF} for a number of projectile–target combinations. The ratios $\sigma_{\mathrm{CF}}/\sigma_{\mathrm{R}}$ plotted against A_{CF} for various projectiles of 5 and 7.5 MeV/amu are also shown in Fig. 10. One sees that the value of $\sigma_{\mathrm{CF}}/\sigma_{\mathrm{R}}$ can become small for the heavier projectiles, though not so small that (HI, $xn\gamma$) reactions should be impossible to study.

Another aspect of the high angular momentum concerns the way the compound system decays once it has been formed. A detailed study of this is very complicated but the general features can be simply understood. For recent work, see Richter, Chapter IV.D.1, Thomas (1968), Grover and Gilat (1967a, b, c), Grover (1967), Gilat (1970a), and Gilat and Grover (1971). The nuclear level density for angular momentum J and energy

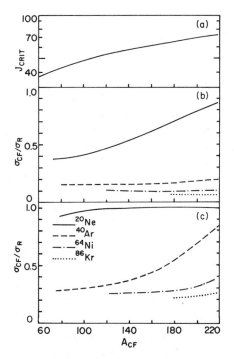

Fig. 10. Empirical trend of J_{crit} (a) and calculated values of σ_{CF}/σ_R for (b) 5 and (c) 7.5 MeV/amu projectiles as functions of the mass of the compound system (estimated from Natowitz, 1970a, b).

$E \gg E_J$ is roughly given by

$$\varrho(E, J) \propto (2J + 1)(E + \delta)^{-2} \exp 2[a(E - E_J + \delta)]^{1/2} \qquad (1)$$

The quantity δ has the value zero for even-even nuclei, δ_e for odd nuclei, and $2\delta_e$ for odd-odd nuclei. It is a correction for the nuclear pairing energy; δ_e has a value of about 1 MeV for rare earth nuclei. The parameter $a \simeq A/8$ MeV^{-1}, and E_J can be interpreted as the energy taken up by the rotation of the nucleus as a whole and therefore not available for excitation. It is given by $E_J = \hbar^2(2\Im)^{-1}J(J + 1)$, where $\Im$ is an effective moment of inertia, which is of the order of that for rigid-body rotation of the nucleus, $\Im_{rig}$. The expression for $\varrho(E, J)$ becomes invalid as E approaches E_J and must become zero at the yrast energy. According to a number of models, the yrast energy is approximately equal to E_J (Thomas, 1968; Gilat, 1970b). In practice, it is given only very roughly by a rotational energy with a constant moment of inertia; $\mathfrak{I}$ is a function, not necessarily smooth, of J and is shell dependent (Gilat and Grover, 1971).

The average change of angular momentum in particle emission to states of energy E in the residual nucleus depends on the initial spin. If this is low, we would expect the spin to increase on the average because the level density increases with J $(E \gg E_J)$. If the initial spin is large, such that E is still larger than E_J but the level density is now decreasing with increasing J, the average spin should decrease. In neither case would one expect a large change because large orbital angular momentum changes are inhibited by the centrifugal barrier E_l. For orbital angular momentum lh, E_l is given by $E_l = \hbar^2 l(l+1)/2MR^2$, where M is the mass of the emitted particle and R the interaction radius. This is already as high as about 10 MeV for neutrons or protons with $l = 5$. However, because the centrifugal barrier decreases with increasing M, heavier particles such as alphas can carry away higher angular momenta than protons or neutrons. These qualitative arguments are in accordance with calculations.

A different situation occurs when the particles are emitted from states near the yrast line. The two cases for emission of neutrons are compared in Fig. 11. Here the excitation energies and yrast lines for both decaying and final nucleus are shown on the same diagram. Although the yrast levels will not be identical for the two nuclei, they should not differ greatly. The vertical arrows in the figure represent the neutron binding energies and the components of the sloping arrows are the kinetic energies and angular

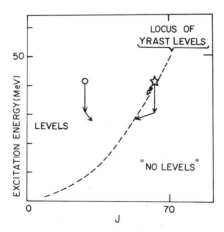

Fig. 11. Schematic diagram featuring the decay of highly excited nuclei with large angular momentum J (redrawn from Grover and Gilat, 1967b).

momenta carried away by the neutrons. The circle and star refer to two different initial states and the tips of the arrows to the corresponding final states. For the state represented by the circle, there is no restriction on the states which may be populated. However, from the state represented by the star, the neutron is forced to carry away a large amount of angular momentum, which will seriously reduce its emission probability. This reduction can be so severe that γ emission, which normally can be neglected when in competition with particle decay, can be favored over neutron decay. Sometimes α decay can be more probable, since α particles can carry away more angular momentum than neutrons.

The region important for γ decay should extend to about one neutron binding energy above the yrast levels, as shown in Fig. 12 (Grover and Gilat, 1967b). There are two distinct regions of γ decay. Well above the yrast line, the γ rays will cascade down, probably by electric dipole transitions, carrying away rather little angular momentum on the average. This region will give rise to the familiar evaporation spectrum. The second type of γ decay occurs when the region of the yrast line is reached. Now the γ rays are forced to cascade down from one yrast level to the next, each reducing the angular momentum of the nucleus. Thus the γ decay down the yrast region must carry away the major fraction of the angular momentum of the original compound system, since neither the preceding γ rays nor the evaporated

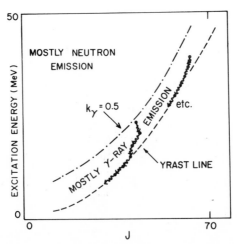

Fig. 12. Schematic illustration of the region of excitation energy and angular momentum in which most deexcitation is by γ radiation. The fraction of nuclei decaying by γ emission is indicated by k_γ (redrawn from Grover and Gilat, 1967b).

neutrons did. The spectrum of these γ rays is quite different from that of an evaporation spectrum and the average energy much lower. An yrast level, or more likely a few nearby levels with the same angular momentum, collects nearly all of the population coming from γ-decaying states with angular momenta higher than its own. (The populated levels near the yrast line are called the yrast region.) Its decay need not necessarily be by dipole radiation since the yrast levels may not always rise in energy with J. One such example is the ground-state collective band (gsb) of a deformed even-even nucleus, which forms the yrast levels for lower values of J. Here successive levels increase by two units of angular momentum and decay has to be by electric quadrupole transitions.

These phenomena have important effects on the (HI, xn) reactions. The threshold energy for the (HI, xn) reaction leading to a particular final nucleus depends on the angular momentum of the state fed in that nucleus. This results in broadening of the peaks in the excitation functions relating to given final nuclei. Each observed peak can be regarded as being made up of a set of displaced peaks, one for each angular momentum. The peaks for alpha-induced reactions, which bring in relatively little angular momentum, may typically have widths at half-maximum of 10–15 MeV, while those from heavier ions may have widths of 20–30 MeV (see Fig. 6). Further, high angular momentum affects the population of the levels from which the observed discrete γ rays arise. The case of a final even-even nucleus with $A \simeq 160$ is illustrated in Fig. 13. The thin line represents the non-gsb yrast levels and the dots and dashes indicate gsb energies for a vibrational and rotational nucleus, respectively. Below the angular momentum where the gsb and non-gsb yrast lines intersect, the members of the gsb form the yrast levels. The average excitation energies and angular momentum ranges following production of this nucleus by (^{4}He, 4n) and (^{40}Ar, 4n) reactions are indicated. For the former, it is assumed that all of the cross section goes into the 4n reaction and that the neutrons carry off a negligible amount of angular momentum. In the (^{40}Ar, 4n) case, the higher angular momentum components are likely to result in surface, (^{40}Ar, 3n), and (^{40}Ar, αxn) reactions and the lower components in (^{40}Ar, 5n) reactions. In both cases, the average excitation energies have been taken to be about one neutron-binding energy above the yrast levels.

In the case of the (^{40}Ar, 4n) reaction, most of the population eventually reaches the non-gsb yrast line region and proceeds down until the point where the gsb intersects it. The population then enters the gsb. Practically all of the gsb population enters at the top when high angular momentum is

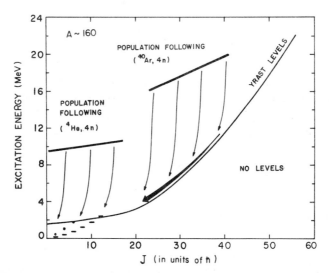

Fig. 13. Schematic illustration of energy levels in a nucleus of mass ≈ 160 versus angular momentum. Indicated are the non-gsb yrast levels, the regions of states populated in (⁴He, 4n) and (⁴⁰Ar, 4n) reactions, and the gsb levels of a typical vibrator (dots) and rotor (dashes) (Newton *et al.*, 1970a, with permission of North-Holland Publishing Co., Amsterdam).

brought into the nucleus and hence all of the observed γ rays below the feeding point have the same intensity. Further, members of the gsb above the intersection point, whether or not they exist in recognizable form there, will not receive significant population in the reaction. This is in good accordance with observation, all but the last one or two observed members having the same intensity (Ward *et al.*, 1967).

Measurements on six (⁴⁰Ar, 4n) reactions leading to final rotational and quasirotational even-even nuclei have shown that the mean time interval between the formation of the compound nucleus and the population of the higher members of the gsb is about 10 psec; only a few percent of the feeding could be slower than this (Diamond *et al.*, 1969; Newton *et al.*, 1973). With scanty experimental data, it is not certain that this result is a general one, but any satisfactory model must be able to explain it. The dipole decay to the yrast region should be fast so we need to show that the decay down the yrast region and the transfer from it to the gsb can occur in a time of less than 10 psec. Since the yrast cascade carries off a large amount of angular momentum, the average energy of the γ rays in it can only be several hundred keV (see Fig. 13). These γ rays can only be E1, M1, or E2 if the feeding time is not to exceed 10 psec. The limited evidence so far (Mollenauer, 1962) suggests that many are of E2 type. Newton *et al.* (1970a) pointed out that the variation of level energy with angular momentum in the yrast region must be very regular, otherwise the rapid variation of γ-transition probability with energy (E_γ^5 for E2 transitions) would produce traps and hence longer feeding times. In addition, the E2 transitions would have to be about ten times faster than the single-particle estimates.

Their simple explanation for rotational nuclei, which may also be appropriate for vibrational ones at high angular momenta, is that the non-gsb yrast region, which is two or three hundred keV wide, consists of rotational states based on two- and, at higher angular momenta, four-quasiparticle states (see Fig. 14). These are expected to have larger moments of inertia than the gsb. This would explain the enhanced E2 transitions, the regular spacing, and the nonobservance of discrete γ lines in the yrast cascade. However, many of these bands, and particularly the lowest ones, have a large value of K, the quantum number of the component of angular momentum along the nuclear symmetry axis. The gsb has $K = 0$ and any transitions to it from other bands will be seriously hindered unless they obey the selection rule $\Delta K \leqslant \lambda$, where λ is the multipolarity of the transition (Nathan

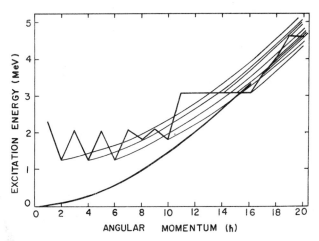

Fig. 14. Energy levels in a rare earth rotational nucleus plotted against angular momentum J. The heavy jagged line is an estimate of the lowest two-quasiparticle state for given values of J and the light lines correspond to rotational bands built on these states. The heavy smooth line represents the gsb levels and, to avoid confusion, is not drawn beyond $J = 16$ (redrawn from Newton *et al.*, 1970a, with permission of North-Holland Publishing Co., Amsterdam).

and Nilsson, 1965). Therefore rapid decay into the gsb can only occur if the bands with different K are mixed. At high angular momenta, such mixing can occur through the Coriolis operator, which mixes states differing in K by ± 1 (for discussion, see Rasmussen, Chapter IX.B). Stephens and Simon (1972) have considered this (both from the point of view of physical insight and detailed calculation) for the case of rare earth rotational nuclei. They find that the levels of the yrast region mostly arise from levels originating from the $i_{13/2}$ shell model orbital, for which the Coriolis operator has a particularly large value, and that sufficient mixing occurs at the transfer point. However, further work, both experimental and theoretical, needs to be done on this subject. No experimental data are available on the feeding times for final vibrational nuclei which intersect the non-gsb at around $J = 6$.

In reactions such as (^{4}He, 4n), much less angular momentum is brought into the compound system. Thus very little of the population reaches the non-gsb yrast region above the intersection point (Figs. 13, 14). In contrast to the (^{40}Ar, 4n) reaction, where the gsb was fed only at the intersection point with the non-gsb yrast region, all members of the gsb below this point will now be fed. Hence the intensities of the γ rays from successively higher members of the band decrease. Experiments show that when the average angular momentum J_i brought into the compound system is small (e.g., $J_{max} \leqslant 4$), it is almost the same as that put into the gsb J_f. However, when J_i is larger, $J_i > J_f$ (Williamson et al., 1968; Mills and Rautenbach, 1968; De Jesus et al., 1971). This can be understood on the basis of K-selection rules. The non-gsb yrast region below the intersection point is likely to be strongly populated because there are many more states of a given angular momentum in it than in the gsb. This can outweigh the energy advantage of direct transitions from higher states to the gsb. When a state in the non-gsb yrast region is populated, it can decay either to a lower rotational state of its band or to the gsb, which forms the true yrast levels below the intersection point. If the K-selection rules allow an E1, M1, or E2 transition to the gsb, then this is likely to be preferred because of greater decay energy. However, if this is not the case, and if K is a good quantum number for the decaying state, then a rotational transition may be more probable.

At the relatively low angular momenta involved in these reactions, K mixing due to the Coriolis force is likely to be small and hence the higher-K states decay mainly by stretched rotational transitions to the band heads (Ferguson et al., 1972). These either decay by K-forbidden transitions to the gsb or, in special cases, a K-allowed transition to a low-lying isomeric state of high K may be favored. Sometimes such states can collect nearly half of the total population which enters the gsb. Hence when little angular momentum is brought into the compound system, states of the gsb and states of low J and K in the non-gsb yrast region are populated from the decay of the higher states. Rotational transitions in the non-gsb yrast region are rare and most of the initial angular momentum appears in the gsb. When the incoming angular momentum is higher, rotational transitions of stretched E2 or M1 type occur in the non-gsb yrast region and carry away angular momentum; hence that reaching the gsb is less than that brought into the compound system. Some K-forbidden transitions to the gsb are predicted and these should result in long feeding times for part of the gsb population. Ferguson *et al.* (1972) have reported a number of cases where around 5% of the transitions in the gsb are delayed by more than 30 nsec, in agreement with this prediction.

We see that the model briefly described here explains very well the present experimental data on (xn) reactions induced by light or heavy ions leading to final rotational even-even nuclei. Further data are desirable, particularly for final nonrotational nuclei. A similar model can probably be used for final odd-mass nuclei. In these, we might expect that the non-gsb yrast region would consist of rotational states based on three- and five-quasiparticle states. In even-even nuclei, the pairing energy gap contains collective states based on the ground state. However, in odd-mass nuclei, there are a number of one-quasiparticle states in the gap, each with its collective band. One expects, and usually observes, population of several of these bands, that band forming the yrast levels receiving most. One or more of these bands is likely to have a K value much greater than zero. Thus in cases such as ^{4}He-induced reactions, decay from the non-gsb yrast region to the lower bands can often occur through K-allowed transitions even for higher incoming angular momenta. Rotational transitions in this region are therefore likely to be much more rare than for the even-even nuclei and most of the incoming angular momentum may appear in the final states. This is in agreement with the limited evidence (Ryde, 1971). The case of odd-odd nuclei is much less clear cut and few experiments have been reported.

Although generally the non-gsb yrast line may be rather smooth for high J, this is less likely to be true when J is small. Sometimes a two- or three-quasiparticle state may have an unusually low excitation energy, and be nearly or actually an yrast level. In this case, it may collect a lot of the population which would otherwise go directly to the gsb. Such a state has a high value of K and consequently a long lifetime, i.e., it is an isomeric state. The (HI, $xn\gamma$) reaction is admirably suited for the production and study of complex isomeric levels because of its lack of sensitivity to the detailed structure of the populated states. Such studies form an important part of the work in this field.

D. EXCITATION FUNCTIONS

In order to make a particular final nucleus in a (HI, $xn\gamma$) reaction, one has to estimate the optimum bombarding energy $E_{pk}(x)$. An empirical expression which gives quite good results (particularly for $A > 100$) is

$$E_{pk}(x) = (1 + A_1/A_2)(-Q_x + \alpha x) \tag{2}$$

where Q_x is the Q value for the given (HI, $xn\gamma$) reaction and α is about 6 MeV (Alexander and Simonoff, 1964; Neubert, 1971). Since successive values of Q_x differ by the neutron binding energies, which typically have values of about 8 MeV, successive peaks in the excitation functions differ in energy by about 15 MeV. Of course, the reaction only has a usable cross section if $E_{pk}(x) \gtrsim E_B$ (Fig. 6). For projectiles heavier than ^{4}He, the minimum number of emitted neutrons varies from two to four depending on the particular case.

Absolute cross sections for (HI, xn) reactions can be computed with a program due to Sikkeland (1967). This program takes into account grazing reactions, fission competition, and angular momentum effects, but it contains many simplifications and neglects γ decay. Nevertheless, using empirical values for the various parameters, it often gives very good results in the region of rare earth and heavy nuclei. Since competition from charged-particle emission is not included, one would not expect good results for the lighter nuclei or for very neutron-deficient nuclei.

The excitation function of an individual γ ray can be useful to study because it can give an indication of the spin of the state from which the γ ray originates (Newton, 1968). For example, if the yields of the γ rays from various states of the gsb divided by that from the lowest excited state of the band are plotted against bombarding energy, the slopes of the curves increase with the angular momentum of the decaying state. This is because the threshold increases with angular momentum and because the angular momentum brought into the compound system increases with bombarding energy. One would expect less effect in reactions induced by heavier ions, where more of the population of the states comes via the non-gsb yrast region beyond the intersection point with the gsb than in, say (^{4}He, xn)

reactions. However, one has to be careful in the interpretation of this method (Mills and Rautenbach, 1968; De Jesus *et al.*, 1971).

E. ANGULAR DISTRIBUTIONS OF DEEXCITATION γ RAYS

It is found experimentally that the observed γ rays from (HI, $xn\gamma$) reactions often have very pronounced angular distributions, with anisotropies near the maximum possible for the particular transitions. This implies that the angular momentum vectors of the decaying states must be strongly aligned. The compound system itself is usually strongly aligned, since the incoming orbital angular momentum is large and perpendicular to the projectile's direction of motion, i.e., in the $m = 0$ substate. Each emitted neutron carries away little angular momentum, oriented in a roughly random direction, and the same applies to the dipole γ rays emitted before the non-gsb yrast region is reached. Thus at this stage, the angular momentum has its magnitude almost unchanged and its orientation only slightly disturbed (see Fig. 15). The γ transitions down the yrast region are stretched, i.e., classically their angular momentum vectors are parallel to those of the

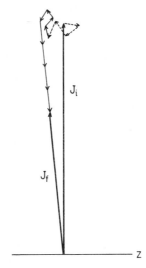

Fig. 15. Classical vector diagram showing how the total angular momentum vector of the compound system (initially vertical) is tilted in the emission of neutrons (dashed lines) and γ-rays (thin lines). The initial angular momentum is taken to be $20\hbar$ and each neutron is assumed to carry away $1.5\hbar$. The γ cascade preceding entry into the gsb is taken to consist of four E1 photons followed by four stretched E2 photons down the non-gsb yrast region.

decaying states, so they reduce the nuclear angular momentum but do not change its alignment (not quite true quantum mechanically). It is not easy to calculate precisely the alignment (substate distribution) for a particular state because this depends on the details of how it is fed. For example, a non-stretched transition occurring near the end of the feeding cascade can markedly reduce the alignment.

The strong alignment obtained in $(HI, xn\gamma)$ reactions gives a useful spectroscopic tool for estimating spins, moments, and transition mixing ratios, without the need for coincidence measurements (Diamond et al., 1966). The angular distribution of the γ rays arising from the decay of an initial state of spin J to a final state has the form

$$W(\theta) = 1 + \sum_{k \text{ even}} G_k A_k P_k(\cos \theta) \tag{3}$$

where the A_k are the theoretical coefficients (Yamazaki, 1967) for the initial state being in the substate $m = 0$ (J integral) or $m = \frac{1}{2}$ (J half-integral). The G_k are attenuation coefficients related to J and to the alignment.

Consider the case of the gsb of an even-even nucleus where the γ rays cascade down the band by stretched E2 transitions. It is a well-known result (Biedenharn, 1960) that if the band is fed only at one level [as would be roughly true in an $(^{40}Ar, 4n)$ reaction], the angular distributions of all the subsequent γ rays are identical. Since the A_k increase as J decreases, the G_k must decrease so that the product $G_k A_k$ remains constant. When the band is fed at many points, as for reactions induced by lighter ions, it is no longer true that all of the γ rays have the same angular distribution. Nevertheless it is roughly correct that the values of $G_k A_k$ are constant even for different reactions, as the empirical results of Fig. 16 show. Here the experimental values for $A_2' \equiv G_2 A_2$ lie within the range 0.30 ± 0.09 and for $A_4' \equiv G_4 A_4$ within -0.09 ± 0.05. Hence, even if no information is available about a given decay scheme, empirical alignments together with angular distribution data can be a useful guide in determining spins, mixing ratios, etc.

If two or more γ rays depopulate a state, it may be possible to obtain more precise information on the alignment. Empirical values for the G_k as functions of J for projectiles ranging from ^4He to ^{19}F are given in Fig. 17. In interpreting angular distributions, it is necessary to know whether levels have significant populations coming through long-lived isomeric states. If so, their alignment may be perturbed due to interactions with atomic or crystalline fields (Frauenfelder and Steffen, 1965). Further, it is essential to

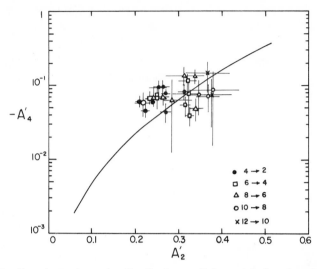

Fig. 16. Experimental angular distribution coefficients A_4' plotted against A_2' for a number of rotational and vibrational γ rays observed in (HI, $xn\gamma$) reactions. The coefficients for $2 \rightarrow 0$ transitions are excluded since they may be attenuated by extranuclear effects. The line represents the relationship expected for a Gaussian population distribution of magnetic substates (redrawn from Diamond et al., 1966).

ensure that the nuclei do not recoil into vacuum, since perturbing fields of 50 MG can be produced by unpaired electrons in the partially stripped and excited ions (Ben-Zvi et al., 1968; Nordhagen et al., 1970; also see discussion by Häusser, Chapter VII.B). For more details on the angular distributions, see Halpern et al. (1968), Newton (1969) and Draper and Lieder (1970).

F. Choice of Projectiles and Targets

A correct choice of projectile–target combination for producing a given final nucleus can have a major effect on the success of an experiment. The following factors, which may sometimes be conflicting, have to be considered:

(i) Availability of targets and projectiles.

(ii) Types of levels to be studied. If they are of low spin, e.g., β-vibrational states, then as little angular momentum as possible should be brought into the system. This usually requires projectiles with $A_1 \leqslant 4$, whereas for studying high-spin levels, one uses projectiles with $A_1 \geqslant 4$.

(iii) Cleanliness of spectra. The fraction of the total cross section leading to direct or incomplete-fusion reactions should be minimized. This increases with projectile velocity and, at high incoming angular momenta, with angular momentum (Section II.C). For this reason and to maximize the fraction of the cross section for neutron evaporation leading to the desired nucleus, it is usually best to form it in the first reaction with reasonable yield above the Coulomb barrier energy (e.g., the 4n reaction in Fig. 6). When making very neutron-deficient nuclei, where competition from charged-particle decay

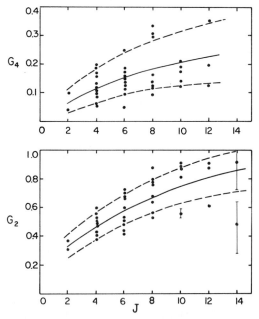

Fig. 17. Experimental values of attenuation coefficients G_2 and G_4 versus angular momentum of decaying state. The continuous lines indicate the average trends of the data. The broken lines indicate the standard errors which might safely be used when using these results to interpret other data (redrawn from Newton *et al.*, 1970b, with permission of North-Holland Publishing Co., Amsterdam).

is serious, best results should be obtained with a target–projectile combination which minimizes x (Section II.B). As the angular momentum brought into the system increases beyond about $16\hbar$, the number of unresolved γ rays from the non-gsb yrast region per observed discrete γ ray increases (see Fig. 13). Thus spectra with the best peak-to-background ratios are likely to be obtained with the lightest projectiles consistent with the other requirements.

(iv) Lifetime measurements (Section III.B and Fossan and Warburton, Chapter VII.H) can be readily performed since the recoiling ions have angular spreads of only a few degrees resulting from neutron emission. They are most easily and accurately done when the recoil velocities are high. For this, the heaviest possible projectiles are best. Heavy projectiles with $A \gtrsim 40$ are also advantageous because the levels are fed in a definite manner from the non-gsb yrast region above the intersection point. For example, when measuring the lifetime of one of the rotational states of the gsb, it is easy to allow for the lifetimes of the preceding rotational states through which the γ-ray cascade passes. However, with lighter projectiles, the state will be populated not only from the rotational state above but from many other states with unknown lifetimes. Heavier projectiles usually give poorer spectra, so one should try to achieve the best compromise taking into account all of the given factors.

G. Heavy-Ion Reactions with Light Target Nuclei

Except in a few special cases, where one uses neutron-rich targets such as $^{48}_{20}\text{Ca}$, the (HI, $xn\gamma$) reaction is unlikely to be dominant when making nuclei with $Z \lesssim 30$ (Section II.B). Many reactions such as (HI, $2pn\gamma$), (HI, $\alpha n\gamma$), etc., may occur with equal or greater probabilities than the (HI, $xn\gamma$) reaction. The problem of disentangling the results is therefore much more complicated. Multidimensional coincidence measurements are essential and some prior knowledge of the lower levels of the product nuclei is very useful. Sometimes observation of the profiles of Doppler-broadened γ-ray peaks can help to distinguish different reactions (Branford, 1972). However, when one goes to reactions between very light nuclei, experiments again become easier. Because of lower excitation energy in the compound system (Fig. 3), reactions in which only a single charged particle is emitted occur with good probability. A number of different reactions still occur but they can be sorted out by particle–γ coincidence measurements.

Many features of these reactions can be understood in essentially the same way as for the (HI, $xn\gamma$) case, though with some differences. Although the decay of the compound system may be statistical (there is some evidence that this is not always so), one is now observing direct particle decay to a given state. Hence no unobserved cascade of γ rays with consequent delay time precedes the observed radiation and lifetimes as short as a few fsec can be measured. A single state can be populated with reasonable probability because the level densities in light nuclei are not very high, particularly for states of high spin. The number of independently fluctuating channels is very limited, particularly if target, projectile, and emitted particle all have spin zero and if the particle is observed along the beam direction. This is in marked contrast to the (HI, xn) reaction and Ericson fluctuations can be very prominent (see Richter, Chapter IV.D.1; Ericson and Mayer-Kuckuk, 1966). An example of a measured excitation function is shown in Fig. 18.

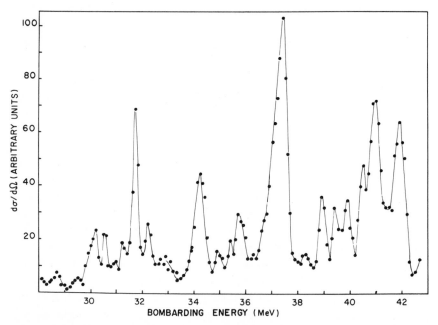

Fig. 18. Excitation function of α-particles leading to the 8$^+$, 13.2-MeV state in ^{24}Mg, taken at 0° to the beam direction [^{12}C(^{16}O, $\alpha_{13.2}$) ^{24}Mg*] (Branford *et al.*, 1972a).

Correct choice of bombarding energy is clearly critical. The usual rules for angular distributions and correlations apply (Ferguson, Chapter VII.G; Devons and Goldfarb, 1957; Litherland and Ferguson, 1961; Litherland, 1961). Isobaric-spin selection rules are also of importance (see Temmer, Chapter IV.A.2). The particle–γ coincidence method is particularly convenient for studying high-spin states a few MeV above the first particle threshold. For these, γ decay competes against particle decay, which is inhibited by the centrifugal barrier, though no such inhibition exists for low-spin states. Therefore, if γ rays from two-particle decay (usually of relatively low energy) are excluded, the coincident particle spectrum in this

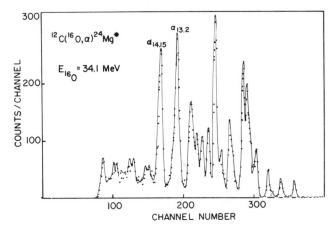

Fig. 19.   Spectrum of α particles at 0° to the beam direction, taken in coincidence with γ-ray pulses in the range 2.5–5.5 MeV from a 5-in.-diameter, 4-in.-thick NaI(Tl) detector. This range of γ energy was chosen in order to enhance α groups leading to states which decay by cascade transitions. The 8+, 13.2- and 14.15-MeV states are well above the thresholds for α decay (9.31 MeV) and proton decay (11.69 MeV) (Branford et al., 1972a).

region shows a few isolated peaks from the high-spin states with little background underneath (see Fig. 19).

The reaction Q values, and hence the modes of decay, vary considerably from case to case with light nuclei. Since pairing energies are large, the excitation energies differ widely when a given compound system is made by different reactions (Fig. 3). For example, excitation energies in ^{28}Si, for incoming particles of the Coulomb barrier energy, are 26.7 MeV for ^{12}C $+\,^{16}O$ and 37.5 MeV for $^{14}N + \,^{14}N$. In the first reaction, mainly one-particle decay occurs, while in the second, mainly two-particle decay. Even if the same compound system at the same excitation energy is made by two different reactions, its decay may not be identical because the angular momenta brought into the system differ. Thus to give an excitation energy of 42 MeV in ^{28}Si, the average incoming angular momentum is of the order of $11\hbar$ for the $^{16}O + \,^{12}C$ reaction, while for the $^{14}N + \,^{14}N$ reaction, it is only about $6\hbar$. Hence, for example, the $^{12}C(^{16}O, \alpha\gamma)^{24}Mg$ reaction is clearly superior to the $^{14}N(^{14}N, \alpha\gamma)^{24}Mg$ reaction for studying high-spin states in ^{24}Mg. Roughly speaking, α emission is slightly favored over proton emission for the lighter even-even nuclei with $N = Z$ and vice versa for odd-odd nuclei (Nomura et al., 1969); neutron emission is less probable. However, every case should be examined on its merits. Background material relevant to this type of experiment can be found in papers by Branford *et al.* (1971, 1972b), Bromley *et al.* (1971), and references therein, Cosman *et al.* (1971), Gobbi *et al.* (1971), Zioni *et al.* (1970) and Nomura *et al.* (1969).

III. Experimental Methods

A. Beam Requirements

The minimum bombarding energy required to perform a (HI, $xn\gamma$) reaction is a little greater than that of the Coulomb barrier in the laboratory system. This energy is of the order of 5 MeV/amu for the heaviest nuclei which can be studied. Good energy resolution is not required, as can be seen from Fig. 6, hence machines such as heavy-ion linear accelerators (HILACs) and cyclotrons are quite satisfactory (see Harvey, Chapter I.B). It is even possible to do experiments with fixed-energy machines, varying the bombarding energy with absorbers, provided that suitable beam handling and shielding facilities are provided. The required beam currents are small and of the order of a few nA for ions of charge one. So far most of the work with the heavier ions up to ^{40}Ar has been done with HILACs which, although designed as fixed-energy accelerators, are easily variable in energy. Much heavier ions than ^{40}Ar with suitable energies will soon be available from new HILACs in the USA and West Germany and from the HILAC/cyclotron combination in France (see Bock, Chapter I.C.1). Present tandem Van de Graaff accelerators, which reach terminal voltages of about 10 MV, satisfy the energy requirements for ^{16}O and lighter ions (see Allen, Chapter I.A). The new generation of these machines should reach terminal voltages of 14–20 MV, thus extending the range considerably. The excellent energy resolution of the tandem Van de Graaffs is of little advantage in (HI, $xn\gamma$) experiments but is essential when studying cases such as the ^{16}O(^{12}C, $\alpha\gamma$)^{24}Mg reaction. This is because one has to detect and resolve the emitted particles and also because the excitation functions fluctuate rapidly with bombarding energy. For (HI, $xn\gamma$) reactions, it is desirable to have pulsed beams so that one can measure during (in-beam) or outside (out-of-beam) the beam pulses. In this way, the contributions of long-lived products, radioactive nuclei or isomeric states, can be subtracted from the in-beam spectra and studied in their own right. Some accelerators such as cyclotrons and HILACs are intrinsically pulsed, and this feature has been exploited in measurements of lifetimes and magnetic moments for isomeric states (Yamazaki and Ewan, 1968; Maier *et al.*, 1972). However, some form of external pulsing is necessary if a large range of lifetimes is to be covered.

B. (HI, $xn\gamma$) REACTIONS

For achieving a satisfactory peak-to-background ratio and for resolving nearby lines, good γ resolution is essential. The importance of the former cannot be overemphasized. The background under a peak arises from the tails of the Compton distributions of higher-energy γ rays and from the photopeaks of weak, unresolved γ rays. Small fluctuations are therefore superimposed on the generally smoothly varying background as a function of channel number. To identify positively a peak, one must be able to make it stand out above the fluctuations (see Fig. 20). Also, when investigating peaks which are smaller than the background, the poorer the resolution is, the greater is the statistical accuracy required to establish their presence. Currently only Ge(Li) detectors satisfy the resolution requirement (see Goulding and Pehl, Chapter III.A). For best results, a number of different types are required. Roughly, for γ energies above about 500 keV, a large-volume, true-coaxial detector is best, though it can be used satisfactorily down to about 100 keV. Sometimes these detectors give better results for low-energy γ rays if they enter the side, rather than the front face, of the crystal. Thick ($\gtrsim 1$ cm) planar detectors with volumes of about 10 cm^3 give good results from very low energies up to about 1 MeV. They have ad-

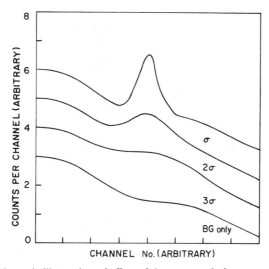

Fig. 20. Schematic illustration of effect of detector resolution on peak-to-background ratio. Curves, displaced vertically for clarity, are shown for resolutions (FWHM) of σ, 2σ, and 3σ, and for the background only.

vantages over the coaxial detectors of better timing and better resistance to radiation damage. They should always be used when fast timing is essential. For energies below about 150 keV, small Ge(Li) planar X-ray detectors give by far the best results both in resolution and in peak-to-background ratio, since they are very insensitive to higher-energy radiation.

Since the peaks of the excitation functions are wide and separated by about 15 MeV (Fig. 6), for normal spectroscopic studies, targets need not be thinner than about 5 MeV, which is equivalent to a few mg/cm^2 for heavy ions (Northcliffe and Schilling, 1970). Metallic targets can often be made easily by rolling. Separated isotopes of the rare earths, in oxide form, can be reduced to the metal with high efficiency (Westgaard and Bjornholm, 1966). Thin foils of highly reactive metals such as the latter should be handled only with nonmetallic tweezers, since a spark will instantly ignite them. When rolling is not possible, targets can be made from powdered material by fixing it to thin Mylar foil with dilute cellulose–cement solution (Lederer et al., 1971) or in some cases to a lead backing. Another method is to make a suspension of the powder in Formvar (Evans et al., 1971). Successful results can be obtained even if the powder is in oxide form because of the large cross section ($\sim$ 1 b) for the (HI, $xn\gamma$) reaction. Most of the reactions with light-element impurities produce high-energy γ rays and, with a few exceptions, do not spoil the spectra from the (HI, $xn\gamma$) reaction, except for increasing the continuous background. With heavier ions such as ^{40}Ar, the ranges of the recoiling nuclei can be several mg/cm^2. Hence all or a large fraction of them may recoil out of the target and emit γ rays when moving at high velocity. This will result in the observed lines being broadened and shifted in energy. If one is studying the reaction which takes place just above the Coulomb barrier, this difficulty can be overcome by using a thick target, the only disadvantage being a relative enhancement of Coulomb excitation and X-ray lines. For reactions at higher energies, one can use a target backing, which should be chosen to produce as small a number of discrete transitions and as low a continuous background as possible. For not too high bombarding energies, a heavy element such as lead which has a high Coulomb barrier may be best.

Measurements with the recoil-distance Doppler shift method (Diamond et al., 1969; Fossan and Warburton, Chapter VII.H) require thin, uniform targets of about 0.5–1 mg/cm^2. Such targets are best made by evaporation, or possibly by electrodeposition, and if possible should be self-supporting. In order to define the target/plunger distance to a precision of ± 0.02 mm or better, they have to be stretched. The plunger material must be chosen so as to minimize background radiation, which can be serious with thin targets. If the lowest (HI, $xn\gamma$) reaction is being studied, lead is a suitable material.

It is often useful to measure the conversion-electron spectrum in addition to that of the γ rays, since a comparison of the two gives multipolarity information. A suitable spectrometer of adequate resolution is one of the wedge type due to Kofoed-Hansen et al. (1950). This is convenient, since both target and detector [which should be of the Si(Li) type] are outside of the magnetic field. Since data have to be taken point by point, much larger beam currents are required than in γ-ray experiments. For conversion-electron spectroscopy, it may be necessary to use thinner targets in order to avoid degrading the energy resolution. The actual thickness depends on the electron energies of interest. Care must be taken that a large fraction of the decaying nuclei do not recoil out of the target, thus causing large kinematic spreads in the electron energies. An alternative arrangement, which takes advantage of the high rate of energy loss of heavy ions, is to use a thick target inclined at a grazing angle of 5–10° to the beam direction and observe the electrons normal to the target (Diamond et al., 1963).

When thin targets are used, a beam dump is required. To minimize background, the material which the beam hits should be a heavy element and the dump, well shielded with concrete and lead, should be about 6 ft from the target. When studying reactions other than the first above the Coulomb barrier, large backgrounds may come from collimators since the bombarding energy may exceed the barrier for any stable nucleus. Thus collimators near the target must be well shielded or better still avoided altogether by using a quadrupole magnet to focus the image of a distant aperture on the target.

Other techniques used are standard. Although in early work much of the interpretation of experiments rested on systematics and γ-ray energy sums and differences, this can sometimes lead to error. Multidimensional coincidence measurements, using event-by-event collection, are now a necessity.

C. REACTIONS WITH LIGHT NUCLEI

In these reactions, lifetimes of decaying levels are often less than the slowing-down time of the ions (~ 1 psec). Therefore many γ-ray peaks are seriously Doppler-broadened and in order to observe them at all, it is essential to take the γ spectra in coincidence with outgoing particles, thus effectively collimating the recoiling ions. Because the energy levels in light nuclei are relatively widely spaced and because it is possible to select the γ-ray spectra from individual levels by coincidence with the outgoing particles, NaI(Tl) detectors are often useful. They have the advantages over present Ge(Li) detectors of higher efficiency and better response function (ratio of full-energy peak to total spectrum). The response function is of major importance in angular distribution measurements. Doppler broadening at 90° to the beam direction can often be large ($\sim$ 100 keV) because of finite detector angle, and in this case, the performance of a large NaI(Tl) detector is much superior to that of a Ge(Li) detector. However, in studying details of γ branching and in lifetime measurements with the Doppler shift attenuation method (Fossan and Warburton, Chapter VII.H) large-volume, high-resolution Ge(Li) detectors are necessary.

A convenient method is to detect both particles and γ rays at 0° to the beam direction. This angle is chosen in order to minimize Doppler and kinematic broadening. A foil thick enough to stop the beam but thin enough to let the reaction particles through is either used as a target backing or is placed just behind the target. In the first case, γ decay takes place in the foil and in the second, in vacuum. Comparison of the two spectra gives information on the lifetime (Branford *et al.*, 1972b). The particles are detected either by a single surface-barrier detector or by a particle identification system. Particle energy resolution is determined mainly by the foil and by kinematic broadening and the latter problem can be overcome to some extent by using a radially position-sensitive detector. The large recoil velocities ($v/c \simeq 3\%$) are ideal for lifetime measurements. The stopping power is near its maximum value and shell effects are unimportant. Multiple scattering is also less serious than at lower velocities.

When reactions with light nuclei are studied, the targets have to be relatively thin, 10–50 μg/cm² being typical. This is because the excitation functions can be rapidly fluctuating and because particle spectra are measured. Carbon buildup is a serious problem since the targets are thin and reactions with carbon have large yield. This is in contrast to the (HI, $xn\gamma$) case, where light-element impurities rarely cause serious difficulties. To minimize this problem, the target should be surrounded by a shroud cooled with liquid nitrogen or, better still, a "clean" beam line and target chamber should be used (see discussion by Rolfs and Litherland, Chapter VII.D).

IV. Nuclear Structure Information from Heavy-Ion Reactions

The heavy-ion-induced reactions have in the past few years greatly enriched our understanding of nuclear behavior under rapid rotation. In fact, before heavy ions became available practically nothing was known about this subject. We have indicated how a study of these reactions has given us some understanding of the levels in the yrast region and no doubt there is a great deal more to be learned. However, most of the work has been directed toward a study of the well-separated levels below the energy gap. One of the great virtues of the (HI, $xn\gamma$) reaction is that it enables measurements on very large ranges of neutron-deficient nuclei with constant Z and varying N. Thus one can determine how level energies and properties vary when one nucleon number is fixed and the other is varied. Such systematic studies, which can rarely be done using other reactions, can lead to much greater insight than detailed data on a single nucleus. The large, uniform, and well-directed recoil velocities are a valuable feature of the (HI, $xn\gamma$) reactions which has been used for lifetime measurements. This feature, in conjunction with the strong alignment of angular momentum produced in these reactions, can be used to recoil the decaying nuclei into suitable environments for performing magnetic moment measurements. Neither this possibility nor that of measuring lifetimes has been much exploited so far, but they undoubtedly will be in the future. Here we only refer briefly to a small number of recent illustrative experiments. Work in this field up to the beginning of 1968 has been reviewed by Newton (1969) and later work by Stephens (1969) and Diamond (1970). Many of the models discussed here are covered in detail by Rasmussen, Chapter IX.B.

A. EVEN-EVEN NUCLEI

Much work has been done on the rotational and quasirotational gsb of even-even nuclei. Until recently, the results suggested that the moments of inertia $\mathfrak{J}$ of these bands increased smoothly with increasing J, reaching values of about $0.75\mathfrak{J}_{rig}$ for the highest observed values of $J(\sim 18)$. However, Johnson *et al.* (1971, 1972), with (α, xn) reactions, have reported three cases, ^{158}Dy, ^{160}Dy (Fig. 1), and ^{162}Er, where $\mathfrak{J}$ appears to change quite abruptly at about $J = 16$. This is shown in Fig. 21, where $\mathfrak{J}$ is plotted against the rotational frequency ω, which is related to $\mathfrak{J}$ by $\mathfrak{J}\omega = \hbar[J(J+1)]^{1/2}$. The data for ^{168}Yb, also shown, exhibit "normal behavior."

The phenomena have been understood as follows. A nucleus consisting of a set of nucleons moving independently in an average nonspherical potential would have a moment of inertia equal to $\mathfrak{J}_{rig}$ (Bohr and Mottelson, 1955). In a deformed nucleus, the principal residual force is the pairing force, the effect of which is to make $\mathfrak{J} < \mathfrak{J}_{rig}$. Under rotation, two effects are important. The nucleus can stretch under the influence of the centrifugal force, thus increasing $\mathfrak{J}$. Models based on this effect alone are known as centrifugal stretching models (Diamond *et al.*, 1964; Mariscotti *et al.*, 1969).

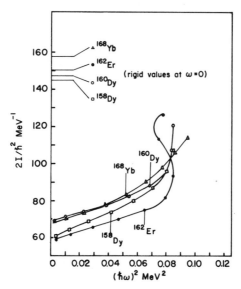

Fig. 21. Moment of inertia as a function of the square of the angular velocity for four rotational nuclei (redrawn from Johnson *et al.*, 1972, with permission of North-Holland Publishing Co., Amsterdam).

The other effect is related to the Coriolis force, which classically has the form $F_C = 2m\omega \times \mathbf{v}$, $\mathbf{v}$ being the velocity of the particle of mass m (Feynman, 1963). Now the pairing force produces a strong coupling between the time-reversed orbits, i.e., in a deformed nucleus between orbits having components on the nuclear symmetry axis of Ω and $-\Omega$. Since these orbits have nucleons traveling in opposite directions, F_C tends to push them apart, i.e., to reduce the interaction between them and hence to increase $\mathfrak{J}$. This is known as the Coriolis antipairing effect (CAP). In practice both of these effects should be considered, though their relative importance varies with the case. Both give a steadily increasing value of $\mathfrak{J}$ for increasing J up to about 14. However, at very large angular momenta, CAP produces a different result. The pairing-correlation effect is an example of a cooperative phenomenon such as ferromagnetism or superconductivity, to which it has similarities. When the pairing interaction is sufficiently reduced by CAP, the pair-correlated state is expected to disappear completely and hence the moment of inertia should become $\mathfrak{J}_{rig}$ quite suddenly. This effect, first predicted by Mottelson and Valatin (1960), may be smoothed out to some extent because neutrons and protons have different pairing interactions.

Johnson *et al.* (1971, 1972) have suggested that their results show the first example of this phenomenon.

An alternative explanation has been given by Stephens and Simon (1972) (Section II.B). When the γ-ray population leaves the non-gsb yrast levels to go into the gsb, there will be a sudden change of $\mathfrak{J}$, from something presumably near $\mathfrak{J}_{rig}$ to that of the appropriate point in the gsb. The value of ω^2 can also increase as one transfers to the gsb so that a discontinuous curve, somewhat like that for ^{162}Er, can be obtained. The discontinuity can be smoothed out by introducing interactions between the gsb and the non-gsb yrast levels and Stephens and Simon claim that curves similar to any of those in Fig. 21 can be obtained with their model. Although both explanations depend on the Coriolis force, they are not identical and it is of great interest to determine which, if either, of them is correct.

Work has not been confined to rotational or vibrational nuclei only.

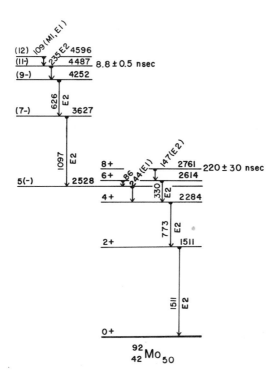

Fig. 22. Levels in ^{92}Mo from the ^{90}Zr(^{4}He, 2nγ)^{92}Mo reaction (Lederer *et al.*, 1971).

In Fig. 22, we show an example of the decay scheme of a nucleus with a closed neutron shell obtained with an $(\alpha, 2n)$ reaction (Lederer *et al.*, 1971).

B.   ODD-MASS AND ODD-ODD NUCLEI

Much less work has been done on these nuclei. Spectra are usually more complex than for the even-even nuclei (Section II.C). A typical decay scheme derived by Leigh *et al.* (1972) for a deformed nucleus from the ^{169}Tm(^{12}C,

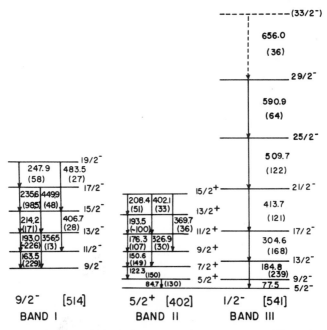

Fig. 23.   Rotational bands in decay scheme of ^{177}Re from the ^{169}Tm(^{12}C, 4nγ)^{177}Re reaction. The γ ray depopulating the $\frac{9}{2}^-$(514) state was not seen. Transition intensities are shown in parentheses (redrawn from Leigh *et al.*, 1972, with permission of North-Holland Publishing Co., Amsterdam).

4nγ)^{177}Re reaction is shown in Fig. 23. Three rotational bands are popu-lated. The one on the right is of interest in that it looks similar to the ro-tational band of an even-even nucleus, successive levels differing by two in J. It is in fact a $K = \frac{1}{2}$ band arising from the $h_{9/2}$ shell model orbital. Such bands from an orbit of high spin have large decoupling parameters and are strongly Coriolis-mixed to the other states arising from it. The levels with $J = \frac{1}{2}$, $\frac{3}{2}, \frac{7}{2}, \frac{11}{2}$, etc., are displaced upward in energy relative to the $\frac{5}{2}$ level and receive negligible population in the reaction. Coriolis mixing calculations with reasonable parameter values give good agreement with the level spacings in this and other cases (Hjorth *et al.*, 1970; Winter *et al.*, 1970). A notable effect of the mixing in such cases is the unusually large apparent moment of inertia for the band. In the case of the odd-odd nuclei ^{160}Ho and ^{162}Ho, which have a low-lying state in which both the odd-proton and odd-neutron state arise from high-spin orbitals, the mixing, as might be expected, is even stronger and $\mathfrak{J} = 2\mathfrak{J}_{rig}$ (Leigh *et al.*, 1970). In all cases of this type, the strong Coriolis interaction between states from a high-spin orbital passes on the irregular spacing of the $K = \frac{1}{2}$ band (which itself arises from the same interaction, Nathan and Nilsson, 1965) to the states of higher K.

The value of systematic investigations on a range of nuclei is shown, for example, by the experiments of Leigh *et al.* (1972) which suggested a relationship between the quasiparticle energies and the nuclear hexadecapole moments of the odd-mass rhenium nuclei. Systematic experiments on the odd-mass thallium isotopes (Newton *et al.*, 1970b) suggested oblate deformation for a band of excited states in these nuclei, which have one hole in the 82-proton closed shell.

C. Isomeric States

A wide variety of data on the decay of high-spin multi-quasiparticle states in deformed and spherical nuclei has been derived from out-of-beam (HI, $xn\gamma$) experiments. The type of information obtained is illustrated in Fig. 24, from the work of Bergström *et al.* (1970) and Maier *et al.* (1971), who used reactions induced by ^{4}He, ^{7}Li, and ^{11}B ions to study ^{211}At. The isomer at 4.816 MeV probably has a spin of $\frac{39}{2}$ and is likely to be a four-particle, one-hole state. This type of work is of great interest since little is known about such complex nuclear states. The (HI, $xn\gamma$) reactions provide the best method of producing them cleanly and with large cross section. Of course, information is also obtained about states populated in the decay of an isomer. Although γ rays from its decay are usually seen in-beam, the out-of-beam spectra are normally of much superior quality due to the absence of the continuum γ rays and those from states not populated in the isomeric decay.

D. Light Nuclei

For practical reasons, most of the work has been directed toward studies of the deformed even-even nuclei ^{20}Ne (Häusser *et al.*, 1971) and ^{24}Mg (Branford *et al.*, 1971, 1972a, b). Rotational bands based on the ground

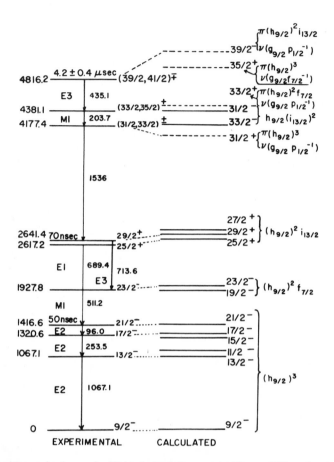

Fig. 24. Isomeric decays in ^{211}At derived from out-of-beam (HI, $xn\gamma$) experiments (redrawn from Maier *et al.*, 1971, with permission of North-Holland Publishing Co., Amsterdam).

state and some excited states have been seen in these nuclei up to $J = 8$. These bands are of particular interest because of the small number of nucleons involved in them. One cannot expect, nor does one get, the very regular behavior of the bands that one usually (but not always!) gets for heavy nuclei. A further point of interest concerns whether or not the bands cut off at a certain spin value, as would be expected if they arose from one configuration, in this case the s-d shell. The lifetime information obtainable in these experiments is of crucial importance for a proper understanding of the rotational bands.

ACKNOWLEDGMENTS

Comments on the manuscript from Drs. D. Branford, S. M. Ferguson, and J. R. Leigh are greatly appreciated. I am particularly indebted to Dr. S. D. Cirilov for help throughout the preparation of this chapter.

References

Alexander, J. M., and Simonoff, G. N. (1964). *Phys. Rev.* **133**, B93.

Ben-Zvi, I., Gilad, P., Goldberg, M., Goldring, G., Schwartzschild, A., Sprinzak, A., and Vager, Z. (1968). *Nucl. Phys.* **A121**, 592.

Bergström, I., Fant, B., Herrlander, C. J., Wikström, K., and Blomqvist, J. (1970). *Physica Scripta* **1**, 243.

Beringer, R., and Knox, W. J. (1961). *Phys. Rev.* **121**, 1195.

Biedenharn, L. C. (1960). In "Nuclear Spectroscopy" (F. Ajzenberg-Selove, ed.), Part B, pp. 732–810. Academic Press, New York.

Blatt, J. M., and Weisskopf, V. F. (1952). "Theoretical Nuclear Physics," Wiley, New York.

Bohr, A., and Mottelson, B. R. (1955). *Mat. Fys. Medd. Dan. Vid. Selsk.* **30**, No. 1.

Branford, D. (1972). Private Communication.

Branford, D., Gardner, N., and Wright, I. F. (1971). *Phys. Lett.* **36B**, 456.

Branford, D., Nagorcka, B. N., Newton, J. O., and Robinson, J. M. (1972a). (To be published).

Branford, D., McGough, A. C., and Wright, I. F. (1972b). *Nucl. Phys.* (To be published).

Bromley, D. A., Chua, L., Gobbi, A., Maurenzig, P. R., Parker, P. D., Sachs, M. W., Shapira, D., Stokstad, R. G., and Wieland, R. (1971). *J. Phys. (Paris)* **32**, C6 Suppl. 11–12, 5.

Cohen, S., Plasil, F., and Swiatecki, W. J. (1963). *Proc. Conf. Reactions between Complex Nuclei 3rd, Asilomar, California,* p. 325. Univ. of California Press, Berkeley.

Cosman, E. R., Sperduto, A., Moore, W. H., Chin, T. N., and Cormier, T. M. (1971). *Phys. Rev. Lett.* **27**, 1074.

De Jesus, A. S. M., Mills, S. J., and Rautenbach, W. L. (1971). *Nucl. Phys.* **A172**, 323.

Devons, S., and Goldfarb, L. J. B. (1957). In "Encyclopedia of Physics" (S. Flügge ed.), Vol. 42, p. 362. Springer, Berlin.

Diamond, R. M. (1970). *Conf. Properties Nuclei far from Region Beta Stability, Leysin, Switzerland.* Univ. California Lawrence Radiat. Lab. Rep. UCRL-19961.

Diamond, R. M., Elbek, B. and Stephens, F. S. (1963). *Nucl. Phys.* **43**, 560.

Diamond, R. M., Stephens, F. S., and Swiatecki, W. J. (1964). *Phys. Lett.* **11**, 315.

Diamond, R. M., Matthias, E., Newton, J. O., and Stephens, F. S. (1966). *Phys. Rev. Lett.* **16**, 1205.

Diamond, R. M., Stephens, F. S., Kelley, W. H., and Ward, D. (1969). *Phys. Rev. Lett.* **22**, 546.

Draper, J. E. and Lieder, R. M. (1970). *Nucl. Phys.* **A141**, 211.

Ericson, T. (1960). *Advan. Phys.* **63**, 479.

Ericson, T., and Mayer-Kuckuk, T. (1966). *Ann. Rev. Nucl. Sci.* **16**, 183.

Evans, M., Ellis, A. E., Leigh, J. R., and Newton, J. O. (1971). *Phys. Lett.* **34B**, 609.

Ferguson, S. M., Ejiri, H., and Halpern, I. (1972). *Nucl. Phys.* **A188**, 1.

Feynman, R. (1963). *In* "Lectures in Physics," Vol. 1, p. 19–28. Addison-Wesley, Reading, Massachusetts.

Frauenfelder, H. and Steffen, R. M. (1965). *In* "α, β and γ-ray Spectroscopy" (K. Seigbahn, ed.), Vol. 2, p. 997. North-Holland Publ., Amsterdam.

Gilat, J. (1970a). Brookhaven Nat. Lab. Rep. No. BNL 50246 (T-580) (unpublished).

Gilat, J. (1970b). *Phys. Rev.* **C1**, 1432.

Gilat, J., and Grover, J. R. (1971). *Phys. Rev.* **C3**, 734.

Gobbi, A., Maurenzig, P. R., Chua, L., Hadsell, R., Parker, P. D., Sachs, M. W., Shapira, D., Stokstad, R., Wieland, R., and Bromley, D. A. (1971). *Phys. Rev. Lett.* **26**, 396.

Grover, J. R. (1967). *Phys. Rev.* **157**, 832.

Grover, J. R., and Gilat, J. (1967a). *Phys. Rev.* **157**, 802.

Grover, J. R., and Gilat, J. (1967b). *Phys. Rev.* **157**, 814.

Grover, J. R., and Gilat, J. (1967c). *Phys. Rev.* **157**, 823.

Halpern, I., Shepherd, B. J., and Williamson, C. F. (1968). *Phys. Rev.* **169**, 805.

Häusser, O., Alexander, T. K., McDonald, A. B., Ewan, G. T., and Litherland, A. E. (1971). *Nucl. Phys.* **A168**, 17.

Hjorth, S. A., Ryde, H., Hagemann, K. A., Løvhøiden, G., and Waddington, J. C. (1970). *Nucl. Phys.* **A144**, 513.

Johnson, A., Ryde, H., and Hjorth, S. A. (1972). *Nucl. Phys.* **A179**, 753.

Johnson, A., Ryde, H., and Sztarkier, J. (1971). *Phys. Lett.* **34B**, 605.

Kalinkin, B. N., and Petrov, I. Z. (1964). *Acta. Phys. Polon.* **25**, 265; Univ. of California, Lawrence Radiat. Lab. Rep. No. UCRL Trans – 1151.

Kofoed-Hansen, O., Lindhard, J., and Nielsen, O. B. (1950). *Mat. Fys. Medd. Dan. Vid. Selsk.* **25**, No. 16.

Lederer, C. M., Jaklevic, J. M., and Hollander, J. M. (1971). *Nucl. Phys.* **A169**, 449.

Leigh, J. R., Stephens, F. S., and Diamond, R. M. (1970). *Phys. Lett.* **33B**, 410.

Leigh, J. R., Newton, J. O., Ellis, L. A., Evans, M. C., and Emmott, M. J. (1972). *Nucl. Phys.* **A183**, 177.

Litherland, A. E. (1961). *Can. J. Phys.* **39**, 1245.

Litherland, A. E., and Ferguson, A. J. (1961). *Can. J. Phys.* **39**, 788.

Maier, K. H., Leigh, J. R., Pühlhofer, F., and Diamond, R. M. (1971). *Phys. Lett.* **35B**, 401.

Maier, K. H., Nakai, K., Leigh, J. R., Diamond, R. M., and Stephens, F. S. (1972). *Nucl. Phys.* **A186**, 97.

Mariscotti, M. A. J., Scharff-Goldhaber, G., and Buck, B. (1969). *Phys. Rev.* **178**, 1864.

Mills, S. J., and Rautenbach, W. L. (1968). *Nucl. Phys.* **A124**, 597.

Mollenauer, J. F. (1962). *Phys. Rev.* **127**, 867.

Morinaga, H., and Gugelot, P. C. (1963). *Nucl. Phys.* **46**, 210.

Mottelson, B. R., and Valatin, J. G. (1960). *Phys. Rev. Lett.* **5**, 511.

Myers, W. and Swiatecki, W. J. (1965). Univ. of California Rep. UCRL - 11980, Berkeley.

Nathan, O., and Nilsson, S. G. (1965). In "α, β and γ-ray Spectroscopy" (K. Seigbahn, ed.), Vol. 1, pp. 601–700. North Holland Publ., Amsterdam.

Natowitz, J. B. (1970a). *Phys. Rev.* **C1**, 623.

Natowitz, J. B. (1970b). *Phys. Rev.* **C1**, 2157.

Neubert, W. (1971). *Nucl. Instrum. Methods* **93**, 473.

Newton, J. O. (1968). *Nucl. Phys.* **A108**, 353.

Newton, J. O. (1969). *In* "Progress in Nuclear Physics" (D. M. Brink and J. H. Mulvey, ed.), Vol. XI, pp. 53–144. Pergamon, Oxford.

Newton, J. O., Stephens, F. S., Diamond, R. M., Kotajima, K., and Matthias, E. (1967). *Nucl. Phys.* **A95**, 357.

Newton, J. O., Stephens, F. S., Diamond, R. M., Kelley, W. H., and Ward, D. (1970a). *Nucl. Phys.* **A141**, 631.

Newton, J. O., Cirilov, S. D., Stephens, F. S., and Diamond, R. M. (1970b). *Nucl. Phys.* **A148**, 593.

Newton, J. O., Stephens, F. S., and Diamond, R. M. (1973). *Nucl. Phys.* **A210**, 19.

Nomura, T., Morinaga, H., and Povh, B. (1969). *Nucl. Phys.* **A127**, 1.

Nordhagen, R., Goldring, G., Diamond, R. M., Nakai, K., and Stephens, F. S. (1970). *Nucl. Phys.* **A142**, 577.

Northcliffe, L. C., and Schilling, R. F. (1970). *Nucl. Data Tables* **A7**, 233.

Ryde, H. (1971). Private communication to I. Halpern.

Sikkeland, T. (1967). *Arkiv. Fysik* **36**, 539.

Stephens, F. S. (1969). *Proc. Int. Conf. Properties Nucl. States, Montreal*, p. 127. Les Presses de l'Univ. de Montreal.

Stephens, F. S., and Simon, R. S. (1972). *Nucl. Phys.* **A183**, 257.

Stephens, F. S., Leigh, J. R., and Diamond, R. M. (1971). *Nucl. Phys.* **A170**, 321.

Thomas, T. D. (1968). *Ann. Rev. Nucl. Sci.* **18**, 343.

Ward, D., Stephens, F. S., and Newton, J. O. (1967). *Nucl. Phys.* **19**, 1247.

Westgaard, L. and Bjornholm, S. (1966). *Nucl. Instrum. Methods* **42**, 77.

Williamson, C. F., Ferguson, S. M., Shepherd, B. J., and Halpern, I. (1968). *Phys. Rev.* **174**, 1544.

Winter, G., Funke, L., Kaun, K. N., Kemnitz, P., and Sodan, H. (1970). *Phys. Lett.* **33B**, 161.

Yamazaki, T. (1967). *Nucl. Data Tables* **A3**, 1.

Yamazaki, T., and Ewan, G. T. (1968). *Nucl. Instrum. Methods* **62**, 101.

Zioni, J., Jaffe, A. A., Friedman, E., Haik, J., and Schectman, R. (1970). *In* "Nuclear Reactions Induced by Heavy Ions" (R. Bock and W. R. Hering ed.), p. 693. North Holland Publ., Amsterdam.

VII.F DETAILED SPECTROSCOPY FROM FISSION

E. Cheifetz[†] and J. B. Wilhelmy[‡]

LAWRENCE BERKELEY LABORATORY
UNIVERSITY OF CALIFORNIA, BERKELEY, CALIFORNIA

[†] Current address: Weizmann Institute of Science, Rehovoth, Israel.
[‡] Current Address: Los Alamos Scientific Laboratory, Los Alamos, New Mexico, 87544.

I.   Introduction

The fission of a heavy element produces over 200 neutron-rich short-lived isotopes with a wide mass distribution varying from $A = 80$ to 165 and having a neutron-to-proton ratio close to that of the fissioning nucleus. Most of the products are far removed from the line of beta stability and can be produced in the laboratory only by the fission process. Thus all the studies concerning the radioactive properties and the nuclear spectroscopy of these isotopes are associated with experimental methods of rapid separation of the fission products and various techniques of singling out specific radiations from the decay of many nuclei. A chart of nuclides is shown in Fig. 1. The regions of isotopes which are produced in the fission of ^{235}U induced by thermal neutrons are indicated. The two separated regions of high yields are due to the asymmetry in the mass distribution of fission. The yield in the region of symmetric mass division can be raised if medium- or high-energy

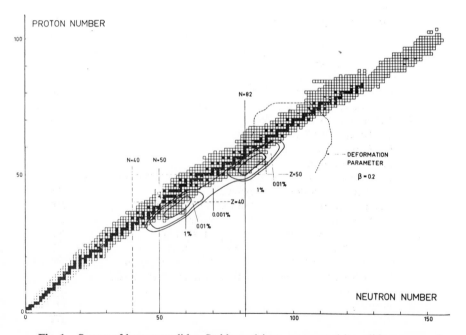

Fig. 1.   Survey of known nuclides. Stable nuclei are represented by solid squares and radioactive ones by open squares. The contours of the mass yield distribution populated in fission of ^{235}U are indicated (from Borg *et al.*, 1971, with permission of North-Holland Publishing Co., Amsterdam).

projectiles induce the fission. However, use of very high-energy projectiles to induce fission has limited advantages because many neutrons are emitted either before or after fission, resulting in products that are not neutron rich and thus not unique to the fission process. The yield of products of the easily available spontaneous fission source ^{252}Cf is similar to that of ^{235}U + thermal n with a shift in the relative yields of the light fission products toward heavier isotopes.

The fragments are produced with high excitation and deexcite first by emission of neutrons in a time duration of 10^{-18}–10^{-14} sec and then deexcite further by emission of γ rays, conversion electrons, and X rays. The time scale of the latter process is generally 10^{-13}–10^{-7} sec with longer half-lives associated with a few specific isomeric states. For an extended review and details concerning the process of fission and the questions associated with fission kinetic energies, fission yields, and neutron emission, the reader is referred to Hyde (1964).

Measurements of the broad aspects of the prompt γ-ray decay of the fragments have been carried out by several authors and the results for the total energy, the average multiplicity, and the spectral shape of the γ rays which were published prior to 1964 have been summarized in the review by Johansson and Kleinheinz (1965). More recent results have been reported by Peelle and Maienschein (1971), who have determined that the total prompt gamma-energy release in the fission of ^{235}U induced by thermal neutrons is 7.25 ± 0.26 MeV with a multiplicity of 8.13 ± 0.35 γ rays per fission. This is of course shared by the two fragments. In addition to that, 350 ± 100 keV of energy is emitted in the form of delayed gamma radiation from the fragments and an estimated 45 keV is emitted as conversion and Auger electrons. Similar emission of γ rays occurs in other fissioning nuclei. The γ rays are also emitted anisotropically with respect to the fission direction with $[N(0°) - N(90°)]/N(90°) \sim 0.12$.

The broad aspects of γ-ray emission have also been investigated as a function of the mass of the fragments by Maier-Leibnitz et al. (1965) and also by Armbruster et al. (1969). They found that the anisotropy, number, and total energy of the γ rays depend only weakly upon the mass ratio of the two fragments; however, the mass dependence of the number of quanta emitted as a function of specific fragment mass indicates a saw-tooth dependence similar to that observed in neutron emission from fission fragments. The angular momentum of the primary fission fragments has been interpreted to be 6–10$\hbar$ from studies of gross γ-ray anisotropy (Strutinskii, 1960; Kapoor and Ramanna, 1964; Hoffman, 1964; Skarsvåg and Singstad,

1965; Skarsvåg, 1967; Graff *et al.*, 1965; Val'skii *et al.*, 1967, 1969; Armbruster *et al.*, 1969), from yields of isomeric levels and ground states in prompt fission products (Croall and Willis, 1963; Warhanek and Vandenbosch, 1964; Sarantities *et al.*, 1965; Loveland and Shum, 1971), and from intensities of ground-state band γ rays emitted deexciting prompt even-even fission products (Wilhelmy *et al.*, 1972). Most of the angular momentum is dissipated through a cascading series of γ-ray transitions as the primary products deexcite.

Once the nuclei reach the ground state, or perhaps in some cases a long-lived isomeric state, beta decay takes place with typical half-lives on the order of 1 sec. Most of the fragments are far removed from the line of beta stability and have high Q_β values. For example, the Q_β values for the decay of ^{94}Rb to ^{94}Sr and ^{96}Rb to ^{96}Sr are 8.59 ± 0.30 and 10.82 ± 0.50 MeV, respectively (Macias-Marques *et al.*, 1970). These beta decays thus populate many excited states in the daughter nuclei and result in very complex γ-ray spectra. The detailed study of these decay schemes will extend accurate nuclear level information to much higher energies than have previously been obtained from decay studies on nuclei closer to stability.

From the spectroscopy of fission products, information is obtained about three particularly interesting regions of isotopes: (a) the $A = 100$ region, for which there is evidence for large deformation in the neutron-rich isotopes, (b) the neutron-rich isotopes around the doubly magic $^{132}_{50}$Sn isotope, and (c) the mass region around $A = 140$–150, where there is a transition from spherical to deformed nuclei.

The recent advances in the production of high-resolution photon detectors and in the development of on-line mass separators and the improvement in rapid radiochemistry have opened the field of fission product spectroscopy to many new possibilities which were impractical earlier. Much research effort is currently being directed to this field and consequently much new information has been obtained. The experimental techniques that are used in studies of fission product spectroscopy are divided into two types: (a) studies of prompt and delayed radiations from the fragments prior to beta decay in which the fragments are characterized by their kinetic energies and the identification of the isotopes is made by determining the properties of their particular decay, and (b) studies of post-beta-decay deexcitation, where product characterization is either by mass separation or fast chemical isolation.

In this chapter, we will only discuss the study of those isotopes for which the fission process plays a dominant role, i.e., prompt fission radiation measurements, and those beta-decay studies which observe the high-fission-yield, short-lived, prompt products. In the discussion part, an attempt will be made to correlate the results obtained from prompt decay with the results from beta decay and to summarize the nuclear information obtained in the various regions.

II. Spectroscopy of Prompt Fission Products

A. MASS IDENTIFICATION BY KINETIC ENERGY MEASUREMENTS

The prompt fission radiations are those resulting from the deexcitation of the primary fission products. The fission fragments are formed in highly excited states and dissipate this energy utilizing the standard method—neutron evaporation followed by γ-ray emission. Our primary concern will not be with the average radiation properties but with the ability to assign transitions to specific isotopes in order to eventually establish nuclear structure information for those isotopes which are only accessible through the fission process. The major obstacle to detailed spectroscopic studies is the complexity of the radiations emitted. For example, in the spontaneous fission of ^{252}Cf, there are ~ 100 isotopes having independent yields of 0.5–3.5% per fission (Watson and Wilhelmy, 1969). Since each isotope emits 3–4 MeV in γ-ray energy, the prompt spectra are very complex. For transitions with life-times shorter than ~ 200 nsec, the only technique available consists in determining the masses of the isotopes from the kinetic energy measurements of the separating fission fragments and determining their atomic numbers from measurements of the characteristic K X rays.

If we ignore for a moment the effect of neutron emission, then the conservation of momentum would require

$$M_1 V_1 = M_2 V_2 \quad \text{or} \quad M_1 E_1 = M_2 E_2 \tag{1}$$

where M, V, and E are the mass, velocity, and kinetic energy of the fragments 1 and 2. Therefore a ratio of the measured kinetic energies would give a ratio of masses and a simple conservation of the number of particles $(M_1 + M_2 = M_f)$ would uniquely determine the masses of the isotopes formed.

The neutron evaporation introduces two major effects in attempting to determine the masses of the isotopes from the measured kinetic energies. The first effect is the resultant dispersion in the measured kinetic energy through the evaporation of neutrons. In order to determine the mass of isotopes formed, it is necessary to have a measurement of the number of neutrons evaporated from each fragment as a function of the measured variables—the two kinetic energies of the fragments. Such arrays have been established for the spontaneous fission of ^{252}Cf by Bowman *et al.* (1962) and for the thermal-neutron-induced fission of ^{235}U by Apalin *et al.* (1965) and Maslin *et al.* (1967). With these arrays and utilizing energy and momentum conservation, it is possible to use an iterative procedure such as that described by Watson (1966) to determine the mass of the isotope. However, this correction is for the average kinetic energy distribution for a fragment pair. The dispersion in the kinetic energy for a given fragment due to the isotropic evaporation of the neutrons will be on the order of the kinetic energy of the neutrons. This means there is an $\sim$ 1–2% energy dispersion introduced by neutron evaporation. Since the mass is inversely proportional to the kinetic energy, a 2% energy dispersion results in a 2% dispersion in the determined mass. The actual energy resolution of the solid-state detectors used to measure the kinetic energies is $\sim$ 1 MeV and therefore contributes the same, or less, to the total dispersion than the inherent width introduced by the neutron evaporation. In practice the dispersion can best be determined from the experimental data.

Figure 2 is taken from Watson (1966) and shows the observed intensity of a 162-keV conversion-electron line as a function of the mass calculated from the measured fragment kinetic energies. The measured 4.8 amu

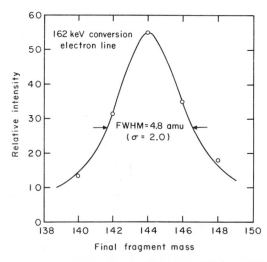

Fig. 2. An example showing the direct determination of the experimental mass resolution. The intensity of a discrete conversion-electron transition is plotted as a function of the average mass value with which it is observed (from Watson, 1966).

FWHM is fairly typical of results found in other experiments. This poor mass resolution is the primary limitation in obtaining detailed spectroscopic properties of the fission isotopes. The spectra associated with specific mass regions contain contributions from a large number of isotopes, thus adding to the complexity and limiting the sensitivity in the spectral analysis.

The large mass dispersion is not the only problem introduced from the neutron evaporation. The other main difficulty is in the determination of the centroid of the mass distribution for any specific line. The neutron corrections which go into the mass calculations are for the average number of neutrons emitted by the fragments for a given kinetic energy division. The difficulty arises in that these average properties are applied to give a neutron mass correction for a specific isotope. Such a procedure tends to be correct for "average" isotopes, but for those which lie away from the centroids of the mass distribution, the calculated mass tends to be biased in such a manner as to shift the calculated mass toward the average value of the distribution. The introduced error can be over 1 amu. For the study of average properties of fission radiations, this error may not be serious, but for spectroscopic purposes, an error of 1 amu makes the difference between, for example, an even-even or an even-odd isotope. These will have tremendously different nuclear structure properties.

This problem could in principle be eliminated by measuring neutron multiplicities in coincidence with specific γ rays. In lieu of such data, however, it is necessary now to rely on nuclear systematics of even-even nuclei in establishing the true masses of isotopes. As will be discussed, ground-state bands of even-even nuclei are strongly populated in the deexcitation of the prompt fission products and thus can be readily identified. Once the calculated mass systematics of the even-even isotopes are established, odd-A and odd-odd isotopes can be accurately determined by interpolation between known even-even isotopes. An example of this is shown in Table 1

TABLE 1

AN EXAMPLE OF CALCULATED MASS SYSTEMATICS FROM KINETIC ENERGY MEASUREMENTS

Isotope	^{108}Ru[a]	^{109}Ru	^{110}Ru[a]	^{111}Ru	^{112}Ru[a]
Measured mass (amu)	108.99	109.48	110.15	110.89	111.85
Predicted mass from e–e	—	109.57	—	111.00	—

[a] Cheifetz et al. (1970a).

for transitions from five adjacent Ru isotopes. Intensities of gamma rays from the various Ru isotopes were evaluated as a function of the calculated masses. The centroids of these distributions (similar to Fig. 2) were determined to give the "measured mass" of each isotope. It is seen that the measured mass distribution is narrower than the true distribution (since the tendency is to squeeze the distribution in toward the average) and that for one isotope (^{108}Ru) the measured mass is ~ 1 amu in error. The transitions from the odd-A isotopes have measured mass distributions which are not the integer values they should be, but do lie between the adjacent even-even values. A simple interpolation between the even-even measured mass values gives a predicted mass for the odd-A isotopes. It is seen that these predicted values differ from the measured value by only ~ 0.1 amu and therefore enable confident assignment of the transitions to the specific isotopes.

With these techniques, it is possible to accurately determine the mass of given transitions using fission-product kinetic energy measurements. However, it should be emphasized that their determinations require relatively strong, spectrally isolated transitions for which systematics in neighboring even-even nuclei have been established. Unfortunately, these requirements eliminate a large number of the observed transitions from accurate assignment to specific isotopes.

B.   Element Identification by X-Ray Coincidence Measurements

Continued improvements in X-ray detector resolution have enabled reliable assignments of fission γ-ray transitions to specific elements. The energy spacing between adjacent $K\alpha_1$ X rays of fission product nuclei varies from ~ 700 eV for the lightest fission product elements ($_{33}$As) to ~ 1500 eV for the heaviest ($_{63}$Eu). The resolution of the current generation of Si(Li) X-ray detectors is between 250 and 500 eV, which permits easy separation of X rays from adjacent elements. Four types of X-ray coincidence experiments have been performed which yield information on prompt fission products: (a) X rays measured in coincidence with fragment kinetic energies, (b) X rays measured in coincidence with γ rays from a fission source, (c) X rays measured in coincidence with γ rays and fission-product kinetic energies, and (d) X rays measured in coincidence with conversion electrons and fragment kinetic energies.

Of the various methods, (a) has probably been the most studied (Glendenin and Unik, 1965; Kapoor et al., 1965; Atneosen et al., 1966; Watson et al., 1970a; Reisdorf et al., 1971). However, this technique does not give specific spectroscopic properties of individual isotopes. The measured yield per element is averaged over several isotopes. These data have been analyzed to extract fission properties such as the product charge and mass distribution. They have also been analyzed to give some information on nuclear systematics. There is an observed odd-even fluctuation in the X-ray yield with the odd-Z isotopes having the higher yield. This is indicative of odd-Z isotopes having a preponderance of lower-energy transitions which in turn emit conversion electrons and K X rays. The even-Z X rays include yields from even-even isotopes which have in general wider-spaced energy levels and therefore fewer conversion electrons and subsequent K X rays. The other predominant systematic feature is that the X-ray yield increases for the heavier fission products. This is consistent with the expected onset of nuclear deformation for these rare earth isotopes. Permanent deformation results in low-lying, closely spaced bands which are conducive to high X-ray yield.

X rays measured in coincidence with γ rays from a fission source have yielded the best data for assignment of transitions to specific elements. The original work of Ruegsegger and Roy (1970) and Eddy and Roy (1971) has been repeated in a more general analysis with much improved resolution by Hopkins et al. (1971, 1972). In these experiments, a fission source (^{252}Cf) is placed between solid-state γ-ray and X-ray detectors. X-ray–γ-ray coincidences are recorded in a two-dimensional format. Digital windows are set on energy intervals containing the K X rays of the fission product elements. The gated γ-ray spectra are thus associated with specific elements. The major advantage of this approach is that very high geometries can be obtained for the coincidence studies. The two photon detectors are placed as close together as practical on either side of a thin, encapsulated fission source. The fission fragments are thus stopped rapidly ($\sim 10^{-12}$ sec) and transitions emitted from the stopped fission products are non-Doppler-shifted and appear as sharp, discrete lines.

There are, however, two primary limitations to this type of analysis. The first is that there is no fission coincidence signal to ensure that the transitions observed are prompt. The second is that, if prompt, the coincident γ rays could be associated with either the element (Z) for which the K X-ray window is set or for the complementary element $(Z_f - Z)$. These ambiguities cannot be resolved with only these types of data. However, by combining these results with experiments in which fission-product mass identification is performed through kinetic energy measurements, it is possible to overcome these difficulties and make assignments of transitions to specific elements. An example of the spectra obtained in coincidence with X rays of the complemen-

tary fission elements Mo–Ba is shown in Fig. 3. Several discrete transitions
are readily discernible above the general background. Some transitions,
such as the 94.8–94.9-keV line, are seen in both spectra. This is not an ac-
cidental coincidence but arises because both the isotope from which it is
emitted and at least one of the complementary isotopes have other transitions
which are decaying by internal conversion to give K X rays. To establish with

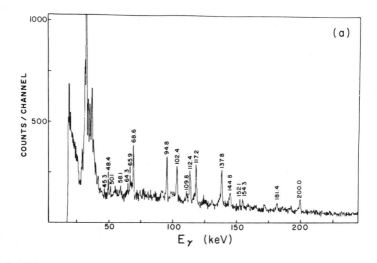

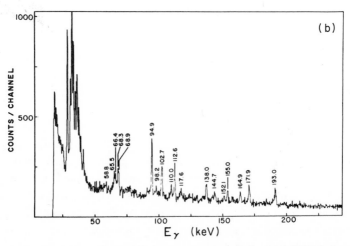

Fig. 3.   Low-energy portions of gamma-ray spectra obtained for ^{252}Cf fission products
measured in coincidence with $K\alpha$ X rays of the complementary elements (a) Mo gate and
(b) Ba gate (from Hopkins *et al.*, 1971).

which element the transition is associated, it is necessary to go to experiments in which mass-sorted information is obtained. In this case, it is found that the transition is from mass 105, which is a light fission product and therefore the γ ray is from ^{105}Mo (Cheifetz et al., 1971a).

It is important to note that even if a transition is observed only with the X ray of one element, it is still possible to associate it with the complement of that element. For example, the indicated 181.4-keV and 200.0-keV γ rays in the Mo spectrum are in fact the $2^+ \rightarrow 0^+$ transitions in ^{146}Ba and ^{144}Ba, respectively (Wilhelmy et al., 1970a), while the 171.9-keV and 193.0-keV γ-ray lines in the Ba-gated spectrum are the $2^+ \rightarrow 0^+$ transitions in ^{106}Mo and ^{104}Mo, respectively (Cheifetz et al., 1970a). In general, the $2^+ \rightarrow 0^+$ transitions in even-even nuclei have only small probability of being in coincidence with X rays from the element with which they are associated. Such a coincidence would require internal conversion of a transition in the same cascade. The other ground-state band transitions are usually of higher energy than the $2^+ \rightarrow 0^+$ transition and thus have small probability of decaying by internal conversion. Isotopes of the complementary element may have a transition with a substantial probability for internal conversion, thus greatly enhancing the X-ray–γ-ray coincidence rate.

The third technique, in which X rays are measured in coincidence with γ rays and fission-product kinetic energies, in principle overcomes the limitation of the two-parameter coincidence method. These experiments, however, have practical limitations of their own. They are electronically more complex, requiring simultaneous measurement of additional parameters. The necessity of placing fragment detectors in the experimental configuration results in poorer total coincidence geometry. Figure 4 is a schematic representation of an experimental configuration used for such studies. Doppler shifts are observed for γ rays emitted from fission products in flight, and this introduces additional complexity in the spectra. These limitations are counterbalanced by the positive aspects associated with the removal of the ambiguities of whether the observed γ-ray is prompt and whether it is from a light or heavy fission product. This technique has been instrumental in establishing the systematics of the even-even fission products (Cheifetz et al., 1970a, b; Wilhelmy et al., 1970a).

The measurement of X rays in coincidence with conversion electrons permits the direct determination of the element with which the observed electron is associated. If the conversion electron is emitted from the K electron shell, then the resultant vacancy can be filled with the emission of K X rays. If (as in the techniques mentioned earlier) a γ ray is measured in

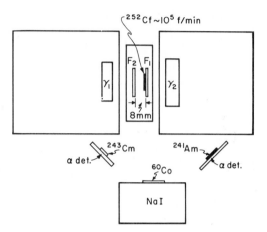

Fig. 4. A schematic representation of an experimental configuration used for four parameter (F_1, F_2, γ_1, γ_2) coincidence measurements of prompt radiations from the fission of ^{252}Cf. The sources and detectors indicated on the bottom of the figure are used for external gain stabilization of the photon detectors (from Cheifetz *et al.*, 1970a).

coincidence with a K X-ray, it is necessary that a coincident transition decay by internal conversion. Therefore, the yield of the X-ray–γ-ray coincidence is dependent on the conversion probability of these coincident transitions. Watson *et al.* (1970b) have performed multiparameter X-ray–conversion-electron–fragment-kinetic-energy coincidence studies on the products formed in the spontaneous fission of ^{252}Cf. From their analysis, they have been able to assign several discrete transitions to specific isotopes. The difficulty in these studies is the relatively poor resolution in the conversion-electron spectra (4.0 keV FWHM at 19 keV to 7.4 keV FWHM at 161 keV). This should be compared with a γ-ray resolution of better than 1 keV over the same energy region. Since the spectra are very complex, good resolution is important for confident assignment of transitions to specific isotopes.

C. Prompt Gamma-Ray Emission

With the development of solid-state detectors, multiparameter pulse-height analyzers, and small on-line computers, many studies have been performed on γ rays associated with specific fission products. In early work, Bowman *et al.* (1964, 1965) and Johannson (1964, 1965) studied the spontaneous fission of ^{252}Cf and attempted to assign the observed γ rays to specific masses. The former authors have tabulated 31 transitions covering an energy range of 76–586 keV which they tentatively assign to 16 specific isotopes, but claim an uncertainty of ± 1 amu in the mass determination and $\pm 1Z$ in the charge determination. The mass determinations were obtained using the measured fragment kinetic energies but the charge was only inferred as being the most probable charge (Z_p) for the calculated mass. No level assignments were obtained. Johansson did not present detailed tabulations of transitions associated with specific isotopes but did present some γ-ray spectra associated with mass intervals. In the light fission product region around mass 110, he saw systematic structure similar to that observed in the deformed rare earth mass region around mass 152 and concluded that this was strongly suggestive that the light fission products were permanently deformed. He speculated that the γ rays at ~ 130 keV and ~ 300 keV were the $2^+ \to 0^+$ and $4^+ \to 2^+$ ground-state band transitions in ^{110}Ru. These levels are not now believed to be correct, but nevertheless the considerations regarding the feasibility of light fission product nuclei being deformed were substantially valid and generated interest in the development of fission spectroscopy so that this region could be studied.

In more recent investigations, the prompt spectroscopy of products formed in the thermal neutron fission of ^{235}U have been studied by Horsch and Michaelis (1969), Horsch (1970), and Khan et al. (1973) and of ^{235}U and ^{239}Pu by Schindler and Fleck (1973). The previous work had been done on the spontaneous fission of ^{252}Cf and it is important for two reasons that similar measurements be extended to the fission of other nuclei: The first is that there should be a large number of the same isotopes produced abundantly by all processes and therefore a higher degree of confidence could be placed in the assignments if independent methods gave the same results. The second point is that, especially in the light fission product region, some different isotopes are produced. In U fission, the light fission product masses having $> 1\%$ fission yield lie between $84 \leqslant A \leqslant 105$, while for Cf, the range is $93 \leqslant A \leqslant 118$. This implies the Cf fission data would be most suitable for studying the possible region of deformation around $A = 110$, while the U data would give unique results on nuclei having close to the 50-neutron closed shell.

In the most recent U work, Khan et al. (1973) tabulate 131 transitions assigned to specific isotopes. Many of the transitions are believed to be the same as those determined in the fission of ^{252}Cf. Based on nuclear systematics, conversion-electron measurements, and strong yield in the prompt γ-ray spectra, they tentatively assign $2^+ \to 0^+$ transitions in ^{90,92}Kr and ^{94,96,98}Sr.

In these studies, which have observed prompt transitions from fragments in flight, serious effects due to Doppler shifting are encountered. The fragments have high velocities $(v/c \sim 0.02–0.05)$ and therefore the laboratory transition energy depends on the fragment velocity and the angle at which the observation is made relative to the fragment motion. If the source is observed directly, each transition will appear twice, once with the fragment moving toward the detector and once with it moving away. This doubles the complexity of the already very complex spectra. The other problem is that, due to the finite acceptance angle for fragment detection (which under normal conditions would be made large to enhance coincidence efficiency), the observed transitions are not only energy-shifted but can appear substantially broadened. This results in an effective worsening of experimental resolution which again is very serious for accurate analysis of the spectra.

To overcome these difficulties and still be able to observe all prompt transitions, a technique was developed for electrodepositing the ^{252}Cf source on a phosphorous-diffused Si surface barrier fragment detector. Using this method, fragments which enter the detector are stopped rapidly ($\sim 10^{-12}$ sec) and they produce sharp γ-ray lines having yields representative of the transition yield per fission. If the complementary fragment detector is close to the source-plated detector (as shown in Fig. 4), then transitions coming from fragments in flight are Doppler-shifted but the wide acceptance angle causes broadening to such a degree that these transitions substantially contribute only to the general background. Using this experimental configuration, specific level assignments have been made to all of the high-yield prompt even-even fission products (Cheifetz et al., 1970a, b; and Wilhelmy et al., 1970a).

The establishment of ground-state bands in even-even fission product nuclei has been possible for two main reasons. The first is that they are populated very strongly in the fission process. The separating fragments carry $\sim 7\hbar$ units of angular momentum, and for even-even fission products this angular momentum is primarily dissipated through a cascade of transitions in the ground-state band. It has been shown that the $2^+ \rightarrow 0^+$ transitions in even-even fission products represent 95–100% of the independent yields of these isotopes (Cheifetz et al., 1971b), which results in these transitions being the most intense that can be observed in fission. The second feature permitting the assignments is the tremendous regularity of ground-state band levels in even-even nuclei. Figure 5 is a so called "Mallmann" plot in which the energy of the spin-I level of the ground-state band relative to the energy of the 2^+ level is plotted against the energy ratio of the 4^+ and 2^+ levels. Presented as points on the plot are all known data as tabulated by Mariscotti et al. (1969) for even-even nuclei for which the ratio E_4/E_2 > 2.23. Also shown as open squares are the results of analysis of the prompt fission products. It is seen that they are very consistent with the smooth trends observed for other nuclei and this greatly enhances the confidence of the assignments. To further substantiate assignments, accurate atomic numbers were determined as described earlier using X-ray–γ-ray coincidence

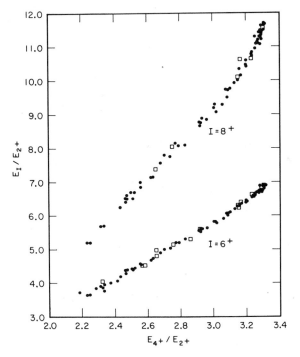

Fig. 5. Plot of the energy ratios $E8^+/E2^+$ and $E6^+/E4^+$ versus $E4^+/E2^+$ for ground-state band levels of even-even nuclei. The data presented as solid circles represent all of the previously known data on these isotopes as tabulated by Mariscotti *et al.* (1969). The data presented as open squares represent the recent experimental results obtained for even-even fission product nuclei (taken from Cheifetz *et al.*, 1970b).

measurements, and γ–γ coincidence studies yielded information on cascade relationships.

This success has not been extended, however, to the odd-A and odd-odd nuclei produced in fission. For many strong transitions, mass and charge determinations can be made, but so far there has not been enough correlated information obtained to make specific level assignments. The reason detailed spectroscopic properties have not been obtained for these nuclei is the absence of the two strong advantages which were present in the even-even nuclei. In the deexcitation of the odd-A and odd-odd fragments, the dissipation of the angular momentum can possibly occur through a number of different paths, whereas in the even-even products, the deexcitation path strongly favors cascading through the ground-state band. This fragmentation of the γ-rays leads to transitions having lower intensities, which makes them more difficult to observe. The second point is that the systematics' which are strongly present in even-even nuclei are to a large extent absent in the non-even-even isotopes. To build reliable level assignments in the odd-A nuclei, it is usually necessary to observe crossover and interband cascade transitions. These transitions may have low relative yields compared with more favorable decay branches. The complexity of the spectra usually limits the observable transition intensity range to approximately a factor of ten. That is, in general, transitions having a factor of ten less intensity than the strongest lines in the spectrum cannot be determined with a great enough accuracy to make positive isotopic assignments. Without substantial improvements in both fragment and photon detector resolution, it seems unlikely that substantial spectroscopic information will be obtained strictly from prompt γ-ray spectroscopy. Perhaps the best information will come from a combination of this technique with results obtained in post-beta-decay studies using on-line isotope separators and/or fast radiochemistry. The post beta decay and prompt deexcitation will usually have different level population patterns. Correlation of these two sets of data should be very helpful in studying the odd-A and odd-odd isotopes.

D. CONVERSION-ELECTRON EMISSION STUDIES

Conversion-electron emission studies have also contributed to our spectroscopic knowledge of prompt fission products. The identification of conversion-electron lines as a function of mass determination from kinetic energy measurements for ^{252}Cf has been carried out by the Berkeley group (Bowman et al., 1965; Watson, 1966; Watson et al., 1967, 1970b) and by Atneosen et al. (1966). Accurate Z determinations of conversion electrons have been made by Shapiro et al. (1971) using X-ray–conversion-electron studies. Conversion electrons from ^{235}U have been recently reported by Khan et al. (1973). In general these studies give complementary information to that obtained in γ-ray investigations of transition energies. The data are however, biased toward conditions which favor internal conversion. The gross spectrum of conversion electrons decreases almost exponentially with increasing energy (Atneosen et al., 1966). This strongly favors the study of low-energy transitions. Also, there is a favoring of isomeric transitions since these are cases which generally decay with a higher multipolarity and/or with lower energies. Both of these situations are conducive to higher conversion-electron yield.

There are two additional problems in these studies. The first is the necessity of eliminating photon background from the electron spectrum. This can be accomplished by recording the data once with no shielding between the detector and source (enabling both conversion electrons and photons to be detected) and once with thin shielding to absorb the electrons (enabling only detection of the photons). The electron spectrum is then determined by subtracting the latter spectrum from the former. In the studies by Watson et al. (1966, 1967, 1970b), the electrons were separated from the photons by using a magnetic steering device and detected in a shielded Si(Li) electron detector. The photon background problem is not as serious as that due to the relatively poor resolution for the electron detection systems. This is partially due to the inherently poorer resolution for the detection of electrons caused by surface dead-layer effects and partially due to Doppler effects caused from observing electrons from fragments in motion. The Doppler shifts are dependent on the transition energy and therefore the observed resolution is also energy dependent. To eliminate the Doppler shift problem, Shapiro et al. (1971) observed electrons emitted from fission products stopped in thin plastic scintillators. Such a procedure can only be a partial solution since the effective electron resolution in this case is strongly affected by energy losses and dispersions introduced as the electrons pass through the fragment-stopping material. This approach is limited to studies of transitions greater than 100 keV.

One important quantity which can be derived from conversion-electron studies is the multipolarity of the transitions. From observed intensities of K and L conversion-electron lines, Watson and Khan et al. have been able to assign several transitions as having E2 multipolarity and one having M1 multipolarity. These determinations are especially important for confirmation of the $2^+ \rightarrow 0^+$ transition assignments made in even-even fission product nuclei. Of the E2 γ rays, twelve have been consistently established as being $2^+ \rightarrow 0^+$ transitions.

As the electron-measuring experiments are improved, it will be possible to ratio the measured K-electron absolute intensities with the tabulated γ-ray intensities to obtain more information on transition multipolarities. This should be very helpful in determining nuclear structure properties in odd-A and odd-odd nuclei.

E. TIME DISTRIBUTION OF GAMMA RAYS

Experiments measuring the time distribution of gross radiation from fission (Johansson and Kleinheinz, 1965; Johansson, 1964, 1965; Skarsvåg, 1970) have shown that most of the radiation is emitted in times on the order of 10^{-11} sec after fission. The average energy per photon is ~ 0.5–1.0 MeV and angular correlation and angular momentum studies have shown them to be predominantly E2 in multipolarity. This implies that the transitions are collective (10–100 single-particle units) which is consistent with the assumptions of cascading band transitions. Conventional electronic timing techniques have shown that less than 10% of the yield of γ rays have lifetimes in the 10^{-9}–10^{-5}-sec range. Even though this is a small portion of the total radiation spectrum, it is the most amenable to spectroscopic analysis.

By shielding the fission source from the γ-ray detector and observing only transitions having $t_{1/2} \gtrsim 10^{-9}$ sec, a significant simplification in the spectra can be obtained. This is due to the avoidance of the bulk of the transitions which have lifetimes shorter than this and also to observing only transitions from stopped fragments, which eliminates Doppler shift problems. Using the experimental configuration indicated in Fig. 6, John *et al.* (1970) were able to tabulate the γ-ray energies, lifetimes (3 nsec $\leqslant t_{1/2} \leqslant 2000$ nsec), and masses (from fragment kinetic energies) for 144 transitions. Many of these assignments have been confirmed by similar studies by Ajitanand (1971)

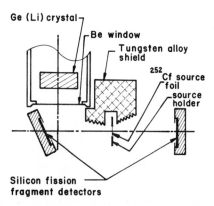

Ge (Li) crystal
Be window
Tungsten alloy shield
^{252}Cf source foil
source holder
Silicon fission fragment detectors

Fig. 6. A schematic representation of an experimental configuration used for studying isomeric transitions from isotopes formed in the spontaneous fission of ^{252}Cf. The photon detector is shielded from the source to minimize prompt radiation detection. The flight time for fission products to reach the desired γ-ray detection position (the fragment detector on the left) is ~ 3 nsec (from John *et al.*, 1970).

and by Clark *et al.* (1973). Substantial groupings of transitions have been found having the same lifetime and being associated with the same mass, thus indicating a cascade of prompt transitions following deexcitation of an isomeric level.

This information has been used to determine definitive level structure information on the even-even nuclei ^{134}Te and ^{136}Xe. Both of these nuclei have $N = 82$ closed neutron shells and presumably ground-state proton configurations predominantly made out of couplings of particles in the $g_{(7/2)}$ proton level. Such near-doubly-magic nuclei should have a low-lying level structure which favors an isomeric $[g_{(7/2)}]^2$ 6^+ state. Such a level is, from fission angular momentum considerations, populated strongly in the prompt deexcitation. Observation of these predicted levels has given some of the first concrete information on level structures for nuclei near the doubly magic ^{132}Sn nucleus.

Jared *et al.* (1973), using a modification of the recoil technique, have measured lifetimes in the 0.05–7 nsec range for several transitions deexciting prompt spontaneous fission products of ^{252}Cf. The Cf source was placed on an arm whose distance from a solid state fission detector was controlled with a precision micrometer. Intensities of non-Doppler-shifted γ rays were measured at six source positions. This information, when coupled with the knowledge of the kinetic energies of the specific fragments, leads to an accurate determination of the transition lifetimes.

F. Angular Distribution of Specific Gamma Rays

Angular distribution studies of prompt gamma transitions give information on the alignment and magnitude of the angular momentum in fission and, if performed on specific γ rays, give spectroscopic information on the multipolarity of the transition. Various experiments have demonstrated that the fragment angular momentum is preferentially aligned in a plane perpendicular to the axis of the separating fission fragments. This situation is somewhat analogous to that encountered in charged-particle-induced reactions, in which the angular momentum is found to be aligned in a plane perpendicular to the beam axis. By measuring the intensity of the transition relative to the axis of alignment and knowing something about the spin history of the deexcitation, it is possible to make multipolarity assignments.

As always, the situation is more difficult with fission, due to spectral complexity and Doppler shift effects. However, the high-resolution γ-ray detectors have made such studies feasible for some of the stronger transitions seen in fission. Figure 7 shows the experimentally determined angular distri-

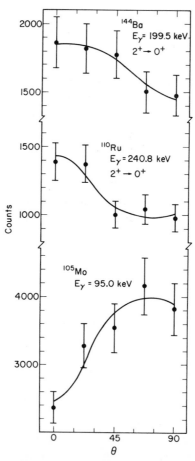

Fig. 7. Angular distribution of three prompt gamma rays relative to the fission fragment separation axis. The solid lines represent a least-squares fit of the data to the angular distribution expression $W(\theta) \propto 1 + a_2 P_2(\cos\theta) + a_4 P_4(\cos\theta)$ (from Wilhelmy *et al.*, 1972).

butions of three prompt fission transitions. The top two transitions are $2^+ \rightarrow 0^+$ transitions in even-even nuclei and these are seen to be forward peaked (i.e., Intensity 0°/Intensity 90° > 1). This is what would be expected for a stretched E2 transition. The bottom angular distribution is for a transition from the odd-A isotope ^{105}Mo and is seen to be peaked around 90°. Though the spin and parity of the emitting and residual levels are not known in this case, the observed angular distribution is consistent with this being a stretched, predominantly M1, transition deexciting a collective band.

Quantitative interpretation of the magnitude of the anisotropy will be dependent on possible nuclear reorientation effects due to magnetic field perturbations present in the stopping material. Such effects have not been corrected for in these experiments, but only have the ability to reduce the anisotropy and not to change the peaking angle of the distributions. Again, as more detailed properties become known, the angular distribution studies will become more valuable in making accurate nuclear structure assignments.

III. Post-Beta-Decay Spectroscopy

The fission process has always been an important source of radioactive nuclei for spectroscopic studies. Much of the understanding of nuclear level structure has come from analysis of radiation spectra following beta decay of isotopes produced in fission. Our emphasis, however, will not be on the more standard aspects of studying radioactive isotopic decay, but on those aspects which are still intimately related to the fission process as the source of the isotopes. In practice, this means analysis of the abundant short-lived activities produced as primary fission products. In these studies, as in all fission radiation studies, the most important objective is the separation of the desired radiation or isotope from a multitude of others. Since the high-yield fission products typically have half-lives on the order of 1 sec, this requires rapid and/or continuous separation techniques. We look at three of the methods employed: (a) the analysis of mass and chemically unseparated fission products, (b) rapid radiochemical separations, and (c) on-line mass separators. This section is concluded by briefly comparing the types of results obtained from prompt fission radiation analysis and those obtained in beta-decay studies. These comparisons are made to demonstrate the complementary nature of the two methods in giving nuclear structure information.

A. Unseparated Fission Products

The development of the high-resolution Si(Li) and Ge(Li) photon detectors has permitted meaningful analysis of X-ray and γ-ray spectra from unseparated fission products. One aspect of these studies has been the development of their analytic capabilities for determining yields of fission-produced isotopes (Gordon et al., 1966). If the gross products are studied at times $\gtrsim 1$ hr after fission, there is an appreciable reduction of spectral complexity. Many mass chains have decayed to stability and those that have not reached stability usually contain isotopes whose decay properties have been well studied. Therefore, by measuring the intensities of well known γ-ray transitions, it is possible to determine the yields of a large number of isotopes.

Our primary interest is not in the development of these analytic aspects but in those which can contribute new nuclear structure information to less studied isotopes. Using a bent crystal spectrometer to observe X rays and γ rays from a ^{235}U source irradiated in a nuclear reactor, John et al. (1967) and John (1971) were able to tabulate 257 transition energies from 30 to 353 keV. No fission coincidence separation was performed in the experiment so both prompt and post-beta-decay transitions were recorded. The value of these measurements lies in their accurate transition energy determinations. Experimental uncertainties ranged from ~ 1 eV at the lower energies to ~ 100 eV for the weaker, higher-energy transitions. This accuracy may be important for positive assignments of cascading and crossover transitions de-exciting the isotopes.

Another experiment used to study mass or chemically unseparated fission products was performed by Wilhelmy (1969) and Wilhelmy et al. (1970b). In these studies, conventional Ge(Li) detectors were used for the γ-ray measurement. Some "physical" separations were obtained by recording X-ray–γ-ray coincidence measurements to determine the charges with which the γ rays were associated. Half-life determinations were obtained by observing γ-ray intensities from fission products which were embedded in a variable-speed moving-belt system. In these studies, half-lives (as short as 0.065 sec) and yields were determined for 617 individual γ rays having energies between 40 and 720 keV. This experiment gave information on the shortest-lived beta-decay products produced in fission, independent of their chemical properties. Since no mass or chemical separations were performed, accurate information was obtained on the yield of the transition per fission without, as in the other methods, having to know the efficiency of the sep-aration technique. However, the measurements of transition intensities and beta-decay half-lives do not give nuclear structure information. In general, they do not even uniquely identify the isotope from which the transition is emitted. If something is known about the nuclear systematics, it is possible to obtain more information. Since ground-state band data on even-even nuclei present the most regular systematics, most of the information from these studies has been for the beta decay of odd-odd parents. Half-lives for 33 odd-odd nuclei ranging from ^{100}Nb to ^{154}Pm were identified by quanti-tatively observing $2^+ \to 0^+$ transitions in the resultant even-even daughter nuclides. Of this number, 17 related to isotopes which had not previously been identified. In addition to these odd-odd nuclei, some 15 other new isotopes were identified from beta-decay systematics using the X-ray–γ-ray coincidence data to establish the atomic number and/or genetic decay relations to identify grandparent activities when the properties of the parents were known.

B.   RAPID RADIOCHEMICAL SEPARATION

The study of fission product radiation has primarily been accomplished by the radiochemical isolation of desired isotopes. In the evolution of these analyses, much sophistication has been introduced in the development of rapid, many-step chemical separations. Now that there is much current interest in the developments of isotopic isolation based on prompt radiation studies and on-line mass separators, it is only fair to point out that much of the most detailed information on fission product radiation has come, and is still coming, from radiochemical analysis. It is not our purpose to go into the specific details necessary for the chemical separation steps. Clearly each element requires its own set of procedures. We will only mention some of the techniques employed and the types of results they yield. For a detailed survey of rapid radiochemical techniques, the review article by Herrmann and Denschlag (1969) should be consulted.

As we have previously stated, if the highest-yield fission products are to be studied, it is necessary to have separation techniques which can be performed in times on the order of one to several seconds. The simplest and one of the most successful techniques is one in which no chemical separation need be performed, i.e., one in which physical properties can be used for the desired isotopic isolation. The elements readily accessible to this method are the two inert gases produced in fission—Kr and Xe. It has been long known that these gases emanate rapidly out of uranyl stearate compounds and can be rapidly isolated by using a gas-flow-sweeping procedure. Half-lives and spectroscopic properties of the decaying Kr and Xe isotopes and their genetic decay products have been studied in this way (Wahlgren and Meinke, 1962; Wahl et al., 1962; Patzelt and Herrmann, 1965).

Another very rapid technique for isolating short-lived products can be performed for the fission-produced halogen isotopes Br and I. By allowing fission products to recoil into methane, high-yield hot atom exchange reactions occur of the type

$$Br^* + CH_4 \rightarrow CH_3Br + H\cdot$$

The methyl halides produced by this method can be rapidly transported to a counting position for γ-ray and/or delayed neutron measurements (Silbert and Tomlinson, 1966; Schüssler *et al.*, 1969; Herrmann *et al.*, 1970). With this technique, the Mainz group has been able to identify bromine isotopes as heavy as ^{92}Br ($t_{1/2} = 0.25 \pm 0.07$ sec) and iodine isotopes as heavy as ^{141}I ($t_{1/2} = 0.45 \pm 0.10$ sec) (Kratz, 1970).

Where physical methods can be applied, great improvement can be obtained in the ease and rate at which separations can be performed. However, for the majority of the elements produced in fission, no such methods

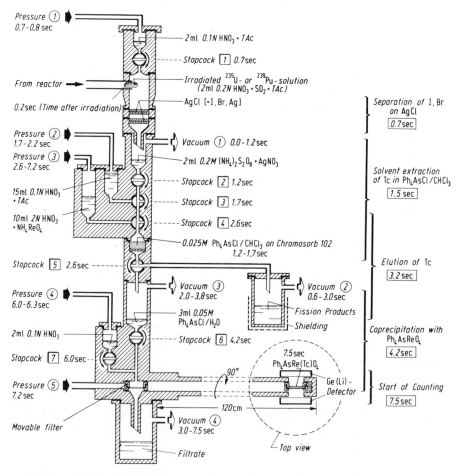

Fig. 8. Experimental chemical procedure used to rapidly isolate Tc fission products (from Trautmann *et al.*, 1972).

are available and therefore it is necessary to rely on chemical procedures. The specific procedure used, of course, depends on the element to be studied. There has been a continuing effort in the development of rapid chemical procedures to allow the study of the shortest-lived products. One example will be used to demonstrate the type of procedures which are being developed. The isolation of the element Tc requires a multistep chemical separation. Trautmann *et al.* (1972) have perfected a procedure in which a neutron-

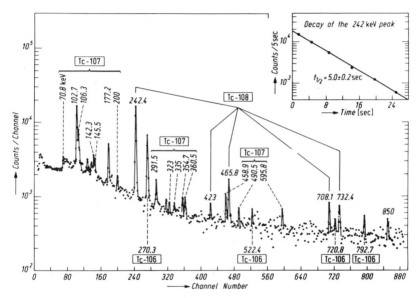

Fig. 9. Gamma-ray spectrum obtained for short-lived isotopes following chemical separation of Tc fission products. Transitions from three separate decays are observed; 36-sec ^{106}Tc, 21-sec ^{107}Tc and 5-sec ^{108}Tc (from Trautmann *et al.*, 1972).

irradiated sample of ^{239}Pu is pneumatically transported to a vessel where several steps of washing, filtering, solvent extracting, and precipitating are performed to isolate the desired technetium isotopes (Fig. 8). The total elapsed time for this procedure is ~ 7.5 sec and gives a 30–40% yield of chemically well-separated technetium isotopes. A portion of the γ-ray spectrum obtained from this method is shown in Fig. 9. Detailed decay information is obtained on the decay of technetium isotopes as heavy as ^{108}Tc ($t_{1/2} = 5.0 \pm 0.2$ sec). Such studies are being applied to a wide variety of elements and are resulting in a substantial increase in our knowledge of the nuclear properties of these very neutron-rich fission product isotopes.

C.   Use of On-Line Mass Separators[†]

On-line mass separators have contributed a great deal to the spectroscopy of fission products by providing a very effective rapid separation method. The principle of operation of an on-line mass separator is to stop the fragments in gas or solid matter and then ionize them in an ion source to charge $+1$. The ionized products are accelerated in an electric field of 50–100 kV and analyzed by a magnet. In these systems, absolute identification of the masses is obtained and the high resolving power of $A/\Delta A \sim 1000$ is such that even isotopes which are produced in low yields can be identified clearly. The time that elapses from the occurrence of the fission until the desired isotope is brought to the collector depends very critically on the transport time from the stopping location to the ion source. In systems which are operating at present, this time is as short as about 0.3 sec. Future systems will perhaps be able to produce separated products in about 1 msec by having the fragments stop in a flow of gas with rapid ionization following that. Transit times of this order permit the study of radiations following beta decay of the prompt products and their daughters and perhaps even radiations from some of the longest-lived prompt isomeric levels. The transport and ionization properties of the products are linked to their chemical nature; thus the systems are selective with respect to atomic numbers, with certain elements such as noble gases being easy to transport and other such as alkalis being easy to ionize. Chemical separation can be made in principle during the passage of the product to the ion source or through specific ionization techniques; nevertheless, great difficulties are encountered with analyzing certain fission products such as Y, Zr, Nb, Mo, and Ru in these systems.

There are two general types of on-line isotope separators that are currently being used: (a) The Isolde type, which was developed at CERN and described by Kjelberg and Rudstam (1970). It has a wide mass range ($\pm 15\%$ of the central mass) and employs fringe-field vertical focusing to give spot focusing. (b) The Scandinavian-type separator, which has a 90° sector magnet generally with no magnetic vertical focusing. The mass range at the collector is usually $\pm 8\%$.

In these separators, the masses are usually separated by about 1 cm at the collector and a slit system is employed to select a specific mass. Some separator systems employ secondary beam handling beyond the

[†] This subject is covered in detail by Klapisch, Chapter II.B, and will only be briefly covered here.

focal plane of the separator to allow several detection stations and to remove the detection area from the high background radiation at the collector.

Another type of separator being actively used is the "gas-filled mass separator." These separators isolate the products in about 1 μsec from the time of fission so that, in addition to their performance as mass analyzers for post-beta-decay studies, some isomeric transitions from the prompt products can be identified. This application was first proposed by Fulmer and Cohen (1958) and was later developed and used successfully by the group at the Jülich reactor in West Germany (Armbruster et al., 1967, 1969, 1971). In this technique the fission fragments are emitted with their initial velocities of about 10^9 cm/sec into a magnetic analyzer filled with gas at low pressure, which means they undergo many atomic collisions in their path. The mean value of the ionic charge of a fragment in the gas has an almost linear relationship with its velocity; thus the mean trajectory in the magnet depends only on the fragment mass and is independent of its velocity. In practice, a specific trajectory is obtained for a definite curve in the (Z, A) plane for any value of the magnetic field for a wide range of fragment velocities.

The resolving powers of these separators are determined by several factors, of which the charge changing statistics and the scattering in the gas are very important. To these are added effects such as the velocity dispersion and the source size. The resolution depends also on the luminosity of the separator (luminosity = source area × transmission) and is in principle not as high as that of conventional mass separators. In operation, for example, with a luminosity of 2×10^{-5} cm^2, a resolution of 4.2% $[\Delta(B\varrho)/B\varrho]$ was obtained for studies of ^{98}Zr (Grüter et al., 1970). This resolution is therefore comparable to that obtained in direct fragment counting experiments. Its advantage, however, is in the selectivity of a specific mass region and the elimination of interfering background radiation from the collector. This aspect is especially useful for the study of microsecond isomeric states which are produced in the fragments during the prompt deexcitation and are not formed by beta decay. Another important feature of this separator is that no chemical selection occurs in the analysis, thus products which are very hard to ionize in ion sources such as Y, Zr, Nb, Mo, Tc, and Ru can be studied.

The contributions of both types of spectrometers have been and will continue to be very important in the detailed analysis of isotopes produced in fission.

D. COMPARISON OF STATES POPULATED AFTER BETA DECAY WITH THOSE
   POPULATED BY PROMPT DEEXCITATION

The field of fission product spectroscopy is in a state of rapid evolution
with regard to both prompt and post-beta-decay studies. Complete detailed
comparisons between levels populated by the two methods are not possible
at this time but continued improvements in procedures will permit more
comparison in the near future. The possibility of good overlap in the data
is increased if high-fission-yield isotopes can be studied in both cases.
The difficulty in achieving this is due in part to the short half-lives of prompt
products and in part to the relatively small isotopic dispersion of any
isobaric chain of fission products [$\sigma = 0.59$ atomic charge units (Norris
and Wahl, 1966)]. The short half-lives of the high-yield fission products make
their isolation difficult and have limited the completeness of the available
decay studies. The small isotopic dispersion has two undesirable effects. For
the beta-decay studies to give information on the highest-yield prompt
products, it is necessary to study the decay of their even more neutron-rich
parents. These are one charge unit further removed from the most probable
isotope production region and thus have smaller initial yields and shorter
half-lives, both of which increase the difficulty of their study. For the prompt
product measurements, the narrow mass dispersion limits the number of
isotopes on which information is obtained. Since the sensitivity of the prompt
studies is very poor compared to most of the post-beta-decay studies, it is
only possible to obtain data for the higher-yield isotopes. Those isotopes
which are very far removed from the most probable ones are not easily
amenable to specific level assignments.

The other limitation imposed by the prompt studies is that essentially all
of the detailed information currently available is on even-even isotopes.
Thus, for a comparison to be performed for these cases, it is necessary to
study the beta decay of the odd-odd parents.

As an example of these comparisons, we will use the levels in ^{142}Ba. A
partial decay scheme has been determined by Larsen et al. (1971) following
beta decay of ^{142}Cs. These studies are part of the analysis of the decay chain
for isobars of $A = 142$ which have been isolated using an on-line mass
separator to extract the gaseous ^{142}Xe produced in the thermal neutron
fission of ^{235}U. The levels in ^{142}Ba are populated following the beta decay
of ^{142}Cs, which has a half-life of 1.67 sec, which is in turn produced through
the decay of the 1.24 sec ^{142}Xe. Using a moving-tape collector system to
separate transitions from these two isotopes of similar half-life and measuring

γ rays in singles and coincidence modes, Larsen *et al.* (1971) were able to identify 144 transitions following the beta decay of ^{142}Cs. These transitions had energies from 140 to 5394 keV. From sums and differences of energies and γ–γ coincidences, they were able to assign only 15 of these lines to nine excited levels. Their tentative partial decay scheme is shown in Fig. 10a. They were not able to make spin and parity assignments to many of the levels. The spin of the decaying ^{142}Cs is apparently low and therefore only

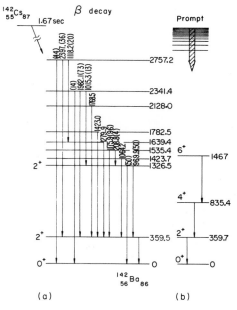

Fig. 10. A comparison of assignable transitions observed in ^{142}Ba following (a) beta decay of 1.67-sec ^{142}Cs (Larsen *et al.*, 1971) and (b) prompt deexcitation of ^{252}Cf fission products (Wilhelmy *et al.*, 1970a).

low-spin states are appreciably populated in the decay. Nuclei in this region are in a transitional area between the spherical nuclei close to the ^{132}Sn core and the permanently deformed isotopes in the rare earth region. It is therefore desirable to have information on the collective higher angular momentum levels so that a more detailed understanding of this transition phase can be obtained. From analysis of the prompt data, Wilhelmy *et al.* (1970a) were able to establish the 2^+, 4^+, and 6^+ members of the ground-state band of ^{142}Ba (shown in Fig. 10b). The dissipation of the angular momentum of the primary fragments favors the observation of the higher-spin members of the ground-state band. However, in the prompt analysis, only three levels have been established and no information is available on intrinsic levels or other collective states except those in the ground-state band.

In the study by Larsen *et al.* (1971) of the beta decay of ^{142}Cs, one of the weaker transitions they observed was of 475.0 keV. There was some indication that this γ ray was in coincidence with the 359.5-keV $2^+ \rightarrow 0^+$ transition. They speculated that this γ ray might be the $4^+ \rightarrow 2^+$ transition but did not have sufficient data to establish the level. In the prompt studies, the $4^+ \rightarrow 2^+$ transition was assigned as being 475.5 keV. This then complements the beta-decay studies in confirming this important level. With this established, the intensity of the transition observed in the post-beta-decay studies then gives information on the beta-decay branching strength, which in turn gives additional information on the spin of the decaying ^{142}Cs.

IV. Nuclear Properties of Fission Products

A. SYSTEMATIC TRENDS OF LEVELS OF EVEN-EVEN PRODUCTS

One of the most successful aspects of fission product spectroscopy has been the systematic information obtained on the energy levels of the even-even products. A good portion of this information comes from the study of the prompt deexcitation of the products. Many of these data have been confirmed and additional even-even isotopes identified from radiochemical or on-line mass separated studies. Data concerning low-lying states of even-even fission products are summarized in Table 2.

The table contains the energies of the $2^+ \rightarrow 0^+$, $4^+ \rightarrow 2^+$, $6^+ \rightarrow 4^+$, and $8^+ \rightarrow 6^+$ ground-state band transitions and the first excited 0^+ (denoted $0'^+$) and the second excited 2^+ (denoted $2'^+$) levels. The half-life of the first 2^+ states are given when known. The prompt intensities and the yield of the $2^+ \rightarrow 0^+$ transitions associated with the beta decay of the parent isotope in spontaneous fission of ^{252}Cf are also given as well as the half-life of the beta-decaying parent.

The information of Table 2 can be compared with the enormous amount of data on ground-state bands of isotopes produced mostly by (particle, xn) reactions (see Newton, Chapter VII.E). A partial list of such data, mostly for neutron-deficient and stable isotopes, has been compiled by Mariscotti *et al.* (1969). The fission products represent an addition of over 40 isotopes to that list. A regular behavior for levels of most ground-state bands is found. An illustration of this phenomenon in the even isotopes of $_{58}$Ce is

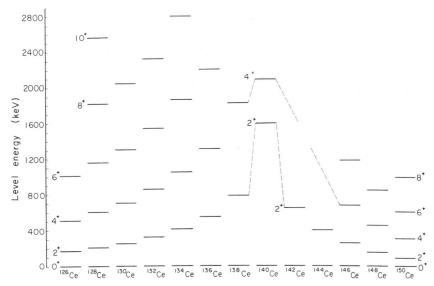

Fig. 11. Ground-state band energy levels in the even cerium isotopes (from Diamond, 1970).

shown in Fig. 11. The states of the neutron-deficient isotopes were observed in heavy-ion reactions and reported by Diamond (1970) and the states of isotopes with $A \geqslant 142$ were found in fission product studies (see Table 2). At each end of the mass scale, a region of deformed nuclei appears and in the center there is a definite increase in the height of the levels due to singly magic ^{140}Ce $(N = 82)$.

In most regions, the energies of the 2^+ state change smoothly both along constant-Z and constant-N lines. This feature can be used to predict in a rough manner the locations of the lowest 2^+ states of nearby even-even nuclei. A few rare but important exceptions will be discussed later.

A quantitative systematic behavior appears in the ratio of the energies of the levels of the ground-state band. The Mallmann (1959) plots show a very smooth trend for the ratio of $E(I^+)/E(2^+)$ when plotted as a function of $E(4^+)/E(2^+)$, where $E(I^+)$ is the energy for spin level I in the ground-state band (Fig. 5). These empirical systematics have been successfully interpreted using a variety of two-parameter theoretical models (Diamond *et al.*, 1964; Harris 1965; Mariscotti *et al.*, 1969). More recently, Ma and Rasmussen (1971) have shown that these two-parameter models can be presented on a more microscopic basis. In their calculations, they have included effects of

TABLE 2
Properties of Even-Even Fission Products

Isotope	$2^+ \rightarrow 0^+$ E (keV)	$2^+ \rightarrow 0^+$ $t_{1/2}$ (nsec)	$4^+ \rightarrow 2^+$ E (keV)	$6^+ \rightarrow 4^+$ E (keV)	$8^+ \rightarrow 6^+$ E (keV)	2^+ E (keV)	0^+ E (keV)	2^+ Yield from ^{252}Cf (%/fission) Prompt	Beta decay	$t_{1/2}$ of Parent (sec)	Ref.
^{88}Kr	(869)	—	—	—	—	—	—	—	—	—	a
^{90}Kr	(706)	—	—	—	—	—	—	—	—	—	a
^{92}Kr	(956)	—	—	—	—	—	—	—	—	—	a
^{90}Sr	831	—	—	—	—	—	—	—	—	156;258	b,c,d
^{92}Sr	813	—	—	—	—	—	—	—	—	4.4	b,c,d
^{94}Sr	835	—	—	—	—	—	—	0.51	—	2.7	b,c,d
^{96}Sr	813	—	—	—	—	—	—	0.34	—	0.23	b,c,d
^{98}Sr	(193)	—	—	—	—	—	—	—	—	—	w
^{98}Zr	1223	—	—	—	—	—	854	~0.3	—	—	c,e,f
^{100}Zr	212.7	0.71	352.1	497.9	—	—	—	1.80	0.3	0.85	c,g,v
^{102}Zr	151.9	2.11	326.6	486	(587)	—	—	1.43	(0.09)	(0.06)	c,g,v
^{102}Mo	296	—	—	—	—	—	698	0.46	2.6	3.0	c,g,h,k
^{104}Mo	192.3	0.91	368.7	520	—	—	—	3.37	1.2	1.0	c,g,v
^{106}Mo	171.7	1.25	350.8	(511.8)	—	—	—	3.37	0.23	1.1	c,g,v
^{108}Mo	(158)	—	—	—	—	—	—	—	—	—	c
^{106}Ru	270	—	444	—	—	793	991	0.16	3.5	36.	c,g,i,j
^{108}Ru	242.3	0.34	423	—	—	—	—	1.94	2.6	5.2	c,g,v
^{110}Ru	240.8	0.34	423	576.1	(708)	—	—	3.49	0.6	0.8	c,g,v
^{112}Ru	236.8	0.32	408.9	—	—	—	—	0.97	—	—	c,v
^{112}Pd	348.8	—	535.8	(644)	—	—	—	0.77	3.4	4.7	c,g,k
^{114}Pd	332.9	0.20	520.7	649.3	—	—	—	1.48	0.8	1.7	c,g,v
^{116}Pd	340.6	0.11	538.0	—	—	—	—	0.87	0.07	0.6	c,g,v
^{114}Cd	558.3	—	724.4	—	—	1208	1134.2	—	1.2	<10	g,l
^{116}Cd	513.5	—	705.3	—	—	1212.8	1154.0	—	1.1	126	g,l

Nuclide											Ref.
118Cd	487.1	—	677.3	(805)	—	1269	—	0.32	0.4	3.5	c,g,l
120Cd	505.5	—	(697.6)	—	—	1324	—	~0.3	—	1.3	c,l
122Cd	(569)	—	—	—	—	—	—	—	—	—	l
132Te	974.6	—	697.1	103.3	—	—	—	>0.2	—	—	m,x
134Te	1278	—	297	115	—	—	—	>1.5	—	11	c,n,s
136Xe	1313.3	—	381.5	197.5	—	(~2640)	—	>0.75	—	83	g,n,p,q
138Xe	589.5	—	482	—	—	—	—	2.3	1.0	6.3	c,g
140Xe	376.8	—	457.9	—	—	—	—	1.5	0.14	0.7	c,g
140Ba	602.2	0.07	—	—	—	—	—	0.52	1.5	66	c,g,q,r
142Ba	359.7	0.70	475.9	632	—	(1326.5)	—	2.90	1.35	1.67	c,g,q,s,v
144Ba	199.4	0.86	331.0	431.7	510.8	—	—	3.60	0.3	1.0	c,g,v
146Ba	181.0	—	333	—	—	—	—	1.01	—	0.2	c,t,v
142Ce	641.2	—	578.1	—	—	1536.1	—	—	3.8	555	g,r,s
144Ce	397.5	—	—	—	—	—	—	0.2	5.5	40	c,g
146Ce	258.6	0.26	410.1	502.3	—	—	—	1.04	2.5	8.3	c,g,v
148Ce	158.7	1.06	295.7	386.5	—	—	—	2.31	0.45	1.3	c,g,v
150Ce	97.1	3.60	209.0	300.7	376.4	—	—	>0.98	—	—	c,v
152Nd	75.9	>2	164.7	247.3	322.1	—	—	>0.6	—	—	c
154Nd	72.8	7.7	162.4	243.7	328.1	—	—	>0.4	—	—	c,v
156Sm	76.0	>2	174.2	258	—	—	—	>0.1	—	—	c,u
158Sm	72.8	>2	167.5	258.2	346	—	—	>0.15	—	—	c

References:

a Horsch (1970)
b Macias-Marques et al. (1971)
c Cheifetz et al. (1970a, b, 1971b)
d Amarel et al. (1967)
e Blair et al. (1969)
f Fogelberg (1971)
g Wilhelmy (1969), Wilhelmy et al. (1970b)
h Trautmann et al. (1972)
i Herrmann et al (1970)
j Casten et al. (1970)
k Casten et al. (1972)
l Bäcklin et al. (1970)
m Kerek et al. (1970)
n John et al. (1970)
o Holm et al. (1970)
p Monnand et al. (1970)
q Lederer et al. (1967)
r Alväger et al. (1968)
s Larsen et al. (1971)
t Tracy et al. (1971)
u Bjerrgaard et al. (1966)
v Jared et al. (1973)
w Khan et al. (1973)
x McDonald and Kerek (1973)

centrifugal stretching and neutron and proton Coriolis antipairing as well as second- and fourth-order cranking effects. They were able to show that the "two-parameter" models can be derived from this basis.

These models are powerful tools in fission product studies since they provide predictions regarding energy locations and transition lifetimes for levels in the ground-state band. Establishing experimental states which satisfy the systematics greatly increases the confidence for level assignments in even-even fission product nuclei.

B.   THE REGION OF DEFORMATION AROUND $A \approx 100$

When isotopes are presented in a chart of nuclides as a function of neutron and proton number, the regions of deformation occur in the center of the rectangles defined by lines joining the magic numbers. If this is to be used as a criterion for deformation, then many deformation regions should exist. The nuclei around $A = 25$, the rare earth, and the actinide isotopes have established stable deformation properties. The reasons those regions have been identified is that they lie along the valley of beta stability and are therefore relatively accessible. As experimental techniques are developed, other possible permanently deformed areas of nuclei will be produced and studied. One of these is a region populated by light fission products where there is a substantial yield of isotopes lying within the rectangle defined by $28 < Z < 50$ and $50 < N < 82$. Information has been limited in this area because of experimental difficulties associated with identifying properties of the short-lived fission products. Studies by Johansson (1965) and Zicha et al. (1969) indicated a large deformation in the neutron-rich $_{44}$Ru isotopes but these results were not confirmed by other experiments.

A general survey of this region showing rotational-like spectra in the even-even products emerged from the studies of prompt ^{252}Cf fission products by Watson et al. (1970b) and Cheifetz et al. (1970a, b). An example of the behavior of ground-state band levels for light even-even fission product nuclei is shown in Fig. 12. The energies of the lowest 2^+ states decrease for neutron numbers greater than $N = 56$. In $_{46}$Pd and $_{44}$Ru, this occurs in a very smooth manner with the exception that the 2^+ state in ^{116}Pd is slightly higher than the 2^+ state in ^{114}Pd, probably because of the approach to the $N = 82$ shell. More abrupt decreases in $E2^+$ are seen in $_{42}$Mo and $_{40}$Zr isotopes. The change in the lowest 2^+ from ^{98}Zr ($E2^+ = 1223$ keV) to ^{100}Zr ($E2^+ = 212.7$ keV) is the most abrupt change for all the known adjacent even-even isotopes. The ratios $E4^+/E2^+$ follow the behavior of the $E2^+$ value since they also change smoothly in Pd and Ru nuclei and more

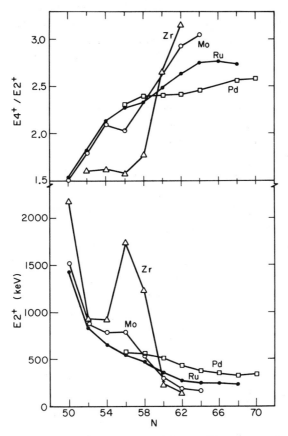

Fig. 12. Energies of the lowest 2^+ levels and the $E4^+/E2^+$ energy ratios for the heavier Zr–Pd isotopes (from Cheifetz *et al.*, 1970b).

abruptly in isotopes of Mo and Zr. The ratios for ^{102}Zr (3.15) and ^{106}Mo (3.06) are approaching the theoretical rotor limit of 3.33.

Additional information concerning the deformation of these light fission-product nuclei has been obtained from transition lifetime measurements. The experimentally determined reduced transition probabilities are related in a model-dependent manner to the nuclear quadrupole moment and the deformation (Stelson and Grodzins, 1965). Extracting these quantities for the light-fission-product region indicates that ^{102}Zr has the largest deformation. It has a B(E2) 91 times the single-particle value and a calculated deformation of $\beta = 0.38$ (Jared *et al.*, 1973). The other new isotopes in the light-fission-product region have smaller values for their deformation parameter than ^{102}Zr but even they have, in several instances, values that are equal to or larger than those found in the rare earth and actinide regions, and in all cases the values are larger than those typically found in spherical nuclei near closed shells.

The abrupt change in $E2^+$, $E4^+/E2^+$, and the B(E2) values between ^{98}Zr and ^{100}Zr is much sharper than the corresponding change between ^{150}Sm and ^{152}Sm, which shows the well-known discontinuity for $N = 88$ and $N = 90$ isotopes. This indicates perhaps a phase change which could be correlated with an abrupt change in the ground-state masses of these nuclei.

Theoretical calculations regarding deformations in the light-fission-product region have been carried out by Arseniev et al. (1969) prior to the publications of detailed experimental results. The authors used the Mottelson and Nilsson (1959) method as modified by Bes and Szymanski (1961) for calculating equilibrium deformations. In this method, the deformation-dependent single-particle energies, with pairing and Coulomb effects included, are summed to give the potential energy as a function of deformation. The equilibrium deformation is assumed to occur at the location of the potential energy minimum. The calculations predict potential energy minima for both prolate and oblate deformation, with the oblate being somewhat deeper. However, the magnitude of either the prolate or the oblate deformation is smaller than that experimentally determined from transition probability measurements. For ^{102}Zr, the calculation gives $\beta_2 = -0.29$ (the model-dependent value calculated from the experimental data is $|\beta_2| = 0.38$) and the minimum in the potential energy is 2.6 MeV below the energy of the spherical configuration.

Ragnarsson (1970) and Ragnarsson and Nilsson (1971) have also performed detailed calculations of the properties of the light fission-product nuclei, using the Strutinskii normalization method. This method accounts for the total energy of the nucleus by considering both the liquid drop energies of the deforming nucleus and the oscillating nature of the shell correction (Strutinskii, 1967). For small deformations, this method gives essentially the same results as the method mentioned earlier. Several of the parameters that affect the results were investigated, in particular two different sets of the potential energy parameters κ (spin–orbit coupling term) and μ (anharmonic correction term) were tried. The results are presented in terms of the deformation coordinates ε and ε_4 and again predict minima for both prolate and oblate deformations. A summary of the results is shown in Fig. 13. The calculations agree with the general trends of the

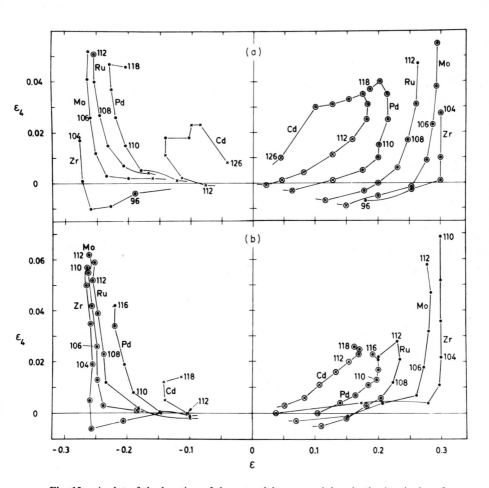

Fig. 13. A plot of the location of the potential energy minima in the $(\varepsilon, \varepsilon_4)$ plane for isotopes of $Z = 40, 42, 44, 46,$ and 48. Each nucleus has a calculated potential energy minimum at both prolate $(+\varepsilon)$ and oblate $(-\varepsilon)$ deformations. The more stable of these minima for each nucleus is marked with a circle. In the cases where the energy of the minimum differs from the spherical-shape energy by less than 0.5 MeV, it is marked by a cross to indicate that these isotopes may in fact not be deformed. (a) Using linearly extrapolated values from the deformed actinide and rare earth regions for the spin-dependent parameters of the potential κ (spin–orbit coupling term) and μ (anharmonic correction term). (b) Calculations for the same nuclei using values of these parameters modified in such a manner as to give better agreement with spectroscopic data in this region (from Ragnarsson, 1970).

experimental results in that the deformations increase from Pd to Zr; however, the absolute values of the calculated deformation parameters are somewhat smaller than the experimental ones.

More recent calculations by Krappe and Nix (1973) are successful in reproducing the observed deformation in ^{102}Zr. Their model replaces the surface energy calculated with the assumption of sharp surfaces, which is inherent in the traditional liquid-drop model, with an effective surface energy calculated from a more realistic short-ranged two-particle Yukawa interaction. This macroscopic potential energy surface is then corrected for microscopic shell effects (Strutinskii, 1967) and results in an overall potential energy minimum occurring for a quadrupole moment of 3.9 b. This compares favorably with the experimental value of 3.6 ± 0.2 b (Jared et al., 1973).

The sign of the deformation cannot be accurately determined from $B(E2)$ values. One possible method for determining the sign would be to establish the low-lying single-particle levels in the odd-A fission products. In the splitting of the degenerate spherical levels due to distortion in shape, the high-spin components decrease in energy for oblate deformations and increase for prolate shapes. Therefore, if the light-fission-product region is oblate as predicted, there is a possibility of having low-lying high-spin levels. Such states are conducive to having distinguishable rotational bands built upon them and also having isomeric levels which can be characterized by their decay to lower-spin levels. For neutron states, the (505) $\frac{11}{2}^{-}$ level should be important and, for proton states, if the deformation is very large ($-\varepsilon \gtrsim 0.4$), the same (505) $\frac{11}{2}^{-}$ proton level may be seen. These high-spin, odd-parity levels should not appear in the low-energy spectra for prolate deformed nuclei. Another possible method for establishing prolate or oblate deformations would be to determine the sign of the quadrupole moment relative to the axis of symmetry. Rotational level spacings (to the lowest order) depend only on the magnitude of the deformation, and therefore cannot be used to distinguish between oblate and prolate shapes.

Grüter et al. (1971) have recently reported a rotational band in ^{97}Y built on a $K = \frac{5}{2}$ level. From the energy level spacings and the mixing ratio determined from cascading and crossover transitions, they concluded the band head was most likely the [422] $\frac{5}{2}^{+}$ Nilsson state. These results suggest a large prolate deformation having a value of $\beta \approx 0.6$. If these results can be substantiated, they would imply a large discrepancy with the theoretical predictions.

The calculations do not predict the existence of a sharp transition region for the Zr isotopes. The shape transition in the calculations is very smooth when compared with the data. Experimentally the nucleus ^{96}Zr appears to have extra stability due to the closing of two subshells: the $2p_{1/2}$ proton level at $Z = 40$ and the $2d_{5/2}$ level of $N = 56$. This stability apparently is quite localized since the addition of four neutrons brings on a rapid onset of deformation. The calculations do not reproduce the increased stability for ^{96}Zr and therefore predict the onset of deformation to be more gradual than observed.

One other conjecture should be mentioned for this region. The calculations have predicted that these isotopes have well-defined minima for both prolate and oblate shapes, with the oblate state usually being somewhat more stable. This raises the possibility of shape isomeric transitions between deformed prolate states and those having oblate deformations. For such isomers to exist, it may be necessary that both the prolate and oblate deformations correspond to true potential energy minima. The possibility exists that one of the localized minima may only be a saddle point and not constrained for asymmetric (γ) deformations. Such a situation would not be conducive to having a stable shape and therefore would reduce the probability of having an experimentally observable isomer. Further calculations including asymmetric deformations in this region wlil be useful in establishing this point.

C.   The Doubly Magic $Z = 50$ and $N = 82$ Region

The doubly magic region $Z = 50$, $N = 82$ lies on the neutron-rich side of the line of beta stability and contains many yet-unidentified, very short-lived isotopes. Most of these can effectively be reached only through the fission process. Using the techniques described in this chapter with particular emphasis on "on-line mass separators," new studies can be anticipated in this interesting region. The major handicap in such studies is the low yields of fission products with $A \leqslant 132$ in either thermal neutron fission of ^{235}U or in spontaneous fission of ^{252}Cf. The situation can, to a certain extent, be remedied by the use of medium- or high-energy projectiles to induce fission in uranium and increase by manyfold the yield of these products.

The attractive feature of spectroscopy in doubly closed shell regions is the relative simplicity of the available levels. Many transitions between low-lying states can be described in terms of pure single-particle or single-hole states. This is particularly apparent in the so-called valence nuclei having one particle or hole added to a doubly magic core, i.e., $^{131}_{50}$Sn, $^{133}_{50}$Sn, $^{131}_{49}$In, and $^{133}_{51}$Sb. The low-energy excited states can be interpreted as a promotion of a single particle or hole to a higher level. An example is the 963-keV γ transition observed in the mass-separated $A = 133$ chain (Holm et al., 1970).

The transition was measured in coincidence with electrons from the beta decay and was shown to be associated with the upper part of the beta spectrum, thus ensuring that it belonged to the low-energy portion of the level scheme. The beta half-life in this case was 1.7 sec, which has not been found earlier in the beta decays of the 133 chain. From these considerations, it was reasonable to assign this new half-life to the previously unidentified decay of ^{133}Sn to ^{133}Sb. The 963-keV transition is proposed to arise in the decay of the first excited state (formed by promotion of a proton to the $d_{5/2}$ state) to the $g_{7/2}$ ground state of ^{133}Sb. The energy of this transition also agrees with the general systematics of the $d_{5/2}$–$g_{7/2}$ energy differences in the $N = 82$, odd–Z nuclei. In such nuclei, odd numbers of protons are filling the $g_{7/2}$ ground state and the excited states are formed by promotion of one of these particles into the $d_{5/2}$ level. The fact that one or more pairs of protons already occupy the $g_{7/2}$ state introduces pairing effects which influence the location of the energy levels, though a smooth systematic behavior still occurs.

In the region of doubly magic shells, there is usually a high probability of having isomeric levels. In the ^{132}Sn region, for example, the $h_{11/2}$ neutron hole states are expected to lie at low excitation energies. Since there is an absence of other compatible spin states near in energy, once populated, these levels will remain as isomers. Unlike other isomeric levels, these may not decay by delayed γ-ray or conversion-electron emission, but instead will beta decay to a more suitable level in the daughter nucleus. This situation will result in multiple half-lives for a single isotope, which adds additional complexity to the unraveling of the nuclear level structure.

Isomeric levels also occur in even-even nuclei when pairs of particles outside the core couple to give a low-lying state with high spin. Such a situation occurs in $^{134}_{52}$Te, which is a high-yield prompt fission product (in both uranium and ^{252}Cf fission) and was found (John et al., 1970) to have an isomeric 6^+ state ($t_{1/2} = 162$ nsec) at 1690 keV, which decays by a 115-keV E2 transition to the 4^+ state. The isomer is most probably made up of two $g_{7/2}$ protons which couple to spin 6^+. Similar isomers occur in other $N = 82$, even-Z nuclei. The calculation of the properties of these states requires a solution of the pairing problem. For a detailed discussion of this aspect, the reader is referred to the review article by Kisslinger and Sorensen (1963). The particular cases of the 6^+ state in ^{134}Te and other $N = 82$, even-Z nuclides were calculated by Heyde et al. (1971) using a single-particle potential fitted to results of even-odd nuclides in this region. They obtained good agreement with experimental energy levels and transition probabilities, and found that the 6^+ state of ^{134}Te could be described by an admixture of 0.988 $(g_{7/2})^2$ and 0.152 $(d_{5/2}, g_{7/2})$. Verification of the purity of this level may be possible by experimentally determining its g factor.

A low-lying 6^+ state appears also in other more neutron-deficient Te isotopes, with these states occurring at about 1600 keV. In ^{132}Te, there is also an isomeric $6^+ \rightarrow 4^+$ transition which is similar in nature to that of ^{134}Te, implying that the two neutron holes do not play a significant role in this case. The low-lying neutron states which are found in the even Te isotopes are of negative parity (5^- and 7^-) and probably arise from the coupling of $h_{11/2}$ and $d_{3/2}$ neutrons (Kerek *et al.*, 1970, 1972).

D. The Onset of Deformation in the Rare Earth Region

The heavy fission products contain a wide range of isotopes, varying from strongly spherical nuclei near the ^{132}Sn closed shell to those well into the deformed rare earth region. Therefore, by studying these isotopes, information can be gained on the transitional phase of nuclear structure from spherical to deformed shapes.

The general smooth trends of the energies of the first excited 2^+ levels as well as the $E4^+/E2^+$ ratios of even-even nuclei have already been discussed. It has been noted, however, that an abrupt change in these quantities does occur between ^{98}Zr and ^{100}Zr. Prior to the discovery of this effect in zirconium, there were two instances of known abrupt changes in both the $E2^+$ value and $E4^+/E2^+$ ratio for adjacent even-even nuclei. One case is the well-established change between $N = 88$ and $N = 90$ for isotopes of Sm and Gd, and the other is in the region of $N = 106$–112 between Os and Pt isotopes. Both of these cases are on the borders of deformation regions and both also occur when the number of protons and neutrons are not close to filled shells. Fission product spectroscopy provides data on the levels of the neutron-rich nuclei with Z below that of samarium in the 88–90-neutron region and therefore yields information on this anomalous discontinuity for lighter isotopes. The $E4^+/E2^+$ ratios of nuclei in this region are shown in Fig. 14. The points which were found in fission studies are marked in the figure and show that the 88–90-neutron discontinuity is localized in the $Z = 60$–66 range. In lighter elements, as well as in heavier ones, the changes in the $E4^+/E2^+$ ratio (and also the $E2^+$ values) proceed in a very smooth way, as is evident in isotopes of $_{56}$Ba and $_{58}$Ce (Fig. 11).

Calculations of the properties of nuclei in the transition region between spherical and deformed shapes (in particular the $Z = 56$–64 region) have been performed by Nilsson *et al.* (1969), Ragnarsson (1970), and Ragnars-

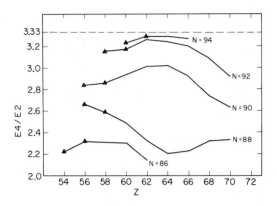

Fig. 14. Systematic behavior of the ratio $E4/E2$ as a function of proton number in the $N = 86–92$ region. Data presented as triangles are from prompt fission product studies (from Wilhelmy *et al.*, 1970a).

son and Nilsson (1971). The calculations give the minimum in the potential energy as a function of deformation and reproduce the general trends of larger deformation with more neutrons for a given number of protons and smaller deformation with decreasing number of protons for a given neutron number; however, here as in the zirconium region, no sharp onset of deformation is calculated. The pairing-plus-quadrupole treatment of Kumar and Baranger (1968) gives similar results, but as Kumar (1970) points out, the deformation of the nucleus is not given by the potential energy shape but by the properties of the ground state which depend both on the potential energy and on the zero-point energy. It is thus possible to have a potential with a minimum at some deformation, such as occurs in ^{150}Sm $(N = 88)$, but with a ground state which is soft and not deformed because the zero-point energy is above the potential energy surface at $\beta = 0$. An apparent sharp shape transition may occur, then, when the zero-point energy becomes confined to a definite deformed region. Many properties of nuclei in the transition region, such as β- and γ-vibrational states and transition probabilities, can be estimated by the pairing-plus-quadrupole calculations. No experimental data concerning these properties in the heavy-fission-product region exist at present, so that detailed comparison with theoretical calculations must wait.

E.   DETAILED SPECTROSCOPY AS A TOOL FOR STUDYING THE FISSION PROCESS

Until now, we have discussed the contribution of the fission process to nuclear spectroscopic information on neutron-rich nuclei. We feel that we should briefly mention that the knowledge of detailed spectroscopic properties of isotopes can in turn yield valuable information with regard to the fission process.

The intensities of transitions deexciting ground-state bands of even-even fission product nuclei are related to the angular momentum present in the fissioning system (Wilhelmy *et al.*, 1972). Since a wide range of even-even products are formed in fission, it is possible, by measuring the γ-ray intensities of known nuclei, to study the angular momentum effects for various mass splits and/or as a function of the fission kinetic energy release. Another application of spectroscopic knowledge has been relating intensities of $2^+ \rightarrow 0^+$ ground-state band transitions in even-even nuclei to the independent yields of the isotopes (Cheifetz *et al.*, 1971b). This gives direct information on the charge and mass dispersion in fission and has an advantage in comparison with radiochemical or mass separation techniques in that the yield can be correlated with other properties of fission, such as kinetic energy release or neutron emission yields. Other possible applications of spectroscopic data still, of course, exist. For example, the correlation of γ rays from known nuclei with their neutron evaporation yields would give information on the excitation energies of specific fragments formed in fission.

V. Summary

Fission product spectroscopy is now in a very active period of development. The continuing improvements of experimental techniques with regard to rapid radiochemistry, on-line mass separation, and prompt radiation studies are now enabling detailed determinations of the nuclear structure properties of isotopes far from the region of beta stability.

Studies following beta decay of high-yield fission products will give information on Q_β values, level energies, level spins and parities, transition multipolarities, mixing ratios, transition probabilities, and delayed neutron emission properties. Prompt radiation studies, which are analogous to "in-beam" spectroscopy, should yield data on ground-state bands in even-even nuclei, stretched cascades in odd-A and odd-odd isotopes, population yields of high-spin isomers, magnetic moments of isomeric levels from perturbed angular correlation studies, transition multipolarities from angular distribution measurements, and transition lifetime information using Doppler shift techniques.

Even if all the technical difficulties associated with the experiments are overcome, it still should be realized that there are some limitations in the type of information which can be obtained in fission product studies. Stripping and pickup reactions, which yield information on spectroscopic factors, as well as Coulomb excitation experiments, which give data on static quadrupole moments, require stable or near-stable isotopes which can be made into targets. The short-lived fission products will not be amenable to such analyses.

Although there are some inherent limitations and numerous technical difficulties for spectroscopic studies of fission products, more than 200 undiscovered neutron-rich isotopes still await investigation.

ACKNOWLEDGMENTS

We would like to thank professors G. Herrmann, W. John, and C. F. Moore for their cooperation in providing information which we presented in this chapter. We would also like to thank Dr. C. F. Tsang for his comments on portions of this work. Special gratitude is, however, due Dr. S. G. Thompson, who has been instrumental in participating and guiding all aspects of our studies.

References

Ajitanand, N. N. (1971). *Nucl. Phys.* **A164**, 300.

Alväger, T., Naumann, R. A., Petry, R. F., Sidenius, G., and Thomas, T. D. (1968). *Phys. Rev.* **167**, 1105.

Amarel, I., Bernas, R., Foucher, R., Jastrzebski, J., Johnson, A., Teillac, J., and Gauvin, H. (1967). *Phys. Lett.* **24B**, 402.

Apalin, V. F., Gritsyuk, Yu. N., Kutikov, I. E., Lebedev, V. I. and Mikaelian, L. A. (1965). *Nucl. Phys.* **71**, 553.

Armbruster, P., Eidens, J., and Roeckl, E. (1967). *Arkiv Fysik* **36**, 293.

Armbruster, P., Hossfeld, F., Labus, H., and Reichelt, K. (1969). *Proc. IAEA Symp. Phys. Chem. Fission, 2nd, Vienna*, p. 545. Int. At. Energy Agency, Vienna.

Armbruster, P., Eidens, J., Grüter, J. W., Lawin, H., Roeckl, E., and Sistemich, K. (1971). *Nucl. Instrum. Methods* **91**, 499.

Arseniev, D. A., Sobiczewski, A., and Soloviev, V. G. (1969). *Nucl. Phys.* **A139**, 269.

Atneosen, R. A., Thomas, T. D., Gibson, W. M., and Perlman, M. L. (1966). *Phys. Rev.* **148**, 1206.

Bäcklin, A., Fogelberg, B., and Hedin, G. (1970). *Proc. Conf. Properties Nuclei Far from the Region of Beta Stability, Leysin, Switzerland, 1970* **2**, 1069. CERN, Geneva.

Bes, D. R., and Szymanski, Z. (1961). *Nucl. Phys.* **28**, 42.

Bjerrgaard, J. G., Hansen, O., Nathan, O., and Hinds, S. (1966). *Nucl. Phys.* **86**, 145.

Blair, A. G., Beery, J. G., and Flynn, E. R. (1969). *Phys. Rev. Lett.* **22**, 470.

Borg, S., Bergström, I., Holm, G. B., Rydberg, B., de Geer, L.-E., Rudstam, G., Grapengiesser, B., Lund, E., and Westgaard, L. (1971). *Nucl. Instrum. Methods* **91**, 109.

Bowman, H. R., Milton, J. C. D., Thompson, S. G., and Swiatecki, W. J. (1962). *Phys. Rev.* **126**, 2120; (1963). *Phys. Rev.* **129**, 2133.

Bowman, H. R., Thompson, S. G., and Rasmussen, J. O. (1964). *Phys. Rev. Lett.* **12**, 195.

Bowman, H. R., Thompson, S. G., Watson, R. L., Kapoor, S. S., and Rasmussen, J. O. (1965). *Proc. IAEA. Symp. Phys. Chem. Fission, Salzburg,* **2**, 125. Int. At. Energy Agency, Vienna.

Casten, R. F., Flynn, E. R., Hansen, O., Mulligan, T., Sheline, R. K., and Kienle, P. (1970). *Phys. Lett.* **32B**, 45.

Casten, R. F., Hansen, O., Flynn, E. R., and Mulligan, T. J. (1972). *Nucl. Phys.* **A184**, 357.

Cheifetz, E., Jared, R. C., Thompson, S. G., and Wilhelmy, J. B. (1970a). *Phys. Rev. Lett.* **25**, 38.

Cheifetz, E., Jared, R. C., Thompson, S. G., and Wilhelmy, J. B. (1970b). *Proc. Conf. Properties Nuclei far from Region Beta Stability, Leysin, Switzerland,* 1970 **2**, 883. CERN, Geneva.

Cheifetz, E., Jared, R. C., Thompson, S. G., and Wilhelmy, J. B. (1971a). *1971 Annu. Rep. Nucl. Chem. Div.* Lawrence Berkeley Lab.

Cheifetz, E., Wilhelmy, J. B., Jared, R. C., and Thompson, S. G. (1971b). *Phys. Rev.* **C4**, 1913.

Clark, R. G., Glendenin, L. E., and Talbert, W. L. (1973). *Proc. IAEA. Symp. Phys. Chem. Fission, 3rd, Rochester.* Int. At. Energy Agency, Vienna.

Croall, I. F., and Willis, H. H. J. (1963). *J. Inorg. Nucl. Chem.* **25**, 1213.

Diamond, R. M. (1970). *Proc. Conf. Properties Nuclei far from Region Beta Stability, Leysin, Switzerland, 1970* **1**, 65. CERN, Geneva.

Diamond, R. M., Stephens, F. S., and Swiatecki, W. J. (1964). *Phys. Lett.* **11**, 315.

Eddy, N. W., and Roy, R. R. (1971). *Phys. Rev.* **C3**, 877.

Fogelberg, B. (1971). *Phys. Lett.* **37B**, 372.

Fulmer, C. B., and Cohen, B. L. (1958). *Phys. Rev.* **109**, 94. See also Cohen, B. L., and Fulmer, C. B. (1958). *Nucl. Phys.* **6**, 547.

Glendenin, L. E., and Unik, J. P. (1965). *Phys. Rev.* **140**, B1301.

Gordon, G. E., Harvey, J. W., and Nakahara, H. (1966). *Nucleonics* **24** (12), 62.

Graff, G., Lajtai, A., and Nagy, L. (1965). *Proc. IAEA Symp. Phys. Chem. Fission, Salzburg,* **2**, 163. Int. At. Energy Agency, Vienna.

Grüter, J. W., Sistemich, K., Armbruster, P., Eidens, J., and Lawin, H. (1970). *Phys. Lett.* **33B**, 474; *Proc. Conf. Properties Nuclei far from Region Beta Stability, Leysin, Switzerland, 1970* **2**, 967. CERN, Geneva.

Grüter, J. W., Sistemich, K., Armbruster, P., Eidens, J., Hübenthal, K., and Lawin, H. (1971). Preprint; see also Grüter, J. W. (1972). Ph. D. thesis. Both from Kernforschungsanlage, Juelich Institut fuer Neutronenphysik.

Harris, S. H. (1965). *Phys. Rev.* **138**, B509.

Herrmann, G., and Denschlag, H. O. (1969). *Annu. Rev. Nucl. Sci.* **19**, 1.

Herrmann, G., Kaffrell, N., Trautmann, N., Denig, R., Herzog, W., Hübscher, D., and Kratz, K. L. (1970). *Proc. Conf. Properties Nuclei Far from the Region Beta Stability, Leysin, Switzerland, 1970* **2**, 985. CERN, Geneva.

Heyde, K., Waroquier, M., and Berghe, G. V. (1971). *Phys. Lett.* **35B**, 211.

Hoffman, M. M. (1964). *Phys. Rev.* **133**, B714.

Holm, G. B., Andersson, G. I., Borg, S., and de Geer, L.-E. (1970). *Proc. Conf. Properties Nuclei Far from the Region Beta Stability, Leysin, Switzerland, 1970* **2**, 1008. CERN, Geneva.

Hopkins, F. F., Phillips, G. W., White, J. R., Moore, C. F., and Richard, P. (1971). *Phys. Rev.* C4, 1927.

Hopkins, F. F., White, J. R., Phillips, G. W., Moore, C. F., and Richard, P. (1972). *Phys. Rev.* C5, 1015.

Horsch, F. (1970). *Proc. Conf. Properties Nuclei Far from the Region Beta Stability, Leysin, Switzerland, 1970* 2, 917. CERN, Geneva.

Horsch, F., and Michaelis, W. (1969). *Proc. IAEA Symp. Phys. Chem. Fission, 2nd* p. 527. Int. At. Energy Agency, Vienna.

Hyde, E. K. (1964). "The Nuclear Properties of the Heavy Elements," Vol. III, Fission Phenomena, Prentice Hall, Englewood Cliffs, New Jersey.

Jared, R. C., Nifenecker, H., and Thompson, S. G. (1973). *Proc. IAEA Symp. Phys. Chem Fission, 3rd, Rochester.* Int. At. Energy Agency, Vienna.

Johansson, S. A. E. (1964). *Nucl. Phys.* 60, 378.

Johansson, S. A. E. (1965). *Nucl. Phys.* 64, 147.

Johansson, S. A. E., and Kleinheinz, P. (1965). *In* "Alpha, Beta and Gamma-Ray Spectroscopy," (K. Seigbahn, ed.), Vol. 1, p. 805. North-Holland Publ. Co., Amsterdam.

John, W. (1971). Private communication.

John, W., Massey, R., and Saunders, B. G. (1967). *Phys. Lett.* 24B, 336.

John, W., Guy, F. W., and Wesolowski, J. J. (1970). *Phys. Rev.* C2, 1451.

Kapoor, S. S., and Ramanna, R. (1964). *Phys. Rev.* 133, B598.

Kapoor, S. S., Bowman, H. R., and Thompson, S. G. (1965). *Phys. Rev.* 140, B1310.

Kerek, A., Bergström, I., Borg, S., Carle, P., Holm, G., and Rydberg, B. (1970). *Proc. Conf. Properties Nuclei Far from the Region Beta Stability, Leysin, Switzerland, 1970*, 2, 1010, CERN, Geneva.

Kerek, A., Holm, G. B., Borg, S., and deGeer, L.-E. (1972). *Nucl. Phys.* A195, 177.

Khan, T. A., Hofmann, D., and Horsch, F., (1973). *Nucl. Phys.* A205, 488.

Kisslinger, L. S., and Sorensen, R. A. (1963). *Rev. Mod. Phys.* 35, 853.

Kjelberg, A., and Rudstam, G. (1970). CERN-70-3, Geneva, Switzerland.

Krappe, H. J., and Nix, J. R., (1973). *Proc. IAEA Symp. Phys. Chem. Fission, 3rd, Rochester.* Int. At. Energy Agency, Vienna.

Kratz, K. L. (1970). Jahresbericht 1970, BMBW-FB K 71-12, p. 7. Univ. of Mainz, Germany.

Kumar, K. (1970). *Proc. Conf. Properties Nuclei Far from the Region of Beta Stability, Leysin, Switzerland,* 2, 779. CERN, Geneva.

Kumar, K., and Baranger, M. (1968). *Nucl. Phys.* A122, 273.

Larsen, J. T., Talbert, W. L., Jr., and McConnell, J. R. (1971). *Phys. Rev.* C3, 1372.

Lederer, C. M., Hollander, J. M., and Perlman, I. (1967). "Table of Isotopes," Wiley, New York.

Loveland, W. D., and Shum, Y. S. (1971). *Phys. Rev.* C4, 2282.

Ma, C. W., and Rasmussen, J. O. (1971). *Phys. Rev.* C2, 798.

Macias-Marques, M. I., Foucher, R., Cailiau, M., and Belhassen, J. (1970). *Proc. Conf. Properties Nuclei Far from the Region Beta Stability, Leysin, Switzerland, 1970* 1, 321. CERN, Geneva.

Macias-Marques, M. I., Johnson, A., Foucher, R. and Henck, R. (1971). *J. Phys.* 32, 237.

Maier-Leibnitz, H., Schmitt, H. W., and Armbruster, P. (1965). *Proc. IAEA Symp. Phys. Chem. Fission, Salzburg,* 2, 143. Int. At. Energy Agency, Vienna.

Mallmann, C. A. (1959). *Phys. Rev. Lett.* **2**, 507.

Mariscotti, M. A. J., Scharff-Goldhaber, G., and Buck, B. (1969). *Phys. Rev.* **178**, 64.

Maslin, E. E., Rodgers, A. L., and Core, W. G. F. (1967). *Phys. Rev.* **164**, 1520.

McDonald, J., and Kerek, A. (1973). *Nucl. Phys.* **A206**, 417.

Monnand, E., Blachot, J., Carraz, L. C. and Moussa, A. (1970). *Proc. Conf. Properties Nuclei Far from the Region Beta Stability, Leysin, Switzerland, 1970*, **2**, 1119. CERN, Geneva.

Mottelson, B. R., and Nilsson, S. G. (1959). *Kgl. Danske Videnskab. Selskab. Mat.-Fys. Skrifter* **1**, No. 8, 1.

Nilsson, S. G., Tsang, C. F., Sobiczewski, A., Szymanski, Z., Wycech, S., Gustafson, C., Lamm, I. L., Möller, P., and Nilsson, B. (1969). *Nucl. Phys.* **A131**, 1.

Norris, A. E., and Wahl, A. C. (1966). *Phys. Rev.* **146**, 926.

Patzelt, P. and Herrmann, G. (1965). *Proc. IAEA Symp. Phys. Chem. Fission, Salzburg*, **2**, 243. Int. At. Energy Agency, Vienna.

Peelle, R. W., and Maienschein, F. C. (1971). *Phys. Rev.* **C3**, 373.

Ragnarsson, I. (1970). *Proc. Conf. Properties Nuclei Far from the Region Beta Stability, Leysin, Switzerland, 1970* p. 847. CERN, Geneva.

Ragnarsson, I., and Nilsson, S. G. (1971). *Prog. Rep. Exp. Theoret. Stud. Transitional Nuclei* Inst. of Nucl. Phys. Orsay, France.

Reisdorf, W., Unik, J. P., Griffin, H. C. and Glendenin, L. E. (1971). *Nucl. Phys.* **A177**, 337.

Ruegsegger, D. R., Jr., and Roy, R. R. (1970). *Phys. Rev.* **C1**, 631.

Sarantities, D. G., Gordon, G. E. and Coryell, C. D. (1965). *Phys. Rev.* **138**, B353.

Schindler, W. J., and Fleck, C. M. (1973). *Nucl. Phys.* **A206**, 374.

Schüssler, H. D., Ahrens, H., Folger, H., Franz, H., Grimm, W., Herrmann, G., Kratz, J. V., and Kratz, K. L. (1969). *Proc. IAEA Symp. Phys. Chem. Fission, 2nd, Vienna*, p. 591. Int. At. Energy Agency, Vienna.

Shapiro, N. L., Wehring, B. W., and Wyman, M. E. (1971). *Phys. Rev.* **C3**, 2464.

Silbert, M. D. and Tomlinson, R. H. (1966). *Radiochim. Acta.* **5**, 217, 223.

Skarsvåg, K. (1967). *Nucl. Phys.* **A96**, 385.

Skarsvåg, K. (1970). *Nucl. Phys.* **A153**, 82.

Skarsvåg, K. and Singstad, I. (1965). *Nucl. Phys.* **62**, 103.

Stelson, P. H. and Grodzins, L. (1965). *Nucl. Data* **A1**, 211.

Strutinskii, V. M. (1960). *Sov. Phys. JETP* **10**, 613.

Strutinskii, V. M. (1967). *Nucl. Phys.* **A95**, 420.

Tracy, B. L., Chaumont, J., Klapisch, R., Nitschke, J. M., Poskanzer, A. M., Roeckl, E. and Thibault, C. (1971). *Phys. Lett.* **34B**, 277.

Trautmann, N., Kaffrell, N., Behlich, H. W., Folger, H., Herrmann, G., and Hübscher, D. (1972). *Radiochim. Acta.* **18**, 86.

Val'skii, G. V., Petrov, G. A. and Pleva, Yu. S. (1967). *Sov. J. Nucl. Phys.* **5**, 521; (1969). **8**, 171.

Wahl, A. C., Ferguson, R. L., Nethaway, D. R., Troutner, D. E., and Wolfsberg, K. (1962). *Phys. Rev.* **126**, 1112.

Wahlgren, M. A. and Meinke, W. W. (1962). *J. Inorg. Nucl. Chem.* **12**, 201.

Warhanek, H. and Vandenbosch, R. (1964). *J. Inorg. Nucl. Chem.* **26**, 669.

Watson, R. L. (1966). *Univ. of California Lawrence Radiat. Lab. Rep. UCRL-16798*, Ph. D. Thesis, unpublished.

Watson, R. L., and Wilhelmy, J. B. (1969). *Univ. of California Lawrence Radiat. Lab. Rep. UCRL-18632*, unpublished.

Watson, R. L., Bowman, H. R., Thompson, S. G., and Rasmussen, J. O. (1967). *Univ. of California Lawrence Radiat. Lab. Rep. UCRL-17289*, unpublished.

Watson, R. L., Jared, R. C., and Thompson, S. G. (1970a). *Phys. Rev.* C1, 1886.

Watson, R. L., Wilhelmy, J. B., Jared, R. C., Rugge, C., Bowman, H. R., Thompson, S. G., and Rasmussen, J. O. (1970b). *Nucl. Phys.* A141, 449.

Wilhelmy, J. B. (1969). *Univ. of California Lawrence Radiat. Lab. Rep. UCRL-18978*, Ph.D. thesis, unpublished.

Wilhelmy, J. B., Thompson, S. G., Jared, R. C., and Cheifetz, E. (1970a). *Phys. Rev. Lett.* 25, 1122.

Wilhelmy, J. B., Thompson, S. G., Rasmussen, J. O., Routti, J. T. and Phillips, J. E. (1970b). *Univ. of California Lawrence Radiat. Lab. Rep. UCRL-19530*, 178, unpublished.

Wilhelmy, J. B., Cheifetz, E., Jared, R. C., Thompson, S. G., Bowman, H. R. and Rasmussen, J. O. (1972). *Phys. Rev.* C5, 2041.

Zicha, G., Löbner, K. E. G., Maier-Komor, F., Maul, J. and Kienle, P. (1969). *Int. Conf. Properties Nucl. States, Montreal* p. 83. Presses de l'Univ. de Montréal, Montréal.

VII.G   ANGULAR CORRELATION METHODS

A. J. Ferguson

NUCLEAR PHYSICS BRANCH, CHALK RIVER NUCLEAR LABORATORIES
ATOMIC ENERGY OF CANADA LIMITED
CHALK RIVER, ONTARIO, CANADA

I. Introduction

A. OUTLINE OF THE THEORY

Angular correlation theory has been described with varying thoroughness and from various points of view by numerous authors. A very complete treatment is the classic paper by Devons and Goldfarb (1957). Goldfarb (1959) presented a similar but somewhat less formal approach. A more physical development with stress on classical analogies has been given by Biedenharn (1960). Angular correlations in the context of radioactive studies have been discussed by Frauenfelder and Steffen (1965) and in the context of bombardment experiments by Ferguson (1965). We will here present an

277

outline of the theory in minimal form that will attempt to display its basic structure, the details of which may be found from the references just cited. In particular, a difference between the modern density matrix method and the traditional wave function method will be stressed. The density matrix method will be seen to exploit symmetries of ensembles that lead to considerable simplifications. A knowledge of the quantum mechanical theory of angular momentum, including angular momentum vector addition, will be assumed.

By virtue of the intimate connection between the eigenvalues of angular momentum eigenfunctions and their properties under rotations of coordinate systems, angular correlation measurements are the natural means of determining these eigenvalues. The angular momenta we are referring to here are the spins of nuclear states and the angular momenta of radiations emitted by them, whose determination is an important part of nuclear structure studies. Equations (8a) and (8b) give the rotational transformations of an eigenstate of spin j. The transformations for each j are distinct, indeed the rotation matrices $D^j_{mm'}$ $(\Im_1 \Im_2 \Im_3)$ for different j's are orthogonal under integration over the Euler angles $\Im_1$, $\Im_2$, and $\Im_3$, so that the rotational properties of a spin eigenstate constitute its signature. However, rotational properties must be used somewhat indirectly to evaluate spins. Nuclear structure is probed by reactions which involve exciting the nucleus by a beam of particles, followed by the emission of the same or other particles. In general these radiations have directions that can be varied, i.e., rotated about the reaction point, and the problem is to extract the required information from the behavior of the reaction under these rotations. The radiations may be particles or gamma rays, and, since the quantum mechanical treatment of both is essentially the same, we will frequently refer to the radiations as "particles" with the understanding that photons are included. The polarization of a state can be included in the study, the polarization axis being another direction that can be used as a rotational variable.

No preferred coordinate frame exists for the description of a nuclear reaction. For this reason, only relative directions have significance and an angular correlation study must involve at least two directions. In the familiar case of a reaction excited by a high-energy particle beam, the directions normally used are those of the incoming and outgoing particles. In a large class of cases, the incoming beam does not establish a direction. An example is excitation by very slow neutrons entering the nucleus with zero orbital angular momentum, therefore having no definable direction. Another example is where the excited nuclei have a long lifetime and lose the orientation by thermal agitation. In such cases, the two directions may be those of two radiations emitted successively from the nucleus, or else of one emitted radiation and a polarization axis for the excited nucleus established by conventional polarizing methods. The measurement of the correlations between three directions is experimentally feasible, though more difficult than between two, and is termed a triple angular correlation.

Angular correlations are expressed in terms of a function $W(\Omega_1, \Omega_2, \ldots)$ which gives the joint intensity of the observed particles in a nuclear reaction as a function of the directions $\Omega_1, \Omega_2, \ldots$. Normally $\Omega_i = (\theta_i, \phi_i)$, where θ_i and ϕ_i are the polar angles for the directions of propagation of the various particles. If polarization is also included, there will be an Ω_i to represent the direction of the polarization axis. The function W is also a function of the spins of nuclear states and the multipolarities of emitted radiations. The object of the theory is to provide a formula for calculating $W(\Omega_1, \Omega_2, \ldots)$. In the case of a reaction, W will be a cross section, expressed as a function of the reaction angle. In the case of a radioactive decay, W is simply a function representing the joint intensity of two or more outgoing radiations. Excepting the case of a simple reaction, an experimental measurement of W requires a time coincidence between the detected particles.

The methodological difference between the traditional and modern approaches may be summed up by asserting that the former deals with amplitudes and the latter with intensities. The traditional approach can be outlined roughly as follows. A wave function Ψ_i, containing factors that represent the positions and spins of the interacting particles, is set up to represent the system in its initial state. The wave function Ψ_f, representing the system in its final state, is given by

$$\Psi_f = H_{int}\Psi_i \tag{1}$$

where H_{int} is the perturbing Hamiltonian that induces the transitions to the final state. The wave functions Ψ_f will be products of sums of outgoing spherical waves representing the final particles of the reaction, which in general will be different from the initial ones. The squared modulus $|\Psi_f|^2$ of the final state is interpreted as the densities of the outgoing particles, from which the required fluxes are obtained by multiplying by the particle velocities. The angular correlations of the particles are given by the dependence of the spherical waves on the polar angles θ and ϕ for each of the outgoing particles. Concealed by this schematic treatment is the fact that the evaluation of $\Psi_f = H_{int}\Psi_i$ consists primarily in performing a number of vector sums as will be discussed in more detail later. The number of coefficients and terms typically occurring in the vector addition is large enough to constitute a tedious problem when the calculations are performed by hand.

To describe the density matrix–statistical tensor method, consider the wave function that represents either the position or the spin of a particle. The wave function can be expanded as a sum of eigenfunctions of some operator,

$$\Psi = \sum_l a_l \phi_l \tag{2}$$

where the a_l are expansion coefficients and the ϕ_l are the eigenfunctions. An example is the Rayleigh expansion of a plane wave in eigenfunctions of the orbital angular momentum. The squared modulus of Ψ is given by

$$|\Psi|^2 = \Psi \Psi^* = \sum_{ll'} a_l a_{l'}^* \phi_l \phi_{l'}^* \tag{3}$$

and can be interpreted as the density of particles at the point under consideration. Equation (3) is clearly an expansion of the particle density in terms of the functions $\phi_l \phi_{l'}^*$ with coefficients $a_l a_{l'}^*$.[†] The coefficient $a_l a_{l'}^*$ is an element of the density matrix that represents the state Ψ. As discussed so far, Ψ is a pure state, that is, composed solely of the coherent sum, Eq. (2). We are usually involved with an ensemble of particles whose wave functions are not identical and which, due to randomness in the formation of the ensemble, are not correlated with each other. When this is so, the total density for the ensemble is the incoherent sum of the densities arising from the different wave functions. The density matrix for the ensemble is then defined as an average over the ensemble.

Using the notation of Devons and Goldfarb for the present example, a density matrix element is

$$\langle l| \varrho |l' \rangle = \langle a_l a_{l'}^* \rangle_{\mathrm{av}} \tag{4}$$

the average being taken over the ensemble. The basic difference between the wave function and the density matrix approaches is that the latter works with coefficients of density expansions, averaged over the ensemble, rather than with amplitude expansions, and thus retains a more direct association with physical observables. It is usually the case that, due to rotational symmetries,

---

[†] In general $\{\phi_l \phi^*_{l'}\}$ will not be an orthogonal set. However, the expansion $\phi_l \phi^*_{l'} = \sum_k a_{ll'k} \phi_k$ will normally be possible and will allow an expansion of $|\Psi|^2$ to be made in terms of the ϕ_l. This is a key step in the theory.

the averaged densities have simpler properties than the component wave functions. A good example of this is the case of an unpolarized spin of magnitude j. In the wave function treatment, each of the $2j + 1$ substates must be traced through the vector sums that lead to the final wave function Ψ_f, and an ensemble average over these is made to obtain the cross sections. In the density matrix treatment, the ensemble of spins is represented by a diagonal density matrix with all diagonal terms equal, i.e.,

$$\langle jm| \varrho |jm'\rangle = \delta_{mm'}/(2j + 1) \tag{5}$$

The off-diagonal terms vanish in this case because the quantities $a_l a_{l'}^*$, $l \neq l'$, of Eq. (4) are completely uncorrelated in magnitude and phase.

The statistical tensors are defined as linear transformations of the density matrix

$$\varrho_{k\kappa}(jj') = \sum_{mm'} (-)^{j'-m'}(jm, j' - m' | k\kappa)\langle jm| \varrho |j'm'\rangle \tag{6}$$

The density matrix and the statistical tensors constitute complete and equivalent sets of parameters that describe the observable properties of a system. Equation (6) can be inverted using the orthogonality property of the Clebsch–Gordan coefficients $(jm, j' - m'|k\kappa)$ to give the density matrix in terms of the tensors

$$\langle jm| \varrho |j'm'\rangle = \sum_{k\kappa} (-)^{j'-m'}(jm, j' - m' | k\kappa)\varrho_{k\kappa}(jj') \tag{7}$$

Linear transformations among the $2j + 1$ substates of a spin j are introduced when the coordinate axes are subjected to an arbitrary rotation. A general rotation $\mathscr{R}$ is described by the three Euler angles consisting of a rotation $\mathfrak{I}_1$ about the z axis, followed by a rotation $\mathfrak{I}_2$ about the new y axis, then followed by a rotation $\mathfrak{I}_3$ about the new z axis. Under such a rotation, the wave function $|jm'\rangle$ relative to the new axes is

$$|jm'\rangle = \sum_m |jm\rangle D^j_{mm'}(\mathscr{R}) \tag{8a}$$

where $|jm\rangle$ is the wave function relative to the old axes, $D^j_{mm'}(\mathscr{R})$ is an element of the rotation matrix, and the argument represents the angles $(\mathfrak{I}_1, \mathfrak{I}_2, \mathfrak{I}_3)$. The wave function $\langle jm'|$, termed a "bra" by Dirac (1947), is contragredient to the "ket" $|jm'\rangle$ and is transformed under rotation by

$$\langle jm'| = \sum_m D^{j*}_{m'm}(\mathscr{R})\langle jm| \tag{8b}$$

Statistical tensors transform similarly to a "bra" of spin k. Relative to new axes obtained by the rotation, they are given by

$$\varrho_{k\kappa}(jj') = \sum_{\kappa'} D_{\kappa\kappa'}^{k*}(\mathscr{R}) \, \varrho_{k\kappa'}(jj') \tag{8c}$$

Systems with high rotational symmetry are represented by simple tensors. A system with complete spherical symmetry, for example, an unpolarized spin, is represented by the tensor $\varrho_{00}(jj)$ alone. If the system has axial symmetry, then only the tensors $\varrho_{k0}(jj)$, $k \leqslant 2j$, are nonzero.

Angular correlations are evaluated in terms of the fluxes of outgoing particles at specified angular positions. Let the density of a particular group of reaction products at the point (r, θ, ϕ) be $\varrho(r, \theta, \phi)/r^2$. The flux of outgoing particles is $v\varrho(r, \theta, \phi)/r^2$, where v is the velocity of the particles. Using the semiclassical radiation theory (Blatt and Weisskopf, 1952), photons can be treated in a similar way, regarding the vector potential as a wave function and defining the flux as the outward energy flow as given by the average Poynting vector divided by the photon energy $\hbar\omega$. A rigorous development requires the use of quantum field theory (Heitler, 1954) but does not change the results from the point of view of angular correlations, so that the semiclassical theory is adequate for our purposes. It may be noted at this point that the physics of the processes has relatively little to do with angular correlations. What is essential is to resolve the wave functions or densities into components with definite rotational properties. These will be components of definite spin and rank, respectively. The magnitudes of the components are determined by reduced matrix elements of the interaction Hamiltonian whose determination constitutes the physics of the problem.

The interpretation of the theoretical result is in terms of the probability that a particular particle will be detected at the point (r, θ, ϕ). The logical implementation of this concept is to regard (r, θ, ϕ) as a point inside a detector for which the efficiency of detection $\varepsilon(r, \theta, \phi)$ is known at every point. Angular correlation theory is usually developed assuming point counters, but it is quite feasible to assume that the detector has finite extension over which $\varepsilon(r, \theta, \phi)$ is known. The probability of detecting an outgoing particle is then $\int \varrho(r, \theta, \phi)\varepsilon(r, \theta, \phi)dV$, where dV is a volume element. Now consider an operator with matrix elements of the form

$$\langle r, \theta, \phi | \varepsilon | r', \theta', \phi' \rangle = \varepsilon(r, \theta, \phi) \, \delta(r - r') \, \delta(\theta - \theta') \, \delta(\phi - \phi') \tag{9}$$

The expectation value for this operator (Dirac, 1947) for a system in a state

$|\Psi(r, \theta, \phi)\rangle$ is given by

$$W = \int \langle \Psi(r, \theta, \phi)| \, \varepsilon \, |\Psi(r', \theta', \phi')\rangle \, dV \, dV'$$

where $dV = r^2 \sin\theta \, dr \, d\theta \, d\phi$ and similarly for dV'. Because of the δ functions $\delta(r - r')$, $\delta(\theta - \theta')$, and $\delta(\phi - \phi')$, the integral simplifies to

$$W = \int \langle \Psi(r, \theta, \phi) \, \| \, \Psi(r, \theta, \phi)\rangle \varepsilon(r, \theta, \phi) \, dV$$

$$= \int \varrho(r, \theta, \phi) \varepsilon(r, \theta, \phi) \, dV \tag{10}$$

Thus we see that the response of the detector is obtained as the expectation value of the operator ε, called the efficiency matrix. This approach was introduced by Coester and Jauch (1953). It has the advantage of tying the response to the general formal rules for the physical interpretation of quantum mechanics. Since ε is an operator, it will undergo the transformations of representation to which the wave functions are subjected. Thus when density matrices are transformed to statistical tensors, the efficiency matrices (or operators) are likewise transformed to efficiency tensors. Efficiency matrices must be introduced for every particle occurring in the final state of the reaction for the reason that the overall efficiency matrix must contain the same set of variables as the wave function. Some of these may be trivial, such as for unobserved particles and for the residual nucleus that remains after all observed emissions have occurred. The efficiency matrices for such radiations have a standard and simple form. A final advantage is that the efficiency matrices for finite counters are obtained by the same formalism as for point counters, the only complication being that an integration over the counter volume is required. The formalism leads straightforwardly to the results required in cases such as triple or polarization correlations where counter size effects are otherwise difficult to evaluate.

Absorption or emission is induced by the perturbing Hamiltonian. We will make the normal assumption that the initial and final wave functions are expanded in eigenstates of angular momentum. In general the notation of Devons and Goldfarb will be used in which italic Roman letters $a, b, c, \ldots$ will denote the magnitude of angular momentum and spin vector operators and the Greek letters $\alpha, \beta, \gamma, \ldots$ the corresponding substate values. Let the initial state of an emission process be $\langle a\alpha|$ and the final state be

$\langle b\beta| \langle l\lambda|$, where $\langle b\beta|$ represents the residual nucleus and $\langle l\lambda|$ the angular momentum of the outgoing particles. The final state is obtained according to Eq. (1) as

$$\langle b\beta| \langle l\lambda| = \sum_{a\alpha} \langle b\beta l\lambda| H_{int} |a\alpha\rangle \langle a\alpha| \tag{11}$$

The matrix elements of H_{int} are given by the Wigner–Eckart theorem

$$\langle b\beta l\lambda| H_{int} |a\alpha\rangle = (b\beta, l\lambda | a\alpha) \langle b| l \| a\rangle \tag{12}$$

where $(b\beta, l\lambda|a\alpha)$ is a Clebsch–Gordan coefficient and $\langle b|l\|a\rangle$ is the reduced matrix element which is independent of the substate values α, β, and λ. Similarly a matrix element for absorption is given by

$$\langle b\beta| H_{int} |a\alpha l\lambda\rangle = (b\beta | a\alpha, l\lambda) \langle b\| l |a\rangle \tag{13}$$

The asymmetric notation $\langle b|l\|a\rangle$ and $\langle b\|l|a\rangle$ was devised by Goldfarb and Johnson (1960) to distinguish between emission and absorption matrix elements. The Wigner–Eckart theorem has several partly related aspects. First, it expresses the conservation of angular momentum; it implies that the total angular momentum before and after the transition is the same. Second, it implies that the matrix elements of H_{int} are invariant under rotations of the coordinate axes. Third, it implies that the initial and final wave functions for the emission process are related by the vector relation

$$\mathbf{a} = \mathbf{b} + \mathbf{l} \tag{14}$$

and for the absorption process by

$$\mathbf{a} + \mathbf{l} = \mathbf{b} \tag{15}$$

However, in contrast to the usual vector sum, the transformations of Eqs. (14) and (15) are not unitary and the normalizations of the final wave functions contain the reduced matrix element as additional factors. We will refer to emission and absorption processes as dynamical changes.

We are now in a position to formulate the calculation represented by Eq. (1). The final state is obtained from the initial state by a succession of vector sums, some of them being the standard unitary transformation of representation and others the dynamical change arising from the interaction Hamiltonian. Consider a reaction in which a beam of particles with spin s_0 is absorbed by a target of spin s_1. The incident beam will be represented by a plane wave traveling in the positive direction of the z axis which will be expanded as a sum of terms having definite orbital angular momentum l_1. Thus the initial state of the reaction involves the three angular momenta

s_0, s_1, and l_1 which we will require to sum to form the spin of the compound state. Angular momenta can be added only two at a time and we will follow the conventional practice of first adding s_0 and s_1 to form the channel spin a,

$$\mathbf{s}_0 + \mathbf{s}_1 = \mathbf{a} \qquad (16)$$

If s_0 and s_1 are unpolarized, as is normally the case, then a will be also. Equation (16) is a normal vector sum and is not a dynamical change of the type of Eq. (15). We next form the sum

$$\mathbf{a} + \mathbf{l}_1 = \mathbf{b} \qquad (17)$$

where b is the spin of the compound nucleus. In this case, the sum represents a dynamical change and the wave functions representing the state b will contain the reduced matrix elements $\langle b \| l_1 | a \rangle$ as factors.

The breakup of the compound state into the reaction products is very similar. Let the outgoing orbital angular momentum and channel spin be l_2 and c, respectively. The breakup is a dynamical change represented by the vector sum

$$\mathbf{c} + \mathbf{l}_2 = \mathbf{b} \qquad (18)$$

Finally, the channel spin c is a normal vector sum of the spins of the outgoing particles s_3 and the final nucleus, s_4:

$$\mathbf{s}_3 + \mathbf{s}_4 = \mathbf{c} \qquad (19)$$

From the wave function point of view, the final state is composed of outgoing waves formed from the incoming waves of the initial state by the four vector sums, Eqs. (16)–(19). Outgoing fluxes and cross sections can be calculated in the way already noted. Corresponding to the wave function transformations Eqs. (16)–(19), there are transformations that enable the statistical tensors and density matrices of the final state to be evaluated from the initial one. If particles of spin a and orbital momentum l are absorbed to form a nucleus of spin b, then the statistical tensors of b are

$$\varrho_{k\kappa}(bb') = \sum_{k_a \kappa_a k_l \kappa_l} \varrho_{k_a \kappa_a}(aa')\, \varrho_{k_l \kappa_l}(ll')\, (k_a \kappa_a, k_l \kappa_l \,|\, k\kappa)$$

$$\times\, \hat{b}\hat{b}'\hat{k}_a\hat{k}_l \begin{Bmatrix} a & l & b \\ a' & l' & b' \\ k_a & k_l & k \end{Bmatrix} \langle b\| l\,|a\rangle \langle b'\| l'\,|a'\rangle^* \qquad (20)$$

where $\varrho_{k_a\kappa_a}(aa')$ and $\varrho_{k_l\kappa_l}(ll')$ are tensors for a and l, $(k_a\kappa_a, k_l\kappa_l | k\kappa)$ is a

Clebsch–Gordan vector addition coefficient, and

$$
\begin{Bmatrix}
a & l & b \\
a' & l' & b' \\
k_a & k_l & k
\end{Bmatrix}
$$

is a Wigner 9-j coefficient. The caret over a spin indicates the function $\hat{b} = (2b + 1)^{1/2}$, etc. If the nucleus of spin b emits particles of spin a and orbital momentum l, then the tensors of a and l are

$$
\varrho_{k_a \kappa_a}(aa')\, \varrho_{k_l \kappa_l}(ll') = \sum_{bb'k\kappa} \varrho_{k\kappa}(bb')\,(k_a \kappa_a, k_l \kappa_l \,|\, k\kappa)
$$

$$
\times\, \hat{b}\hat{b}'\hat{k}_a\hat{k}_l
\begin{Bmatrix}
a & l & b \\
a' & l' & b' \\
k_a & k_l & k
\end{Bmatrix}
\langle a \| l \| b \rangle \langle a' \| l' \| b' \rangle^* \tag{21}
$$

The tensor transformations required when the spins are related by a simple vector sum rather than an emission or absorption process are the same as Eqs. (20) and (21) but with the reduced matrix elements omitted. Equations (20) and (21) are then inverses of each other. The equations present a rather formidable appearance by virtue of being completely general. In practice, substantial simplifications result from one or more spins being either unpolarized or aligned. For an unpolarized spin, only the rank $k = 0$ is finite. If a is unpolarized in Eq. (20), then $k_a = 0$, and the equation becomes

$$
\varrho_{k\kappa}(bb') = \varrho_{00}(aa)\, \varrho_{k\kappa}(ll')\,(-)^{a+k-l-b'}(\hat{b}\hat{b}'/\hat{a})
$$

$$
\times\, W(lbl'b'; ak)\langle b \| l \| a \rangle \langle b' \| l' \| a \rangle^*
$$

where $W(lbl'b'; ak)$ is a Racah coefficient. The reductions involved here are $|\kappa_a| \leqslant k_a = 0$, $(00, k_l \kappa_l | k\kappa) = \delta_{k_l k} \delta_{\kappa_l \kappa}$, and

$$
\begin{Bmatrix}
a & l & b \\
a' & l' & b' \\
0 & k_l & k
\end{Bmatrix}
=
\begin{Bmatrix}
l & b & a \\
l' & b' & a' \\
k_l & k & 0
\end{Bmatrix}
= \delta_{aa'}\,\delta_{k_l k}(-)^{a+k-l-b'}(1/\hat{a}\hat{k})\, W(lbl'b'; ak)
$$

as given, for example, by Devons and Goldfarb (1957) or Ferguson (1965). If the spin a is aligned or polarized along the z axis, then it will be represented by the tensors $\varrho_{k0}(aa')$, i.e., with only the substates $\kappa = 0$ nonzero.

The density matrix transformations corresponding to Eqs. (20) and (21) are very infrequently required. The result for emission only will be given in

Eq. (22). The density matrix for the outgoing particles is

$$\langle a\alpha l\lambda| \varrho |a'\alpha'l'\lambda'\rangle = \sum_{b\beta b'\beta'} \langle a| l \,\|b\rangle (a\alpha, l\lambda | b\beta)$$

$$\times \langle b\beta| \varrho |b'\beta'\rangle (b'\beta' | a'\alpha'l'\lambda')\langle a'| l' \,\|b'\rangle^* \qquad (22)$$

The notation and ordering of factors conform here to Dirac's rules of notation (Dirac, 1947). The reality of the Clebsch–Gordan coefficients gives $(a\alpha, l\lambda|b\beta) = (b\beta|a\alpha, l\lambda)$. Similarly to the statistical tensors, if the transformation is not a dynamical change but simply a vector addition, then the reduced matrix elements are omitted.

The evaluation of an angular correlation through the use of statistical tensors thus has elements of similarity with the wave function method. A tensor representing the final state of the reaction is defined by

$$\varrho^f = \varrho_{k_0\kappa_0}(cc)\, \varrho_{k_1\kappa_1}(L_1L_1')\, \varrho_{k_2\kappa_2}(L_2L_2')\cdots \qquad (23)$$

where $\varrho_{k_0\kappa_0}(cc)$ is a tensor representing the residual nucleus with spin c and $\varrho_{k_1\kappa_1}(L_1L_1')$, $\varrho_{k_2\kappa_2}(L_2L_2')$,... represent the outgoing particles or photons. The angular correlation function is given by

$$W = \sum \varrho_{k_0\kappa_0}(cc)\, \varepsilon^*_{k_0\kappa_0}(cc)\, \varrho_{k_1\kappa_1}(L_1L_1')\, \varepsilon^*_{k_1\kappa_1}(L_1L_1')\cdots \qquad (24)$$

$\varepsilon_{k_0\kappa_0}(cc)$ is an efficiency tensor for detecting the final spin state and has only the component $\hat{c}\, \delta_{k_00}\delta_{\kappa_00}$. $\varepsilon_{k_1\kappa_1}(L_1L_1')$, $\varepsilon_{k_2\kappa_2}(L_2L_2')$,... are efficiency tensors for detecting the radiations $L_1, L_2,...$ and are functions of the positions and sizes of the detectors. Their evaluation is discussed in detail by Ferguson (1965). The summation in Eq. (24) is over all k, κ, L, and L'. The tensor ϱ^f is evaluated in terms of the tensor for the initial state ϱ^i through the use of Eqs. (20) and (21) and will have the form $\varrho^f = \sum K\varrho^i MM'^*$, where K represents the required recoupling coefficients and M and M' the various reduced matrix elements. The tensor ϱ^i will be a product of factors representing the spins and motions of the initial particles. These factors will consist of tensors for polarized or unpolarized spins and for the orbital angular momenta in plane wave states.

B. Signs of Matrix Elements

If a mixture of orbital angular momenta or multipolarities is present in the outgoing radiation from a reaction or if there is a mixture of spins in the compound nucleus, then coherent interference between these components will be found in the differential cross section. In general, interference must be expected unless it is removed in some specific way. An integration over all directions of the outgoing particles will remove the interference so that total cross sections do not show interference in general. The interference arises from terms of the form $a_l a_{l'}^* \phi_l \phi_{l'}^*$, $l \neq l'$, in Eq. (3), where $|\Psi|^2$ represents the density of outgoing particles. The index l may represent the orbital or the total angular momentum of the particles. Clearly the sign of these terms is sensitive to signs of the two factors. Of particular interest in this context is the interference between gamma-ray multipoles $E(L + 1)$ and ML which have the same parity and frequently have comparable strengths. In angular correlation studies, the terms involving multipole interference will contain the factors $\langle b|L\|a\rangle\langle b|L'\|a\rangle^*$. It is customary to divide the complete expression for W by $|\langle b|L\|a\rangle|^2$, where L represents the dominant multipole, to obtain the correlation function in terms of the multipole mixing ratio,

$$\delta = \langle b| L' \|a\rangle/\langle b| L \|a\rangle \tag{25}$$

$\delta^* = \delta$ because of the reality of the reduced matrix elements for gamma rays. Due to the sensitivity of the interference terms to the relative signs of the amplitudes and consequently to the sign of δ, this sign will normally be found in any analysis that extracts δ from measured data.

The evaluation of δ in terms of basic wave functions is a complex calculation that involves arbitrary choices at various stages. In order that a measured δ should agree in magnitude and sign with one calculated from a model, it is essential that the definitions and conventions that determine each be identical. Previous developments have frequently used different conventions and have often failed to be explicit about them. Rose and Brink (1967) have provided a development of angular correlation theory in which particular care has been taken in the basic definitions. Many recent papers have used this work for the evaluation of mixing ratios. Rose and Brink's conventions are the same as those of Goldfarb (1959), Litherland and Ferguson (1961), and Ferguson (1965). The first discussion in which the sign of δ is included in a comparison between experiment and theory was by Alder *et al.* (1956) in connection with multipole mixing in deexcitations through a rotational band. They pointed out that for an odd-A nucleus,

$$\text{sign}\,\delta = \text{sign}[Q_0/(g_\Omega - g_R)]$$

where Q_0 is the intrinsic quadrupole moment and g_Ω and g_R are the g factors for the odd particle and the collective motion, respectively. This suggestion has been exploited by McGowan and Stelson (1958) and Bernstein and de Boer (1960) to determine g_Ω and g_R. Using Rose and Brink's formulation, Häusser et al. (1966) have made a comparison between multipole mixtures measured experimentally and calculated from shell model wave functions for the nuclei ^{14}N, ^{15}N, ^{16}O, and ^{18}F. One E3/M2 and eight E2/M1 mixtures were studied. All showed qualitative agreement in magnitude with no discrepancies in sign between theory and experiment. A similar study with similar conclusions has been made by Poletti et al. (1967) on seven E2/M1 mixtures in 1p-shell nuclei.

The sign of a mixing ratio is dependent on whether the gamma rays are emitted or absorbed. The reduced matrix element for a state a that absorbs radiation l to form a state b is related to the matrix element for emission by a of the radiation l leaving the residual state b by (Devons and Goldfarb, 1957)

$$\hat{b}\langle b\| l |a\rangle = (-)^{a+l-b}\hat{a}\langle b| l \|a\rangle \tag{26}$$

Let δ_e denote the mixing ratio of Eq. (25) based on emission matrix elements and let $\delta_a = \langle b\|L'|a\rangle / \langle b\|L|a\rangle$ be a similar ratio for absorption matrix elements. Then by Eq. (26),

$$\delta_a = (-)^{L'-L} \delta_e \tag{27}$$

In most cases, for example, E2/M1 mixtures, $L' = L + 1$, so that multipole mixtures for absorption have the opposite sign to those for emission.

The angular correlation for a cascade of two gamma rays described by the spin-multipole sequence $a(L_1L_1')b(L_2L_2')c$, where a is unpolarized and c is unobserved, is

$$W(\theta) = \sum_{kr_1r_2} (-)^{a-c-L_1-L_1'} \delta_{e_1}^{r_1} \delta_{e_2}^{r_2} \bar{Z}_1(L_1bL_1'b; ak)$$

$$\times \bar{Z}_1(L_2bL_2'b; ck) P_k(\cos\theta) \tag{28}$$

where the $\bar{Z}_1$ coefficients are defined and tabulated by Ferguson (1965) and the multipole mixtures for emission are used in conformity with current practice. The exponents r_1 and r_2 have the values 0, 1, or 2 according to whether the multipoles involved are (LL), (LL'), or $(L'L')$. An asymmetry between the incoming and outgoing radiation is present in Eq. (28) in the form of the factor $(-)^{-(L_1+L_1')}$, which, however, can be removed by utilizing Eq. (27) to give

$$(-)^{-(L_1+L_1')} \delta_{e_1} = (-)^{-2L_1'+L_1'-L_1} \delta_{e_1} = \delta_{a_1}$$

L_1 and L_1' are integers so that $(-)^{2L_1'}$ is positive. After this substitution,

Eq. (28) becomes

$$W(\theta) = \sum_{kr_1r_2} (-)^{a-c} \delta_{a_1}^{r_1} \delta_{e_2}^{r_2} Z_1(L_1bL_1'b; ak) \bar{Z}_1(L_2bL_2'b; ck) P_k(\cos\theta)$$

(29)

representing the absorption of L_1 followed by the emission of L_2. This is essentially a reaction involving absorption and emission and, as required by time-reversal invariance, has symmetry between the initial and final states. It was proposed by Biedenharn and Rose (1953) in their early development of the theory that Eq. (29) be used in the interpretation of all correlations regardless of their being reactions or sequential decays, by virtue of its simplicity. The problem described by Ofer (1959) was not anticipated at this point. In a cascade of three gamma rays from a radioactive source, it may be possible to measure the angular correlation between the first and second and also between the second and third. On the Biedenharn–Rose convention, the second transition is treated as an emission process in the first measurement and as an absorption process in the second, and, due to Eq. (27), the values of δ resulting from the two measurements will have opposite signs.

No conflict of sign of this kind will arise if emission-type matrix elements are used consistently to represent emission processes. Specifically, if Eq. (28) is used to interpret the two correlations measured by Ofer, then no inconsistency of sign will appear, including the mixing ratio of the middle transition. The appearance of several papers by Pande, Singh, and Dahiya (Singh et al., 1971; Pande and Singh, 1971a, b) which attempt to measure the "real" sign of δ by measuring cascades in triple coincidence indicates that confusion on the question remains. While their papers are obscure on what is meant by the "real" sign of δ, the problem clearly resides in an adherence to the Biedenharn–Rose convention.

From the viewpoint of an experimenter, conformity to the phase conventions of Rose and Brink means the employment of their formulas for the interpretation of experimental data. Extension of their formulas to polarized gamma rays is straightforward and has been given by Twin et al. (1970) and Taras (1971). A convention that can influence matrix element phases is the order in which the terms of the vector sum occur in the definitions of the reduced matrix elements, Eqs. (12) and (13). The noncommutability of vector sums is reflected in the phase rule for Clebsch–Gordan coefficients

$$(b\beta, l\lambda \,|\, a\alpha) = (-)^{b+l-a}(l\lambda, b\beta \,|\, a\alpha)$$

(30)

Thus if, instead of Eq. (12), a reduced matrix element for emission were defined by

$$\langle l\lambda, b\beta| H_{\text{int}} |a\alpha\rangle = (l\lambda, b\beta | a\alpha)\langle b| l \| a\rangle' \qquad (31)$$

then this element would be related to the standard one by

$$\langle b| l \| a\rangle' = (-)^{b+l-a}\langle b| l \| a\rangle \qquad (32)$$

This relation is very similar to Eq. (26) for emission and absorption matrix elements and thus multipole mixing ratios resulting from the two definitions will be related by

$$\delta_e' = (-)^{L'-L} \delta_e$$

and will have opposite signs. The danger of error from this source is present when angular correlation formulas are developed ab initio using the principles outlined in Section I.A of this chapter. Inversion of the vector sum order would result in the replacement of

$$\begin{Bmatrix} a & l & b \\ a' & l' & b' \\ k_a & k_l & k \end{Bmatrix} \quad \text{by} \quad \begin{Bmatrix} l & a & b \\ l' & a' & b' \\ k_l & k_a & k \end{Bmatrix}$$

in Eqs. (20) and (21) and would lead ultimately to an inversion of the sign of the corresponding δ. The order used in Eq. (12) to define emission reduced matrix elements is the same as that of Rose and Brink (1967).

In an interesting paper, Danos (1971) has discussed the problem of phase definition in relation to nuclear shell model calculations. A major problem in the shell model is angular momentum recoupling and this is also a major source of phase errors. Danos proposes a standard form for all recouplings based on the Wigner 9-j coefficient and having no additional sign factors. In this way, recouplings can be carried through their various steps without troublesome sign factors and consequently with minimum chance of error.

II. Some Particular Techniques

A. COMMENTS ON SPIN-FLIP

Spin-flip is the transition of a particle with spin from one magnetic substate to a different one in the course of a reaction. It arises, of necessity, from spin-dependent forces. In many cases, particularly for reactions that form a compound nucleus, these spin-dependent forces are buried among the details of the internucleon interactions. This can be illustrated by the case of elastic proton scattering at medium and low energies. If no spin-dependent forces operated on the proton, it would behave like a spinless particle; in particular, the two angular momentum components $j = l + \frac{1}{2}$ and $j = l - \frac{1}{2}$ would behave identically, even through resonances. No energy region is known where such a degeneracy occurs. In a case such as this, the mere fact that different j's are separated indicates that both m_l and m_s are mixed in the wave functions and are therefore not constants of the motion.

Since the angular distribution of the gamma rays following a reaction is sensitive to the orientation of the excited nucleus before deexcitation, it has been used in several classes of studies of spin-dependent forces. Satchler (1964) has described in detail the formalism of distorted wave reaction theory with spin-dependent forces included. Formulas for the statistical tensors of the residual nucleus are given which enable the angular correlations of subsequent gamma rays to be predicted.

B. Particle–Particle-Gamma Reactions in a Collinear Geometry

Angular correlation methods, in general, require that the reaction used involve principally states having unique (or "sharp") angular momentum. When mixtures occur, the associated amplitudes are normally not predictable but in some cases may be treated as arbitrary parameters to be found from the experiment. However, the amount of information provided by an angular correlation is ordinarily small so that only a small number of unknown mixtures can be tolerated. In many bombardment experiments, particularly those using tandem accelerators, the compound nucleus has an excitation energy in the region of strongly overlapping states which require an excessively large number of parameters for their specification. To avoid this problem, Litherland and Ferguson (1961) have described two procedures, designated as methods I and II, that allow angular correlation measurements to be interpreted with a small number of arbitrary parameters. In both cases, a reaction of the form $A(p_1, p_2)B^*$ is assumed, where A and B* represent initial and product nuclei and p_1 and p_2 represent any particles. In method II, the outgoing particles p_2 are detected in a counter located at either 0° or 180° relative to the beam axis; in the latter case, an annular counter with a hole for the beam is used. Let A, B*, p_1, and p_2 have spins a, b, s_1, and s_2 with magnetic substates α, β, σ_1, and σ_2. The nucleus B* will be in a bound excited state so that b will be sharp. It is then shown that B* is aligned and can be described by a set of population parameters $w(\beta)$, where $0 \leqslant \beta \leqslant a + s_1 + s_2$. The angular correlations of gamma rays emitted by B* relative to the beam axis are determined by the $w(\beta)$ together with the spins of the states and the multiplicities of the radiations. The $w(\beta)$ depend on the reaction and are unpredictable, but if a, s_1, and s_2 are small, then their number will be small and they can be treated as unknown parameters to be found from the gamma-ray angular correlation measurement. The reaction (α, α') on a target of zero spin is ideal for this method because only the parameter $w(0)$ is required and is merged with the normalization factor in the intensity.

Method II is somewhat less effective in (d, p) and (d, n) reactions than in reactions involving particles of smaller spin. In these reactions, if the target has spin zero, then the normal application of the method requires two adjustable parameters for the $\frac{1}{2}$ and $\frac{3}{2}$ substates of the gamma-emitting state. An unknown multipole mixing ratio is usually present, so that in this case, there are three unknown parameters to be determined from a measurement of the three Legendre polynomial coefficients a_0, a_2, and a_4. If the target has spin greater than zero, then more than two population parameters are required and the method cannot be applied because the number of unknowns exceeds the number of independent measurable quantities. These conclusions are founded on the assumption of unrestricted spin-flipping possibilities which are characteristic of compound nucleus reactions. The application of the method could be broadened considerably if means were found to parameterize the correlation with a smaller number of parameters. Goldfarb (1964) has recommended using the method II particle–gamma geometry in cases where the principal mechanism is a direct reaction. Spin-dependent interactions play a minor part here and Goldfarb shows that the number of substate population parameters is then smaller than in the general case.

Let the incoming and outgoing particles have orbital momenta l_1 and l_2, respectively. The total angular momentum of the incoming and outgoing particles is given by

$$\mathbf{j}_1 = \mathbf{l}_1 + \mathbf{s}_1 \tag{33a}$$

$$\mathbf{j}_2 = \mathbf{l}_2 + \mathbf{s}_2 \tag{33b}$$

The reaction process transfers angular momentum j to the target, consisting of the spin s of the captured particle and l, the captured orbital momentum. Thus

$$\mathbf{j} = \mathbf{l} + \mathbf{s} \tag{33c}$$

$$\mathbf{b} = \mathbf{a} + \mathbf{j} \tag{33d}$$

$$\mathbf{j}_1 = \mathbf{j}_2 + \mathbf{j} \tag{33e}$$

The conservation of angular momentum is contained in Eqs. (33d) and (33e);

Eq. (33d) is the condition that the target absorb angular momentum j; and Eq. (33e) is the condition that the incoming and outgoing beams lose it. These two equations represent dynamical processes and reduced matrix elements are associated with each one.

The essence of the method is to parameterize the gamma-emitting state b in terms of the substates m of the transferred angular momentum j. The tensors of the state b are given by

$$\varrho_{k0}(b) = \sum_{jj'} (-)^{a-b} \varrho_{k0}(jj') \, W(bbjj'; ka) \tag{34}$$

This results from Eq. (20) with the assumption that a is unpolarized. A number of unessential factors have been omitted and reduced matrix elements for the absorption of the transferred particle are contained in $\varrho_{k0}(jj')$. The axial symmetry of the system leaves only the tensors $\varrho_{k0}(b)$ nonzero. The tensors for j can be expressed in terms of the substate populations x_m of m,

$$\varrho_{k0}(jj') = \sum_{m} (-)^{j'-m} (jm, j' - m \,|\, k0) \, x_m \tag{35}$$

Finally the angular correlations of the gamma rays emitted by the state b are given by

$$W(\theta) = \sum_{LL'k} \varrho_{k0}(b)(-)^{c-b-L+L'-k-1} \, \delta^r \hat{L} \hat{L}'$$

$$\times \, (L1, L'-1 \,|\, k0) \, W(LbL'b; ck) \, P_k(\cos\theta) \tag{36}$$

where L and L' are interfering multipoles, δ is the multipole mixing ratio, and c is the final spin following the gamma-ray emission. By eliminating $\varrho_{k0}(b)$ from Eq. (36) through the use of Eqs. (34) and (35), $W(\theta)$ is expressed in terms of the x_m, which are treated as arbitrary parameters, similarly to the way the substate population parameters of b are used in method II. If spin-dependent forces are negligible, then the number of parameters will be less in Goldfarb's approach.

A number of simplifying assumptions are made, namely that only one value of j is involved and that no spin-dependent interactions involving the spins a and b are present. Let λ_1 and λ_2 represent the substates of l_1 and l_2, and m_1 and m_2 those of j_1 and j_2. From Eq. (33e), we obtain $m_1 = m_2 + m$, or

$$m = m_1 - m_2 = \lambda_1 + \sigma_1 - \lambda_2 - \sigma_2 = \sigma_1 - \sigma_2 \tag{37}$$

where $\lambda_1 = \lambda_2 = 0$ because the incident and outgoing waves are in the z direction. Due to spin-dependent forces, spin-flip transitions $\sigma_1 \to \sigma_1'$ and

$\sigma_2 \rightarrow \sigma_2'$ may occur near the nucleus. Goldfarb shows that the spin substate σ for the captured particle is given by

$$\sigma = \sigma_1' - \sigma_2' \leqslant s \tag{38}$$

Combining Eqs. (37) and (38), we get

$$m = \sigma + (\sigma_1 - \sigma_1') - (\sigma_2 - \sigma_2') \tag{39}$$

Equations (37) and (39) imply the limitations

$$m \leqslant s_1 + s_2 \tag{40a}$$

$$m \leqslant s + 2s_1 + 2s_2 \tag{40b}$$

Equation (40a) is related to the limitation on which method II is based, namely $\beta = \alpha + m \leqslant a + s_1 + s_2$. Equation (40b) appears less restrictive than Eq. (40a), but it may be noted that the terms $2s_1$ and $2s_2$ arise from spin-flip, so that this equation may be more restrictive if spin-flip is absent in one or both channels. For no spin-flip in both channels, Eq. (40b) gives $m \leqslant s$, which, in the case of stripping, is a very strong restriction.

A theoretical study of the applicability of this procedure to the $^{52}Cr(d, p)^{53}Cr$ stripping reaction has been made by Goldfarb and Wong (1966). Using the published optical model parameters for both entrance and exit channels, they have computed the contributions to the spin-flip factor $\xi = w(\frac{3}{2})/w(\frac{1}{2})$, where $w(\beta)$ is the population of the substate β, due to direct and compound nucleus processes, at $0°$ and $180°$. The compound nucleus contribution was calculated using the Hauser–Feshbach theory. Spin-flip from the direct interaction is found to be small at $0°$ and large at $180°$ and it is large from the compound nucleus process at all angles. The direct reaction cross section is substantially larger than the compound nucleus cross section at $0°$ so that little spin-flip is predicted at this angle.

Critchley et al. (1971) have applied Goldfarb's procedure to the reactions $^{28}Si(d, p\gamma)^{29}Si$ and $^{9}Be(d, p\gamma)^{10}Be$ at incident energies of 4 and 5 MeV. The reaction protons were separated from the deuteron beam at $0°$ using a magnetic spectrograph. The results for the 4.93-MeV $(\frac{3}{2}^-)$ level of ^{29}Si were compatible with the absence of spin flip; the substate populations of this $\frac{3}{2}^-$ gamma-emitting state have the ratio $w(\frac{3}{2})/w(\frac{1}{2}) = 0.025$.

The 2^+ level of ^{10}Be at 3.37 MeV was studied and here, due to the spin of $\frac{3}{2}^-$ for ^{9}Be, all substates β can be populated without spin-flip. For this reason, conclusions regarding spin-flip cannot be drawn. In this case, they find that a channel spin representation is preferable to Goldfarb's and consequently modify the analysis as follows. In place of Eqs. (33c) and (33d), they write

$$\mathbf{S} = \mathbf{a} + \mathbf{s}, \qquad \mathbf{b} = \mathbf{l} + \mathbf{S}$$

where S is the channel spin, s is the neutron spin, $\frac{1}{2}$, and l is the captured orbital angular momentum, assumed to be $l = 1$. The result $(S = 1)/(S = 2)$ $= 0.3 \pm 0.06$ is found for the channel spin ratio. The equivalent matrix elements for $j = l \pm \frac{1}{2}$ have the ratio

$$|\langle b \| \tfrac{3}{2} \| a \rangle|^2 / |\langle b \| \tfrac{1}{2} \| a \rangle|^2 = 23$$

so that $j = \frac{3}{2}$ is transferred much more strongly than $j = \frac{1}{2}$, in agreement with Goldfarb's assumption.

C. SPIN-FLIP IN INELASTIC SCATTERING

Schmidt *et al.* (1964) have developed an ingenious procedure that uses particle–gamma angular correlations to measure spin-flip in reactions involving spin-$\frac{1}{2}$ particles. It is based on a theorem of Bohr (1959) which states that for any two-body reaction which conserves parity and angular momentum,

$$P_i e^{i\pi S_i} = P_f e^{i\pi S_f}$$

where P_i and P_f are the initial and final intrinsic parities and S_i and S_f are the total spin projections of the initial and final states perpendicular to the reaction plane. The quantization axis is chosen normal to the reaction plane so that the total spin projections S_i and S_f are represented by the sums of the z-component substates. We now consider a reaction involving incoming and outgoing particles of spin $\frac{1}{2}$ on a spin-0 target. If the initial and final states have the same parity, then Bohr's theorem shows that $S_f - S_i$ is even. Let σ_i and σ_f represent the initial and final projections of the spin-$\frac{1}{2}$ particle and m the projection of the residual nucleus. Then $\sigma_f + m - \sigma_i = S_f - S_i$ and is even. For no spin-flip, $\sigma_f = \sigma_i$, in which case m is even. Odd values of m arise only if $\sigma_f \neq \sigma_i$ due to spin interactions.

We now assume that a spin-2 state is excited in the inelastic scattering which decays to a spin-0 state. Polar diagrams representing the radiation patterns from the three possible substates $m = 0$, 1, and 2 are shown in Fig. 1. It is evident that the only substate that produces radiation along the z axis is $m = 1$. Thus the degree of spin-flip can be measured by the intensity of coincidences between a particle detector at some arbitrary angle of scat-

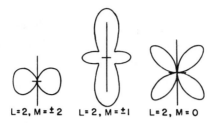

L=2, M=±2   L=2, M=±1   L=2, M=0

Fig. 1. Polar plots for quadrupole radiation. The z axis is the vertical line in the plane of the diagram.

tering and a gamma-ray detector situated on the normal to the plane of scattering.

A number of studies have been made utilizing this procedure, the most recent being by Howell and Galonsky (1972), Wilson and Schecter (1971), Sweeney and Ellis (1971), and McDaniel and Amos (1972). Theoretical evaluations of the results are obtained from DWBA calculations including spin-dependent interactions. In general, moderate agreement between the theory and experiment is obtained, indicating that the optical potentials are generally sufficient to account for the spin-flip process. However, Wilson and Schecter (1971) have obtained anomalously large spin-flip cross sections for $^{32}S(p, p')$, which is probably due to compound nucleus processes. Spin-flip in the reaction $^{12}C(^{3}He, ^{3}He')$ has been studied by Assousa (1968) and of deuterons on Mg, S, Ti, and Ni by Hippelein et al. (1970). In the latter case, population of the $m = 1$ substate only can be observed, while $m = 2$ will be undoubtedly populated by the spin-flip process. For this reason, Hippelein et al. have referred to the process as "half-spin-flip."

D. An Azimuthal Geometry

Goldfarb and Seyler (1968) have proposed a simple geometry that can be applied in the case of resonant reactions to obtain the spin of the resonance and possibly other information. It is assumed that a reaction of the type $(p_1, p_2\gamma)$ or $(p_1, \gamma\gamma)$, with detection of coincidences between the two outgoing radiations, is used. The essence of the method is to find geometries that provide information on the maximum ranks of the statistical tensors that represent the successive states of the reaction. The procedure is to attach the z axis of the coordinate system to either the incoming beam direction or to the first outgoing radiation direction and to examine the azimuthal variation of the final gamma ray about these axes. Let a, b, c, and d represent the spins of the target, compound, and first and second residual states. Let s_1 and s_2 be the spins of the incident and outgoing particles and l_1 and l_2 their orbital angular momenta, respectively. The channel spins S_1 and S_2 are given by $S_1 = a + s_1$, $S_2 = c + s_2$ and the multipolarity of the final gamma ray is assumed to be L. Also let tensors of rank k_1, k_2, and k represent the incoming and outgoing particles and the final gamma ray. These ranks will be limited by the following triangle conditions: $(l_1 l_1' k_1)$, $(b b k_1)$, $(l_2 l_2' k_2)$, $(s_2 s_2' k)$, $(c c k)$, $(L L' k)$, and $(k_1 k_2 k)$. The triple angular correlation is represented by a sum of the invariant functions

$$(4\pi)^{3/2} \sum_{\kappa_1 \kappa_2 \kappa} Y^*_{k_1 \kappa_1}(\Omega_1) Y^*_{k_2 \kappa_2}(\Omega_2) Y_{k\kappa}(\Omega) (k_1 \kappa_1, k_2 \kappa_2 | k\kappa)$$

Here, the arguments Ω_1, Ω_2, and Ω represent the polar angles (θ, ϕ) of the three radiations. If we now consider a geometry in which only the azimuthal angle ϕ of the gamma ray is varied, the correlation function will have the form

$$W(\phi) = \sum_{\kappa=0}^{\kappa_{max}} a_\kappa \cos \kappa\phi \qquad (41)$$

Ω_1 and Ω_2 are taken to be different directions. Goldfarb and Seyler do not comment on a choice for these directions, but a relative angle of 90° between them is probably a reasonable one and has been used by Duray et al. (1969). If now the coordinate z axis is chosen in the outgoing particle direction, then only $\kappa_2 = 0$ will survive, and the Clebsch–Gordan coefficient $(k_1 \kappa_1, k_2 0 | k\kappa)$ requires that $\kappa_1 = \kappa$. By fitting the experimental correlation to Eq. (41), κ_{max} can be determined. It then follows that either $k_{1 \, max} = \kappa_{max}$, implying $l_1 = \kappa_{max}/2$, or $b = \kappa_{max}/2$ or $(\kappa_{max} - 1)/2$, depending on whether b is integral or half-integral, or else $k_{max} = \kappa_{max}$, implying $L = \kappa_{max}/2$ or $c = \kappa_{max}/2$ or $(\kappa_{max} - 1)/2$. Alternatively, if the z axis is chosen in the beam direction, then the same procedure provides information about $k_{2 \, max}$ and l_2.

Duray et al. (1969) have used this technique with the reaction ^{24}Mg(p, p'γ) to obtain a tentative spin of $\frac{3}{2}$ for the 6.92-MeV level of ^{25}Al. Other reports on its use are by Morgan et al. (1971) and Steiner et al. (1971). The restriction to resonance reactions limits its application to the relatively narrow regions of excitation energy where isolated resonances are present and possibly also to isobaric analog resonances provided there is not an excessive background present. By attempting to exploit qualitative information, it represents a departure from current practice in which hypotheses on spins and multipolarities are tested by goodness of fit to fairly compre hensive and accurate data. The latter fail to provide unique assignments frequently and it must be expected that failure for the weaker Goldfarb–Seyler technique will be even more frequent.

III. Perturbations in Vacuum Recoils

Recent studies of gamma-ray angular correlations of reaction products recoiling with high velocities into vacuum (Grodzins, 1970; Berant et al., 1971; and references cited therein) have shown that perturbations of the nuclear orientation due to the hyperfine interactions with the highly stripped atomic electron cloud can cause major attenuation of the correlations. These experiments involve bombardment of foil targets thin enough to allow the reaction product to emerge into the vacuum. In general, no attenuation of the correlation occurs when the recoils are within the solid material of the target, due to the rapid collision rate there. The recoiling atoms enter the vacuum with the loss of many or, in a few cases, all of their electrons, the remaining electrons being in a variety of excited atomic states. Atomic deexcitations proceed rapidly with a period of the order of a few picoseconds. The electron cloud produces at the nucleus a fluctuating magnetic field in the range of one to several hundred megagauss that is the principal cause of the nuclear perturbations. The strongest fields and perturbations arise from the presence of a single $1s_{1/2}$ electron, which, however is improbable unless the recoil velocities $v \geq 0.1c$. A recent experiment by Ward et al. (1972) on gamma-ray correlations from ^{150}Sm nuclei recoiling in vacuum with $v/c = 2\%$ shows significant attenuation of the a_2 and a_4 coefficients with periods of 40×10^{-12} and 16×10^{-12} sec, respectively.

Method II experiments (Litherland and Ferguson, 1961; Ferguson, 1965) are generally performed in circumstances very similar to those just described. Thus care must be taken that hyperfine perturbations in the vacuum do not invalidate the analysis. The results of Ward et al. show that perturbations having periods in the range of 10^{-11} sec will generally be present and precautions to avoid them are required for nuclear lifetimes of this magnitude or greater. An effective precaution is an inert target backing sufficiently thick to ensure that most decays have occurred before the nuclei emerge from the backing. The backing material should not perturb the moving or stopped ions. Nonferromagnetic elements are suitable in this respect. The material also should not introduce excessively strong contaminant peaks into the particle spectrum. The comprehensive range tables of Northcliffe and Schilling (1970) can be used to determine an adequate backing thickness.

IV. Optimization

A principal object of angular correlation studies is the determination of the spins of nuclear levels. However, frequently such efforts are unsuccessful in that, while the results of the experiment unequivocally reject a number of possibilities, there may remain two or more possible choices that provide equally good fits to the results. In this context, it was early thought that triple angular correlation techniques would be much more powerful than double correlation techniques due to the much greater amount of information supplied by the former. This enthusiasm has been somewhat dissipated by the discovery that cases of ambiguity can arise from triple angular correlation studies. Some cases where ambiguity is very conspicuous are the spin sequences 4–2–0, 5–3–1, 6–4–2, etc. A scrutiny of the table of Kaye *et al.* (1968) shows that the triple angular correlation coefficients associated with the lowest-rank tensors are nearly identical for these sequences, whereas appreciable differences are found among those for the highest-rank tensors. While the requirement to distinguish these sequences is probably unlikely, other less obvious ambiguities have arisen. Optimization of some sort will clearly be helpful and possibly critical in such cases. In the example given, this might take the form of optimizing the accuracy with which the coefficients of the highest-rank tensors are measured.

There are two circumstances where optimization may be applied. The first is where there is no prior knowledge about the correlation. Optimization is achieved here by obtaining, in some general sense, the most accurate determination of the expansion coefficients of the correlation. The proposals that have been advanced for this case have been the maximization of the Gram determinant of the normal equations (Ferguson, 1965) and the minimization of a linear function of the variances of the coefficients (Monahan and Langsdorf, 1965; Walraven and McCauley, 1971; Waibel and Grosswendt, 1971). The second circumstance is where some knowledge already exists about the correlation, possibly to the extent that a decision is required between two spin choices. Relatively little work has been done for this case.

A. MAXIMIZATION OF THE GRAM DETERMINANT

The equations for the determination of a set of coefficients in a linear least-squares fit are, in matrix form,

$$Na = PwV$$

where N is the normal matrix; a is a column matrix of the coefficients;

P is a matrix of the set of fitted functions; w is a weight matrix; and V is a column matrix of measured values. The functions will be assumed to be angular correlation functions of one angle θ for the case of a double correlation or of three angles θ_1, θ_2, and ϕ for the case of a triple correlation. We will define the Gram determinant as $G = \det(N)$, where $\det(N)$ signifies the determinant of the normal matrix N. This is equivalent to the classical definition (Margenau and Murphy, 1956) but with the extension that N includes weights that are, in general, different from unity. If $G = 0$, then N is a singular matrix and the equations are insoluble. This condition results from a linear dependence among the fitted functions at the angles selected and implies that the equations contain too many unknowns. It can be resolved by reducing the number of unknowns by fixing some of them at values obtained from other considerations. The selection of coefficients to delete in this way is not entirely arbitrary; they must belong to the functions among which the linear dependences exist. If G is not zero, but small, then the equations will be soluble but ill conditioned and the coefficients obtained from them will have large errors. Maximization of the Gram determinant leads to equations that can be regarded as best conditioned.

The optimum is obtained by maximizing G under variation of the angular positions at which the correlation is measured. The weights may also be included as optimizing parameters, but considerable complication is added to the experimental procedure to realize the proper values. It would further not be applicable to counter array experiments because the weights of the individual points cannot be controlled. The accuracy of a set of parameters obtained from a least-squares fit can be specified by confidence regions. The regions consist of a linear interval for one parameter, an ellipse for two parameters, and a hyperellipse in m dimensions for m parameters, and include the influence of statistical correlations when two or more parameters are involved. For the general case of m parameters, the volume of the confidence hyperellipse in the m-dimensional parameter space is inversely proportional to $G^{1/2}$, so that this volume is minimized when G is maximized. Thus the significance of this optimum is that the confidence region is minimized. To demonstrate this property, let λ_1, λ_2, ..., λ_m represent the eigenvalues of N. Then the principal axes of the hyperellipse are proportional to $\lambda_1^{-1/2}$, $\lambda_2^{-1/2}$, ..., $\lambda_m^{-1/2}$ and the volume V is proportional to $(\prod_{i=1}^{m} \lambda_i)^{-1/2}$. But $\det(N) = \prod_{i=1}^{m} \lambda_i$, so that $V \propto [\det(N)]^{-1/2}$.

The optimum is invariant under any linear transformation of the fitted functions. Let the fitted functions be transformed to a set Q by the transformation $Q = PX$, where X is a nonsingular square matrix. The normal matrix for fitting in terms of Q is $N' = \tilde{Q}wQ = \tilde{X}\tilde{P}wPX = \tilde{X}NX$. Thus $\det(N') = \det(N)\det^2(X)$ and a maximum of $\det(N)$ is also a maximum of $\det(N')$. An illustration of such a transformation is that between the Legendre polynomials $P_k(\cos\theta)$ and the powers of $\cos\theta$. The same set of angles $\{\theta_i\}$ optimizes the fitting of $W(\theta) = \sum_k a_k P_k (\cos\theta)$ and $W(\theta) = \sum_r b_r \cos^r\theta$. The results of the two fits are related by $a = Xb$, where a and b are column matrices of the fitted parameters a_k and b_r and the square matrix X contains the coefficients that express the powers of $\cos\theta$, assumed here to be even, in terms of the Legendre polynomials.

A study has been made (Ferguson, unpublished) of the effectiveness of this optimization in resolving an ambiguity arising in angular correlation measurements attempting to determine the spin of the 4.47-MeV level of ^{22}Ne (Broude and Eswaran, 1964; Buhl et al., 1967). The study was based on the application of method I[†] to the cascade through the $4.47 \rightarrow 1.28 \rightarrow 0$ MeV levels, having the spin sequence $J \rightarrow 2 \rightarrow 0$. Broude and Eswaran used this method in the octant geometries described by Broude and Gove (1963) and found acceptable fits for $J = 2$ and $J = 3$. The same result was obtained from the method II studies of Buhl et al. (1967). The observation of a ground-state transition in this last work eliminates the spin-3 possibility. The procedure was to compute the correlation for several optimal and nonoptimal geometries assuming the parameters of the $J = 2$ and $J = 3$ fits obtained by Broude and Eswaran. The optima were found by maximizing the 19×19 determinant of the normal matrix for fitting the $X^\kappa_{k_1 k_2}(\theta_1 \theta_2 \phi)$ functions (Ferguson, 1965) that describe the triple correlation. An array of seven counters was assumed, set up so that the 42 pairs of possible coincidences were recorded. The computed correlations were then treated as measured data and fits were made to them by the least-squares method assuming the alternate spin assignment. It was found, as was expected, that a greater difference between the two assignments was obtained for optimal than for nonoptimal geometries, the difference being measured by the total squared residuals, or "χ^2," of the fits. The same test was applied to the geometry in which the correlations were measured by Broude and Eswaran with the result that the parameters for the best fit to the synthetic data were different

---

[†]   Method I has been described in Section II.B. It differs from method II in that the particles p$_2$ are not observed, but the angular correlations of two cascade gamma rays from the excited state B* are measured. The measurement is a triple angular correlation involving the directions of the beam and the two gamma rays and is expected to provide sufficient information to determine all of the populations $w(\beta)$ as well as mixing parameters for the gamma rays.

from those for the best fit to the original data. This result is, of course, due to the random fluctuations of the real data and it is concluded that a test of this type cannot be conclusive without an understanding of the effects of fluctuations. The final conclusion was therefore qualitative: The optimal geometries are better, but it has not been shown that they are capable of resolving the ambiguity.

B. Error Minimization

Optimization by minimizing a linear function of the variances of the fitted parameters has been considered by a number of authors, notably Monahan and Langsdorf (1965). The linear function is defined by

$$G = \sum_i \eta_i \sigma^2(a_i)$$

where η_i are an arbitrary set of positive constants, and the $\sigma^2(a_i)$ are the variances of the fitted parameters. Only the case $\eta_i = 1$ for all i has been considered by Monahan and Langsdorf. In the case of ambiguous spin assignments, it may be possible to use a form of G that would exploit the fact that high-rank terms show larger differences than do low-rank terms in such cases. This would be achieved, for example, by choosing $\eta_i = 1$ for high-rank terms and $\eta_i = 0$ for the low-rank terms. For a correlation expressible as a sum of Legendre polynomials $P_k(x)$, such a choice might be $\eta_0 = \eta_2 = 0$ and $\eta_4 = 1$. The correlation function is assumed to be expanded in terms of orthonormal polynomials $p_k(x)$:

$$W(x) = \sum_{k=0}^{n-1} a_k p_k(x)$$

where

$$a_k = \int w(x) \, W(x) \, p_k(x) \, dx$$

and $w(x)$ is the appropriate weight function. We take a and b as the limits over which othogonality holds. We will assume that the $p_k(x)$ are the normalized Legendre polynomials $p_k(x) = (k + \frac{1}{2})^{1/2} P_k(x)$, for which $a = -1$, $b = 1$, and $w(x) = 1$. The set of points $x_1, x_2, ..., x_n$ that are the roots of

$$p_n(x) - c p_{n-1}(x) = 0$$

where c is an arbitrary constant, then minimizes the function G, independently of the errors of the measurements at these points. The constant c is limited by the condition $p_n(a)/p_{n-1}(a) \leqslant c \leqslant p_n(b)/p_{n-1}(b)$, which guarantees that the required roots lie within the interval (a, b). If it is desired to choose one point x_r arbitrarily, then the constant will be given by $c = p_n(x_r)/p_{n-1}(x_r)$.

It is customary in angular correlation work to overdetermine the solution by measuring more points than there are unknown parameters. The analysis here applies to the determination of n coefficients $a_0, a_1, ..., a_{n-1}$ from measurements at the n points $x_1, ..., x_n$. Monahan and Langsdorf have extended their analysis to determine where an additional point x_{n+1} must be located in order to effect the maximum decrease in G, i.e., to effect the greatest improvement in accuracy.

Optimization when the merit function G has a form less simple than that considered here does not appear to be easily soluble and is not considered by Monahan and Langsdorf. Optimal conditions can always be found by purely numerical methods in such cases. Walraven and McCauley (1971) have made a numerical study of optimal conditions for the measurement of the angular correlation function

$$W(\theta) = a_0 + a_2 P_2(\cos \theta) + a_4 P_4(\cos \theta)$$

using the merit function

$$G = \sigma^2(A_2) + \sigma^2(A_4), \quad \text{where} \quad A_2 = a_2/a_0 \quad \text{and} \quad A_4 = a_4/a_0$$

The parameters varied to achieve the optimum were (a) the number of points in the angular range, (b) the angular positions, and (c) the time of counting at each position. In general, the optima depended on the values of A_2 and A_4 and the optimal times of counting at each position are presented as contour plots in terms of these two parameters. Application of these results thus requires a preliminary estimate of the correlation coefficients. General results of the study are: (i) Three points is always the optimal number. Thus overdetermination, which is highly desirable from other points of view, has no value in this context. (ii) If θ is an optimal point, then $270° - \theta$ gives an identical optimum. (iii) Two of the angles were always 90° and 135°, the remaining one being in the range 107–122°, depending on the values of A_2 and A_4.

A similar study has been made by Waibel and Grosswendt (1971). The correlation function $W(\theta) = a_0 + a_2 P_2 (\cos\theta) + a_4 P_4 (\cos\theta)$ measured at three angles has been assumed. Minimization of the individual variances $\sigma^2(a_k)$ and the total variance

$$G = \sum_k \sigma^2(a_k)$$

has been effected first when weights and angles are varied and second when angles alone are varied with the weights equal. The results for these various cases are presented in graphs and tables. This case differs from that studied by Walraven and McCauley (1971) in that the variances $\sigma^2(a_k)$ rather than $\sigma^2(a_k/a_0)$ are considered, and the results are independent of the values of the coefficients.

References

Alder, K., Bohr, A., Huus, T., Mottelson, B., and Winther, A. (1956). *Rev. Mod. Phys.* **28**, 432.

Assousa, G. E. (1968). Thesis, Florida State Univ. Tallahassee, Florida.

Berant, Z. *et al.* (1971). *Nucl. Phys.* **A178**, 155.

Bernstein, E. M., and de Boer, J. (1960), *Nucl. Phys.* **18**, 40.

Biedenharn, L. C. (1960). *In* "Nuclear Spectroscopy" (F. Ajzenberg-Selove, ed.), Part B, pp. 732–810. Academic Press, New York.

Biedenharn, L. C., and Rose, M. E. (1953). *Rev. Mod. Phys.* **25**, 729.

Blatt, J. M., and Weisskopf, V. F. (1952). "Theoretical Nuclear Physics," Wiley, New York.

Bohr, A. (1959). *Nucl. Phys.* **10**, 486.

Broude, C., and Eswaran, M. A. (1964). *Can. J. Phys.* **42**, 1300.

Broude, C., and Gove, H. E. (1963). *Ann. Phys.* **23**, 71.

Buhl, S., Pelte, D., and Povh, B. (1967). *Nucl. Phys.* **A91**, 319.

Coester, F., and Jauch, J. M. (1953). *Helv. Phys. Acta.* **26**, 3.

Critchley, J. B., Calvert, J. M. and Joy, T. (1971). *Nucl. Phys.* **A176**, 129.

Danos, M. (1971). *Ann. Phys.* **63**, 319.

Devons, S. and Goldfarb, L. J. B. (1957). *In* "Handbuch der Physik" (S. Flügge, ed.), Vol. 42, pp. 362–554. Springer, Berlin.

Dirac, P. A. M. (1947). "The Principles of Quantum Mechanics," Oxford Univ. Press, London and New York.

Duray, J. R., Hausman, H. J., Marshall, G. K., Sinclair, J. W. and Steiner, W. S. (1969). *Nucl. Phys.* **A136**, 153.

Ferguson, A. J. (1965). "Angular Correlation Methods in Gamma-Ray Spectroscopy," North-Holland Publ., Amsterdam.

Frauenfelder, H. and Steffen, R. M. (1965). *In* "Alpha-, Beta- and Gamma-Ray Spectroscopy" (K. Siegbahn, ed.), Vol. 2, pp. 997–1198. North-Holland Publ., Amsterdam.

Goldfarb, L. J. B. (1959). *In* "Nuclear Reactions" (P. M. Endt and M. Demeur, eds.), Vol. 1, pp. 159–214. North-Holland Publ., Amsterdam.

Goldfarb, L. J. B. (1964). *Nucl. Phys.* **57**, 4.

Goldfarb, L. J. B., and Johnson, R. C. (1960). *Nucl. Phys.* **18**, 353.

Goldfarb, L. J. B. and Seyler, R. G. (1968). *Phys. Lett.* **28B**, 15.

Goldfarb, L. J. B., and Wong, K. K. (1966). *Phys. Lett.* **22**, 310.

Grodzins, L. (1970). *Proc. Int. Heavy Ion Conf. Heidelberg* (R. Bock and W. R. Hering, eds.), p. 367. North-Holland Publ., Amsterdam.

Häusser, O., Rose, H. J., Lopes, J. S. and Gill, R. D. (1966). *Phys. Lett.* **22**, 604.

Heitler, W. (1954). "The Quantum Theory of Radiation," Oxford Univ. Press (Clarendon), London and New York.

Hippelein, H. H., Jahr, R., Pfleger, J. A. H., Rott, F. and Vieth, H. M. (1970). *Nucl. Phys.* **A142**, 369.

Howell, R. H. and Galonsky, A. I. (1972). *Phys. Rev. C5*, 561.

Kaye, G., Read, E. J. O. and Willmott, J. C. (1968). "Tables of Coefficients for the Analysis of Triple Angular Correlations of Gamma Rays from Aligned Nuclei," Pergamon, Oxford.

Litherland, A. E. and Ferguson, A. J. (1961). *Can. J. Phys.* **39**, 788.

Margenau, H. and Murphy, G. M. (1956). "The Mathematics of Physics and Chemistry," Van Nostrand Reinhold, Princeton, New Jersey.

McDaniel, F. D., and Amos, K. A. (1972). *Nucl. Phys.* **A180**, 497.

McGowan, F. K., and Stelson, P. H. (1958). *Phys. Rev.* **109**, 901.

Monahan, J. E., and Langsdorf, A. (1965). *Ann. Phys.* **34**, 238.

Morgan, J. F., Gerhart, N. L., Hausman, H. J., Sinclair, J. W. and Steiner, W. S. (1971). *Bull. Amer. Phys. Soc.* **16**, 626.

Northcliffe, L. C., and Schilling, R. F. (1970). *Nucl. Data Tables* **A7**, 233.

Ofer, S. (1959). *Phys. Rev.* **114**, 870.

Pande, U. S., and Singh, B. P. (1971a). *Can. J. Phys.* **49**, 2088.

Pande, U. S., and Singh, B. P. (1971b). *Nucl. Instrum. Methods* **97**, 123.

Poletti, A. R., Warburton, E. K., and Kurath, D. (1967). *Phys. Rev.* **155**, 1096.

Rose, H. J., and Brink, D. M. (1967). *Rev. Mod. Phys.* **39**, 306.

Satchler, G. R. (1964). *Nucl. Phys.* **55**, 1.

Schmidt, F. H., Brown, R. E., Gerhart, J. B., and Kolasinski, W. A. (1964). *Nucl. Phys.* **52**, 353.

Singh, B. P., Dahiya, H. S., and Pande, U. S. (1971). *Phys. Rev. C4*, 1510.

Steiner, W. S., Duray, J. R., Hausman, H. J., Kent, J. J., and Sinclair, J. W. (1971). *Bull. Amer. Phys. Soc.* **16**, 132.

Sweeney, W. E., and Ellis, J. L. (1971). *Nucl. Phys.* **A177**, 161.

Taras, P. (1971). *Can. J. Phys.* **49**, 328.

Twin, P. J., Olsen, W. C. and Sheppard, D. M. (1970). *Nucl. Phys.* **A143**, 481.

Waibel, E., and Grosswendt, B. (1971) *Nucl. Instrum. Methods* **93**, 61.

Walraven, R. L., and McCauley, D. G. (1971). *Nucl. Instrum. Methods* **94**, 205.

Ward, D., Graham, R. L., Geiger, J. S., Andrews, H. R., and Sie, S. H. (1972). *Nucl. Phys.* **A193**, 479.

Wilson, M. A. D., and Schecter, L. (1971). *Phys. Rev. C4*, 1103.

VII.H LIFETIME MEASUREMENTS

D. B. Fossan

STATE UNIVERSITY OF NEW YORK, STONY BROOK, NEW YORK

and

E. K. Warburton

BROOKHAVEN NATIONAL LABORATORY, UPTON, NEW YORK

I. Introduction

The radiative lifetimes of nuclear states—or more fundamentally the electromagnetic transition matrix elements we extract from them—are vitally important to our study of the structure of nuclei. These electromagnetic matrix elements provide, after the relative binding energies of the states, the most important touchstone between model wave functions and experiment. Certainly there is no other group of data so complete as well as so varied in its sensitivity to various aspects of the wave functions. It has become routine to ask of a nuclear model its predictions for the electromagnetic transition matrix elements and to judge the model's success by how well these reproduce experiment.

The rate with which nuclear lifetimes are being measured has greatly increased in recent years due to technological advances in electronics, computers, accelerators, and, above all, γ-ray detectors. We can single out the lithium-drifted germanium detector to exemplify the more spectacular innovations; without it, the Doppler shift measurements described in Section IV would hardly be conceivable. One aim of this chapter is to convey a feeling of the vitality and activity in the field of lifetime measurements. We aim to sketch the richness of possible approaches and the scope for ingenuity rather than give an omnibus review of methods and results.

We will concentrate on the three most basic and widely applicable direct timing techniques: the electronic technique, the recoil distance method

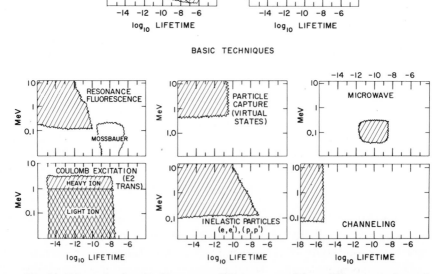

Fig. 1. Schematic of applicable ranges of lifetimes accessible to different experimental techniques. For all methods, the applicability is a function of the energy of the emitted radiation as indicated. The regions are only crudely bounded and the indicated boundaries should not be taken as absolutely excluding nearby regions but rather as a historical indication of the past use of the methods (Schwarzschild and Warburton, 1968).

(RDM), and the Doppler shift attenuation method (DSAM). The time and γ-ray energy regions in which these techniques can be applied are illustrated in Fig. 1. The electronic technique, RDM, and DSAM are used typically in the indicated time regions with time accuracies of 1, 5, and 10%, respectively, but with considerable variation depending on the detailed experimental conditions. Despite the fact that there are lifetimes for which more than one of these techniques can be used, each technique can be applied most favorably in the following approximate timing regions: $\tau > 10^{-10}$ sec for electronic timing; 5×10^{-12} sec $< \tau < 10^{-10}$ sec for the RDM; and $\tau < 5 \times 10^{-12}$ sec for the DSAM. The times 10^{-10} and 5×10^{-12} sec thus form rough time boundaries for the application of these techniques, which enable the experimenter to select the most appropriate technique from an estimate of the lifetime. Assuming an approximate electromagnetic matrix element such as the single-particle values (Blatt and Weisskopf, 1952), summarized in Section II, these time boundaries for a given multipolarity L imply similar boundaries in terms of γ-ray energies. For example, an E2 transition of $E_\gamma = 900$ keV corresponds to $\tau = 10^{-10}$ sec and one at $E_\gamma = 1.7$ MeV corresponds to $\tau = 5 \times 10^{-12}$ sec; therefore, the electronic technique would best be used for $E_\gamma < 900$ keV, the RDM for 900 keV $< E_\gamma < 1.7$ MeV, and the DSAM for $E_\gamma > 1.7$ MeV when the transition is E2. Table 1 summarizes such calculated γ-ray energy boundaries for the various multipolarities using the matrix elements presented in Eq. (7) of II for an intermediate mass of

TABLE 1

Rough Boundaries in Terms of E_γ Which Define the Regions of
Applicability for the Three Basic Timing Techniques[a]

L Electronic $< E_\gamma(10^{-10}\mathrm{sec}) <$ RDM $< E_\gamma(5 \times 10^{-12}$ sec$) <$ DSAM		
E1	15 keV	50 keV
M1	70 keV	200 keV
E2	900 keV	1.7 MeV
M2	2.0 MeV	4.2 MeV
E3	4.5 MeV	7.5 MeV

[a]The techniques considered are electronic, RDM, and DSAM. The E_γ listed, which were obtained from the single-particle matrix elements of Eq. (7), correspond to the rough time boundaries between the techniques of $\tau = 10^{-10}$ sec and 5×10^{-12} sec. Electron-conversion contributions have not been included.

$A = 90$. Although exact matrix elements will cause some variation in the E_γ boundaries, the values in Table 1 represent a rough guide to the choice of experimental technique for the various multipolarities as a function of γ-ray energy. The detailed slow-A dependence of the single-particle values, if desired, can be obtained directly from Eq. (7) (see Wilkinson, 1960).

In addition to these three basic direct timing techniques, there are several indirect timing methods and other special direct techniques which are often competitive with the three basic techniques for the determination of lifetimes. A brief discussion of the specific advantages and the applicability of the indirect techniques—resonance fluorescence, capture cross-section measurements, Coulomb excitation, inelastic electron scattering, and inelastic particle scattering—will be given. The microwave and channeling techniques, which are direct methods, are also discussed. A summary of the lifetime regions where these techniques are applicable is also shown in Fig. 1.

In the next section, we trace the relationship between the lifetimes of nuclear states and the nuclear structure information we obtain from them. In Section III, electronic timing techniques are discussed and in Section IV, the RDM and DSAM are presented. Section V is a brief review of other methods for lifetime determinations.

II. Sensitivity of Lifetime Results to Nuclear Structure

As pointed out in the introduction, the interest in lifetimes of excited nuclear states lies in their relation to the electromagnetic properties of nuclei. The electromagnetic interactions within nuclei are a source of detailed information about nuclear structure. A determination by lifetime measurements of γ-ray transition matrix elements $\langle \psi_b \| \mathscr{M}(L) \| \psi_a \rangle$ that are a measure of this electromagnetic interaction allows a direct and sensitive test of the wave functions ψ_a and ψ_b since the interaction operators $\mathscr{M}(L)$ are to a large extent well understood. Since operators of several multipolarities L are influential in nuclei, the electromagnetic interactions offer a wealth of nuclear information. On the other hand, for cases where the wave functions are known to be very pure, the knowledge of an electromagnetic matrix element gives a measure of small but interesting deviations from the first-order electromagnetic operators. In this section, we will briefly outline the relation and sensitivity of nuclear lifetimes to the nuclear structure and the electromagnetic operators via the transition matrix elements.

The transition probability for a γ transition of multipolarity L from an excited state ψ_a to a final state ψ_b is given by (Blatt and Weisskopf, 1952;

Alder *et al.*, 1956)

$$\lambda(L) = \frac{8\pi(L+1)}{L[(2L+1)!!]^2} \left(\frac{1}{\hbar}\right)\left(\frac{E_\gamma}{\hbar c}\right)^{2L+1} B(L) \qquad (1)$$

where the reduced transition probability $B(L)$ for $a \to b$ is

$$B(L) = (2J_a + 1)^{-1} |\langle \psi_b ||\mathcal{M}(L)|| \psi_a \rangle|^2 \qquad (2)$$

This $B(L)$ represents the sum of squared $\mathcal{M}(L, \mu)$ matrix elements over the m substates of the photon and the final state and an average over the initial m substates. One sees that the transition probability is proportional to the square of an electromagnetic matrix element. The squared matrix element, which contains the nuclear structure information, is most often quoted in terms of the reduced transition probability $B(L)$. The $B(L)$ for the first few electric and magnetic multipoles as calculated from Eq. (1) are

$$B(E1) = 6.29 \times 10^{-16}(E_\gamma)^{-3}\lambda(E1) \quad [e^2F^2]$$

$$B(E2) = 8.20 \times 10^{-10}(E_\gamma)^{-5}\lambda(E2) \quad [e^2F^4]$$

$$B(E3) = 1.76 \times 10^{-3}(E_\gamma)^{-7}\lambda(E3) \quad [e^2F^6]$$

$$\vdots \qquad\qquad (3)$$

$$B(M1) = 5.68 \times 10^{-14}(E_\gamma)^{-3}\lambda(M1) \quad [\mu_0^2]$$

$$B(M2) = 7.41 \times 10^{-8}(E_\gamma)^{-5}\lambda(M2) \quad [\mu_0^2F^2]$$

where E_γ is in MeV and the $\lambda(L)$ are in $\sec^{-1}$. To obtain a $B(L)$ connecting state a to state b, it is thus necessary to measure experimentally the transition probability $\lambda(L)$. When γ decay of multipolarity L to state b is the only decay mode, then the lifetime of state a is directly related to $\lambda(L)$; however, if state a decays to other final states or more than one multipole is involved, then information in addition to the lifetime is needed to extract the $B(L)$. [Although internal electron conversion will not be included in this discussion, allowances for it are required for low E_γ in high-Z nuclei (see Ewan and Graham, 1965).]

A γ transition or branch to a particular final state b may contain more than one multipolarity L, where L is the total angular momentum carried away by the γ ray. The allowed multipolarities L are governed by the relations

$$|J_a - J_b| \leqslant L \leqslant J_a + J_b$$

$$\Delta\pi = (-1)^L \quad \text{for} \quad (EL), \quad \text{and} \quad (-1)^{L+1} \quad \text{for} \quad (ML) \qquad (4)$$

where $\Delta\pi$ is the change in parity in going from state a to state b. The electromagnetic interaction decreases rapidly with L with the result that only the lowest allowed L for the electric and for the magnetic multipoles are important. In addition, for a given L, the magnetic multipole intensity is generally weaker by more than an order of magnitude compared to that of the electric multipole. Thus for transitions where Eqs. (4) require the lowest multipolarity to be electric, the partial transition probability is usually dominated by the electric multipole $E(L)$, while in the reverse situation where the lowest multipole is magnetic, often both the $M(L)$ and $E(L + 1)$ multipoles compete more equally in the γ transition. If a mixture of multipolarities does occur in a γ transition to state b, the transition probability is then equal to the intensity sum $\lambda_b = \lambda(L) + \lambda(L + 1)$, since in the integration of a squared mixed matrix element over all angles for the transition probability, the interference or cross terms cancel out. The magnitude of the mixing ratio δ is defined by $\delta^2 = \lambda(L + 1)/\lambda(L)$, from which the transition probability can be written as $\lambda_b = (1 + \delta^2)\lambda(L)$. The $\lambda(L)$ and $\lambda(L + 1)$ are the quantities that are related to $B(L)$ and $B(L + 1)$, respectively, in Eq. (1). The mixing ratio δ can be obtained experimentally by angular correlation techniques (see Ferguson, Chapter VII.G) or from conversion-electron subshell coefficients or ratios (Ewan and Graham, 1965).

The total transition probability λ for a given initial state a is the sum over the various partial transition probabilities of all γ branches to different final states b, that is, $\lambda = \sum_b \lambda_b$. The fractional γ decay or branching ratios λ_b/λ can be easily obtained from the observation of the γ rays as a function of angle with a detector of known efficiency.

The total transition probability λ is the quantity which is determined in a lifetime measurement of a given excited state. The decay of an excited state is governed by $I(t) = I_0 e^{-\lambda t}$, where $I(t)$ is the number that have not decayed after a delay time t for a population ensemble of I_0; the rate of decay is the differential of this expression $dI(t)/dt = I_0 \lambda e^{-\lambda t}$. These decay relations are used in the various experiments to deduce the lifetime of an excited state which, of course, also determines the total transition probability λ. The lifetime results are usually given as the mean life which is $\tau = 1/\lambda$, with $t_{1/2} = 0.693\tau$. An alternative method of determining the total transition probability or a lifetime is to employ the uncertainty principle $\Gamma\tau = \hbar$ and to measure a radiation width Γ. A radiative width measurement is necessary for the exceedingly short-lived states.

In summary, to obtain a reduced transition probability $B(L)$ or an electromagnetic matrix element, it is first necessary to obtain the total transition probability λ by a lifetime or width measurement. If there are γ branches to more than one final state, the branching ratios λ_b/λ must be measured, and when multipole mixing occurs in any of the branches, the corresponding mixing ratios δ^2 must be determined in order to obtain the various $\lambda(L)$. Thus a given lifetime result along with branching and mixing ratios can lead to several $B(L)$ connecting different final states. Frequently, however, only one branch and one multipole are involved in the decay of an excited state, in which case, the $B(L)$ is directly determined by the lifetime.

The form of the various electromagnetic multipole operators determines the type of nuclear information that is available in the electromagnetic matrix elements. Assuming that the nucleons have point charges with magnetic moments and that the γ-ray wavelength is large compared to the nuclear dimensions, the multipole operators for a nucleus are given by a sum over the single-particle operators for the k nucleons (de-Shalit and Talmi, 1963; Bohr and Mottelson, 1969):

$$\mathcal{M}(\text{EL}, \mu) = \sum_k e(k)\, r_k^{\,L} Y_{L\mu}(\Theta_k, \Phi_k)$$

$$(5)$$

$$\mathcal{M}(\text{ML}, \mu) = \sum_k \{g_s(k)\, \mathbf{s}_k + [2g_l(k)/(L+1)]\, \mathbf{l}_k\} \cdot \nabla_k [r_k^{\,L} Y_{L\mu}(\Theta_k, \Phi_k)]\, \mu_0$$

The $e(k)$ is an effective charge for the kth nucleon, μ_0 is the nuclear magneton, and the $g_s(k)$ and $g_l(k)$ are the spin and orbital g factors for the kth nucleon, respectively. Deviations from these operators under the assumptions given are in general small or at least understood. Corrections that take into account the subtleties of the nuclear exchange interactions are often made in terms of the $e(k)$, $g_s(k)$, and $g_l(k)$.

For a single-particle E1 transition, the corresponding recoil of the remainder of the nucleus makes a major contribution to the transition strength. This effect is accounted for by adding the recoil term $-Ze/A$ to the normal charge; thus, $e(k) = (\frac{1}{2} - t_z)e - Ze/A$, which is Ne/A for a proton and $-Ze/A$ for a neutron. Recoil terms for magnetic multipoles and higher electric multipoles are usually insignificant compared to other uncertainties.

The effective charges are also influenced, in particular for the E2 operators (Bohr and Mottelson, 1969), by a polarization of the nuclear core. In a shell model description, the wave functions of excited states are usually given in terms of a few nucleons outside an inert core. An outer nucleon can induce a core polarization by distorting nucleon orbits in the core; this polarization contributes to the E2 operator an amount roughly comparable to that of the outer nucleon. Considerable experimental information is available on the E2 effective charges (Zamick and McCullen, 1965; Skorka et al., 1966; Marelius et al., 1968; Astner et al., 1972) which are usually expressed as $e(k) = (\frac{1}{2} - t_z)e + (e_{pol})_{E2}$. Similar core polarization contributions to the effective charges for the higher electric moments also exist.

In the magnetic operators, the spin and orbital g factors for free nucleons are $g_s(\text{proton}) = 5.59$, $g_s(\text{neutron}) = -3.83$, and $g_l = (\frac{1}{2} - t_z)$. These values, which correspond to the Schmidt static magnetic moments, have fair accuracy for lighter nuclei. Several corrections to these g factors (Talmi and Unna, 1960) are expected from meson exchange (two-body operators), spin–orbit coupling, core polarization, and possible higher-order effects. For M1 operators, meson exchange and spin–orbit corrections to g_s are insignificantly small. Recent experiments (Yamazaki et al., 1970) and theory (Chemtob, 1969) imply meson exchange corrections to g_l of $\delta g_l(\text{proton}) = +0.09$ and $\delta g_l(\text{neutron}) = -0.06$, which have significant effects for large l values. Core-polarization corrections to g_s (Arima and Horie, 1954; Blin-Stoyle and Perks, 1954) can become important for heavy nuclei where $j = l + \frac{1}{2}$ orbitals are filled as part of the core and the $j = l - \frac{1}{2}$ partner orbitals are not. A particle–hole excitation of $[(j = l + \frac{1}{2})^{-1}(j = l - \frac{1}{2})]1^+$ recoupled to a given particle state represents a configuration admixture that can contribute to g_s to first order in the amplitude. Accurate theoretical information (Blomqvist et al., 1965) is available on the core-polarization corrections. Less information is available on the corrections for the higher magnetic multipoles.

The evaluation of the electromagnetic matrix elements with various nuclear wave functions for a comparison with experimental results is simplified by the fact that the electromagnetic operators are sums of single-particle operators. The particle j_k that is making the transition via the kth single-particle operator can be separated from the remaining nucleons, the so-called parent which does not participate or change in the transition. The closed-shell core that is part of the parent and coupled to $J = 0$ is also inert in the transition. The kth term of the electromagnetic matrix element then reduces to a product of angular momentum coupling coefficients and the reduced single-particle matrix element $\langle j_f \| \mathcal{M}_k(L) \| j_i \rangle$ that represents a single-particle transition. Thus an electromagnetic matrix element involving several particles reduces to a sum over single-particle matrix elements. Formally, this matrix element reduction can be carried out by antisymmetrizing the wave function to obtain the fractional parentage coefficient which multiplies each allowed coupling term (de-Shalit and Talmi, 1963).

Therefore the evaluation of an electromagnetic matrix element is achieved with the knowledge of single-particle matrix elements. The single-particle matrix elements have been calculated and are given in a general form (Bohr and Mottelson, 1969). Simplified expressions for the M1 and E2 operators are as follows:

$$\langle j_f \| \mathscr{M}(M1) \| j_i \rangle = [(3/4\pi) j(j+1)(2j+1)]^{1/2} g_j \mu_0, \qquad j_f = j_i$$

$$= (-1)^{j_f - l - 1/2} [g_l(k) - g_s(k)]$$

$$\times \{(3/4\pi)[2l(l+1)]/(2l+1)\}^{1/2} \mu_0, \qquad j_f \neq j_i$$

$$\langle j_f \| \mathscr{M}(E2) \| j_i \rangle = e(k) [(5/4\pi)(2j_f+1)]^{1/2} \langle j_f | r^2 | j_i \rangle \langle j_f 2\tfrac{1}{2}0 | j_i \tfrac{1}{2} \rangle \tag{6}$$

A simple approximation of these single-particle matrix elements is often used to calculate from Eq. (2) an approximate unit of strength, the so-called Weisskopf unit (Blatt and Weisskopf, 1952):

$$B(EL)_W = (1/4\pi)[3/(3+L)]^2 (1.2A^{1/3})^{2L} \quad [e^2F^{2L}]$$

$$B(ML)_W = (10/\pi)[3/(3+L)]^2 (1.2A^{1/3})^{2L-2} \quad [\mu_0^2 F^{2L-2}] \tag{7}$$

In certain nuclear regions, nuclei exhibit collective properties which are often described in macroscopic parameters rather than by detailed particle wave functions; for example (Bohr and Mottelson, 1953), the deformed nuclei have rotational properties that are related to a deformation, and vibrational nuclei are assigned an oscillator strength. A coupling of single particles to a deformed core as in the Nilsson (1955) model or to a vibration represents a combination of the collective and particle pictures. Electromagnetic matrix elements can also be calculated in terms of these collective parameters for a sensitive comparison with experimental results.

A feature of the electromagnetic operators that in certain ways increases the sensitivity to nuclear structure is the operation of numerous selection rules. In addition to the requirements on angular momentum and parity already mentioned, selection rules for the various multipole operators with regard to isospin, particle configurations, and collective wave functions are very useful. Warburton and Weneser (1969) have recently reviewed the isospin selection rules in electromagnetic transitions (isospin is discussed by Temmer, Chapter IV.A.2); representative examples are: transitions are forbidden unless $\Delta T = 0$ or ± 1; $\Delta T = 0$, E1 transitions in self-conjugate nuclei are forbidden; and $\Delta T = 0$, M1 transitions in self-conjugate nuclei are expected to be weaker by an approximate factor of 100. Regarding particle configurations, an example of an M1 selection rule is: Single-particle electromagnetic transitions are l-forbidden unless $l_i = l_f$ since the M1 operator is diagonal in l. Another selection rule that in certain nuclei gives sensitivity to small admixtures is that M1 transitions in a j^n configuration of identical particles are forbidden (de-Shalit and Talmi, 1963). Similar selection rules apply to the collective wave functions (Alaga *et al.*, 1955); a transition is K-forbidden if $\Delta K > L$, where K is the projection of the angular momentum on the axis of deformation, and there are additional, less restrictive selection rules regarding other parameters of the deformed wave functions. These examples hopefully illustrate the type of nuclear information that can be obtained from the electromagnetic selection rules.

In making theoretical comparisons with electromagnetic matrix elements, a more sensitive check on the nuclear structure of a given state is achieved if several matrix elements of different L connect the same state. In addition to the transition or nondiagonal matrix elements of the electromagnetic multipole operators, a knowledge of the static moments, the magnetic and quadrupole moments, that give the diagonal matrix elements are important in a complete nuclear structure study.

III. Electronic Timing Techniques

This section contains a review of the latest techniques of electronic timing. The intent is to explain these techniques without necessarily giving all of the details and to discuss their present capabilities. The various aspects of electronic timing have been reviewed in the past by Bell (1955, 1965), Schwarzschild (1963), Bonitz (1963), Ogata *et al.* (1968), Meiling and Stary (1968), and Schwarzschild and Warburton (1968). Given the existence of these articles, no attempt will be made at covering or referring to all of the historical developments. This chapter will emphasize in-beam timing applications where the states of interest are populated with nuclear reactions; aspects of electronic timing that are specifically related to radioactivity measurements have been well covered in the previous chapters. The many interesting excited states that are populated with the various nuclear reactions, including the exotic heavy-ion reactions (see Newton, Chapter VII.E), have increased the importance of electronic timing techniques. Although pulsed-beam γ timing will be discussed, the techniques of beam pulsing will not be given. Also, only those electronic timing techniques which apply to the time region of a few psec to a few μsec will be included. This section will be divided into four parts: basic technique, contributions to the time resolution, electronic aspects of timing, and detectors in timing measurements.

A. BASIC TECHNIQUE

The electronic technique or so-called delayed-coincidence technique is widely applicable for the measurement of short nuclear lifetimes. The technique involves the measurement of the distribution of time delays between the formation and subsequent decay of the nuclear state of interest. The distribution of time delays contains the lifetime information through the fact that it is governed by the decay relation $dI(t)/dt = I_0 \lambda e^{-\lambda t}$. The time of formation of the state is usually determined by the detection of particles from a nuclear reaction or other radiations populating the state, while the time of decay is marked by the detection of the decay γ ray (conversion e^-). Pulsed beams can also be used to determine the time of formation. The basic electronic circuit is shown in Fig. 2. Detectors A and B are used for the timing

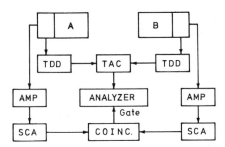

Fig. 2. Schematic diagram of electronic timing technique. A and B are detectors, TDD refers to time derivation device, TAC to time-to-amplitude convertor, and SCA to single-channel analyzer.

determinations just described. For the case of beam pulsing, one of the detectors is replaced with a time signal that is correlated with the beam pulse. Appropriate signals directly from the detectors are fed into the time derivation devices (TDD) which generate standard time pulses. The time intervals between these standard pulses are then measured in a time-to-amplitude convertor (TAC) and recorded in a multichannel analyzer. Additional signal amplitude conditions are often specified by single-channel analyzers (SCA) in order to isolate the state of interest or to improve the signal-to-background ratio and for the capability of time-resolution adjustments. The use of a pulsed beam brings the advantages of a reduced random background between pulses from prompt events as well as higher counting rates, but has the disadvantage of less isolation for a particular state as compared to the two-detector system. These electronic circuits are discussed by Goulding and Landis, Chapter III.D.

The type of detector used depends on the experimental conditions and the method of populating the state. The γ rays can be detected with NaI(Tl) or plastic scintillators coupled to photomultiplier (PM) tubes, or with Ge(Li) or Si(Li) solid-state detectors (these last are discussed by Goulding and Pehl, Chapter III.A). Silicon surface-barrier detectors are usually used for charged particles, while neutrons are detected with liquid or organic scintillators. (For convenience of discussion, the word radiations will be used as a general term for γ rays, β particles, neutrons, and charged particles.) These detectors, in response to the different radiations, produce signals of variable amplitudes with finite rise times; some of the detectors also have variable rise times. To achieve the optimal time information for the different detectors, several types of TDD are used. Details of the detector characteristics and the appropriate TDD will be discussed later.

The electronic timing technique for a given detector combination has a finite time resolution. This can be studied by observing prompt events in the two detectors or by measuring a lifetime short compared to the resolving time. The distribution of time delays for such prompt events is called the prompt resolution function (PRF). The shape of the PRF depends on the type of detector and the radiations detected; a typical PRF is shown by the dashed line in Fig. 3. The figure of merit of a system is usually characterized by the full-width at half-maximum (FWHM) and the approximate logarithmic slope. Contributions to the FWHM from each detector are assumed to combine as the square root of the sum of squares; however, the contributions to the slope do not combine in such a simple manner. The PRF for detectors of unequal responses will in general not be symmetric.

The distribution of time delays $F(t)$ for a state of mean lifetime $\tau = 1/\lambda$ is given by the convolution integral of the normalized PRF, $P(t)$, with the exponential decay

$$F(t) = \int_0^\infty \lambda[\exp(-\lambda t')] P(t - t') \, dt' \tag{8}$$

There are two methods of extracting the lifetime from $F(t)$, namely the slope and the centroid methods; the slope method, which is the most accurate, will be discussed first. A typical delayed curve $F(t)$ is shown by the solid line in Fig. 3. As shown by Newton (1950), the logarithmic slope of the

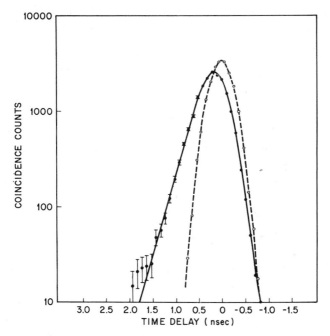

Fig. 3. Distribution of time delays $F(t)$ for the 1.02-MeV state in ^{23}Ne obtained with the $^{22}Ne(d, p\gamma)^{23}Ne$ reaction using a silicon particle detector and plastic scintillator (Fossan *et al.*, 1966). The experimental prompt resolution function (PRF), $P(t)$, is shown by the dashed line. A comparison of $F(t)$ and $P(t)$ illustrates the slope method of timing; in this measurement, the logarithmic slope of $F(t)$ yields a mean lifetime of $\tau = 257$ psec.

delayed curve is given by

$$(d/dt)\,[\ln F(t)] = -\lambda[1 - P(t)/F(t)] \qquad (9)$$

Thus, in the time region where $P(t) \ll F(t)$, the logarithmic slope is a straight line and the mean lifetime can be extracted with an exponential fitting procedure. With proper statistics, the accuracy of the slope method is limited only by the uncertainty in the time calibration of the TAC, which can be kept to less than 1%. Lifetimes that are larger than the PRF slope by $\gtrsim 30\%$ can be determined by this slope method. For lifetimes that are less than this but still greater than the PRF slope, corrections for the PRF are required in order to extract the lifetime from the shape of the delayed curve (Newton, 1950; Boström *et al.*, 1966). Thus the slope of the PRF roughly represents the limiting lifetime that can be extracted from the slope of the delayed curve $F(t)$. The PRF slope is often quoted as the approximate time to fall a factor of two; this figure of merit, termed slope($\frac{1}{2}$), then corresponds to an apparent half-life ($t_{1/2}$).

For lifetimes that are less than the PRF slope, the centroid method can be used. The centroids (first moments) of the delayed curve, $F(t)$, and the PRF, $P(t)$, of Eq. (8) are separated by a time τ (Bay, 1950); thus a careful measurement of the centroids of both functions will give the mean lifetime. This method in general is not as accurate as the slope method, due to large systematic uncertainties in the centroid shifts. In certain special cases, lifetimes of a few percent of the FWHM can be measured by this centroid method. Higher moments (Birk *et al.*, 1962) have also been used in such lifetime evaluations.

B. Contributions to the Time Resolution

The time resolution of the electronic timing technique as characterized by the FWHM and slope($\frac{1}{2}$) of the PRF determines the limiting lifetime sensitivity. For this reason, it is important to understand the various contributions to the time resolution. The resolution contributions can be associated with three phases of the timing process: (1) time variations in the interaction of the radiations with the detectors, (2) time variations in the response of the detectors to the radiation, and (3) time variations associated with the electronic equipment.

Contributions of the first type include the variations in the target-to-detector flight times for the radiations, time characteristics for pulsed beam timing, and time variations resulting from the interaction of the radiation over the detector dimensions. The minimization of these time-resolution contributions such as from, for example, target or source thickness, beam energy resolution, kinematics, and detector geometry are rather straight-forward and will not be discussed further.

The contributions of type 2 relate to time variations in the formation of the detector output signal from the radiation interaction events, such as from the statistics in the charge carrier production, from detector noise, and from the collection of the charge carriers. These effects manifest themselves in the shape of the detector signal, namely the finite rise time, variations in the rise times, and amplitude fluctuations. Contributions of type 3 arise from the intrinsic time resolution of the electronic equipment, the stability of the electronics, time variations introduced by the TDD, and effects of electronic noise (for a related discussion of many of these points, see Chapter III.D).

The time-resolution contributions of types 2 and 3, which are often inter-connected, will be discussed in detail for each type of detector later. For the present, however, we note that there are three general types of these time uncertainties that have application to any detector arrangement and have common usage (Williams, 1971a): time walk, time jitter, and time drift.

To define time walk, consider the most simple TDD, a leading-edge discriminator with a threshold V_T. Consider also the following idealized detector signals $V(t)$ which are to be fed into this discriminator (Fig. 4):

$$V(t) = (V_a/T_r)\, t, \qquad 0 \leqslant t \leqslant T_r$$
$$V(t) = V_a, \qquad\qquad t > T_r \tag{10}$$

V_a is the maximum signal amplitude, which generally is related to the energy of the detected radiation, and T_r is the rise time, where the slope of the rise

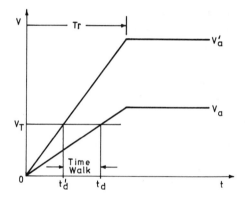

Fig. 4. Illustration of time walk for two idealized signals of different amplitudes that are fed into a leading-edge discriminator of threshold V_T.

$dV/dt = V_a/T_r$ is assumed constant in time. These idealized signals will be used throughout this section to illustrate resolution contributions and the operation of the different TDD. More realistic signals do not change the discussion significantly. For these signals, originating at $t = 0$, the time t_d at which the discriminator threshold V_T is crossed is given by $t_d = T_r(V_T/V_a)$. In the present discussion, the rise time T_r of all the signals is assumed the same. The discriminator output pulses thus are shifted in time by an amount which is inversely proportional to the signal amplitude V_a and directly proportional to the rise time T_r; this time shift is referred to as time walk. Figure 4 illustrates the time walk for two different signal amplitudes. It is easy to see by differentiation that the time walk per unit amplitude change $\Delta t_d/\Delta V_a =$

$T_r V_T / V_a^2$ increases as the signal amplitude decreases. From this discussion on time walk, one can see that for a detector which produces signals with different rise times, as in the case with Ge(Li) detectors, the time walk problem becomes even more severe. There is an additional walk effect due to the fact that the discriminator requires a finite amount of charge

$$Q \propto \int_{t_d}^{t_d + \Delta t_q} [V(t) - V_T] \, dt$$

to trigger once the threshold V_T has been reached. The additional walk time Δt_q needed to collect this charge increases as the signal amplitude decreases, and becomes very large for signals with amplitudes $V_a \approx V_T$. This type of walk can be minimized by selecting with the SCA (see Fig. 2) a region of the signal amplitude that is somewhat above the threshold V_T.

Time jitter refers to the time variations that are due to fluctuations in the detector signal $V(t)$. The two main sources of jitter are the statistical manner in which the detector signal is formed and noise. The processes by which a detector converts the absorbed radiation into elementary charge carriers is statistical in nature. These statistics result in amplitude fluctuations $\Delta V_s(t)$ in the detector signal; the magnitude of $\Delta V_s(t)$, which is time dependent, depends largely on the number of elementary charge carriers. The statistical fluctuations are large for a small number of charge carriers as in the case for scintillator–PM detectors. For a fluctuation $\Delta V_s(t)$ at the discrimination time t_d, the corresponding time variation for the idealized signals is $\Delta t_j = \Delta V_s(t_d) / V_a / T_r$ where V_a / T_r is the signal slope dV/dt. Noise in a detection system also produces a fluctuation ΔV_n in the detector signal amplitude. Noise is generated in the detector or in the electronics associated with the TDD. The magnitude of ΔV_n is generally independent of time. The noise fluctuation ΔV_n corresponds to a time jitter Δt_j defined by the same relation as just given. This noise jitter is very important for semiconductor detectors.

Time drift refers to long-term variations in the timing equipment. The most common time drifts are those associated with the electronics, in particular the TAC and TDD, which result from temperature changes or the aging of components. These effects have been minimized to a large extent by the availability of well-engineered commercial electronics.

This discussion of general time-resolution contributions of types 2 and 3 was based on a leading edge discriminator. These resolution effects can be altered and often reduced with the use of other types of TDD; these will be discussed in Section III.C.

C. ELECTRONIC ASPECTS OF TIMING

The optimal choices among the types, dimensions, and geometry of the detectors for timing measurements are generally determined by the experimental conditions. These choices define the timing contributions of types 1 and 2 which are manifest in the detector signals. Here, the electronic methods used to optimize the time resolution for a given set of detector timing signals will be discussed. As seen in Fig. 2, there are two parts of the timing process that remain, the derivation of the standard timing pulses from the detector signals by the TDD, and the conversion of the time intervals between these pulses into a measurable amplitude by the TAC. The method of time derivation is the more crucial of the two in optimizing the time resolution; the TDD is chosen to minimize the time walk and jitter associated with the detector signals. Time-resolution contributions from the TAC are small.

1. Time Derivation Devices (TDD)

The selection of a TDD depends on the timing characteristics of the detector signals. There are several types of TDD which are presently in use: leading-edge (LE) discrimination, constant-fraction (CF) discrimination, amplitude and rise-time compensation (ARC) timing, and extrapolated-leading-edge (ELE) timing. We give here a brief description of how these TDD work along with their advantages and disadvantages.

The LE or constant-amplitude discriminator is the simplest TDD. It has been explained in Section III.B and illustrated in Fig. 4 for use in the discussion of time walk. Because LE discrimination gives significant time walk effects, it is used generally for narrow dynamic ranges of signal amplitudes. Due to its simplicity, it has wide application where the magnitude of the time walk for the selected dynamic range is small compared to other contributions to the time resolution. The LE discriminator triggers at a fraction $f = V_T/V_a$ for a signal amplitude V_a; this fraction varies over the accepted range of signal amplitudes V_a.

Constant-fraction (CF) discrimination refers to a TDD that triggers at a constant fraction of the amplitude $V(t_d) = fV_a$ for all signal amplitudes V_a. For signals of a definite rise time T_r [see Eq. (10)], the discrimination time for a CF TDD is $t_d = fT_r$, which is independent of V_a. Thus, an ideal CF discriminator has no time walk over any amplitude range of these signals. A circuit designed by Gedcke and McDonald (1968) to achieve CF discrimination is simple in concept and has proved very successful. The method is

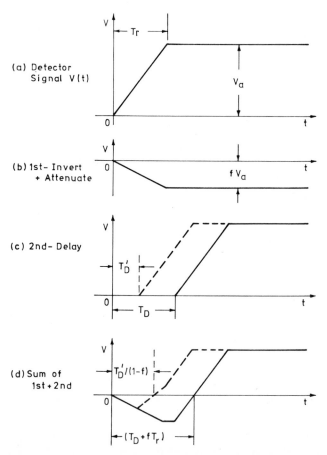

Fig. 5. Illustration of signal processing for constant-fraction (CF) timing and amplitude and rise time compensation (ARC) timing. In (c) and (d), the dashed lines refer to the ARC timing.

illustrated in Fig. 5 with the idealized signals of Eq. (10). The detector signal shown in Fig. 5a is split into two signals; the first is inverted and attenuated by a factor equal to the fraction f giving $-f(V_a/T_r)t$ for $t \leqslant T_r$ and $-fV_a$ for $t > T_r$ as shown in Fig. 5b. The second is delayed by a time $T_D > T_r$ as shown in Fig. 5c and added to the first, yielding, for $T_D < t < T_D + T_r$, the summed signal $V_{sum}(t) = (V_a/T_r)(t - T_D) - fV_a$ as shown in Fig. 5d. The time at which the summed signal passes through zero is easily obtained, $t(V_{sum} = 0) = fT_r + T_D$. Aside from the constant delay T_D, the zero-crossing point of the summed signal yields the CF discrimination time $t_d = fT_r$, and $t(V_{sum}=0)$ is easily determined by a zero-crossing discriminator. The CF TDD thus eliminates time walk over wide amplitude ranges for signals of a definite rise time. It has the additional advantage, especially for scintillator–PM detectors, that the fraction f can be adjusted to the value that optimizes the time resolution for all signal amplitudes and therefore represents an improvement over LE discrimination even for a narrow dynamic range. Another version of the CF TDD (Maier and Sperr, 1970) circumvents the inversion process by using a signal comparison technique with integrated circuits. Integration techniques (unpublished) that replace the zero-crossing discrimination with a special "snap-off" diode have also been used to minimize jitter contributions in CF timing.

Another type of CF discriminator is the so-called crossover timing of, for example, a double delay-line clipped pulse, which has wide use. Since the discrimination fraction for such timing is usually $f \approx 50\%$, the time resolution, at least for scintillator–PM detectors, is not as good as LE timing for a narrow dynamic range, because this fraction gives a large statistical time jitter (Bell, 1966). A fast version of crossover timing (Wieber and Lefevre, 1966) with PM anode signals represents an improvement that is simple; however, again because of statistics, it does not yield the best time resolution.

The ARC timing technique is a time-derivation method which was designed by Chase (1968) for Ge(Li) detector signals that have variable rise times. An ARC TDD works in a manner similar to the CF TDD except for a shorter time delay T_D'. To illustrate the ARC timing method, consider again the idealized detector signals of Eq. (10) as shown in Fig. 5a; however, now both the signal amplitude V_a and the rise time T_r are considered as variables. The detector signal is split into two signals; the first is inverted and attenuated by a fraction f giving $-f(V_a/T_r)t$ for $t \leqslant T_r$ as shown in Fig. 5b. The second is delayed as shown by the dashed line in Fig. 5c and added to the first. The sum as shown by the dashed line in Fig. 5d is described by $V_{sum}(t) = (V_a/T_r)(t - T_D') - f(V_a/T_r)t$ for $T_D' < t < T_r$. From a solution of this equation for the time at which $V_{sum}=0$, one obtains $t(V_{sum}=0) = T_D'/(1-f)$. This zero-crossing time, which again is determined by a zero-crossing discriminator, is independent of both the amplitude V_a and the rise time T_r. Thus with the ARC TDD, time walk has been eliminated for signals of variable rise times over a broad range of amplitudes. This is exactly true if the signals rise linearly with time up to the discrimination time as is the case for the idealized signals. The discrimination time in the ARC method is no longer, however, at a constant fraction of the signal amplitude for variable rise times. For the solution to be valid, the zero-crossing time must occur when the attenuated signal is still rising, before the time $t = T_r$, and thus T_D' must be less than $(1 - f)T_r$. {It should be pointed out that for linear signals with a fixed rise time, the ARC TDD also triggers at a constant fraction of the signal amplitude; this fraction is $T_D'/[T_r(1 - f)]$.} Electronic limitations in the ARC TDD place a limit on the dynamic range for ARC timing. White and McDonald (1973) have used dual CF TDD for ARC.

The ELE timing method is a second TDD that has been designed (Gorni *et al.*, 1967; Fouan and Passerieux, 1968) to compensate for variable rise times as from Ge(Li) detectors. This method uses two leading-edge discriminators set at different threshold levels of the detector signal, as illustrated in Fig. 6.

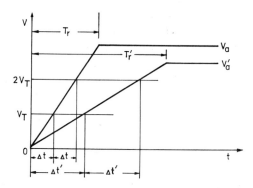

Fig. 6. Illustration of extrapolated-leading-edge (ELE) timing, which employs two leading-edge discriminators.

The idealized signals of Eq. (10) with variable V_a and T_r will again be assumed to explain the method. The first, lower-level discriminator feeds the start of a TAC as in a usual timing measurement (see Fig. 2). A second TAC of identical gain is used to measure the time difference between the triggering of the first discriminator and that of the second, which is set at a threshold $V_T(2) = 2V_T(1)$. For detector signals that rise linearly with time as for the idealized signals, the time difference measured in the second TAC is $t_d(2) - t_d(1) = [V_T(1)/V_a]T_r$. This is also the time difference between the first discriminator trigger $t_d(1)$ and the origin of the signal at $t = 0$. Thus, by adding the second TAC output to the first TAC output, the start signal has been extrapolated to the origin $t = 0$, which contains no time walk. The ELE method, in extrapolating the signal back to its origin, has properly compensated for both variable rise times T_r and variable amplitudes V_a for all signals that trigger the second discriminator. A disadvantage of the ELE TDD is that the time resolution includes the time jitter of both discriminators (Nutt et al., 1970). This effect can be reduced by increasing the threshold difference and lowering the second TAC gain; however, the signal then is required to be linear over a longer period of time.

2. Time-to-Amplitude Converters (TAC)

The last part of the timing process is the conversion by the TAC of the time interval between the standard pulses that are generated by the TDD into a voltage amplitude that can be measured in a multichannel analyzer. There are two basic analog TAC's as reviewed by Ogata et al. (1968), the start–stop converter and the pulse-overlap converter. The start–stop converter works by charging a capacitor from a constant current generator during the time interval between the start and stop pulses. The start pulse from one TDD initiates the charging and the stop pulse from the other terminates it; the amplitude of the capacitor voltage is directly proportional to the time interval between the start and stop pulses. The start–stop converter can operate over wide time ranges of up to 100 μsec and has outstanding drift stability and linearity characteristics. The time-resolution contribution from the TAC has a fundamental limit of less than 10 psec.

The pulse-overlap converter is very simple; two TDD pulses of standard amplitude and width are added and the pulse overlap is integrated, yielding an amplitude which is proportional to the time interval between the two input pulses. The overlap converter is double-valued in time depending on which TDD pulse arrives first. This has the disadvantage of giving twice the random rate; however, with an additional coincidence circuit, the second time region can be used to record random events. The overlap converter has an advantage of a higher input rate capability than the start–stop converter; however, it operates over smaller time ranges and has poorer stability and linearity.

There are also several digital methods for measuring time intervals. Basically the method involves counting pulses from an oscillator during the time interval to be measured. The inherent resolution of ± 1 oscillator cycle can be reduced by several vernier techniques (Lefevre and Russell, 1959; Nutt, 1968). Although the limiting resolution is comparable to the analog converters for many time ranges, stability difficulties and the complexity favor the analog converters, especially for the shorter time ranges. The digital methods are preferable in the time ranges of a few μsec and larger.

D. DETECTORS IN TIMING MEASUREMENTS

1. *Scintillator–PM Detectors*

Scintillator–PM combinations used as detectors have wide application in the detection of nuclear radiations for timing purposes; they are mostly used in the detection of γ rays, neutrons, and β particles (conversion e^-). There are two types of scintillators that are commonly used. The first type, which includes the plastic, liquid, and organic scintillators, is characterized by a fast decay or fast light output, which implies good timing capabilities. For γ rays, these give limited energy information due to the fact that they have essentially no photopeak cross section (unless doped with high-Z material). The second type of scintillator, of which only the NaI(Tl) crystal will be discussed, is characterized by a slower light output, a high light yield, and good energy information in the photopeak because of the high-Z material.

In discussing the timing properties of scintillator–PM detectors, reference will be made to the three kinds of time-resolution contributions discussed in Section III.B. Most of the type 1 contributions are determined by the choice of scintillator dimensions. This choice usually represents a compromise between detector efficiency and time resolution. Resolution contributions of the second type, arising from time variations in the formation of the detector signal, originate both in the scintillator and the PM. The resolution contributions of the scintillator are contained in the time dependence of the photocathode illumination function $I(t)$. This function depends on (a) the light signal generation process and (b) the light collection. The generation of the light signal in the scintillation process involves the production of ionization by the radiation interaction, transfer of excitation energy to the scintillant, photon decay of the scintillant, and often a wavelength shift. The time dependence of $I(t)$ and in particular its rise (Koechlin and Raviart, 1964; Lynch, 1968) are expected to have a complicated relation to the details of these processes; one might expect the function to contain several exponentials including that of the dominant decay time as well as a dependence on the light yield. The light yield, which varies with the energy absorbed, is important in determining the statistical uncertainties in $I(t)$. The collection of the photons from the scintillator adds an important additional time dependence (Cocchi and Rota, 1967) to $I(t)$. Variations in the path length of the photons depend on scintillator geometry and on the reflectivity and diffuseness of the surfaces.

The PM, which converts the photons into a current signal, contributes to the time variation of the detector signal through (a) the statistical uncertainty in the number of photoelectrons, which depends on the conversion efficiency of the photocathode, (b) the transit-time differences of the photoelectrons in traveling from the photocathode to the first dynode due to different path lengths and velocities, and (c) the response function of the dynode gain structure to a single electron at the first dynode (SER).

Numerous theoretical calculations (Gatti and Svelto, 1966; Hyman, 1965; Bertolini et al., 1966; Bengtson and Moszynski, 1970) using these contributions of the scintillator and the PM have been performed to evaluate the time-resolution capabilities of the scintillator–PM detector. The calculations usually give the time resolution as a function of the discrimination triggering fraction of the detector signal (the literature refers to an integrated charge fraction C/R) and as a function of the energy absorbed E in the scintillator. The triggering fraction C/R that yields the minimum time resolution involves a compromise between the statistical uncertainties, which increase with C/R, and the time spread effects, which decrease with an increase in C/R; this optimal fraction varies with the different scintillators and PM. The energy dependence of the time resolution is contained in its inverse proportionality to the statistical uncertainty of the number of photoelectrons emitted from the photocathode. The number of photoelectrons is distributed around the average number N with a spread equal to $N^{1/2}$, which is reflected in the energy resolution and is approximately proportional to $E^{1/2}$; thus the time resolution should be proportional to $E^{-1/2}$.

A number of experiments (Schwarzschild, 1963; Bartl and Weinzierl, 1963; Present et al., 1964; Bertolini et al., 1966; Bengtson and Moszynski, 1970) for various scintillators and PM have been performed to check the theoretical predictions regarding the time resolution. The agreement between theory and experiment has not always been good. The disagreements have in part been due to the difficulty of describing the complicated light generation processes in the scintillator. A difference of two exponentials, one for the rise and the other for the decay, has been used for $I(t)$ by Hyman (1965). Although some parametrization of the rise is necessary, this particular description has not been completely satisfactory. Recently, Bengtson and Moszynski (1970) have obtained quantitative agreement between their experiments and the Hyman theory by describing $I(t)$ by a convolution of the single decay exponential with a Gaussian function. The Gaussian represents the time spread in the light collection (Bertolini et al., 1966) from the scintillator and the time spread of the light signal generation process.

Their experiment also agrees with theory for the energy dependence of the time resolution. They point out the importance, for the energy dependence study, of selecting the same fraction of the distribution of photoelectrons for all energies; thus the energy windows ΔE should be approximately proportional to $E^{1/2}$. With these energy windows, the experimental time resolution is proportional to the predicted $E^{-1/2}$.

Scintillator and PM parameters can be compared with the understanding of the time-resolution contributions of type 2. For the scintillator, a large light output per unit time at the rise of the illumination function $I(t)$ is important according to the theories for best timing. With this function best described by the convolution of a single decay exponential τ with a Gaussian of half-width σ, $I(t)$ at the rise depends on τ, σ, and the total light output $L(E)$ produced by the detected radiation. For no Gaussian spread, i.e., $\sigma \approx 0$, $I(t = 0) = L(E)/\tau$; with the necessary finite Gaussian spread σ, $I(t)$ is more complicated, having a rise time that is somewhat greater than σ. Thus for timing purposes, one should select a scintillator with a large $L(E)$ and small τ and σ. Unfortunately, these quantities are difficult to measure and often depend on the PM used; the quantity σ also depends on the scintillator dimensions.

The fastest plastic scintillators in common use are Naton 136, Pilot B, and NE 111, which have decay constants $\tau \approx 1.7$ nsec. The light output $L(E)$ of NE 111 is about 40% smaller than that of the other two; however, the σ of NE 111 is about a factor of four smaller; this is possibly due to the fact that NE 111 contains no wavelength shifter (Bengtson and Moszynski, 1970). The small σ appears to be more important for timing than the reduced $L(E)$. Stilbene and the liquid scintillators NE 213 and NE 218, with an $L(E)$ about 25% larger than that of Naton 136, have $\tau \approx 3.5$ nsec; as an advantage, they have an additional long-decay component of ~ 280 nsec which can be useful for γ-ray or neutron discrimination. The neutron-induced proton recoil signals excite a larger fraction of the long-lived component than the γ-induced Compton electrons; thus a pulse-shape analysis can be used to distinguish between the γ rays and the neutrons.

NaI(Tl), of the second type of scintillators, has a decay constant of $\tau = 250$ nsec; however, in spite of this long decay time, it is still very useful for fast timing because of a large light output $L(E)$ which is about six times that of Naton 136 (Houdayer et al., 1968).

The PM window to which the scintillator is attached should transmit a large fraction of the scintillator light. Occasionally, quartz windows are used to reduce the light absorption. Also, the optical coupling between the two should be good in order to reduce reflections and the corresponding excessive photon collection times.

The PM time-resolution contributions mentioned here should be considered in selecting a PM. The photocathode should have a high conversion efficiency for photoelectrons since the time resolution is proportional to $N^{-1/2}$. The recently available bialkali photocathodes, as, for example, RCA 8575, have considerably improved quantum efficiencies. The transit-time spread of the photoelectrons between the photocathode and the first dynode is one of the most important properties of a PM for fast timing. The construction, the focus, and the scintillator diameter are important in reducing these time spreads. The SER represents timing contributions from the dynode gain structure. Although it is not as important for timing as the other two, a narrow SER with minimal transit-time spreads yields the best time resolution. First dynodes constructed of gallium phosphide give high multiplication gains and a good SER. The RCA 8850 tube has both a bialkali photocathode and a gallium phosphide first dynode. The PM should also have a low noise level to reduce jitter. In summary, a PM for fast timing should have a high quantum efficiency, a low transit-time spread, a good SER, and low noise; PM tubes in general use for timing are RCA 8575, RCA 8850, XP 1021, and PM 2106.

The time resolution for scintillator–PM detectors can be minimized by use of CF timing. A definite discrimination fraction of the anode signal minimizes for all amplitudes the time variations due to statistical jitter and the PM response. The CF TDD can easily be adjusted to this optimal fraction f, which depends on both the scintillator and the PM (see Present et al., 1964; Bengtson and Moszynski, 1970). Since the rise times T_r for scintillator–PM signals are usually constant, the CF timing drastically reduces time walk over a large dynamic range. The CF timing is preferable for both types of scintillators over any significant dynamic range. For plastic scintillators, the time resolution for a 20% amplitude range is 30% better for CF timing as compared to LE timing. For NaI, a narrow dynamic range is frequently used such as over the photopeak of a single γ ray; in these cases, a LE TDD is often used for reasons of simplicity.

The best time resolution for scintillator–PM detectors to date is from the studies of Bengtson and Moszynski (1970). Using NE 111 scintillators ($d = 2.5$ cm, $h = 1$ cm) and XP 1021 phototubes, they obtained a FWHM = 132 psec and slope($\frac{1}{2}$) = 18 psec, as shown in Fig. 7, with 15% energy windows at 930 keV for ^{60}Co activity. These results imply that the contribution from one detector is FWHM = 94 psec. For these NE 111 scintillators,

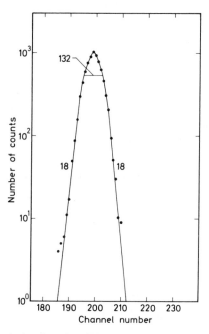

Fig. 7. Prompt resolution function (PRF) from ^{60}Co activity obtained with Ne 111 scintillators ($d = 2.5$ cm, $h = 1$cm) on XP 1021 phototubes, with $\Delta E/E \approx 15\%$ at 930 keV (Bengtson and Moszynski, 1970, used with permission of North-Holland Publishing Co., Amsterdam); times given on curve are in psec.

the resolution limitation is believed to be the PM; from their studies of Naton 136, for which the time resolution was about 15% greater, the scintillator light generation process appears to be the limitation. Figure 8 shows the effectiveness of CF timing with Naton 136 for 90% of the ^{60}Co Compton spectra. Time-resolution studies (Lynch, 1966; Braunsfurth and Körner, 1965) for NaI show a FWHM ≈ 500 psec at 511 keV for one detector; thus NaI is very useful for timing, especially with the added energy resolution and when the extreme time resolution is not needed. A summary of the optimal resolution for plastic–PM and NaI–PM detectors is given in Fig. 9 as a function of absorbed energy in terms of the PRF parameters. The FWHM values represent contributions for one detector only. At low γ energies for plastic scintillators, the absorbed energy E of the Compton edge is small, and thus NaI become preferable to plastic, with comparable time resolution at the higher photopeak energy.

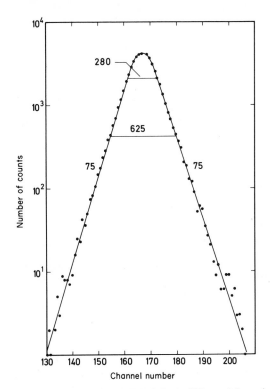

Fig. 8. Prompt resolution function (PRF) from ^{60}Co activity using constant-fraction timing for 90% of the Compton spectra (Bengtson and Moszynski, 1970, used with permission of North-Holland Publishing Co., Amsterdam), Naton 136 scintillators, $d = 2.5$ cm, $h = 2$ cm; times given on the curve are in psec.

There are numerous published timing measurements using scintillator–PM detectors. Several recent examples of in-beam timing measurements for the different radiations and scintillators will be given to illustrate the capabilities of scintillator–PM detectors in electronic timing; charged-particle–γ measurements will be discussed later with respect to silicon-detector timing. First, for the slope method, Lowe *et al.* (1962) and Becker and Wilkinson (1964) have obtained mean lifetimes as short as 250 psec in light nuclei using the slanted-target technique together with pulsed beam–γ (plastic) experiments. Pulsed beam–γ (NaI) timing has been used by Baba *et al.* (1973) to measure lifetimes of high-spin states in ^{91}Nb and ^{91}Zr. The NaI scintillator gives improved γ-energy resolution but with reduced time resolution as compared to plastic. Neutron (NE213)–γ (NaI) timing has been used

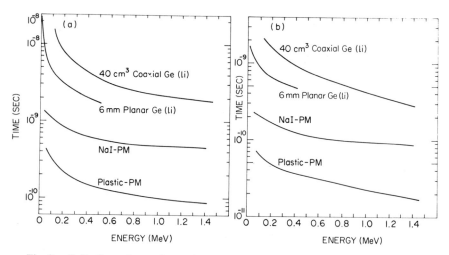

Fig. 9. Collection of experimental prompt resolution function (PRF) parameters, namely (a) the FWHM, and (b) the slope ($\frac{1}{e}$), for different γ-ray detectors. The FWHM values refer to the contribution of a single detector.

by Cochavi and Fossan (1971) to measure lifetimes in ^{92}Nb with the ^{92}Zr(p, nγ)^{92}Nb reaction. In this experiment, the neutron detection avoids confusion from other reactions and the NaI energy resolution allows the isolation of individual γ rays. A beam-shutter β (plastic)–γ (plastic) timing technique has been used by Cochavi et al. (1970a) to measure the $6_1{}^+$ lifetimes in ^{50}Ti and ^{54}Fe following the ^{48}Ca(t, n)^{50}Sc(β^-) and ^{54}Fe(p, n)^{54}Co(β^+) reactions. Jain et al. (1972) have effectively used Pb-loaded (10%) plastic scintillators for low-energy γ rays in an intense β background to measure sub-nsec lifetimes in several Rh and Ag isotopes. The addition of Pb gives a photopeak cross section without drastically affecting the fast decay properties of the plastic scintillator. Centroid-shift measurements with pulsed beam–γ (NaI) timing have been used by Shipley et al. (1969) to measure lifetimes of numerous Coulomb-excited states. For centroid measurements, great care must be taken to avoid any contamination by prompt γ rays; such prompt γ rays usually do not interfere with slope measurements due to the fact that the slope and the prompt regions are separated in time. With plastic scintillators, the centroid method of timing has been used in special cases for exceedingly short lifetimes. For γ–γ measurements where both γ rays trigger the start and the stop detectors, two time spectra are obtained with a centroid separation of 2τ. Barton et al. (1971) have used this technique with the ^{34}S(p, γ)^{35}Cl reaction to obtain a value of $\tau = 37 \pm 4$ psec for the 3.16-MeV level in ^{35}Cl. One of the difficulties in this method is the isolation of the γ rays. NaI detectors have been used in conjunction with one of the plastic detectors to observe the Compton-scattered γ ray, thus yielding the full-energy resolution. This technique along with centroid timing has been used by Sims and Kuhlman (1972) to measure the lifetimes $\tau = 3.8 \pm 0.4$ psec for the 889-keV level in ^{46}Ti and $\tau = 11 \pm 3$ psec for the 612-keV level in ^{192}Pt. Subtle walk and geometric-interaction corrections have to be made for these short lifetimes.

2. Solid-State Detectors[†]

a. Ge(Li) detectors. The outstanding γ-ray energy resolution of Ge(Li) detectors makes their use crucial for timing measurements involving complex γ spectra. The dominant time-resolution contributions for Ge(Li) detectors are the time variations due to charge collection and detector noise of type 2. The electron and hole charges produced by the γ rays exhibit varying collection times depending on the position of the ionizing event and the nonuniformity of electric fields in the detector. An electric field of about 100 V/mm, which is needed for good timing, produces a saturating velocity for both electrons and holes of ~ 0.15 mm/nsec. Collection times for a typical 40-cm^3 true coaxial (cylindrical with both ends open) detector vary from 60 to 120 nsec for these velocities. For modified coaxial (cylindrical with one end closed) and trapezoidal (five-sided) detectors, the collection times vary over even larger values, as expected from geometric considerations (Schmidt-Whitley, 1969). The collection times of planar detectors typically < 7 mm thick, which are used for low-energy γ rays, are somewhat shorter than those of the larger detectors. These variable collection times result in detector signals with a spread in rise times which strongly affects the time resolution. Collection times and signal shapes for coaxial and planar Ge(Li) detectors are given by Cho and Chase (1972) under the assumption of idealized electric field distributions and no velocity saturation effects. This simplified theory appears to be approximately valid only for planar detectors. Jitter from detector noise determines the energy dependence of the time resolution since the statistical jitter is usually small because of the large number of charge carriers in solid-state detectors. The time resolution for Ge(Li) detectors is expected to vary inversely with the energy since the noise is constant (see Section III.B).

[†] Also see Goulding and Pehl, Chapter III.A.

To minimize resolution contributions from the variable rise times of the Ge(Li) detectors, ARC or ELE timing is important. These TDD reduce the time walk of the variable rise times over a large dynamic range as discussed in Section III.C. The preamplifier signal is usually amplified with a fast amplifier before the TDD. A FWHM of 1.9 nsec has been obtained using ARC timing for a 40-cm^3 true coaxial Ge(Li) detector with the 1.33-MeV photopeak of ^{60}Co activity; for a dynamic range from 100 keV to 1 MeV, the time resolution for the same detector was FWHM = 2.8 nsec (Williams, 1971b). Ryge and Borchers (1971) have shown that the time resolution of modified coaxial and trapezoidal detectors is significantly inferior to that of true coaxial detectors. An experimental and theoretical comparison of LE, ARC, and ELE timing for Ge(Li) detectors has been made by Cho and Chase (1972). Optimal time-resolution parameters for a 40-cm^3 true coaxial Ge(Li) detector as a function of energy are shown in Fig. 9. Because realistic detector signals do not have perfectly linear rises, the time walk is not completely eliminated by the TDD; thus, for optimal time resolution, a Ge(Li) detector should be chosen for the shortest rise time and the smallest variation in rise times. There often are wings on the PRF of Ge(Li) detectors due to extreme collection times; these effects, which can be avoided by differentiation of the signal, should be carefully considered for the slope method of timing (Moszynski and Bengtson, 1970). Preliminary results for high-purity, uncompensated Ge show good timing properties for coaxial detectors (Cho and Llacer, 1972). The FWHM and slope($\frac{1}{2}$) values are also shown in Fig. 9 for a 6-mm planar Ge(Li) detector; a FWHM of ~ 4 nsec is typical for 100 keV.

The Ge(Li) detectors have been widely used for timing measurements with pulsed beams, where the outstanding energy resolution provides adequate isolation of excited states without the detection of the associated reaction particles. Representative examples of the slope method of timing with pulsed beams and Ge(Li) detectors are the studies of Yamazaki and Ewan (1969) for Sn nuclei, of Lederer et al. (1971) for mass-90 nuclei, and of Bergström et al. (1970) in the Pb region, where the (α, 2nγ) reaction was used together with natural cyclotron pulsing. The (HI, $xn\gamma$) reactions, when used with pulsed beams and Ge(Li) detectors, have allowed the measurement of many lifetimes of high-spin states in numerous nuclei off the line of stability (Maier et al., 1971; Hageman et al., 1971). Coincidence timing measurements with Ge(Li) detectors are made more difficult because of the low γ-ray detection efficiency. Ellegaard et al. (1971) have measured slope lifetimes in ^{210}Bi using the (d, pγ) reaction. Maier (1970) obtained a result of $\tau = 2.1$

nsec for the 2.17-MeV state in ^{91}Zr using the same reaction. In both of these measurements, constant-fraction discriminators were used for the Ge(Li) and the silicon particle detectors. Centroid timing measurements have also been made with pulsed beams and Ge(Li) detectors by Kim and Milner (1971). Mean lifetimes of the order of 200 psec have been obtained for states in In and Cd isotopes following (p, nγ) and (p, p'γ) reactions; for such short lifetimes, experimental corrections are made for the time walk that remains after ARC timing. With the good energy resolution of Ge(Li) detectors, a two-dimensional analysis of γ-ray energy and time, along with ARC of ELE timing, is very useful in simultaneously studying numerous delayed γ rays following a nuclear reaction.

b. Silicon surface barrier detectors. Silicon surface barrier detectors are widely used for the detection of charged particles with excellent energy resolution. Collection times of the charge carriers are fast, at least for thin detectors, which makes them good for timing. Rise-time calculations (Tove and Falk, 1964; Quaranta, 1965) predict, for a silicon detector of 100 μm thickness, a rise time of $\sim$ 3 nsec. The collection times can be improved by increasing the detector bias, which produces higher electric fields. A thin detector such as the one cited can often be used for α particles and heavy ions which have large stopping powers. The rise times increase with detector thickness as expected and are influenced by the position of the ionization. Plasma discharges, which are most significant for heavy ions, introduce some additional time jitter in these detectors (Moszyński and Bengtson, 1971). Detector noise again influences the energy dependence of the time resolution.

Time resolutions of FWHM = 165 psec with 3-MeV ^{3}He (Sherman *et al.*, 1968) and 180 psec with 5.7-MeV α particles (Moszynski and Bengtson, 1971) using thin surface-barrier detectors and CF timing have been obtained; the PRF slope($\frac{1}{2}$) was $\sim$ 25 psec. Estimate of the FWHM contribution from the silicon detector alone is $\sim$ 100 psec. The FWHM measured for a 2000-μm-thick detector is $\sim$ 650 psec (Williams, 1971b). A LE TDD is often used for simplicity with silicon detectors for measurements involving the narrow amplitude range of a single particle group. An inductive LE time pickoff has been successfully used for measurements of mean lifetimes of about 200 psec with α-decay and reaction particles (Neal and Kraner, 1965; McDonald *et al.*, 1965); capacitive pickoffs are usually more effective for lower energy thresholds. For any significant dynamic range, one of the amplitude-compensation TDD's should be used.

Charged-particle–γ timing measurements are expected to yield exceptionally good time resolution with the use of silicon detectors and plastic scintillators. There are several published slope measurements of mean lifetimes of a few hundred psec for this type of timing. The (d, pγ) reaction has been used to measure lifetimes in ^{23}Ne (see Fig. 3) (Fossan *et al.*, 1966) and in ^{41}Ar (Fossan and Poletti, 1966). For situations when both a prompt and a delayed γ ray fall into the same energy window of a plastic scintillator, the PRF and the decay curve can be measured simultaneously with the different particle groups of the silicon detector. This is especially important for lifetime measurements where the lifetime is not significantly larger than the PRF slope, as in the lifetime measurements of the 0_2^+ states in ^{92}Zr and ^{94}Zr with the (p, p'γ) reaction (Cochavi *et al.*, 1970b). A nice example of the capabilities of this type of timing is the recent result of $\tau = 59 \pm 5$ psec for the ^{40}Ca 3^- state obtained with the (p, p'γ) reaction (Tape *et al.*, 1972). A least-squares fitting procedure of the convolution integral of Eq. (8) was used to properly account for the experimental PRF. The use of NaI scintil-

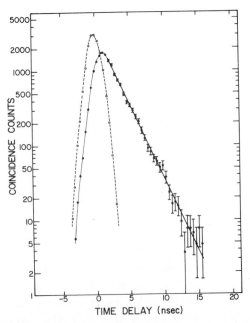

Fig. 10. Experimental delayed curve $F(t)$ (solid line) for the 2527-keV 5_1^- state in ^{92}Mo obtained with the ^{92}Mo(p, p'γ)^{92}Mo reaction using a silicon particle detector and a NaI scintillator (Cochavi *et al.*, 1971). The PRF shown by the dashed curve was measured simultaneously by selecting protons corresponding to the short-lived 2_1^+ state. The logarithmic slope of $F(t)$ gives a mean lifetime of $\tau = 2.24 \pm 0.06$ nsec.

lators in charged-particle–γ timing with its large photopeak cross section increases the yield of these low-counting-rate coincidence experiments and improves the isolation of excited states; the time resolution is then limited, however, by the NaI resolution contributions. NaI scintillators and silicon particle counters have been used for timing measurements with the (p, p'γ) reaction in ^{40}Ca (MacDonald et al., 1969) and in ^{92}Mo (see Fig. 10) (Cochavi et al., 1971). With the good particle energy resolution of silicon detectors, two-dimensional analysis of particle energy and time becomes an effective method for timing studies following nuclear reactions.

IV. Doppler Shift Techniques

The two essentially different Doppler shift techniques for measuring nuclear lifetimes which are in common use today can be characterized as to how the time scale is established. In the recoil distance method (RDM), the time scale is established by the distance a recoil nucleus travels at velocity v, and in the Doppler shift attenuation method (DSAM), it is derived from the slowing-down time of a recoiling nucleus in a stopping medium.

In both techniques, the Doppler shift of electromagnetic radiation emitted from a moving source is utilized. To second order, we have

$$E = E_0[1 + \beta \cos\theta - \tfrac{1}{2}\beta^2 + \beta^2 \cos\theta] \tag{11}$$

where $\beta \equiv v/c$ and θ is the angle between $\mathbf{v}$ and $\mathbf{k}_\gamma$. In both techniques, the production of an ensemble of recoiling excited nuclei that are as unidirectional and monoenergetic as possible is one of the main experimental requirements. For, as is intuitively clear, any spread in θ or v will ultimately be reflected as a spread in time. Thus, we consider first the nuclear excitation processes available for the two methods.

In both, the reactions are of the type

$$M_2(m_1, m_4) M_3{}^*$$

That is, a target nucleus of mass M_2 is bombarded by a beam of particles of mass m_1 to form the nucleus of mass M_3 in an excited state (denoted by an asterisk) and a particle of mass m_4. Simple application of energy and momentum conservation (nonrelativistic) gives us the pertinent relations to guide our choice of reactions and of beam energy: The center-of-mass velocity (in units of the speed of light) is

$$\beta_{CM} = (2m_1 E_1)^{1/2}/(m_1 + M_2) \tag{12}$$

where *all masses and energies are in MeV* and E_1 is the beam energy in the laboratory system. The corresponding z component of velocity for the excited nucleus $M_3{}^*$ at the moment of emission ($t = 0$) is

$$\beta_z(0) = \beta_{CM}(1 + \gamma^{-1} \cos \theta_{CM}) \qquad (13)$$

where θ_{CM} is the center-of-mass angle of the outgoing nucleus $M_3{}^*$, and γ^{-1} is the ratio of the speed of $M_3{}^*$ in the center-of-mass system to the speed of the center of mass in the laboratory system and is given by

$$\gamma^{-1} = \left\{ \frac{M_2 m_4}{m_1 M_3{}^*} \left[1 + \frac{m_1 + M_2}{M_2} \left(\frac{Q}{E_1} \right) \right] \right\}^{1/2} \qquad (14)$$

where the Q value is given by $Q_o - E_x$, with Q_o and E_x respectively the Q value for the ground-state reaction and the excitation energy of the level of M_3 in question. A well-designed RDM or DSAM experiment will have a small spread in $\beta_z(0)$. The problem is to accomplish this without unduly sacrificing intensity. From Eq. (13), we see that the most general way to achieve a narrow spread in recoil velocities is to observe the γ rays in coincidence with particles m_4 corresponding to a unique value of $\cos \theta_{CM}$. For instance, the particles could be detected in an annular counter centered at 180° or, in the case that the particles m_4 have considerably more range than m_1, a counter at 0°. Alternatively, to avoid the great loss of efficiency engendered by a coincidence requirement, the reaction kinematics can be utilized to minimize the spread in $\beta_z(0)$. From Eq. (13), we see that in a singles experiment, there is a total spread in $\beta_z(0)$ of $2\gamma^{-1}\beta_{CM}$ and so we need to minimize γ^{-1}. From Eq. (14), it is clear that for a given reaction we do this by minimizing the term in square brackets. Thus endothermic reactions (Q negative) and near-threshold, neutron-emitting reactions (no Coulomb barrier in the exit channel) are the most commonly used. Actually, the yield close enough to threshold to give an adequately collimated beam of $M_3{}^*$ is high enough in many proton-emitting reactions so that studies with reactions such as (α, p) are quite possible. Another way to minimize γ^{-1} is to make the ratio $M_2 m_4/m_1 M_3{}^*$ small. This can be accomplished dramatically by reversing the reaction, i.e., bombarding a light target with a heavy projectile, the ratio of the γ^{-1} values for the "inverse" and "normal" reactions being M_2/m_1 at the same center-of-mass energy. This procedure is wedded beautifully to the newer electrostatic accelerators and their versatile ion sources. Less dramatic in the minimization of γ^{-1}, but very promising for other reasons, are reactions of the evaporation type, in which two relatively heavy nuclei are welded into a compound nucleus and a few nucleons are then spewed out. In these cases, γ^{-1} is small mainly because m_4 is small compared to m_1.

Finally, we close this short summary of reaction types with the capture reaction in which m_4 is a γ ray. This reaction produces an almost perfectly defined "beam" of M_3^*. It has been extensively utilized for DSAM measurements but usually has too small a Doppler shift ($\beta = \sim 0.01\text{-}1.0\%$) for use with the RDM.

We now consider the two techniques in turn.

A. RECOIL DISTANCE TECHNIQUE

1. *Basic Technique*

If at $t = 0$ a nuclear reaction produces a nucleus in a bound (γ-emitting) excited state recoiling at velocity v, then a time scale is established by the distance D it moves before it emits a γ ray. The use of this time scale in measuring nuclear lifetimes is the basis of the recoil distance method. The range of applicability is determined by the recoil velocities available and by the range of distances which can be measured. This means, at the present time, lifetimes in the range $10^{-8} \gtrsim \tau \gtrsim 10^{-12}$ sec. Early applications of this method were made by Devons and associates (1949, 1954, 1955), Thirion and associates (1953, 1954), Burde and Cohen (1956), and Severiens and Hanna (1956). The method was applied with difficulty and in only special circumstances until the Ge(Li) γ-ray detector was harnessed for its use at the Chalk River Laboratories and applied to lifetime measurements in ^{16}O and ^{17}O (Alexander and Allen, 1965), ^{18}F (Alexander *et al.*, 1966), and ^{14}C and ^{14}N (Allen *et al.*, 1966, 1968). The method developed at Chalk River is termed the plunger technique and was suggested by A. E. Litherland. It has been reviewed by Allen (1966), Warburton (1967), and Schwarzschild and Warburton (1968).

The principle of this technique is illustrated schematically in Fig. 11. We give a zeroth-order description first. A beam induces a reaction in a target thin enough to permit the residual nuclei to recoil freely into vacuum with a component of velocity v along the beam direction. By placing a metal stopper at a distance D, the number of excited nuclei decaying in flight, and those surviving for a time greater than $t = D/v$ and thus having quickly stopped in the metal, can be determined with the aid of the Doppler energy separation of the γ rays. The γ rays detected at an angle θ to the beam from nuclei that decay before reaching the stopper are Doppler-shifted and thus to first order

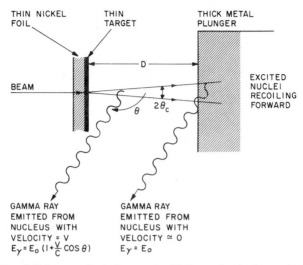

Fig. 11. Recoil distance method of measuring lifetimes of excited states (Schwarzschild and Warburton, 1968).

in v/c their energy is given by

$$E = E_0(1 + \beta \cos\theta) \tag{15}$$

while γ rays from the stopped nuclei have an energy E_0. The two corresponding γ-ray peaks can usually be resolved in the energy spectrum of a Ge(Li) detector. The intensities of the Doppler-shifted peak I_s and the unshifted peak I_0 are given by the well-known formulas for radioactive decay with $t = D/v$:

$$I_s = N(1 - e^{-D/v\tau}), \qquad I_0 = Ne^{-D/v\tau} \tag{16}$$

where N is the total number of reaction-produced γ rays. Hence a measurement as a function of D of the ratio $R = I_0/(I_0 + I_s)$, which is

$$R = e^{-D/v\tau} \tag{17}$$

gives the mean life, since v can be determined from the γ-ray energy spectrum using Eq. (15) or from the kinematics of the reaction. Alternately, I_0 ($= Ne^{-D/v\tau}$) can be determined relative to some quantity other than $I_0 + I_s$ such as integrated beam charge or some other γ-ray peak. This procedure is used when the intensity of I_s is hard to obtain accurately for some reason or another. Application of the RDM to the $^{49}\text{Ti}(\alpha, n)^{52}\text{Cr}$ reaction is illustrated in Fig. 12 (Brown et al., 1973).

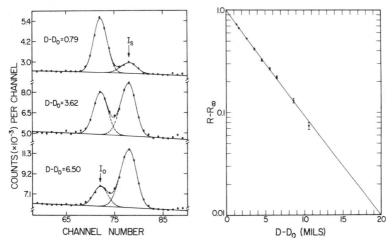

Fig. 12.   The lifetime measurement of the ^{52}Cr 6^+ 3113-keV level (3113 → 2369 keV).
The ^{49}Ti(α, n)^{52}Cr reaction ($Q_0 = 1.206$ MeV) was used with $E_\alpha = 14.5$ MeV. The 744-keV
$6^+ \rightarrow 4^+$ γ ray was detected at $0°$ to the beam. The line shape of this γ ray for three different
plunger distances (in mils) is shown on the left; D_0 is the experimentally determined plunger
"zero." The decay on the right yields $\tau = 59.5 \pm 3.4$ psec. R_∞ is the background value of
$R[= I_0/(I_0 + I_s)]$ at very large distance; it results mainly from large-angle scattering in the
target (Brown *et al.*, 1973).

The velocities necessary to give good separation of the two peaks I_0 and I_s
depend, of course, on the detector resolution and the γ-ray energy. For a
detector with a response of 3-keV FWHM and with 1-MeV γ rays, $\beta \approx 0.5\%$
would be adequate. However, high velocities are also desirable since for a
given lifetime, the appropriate distances for a measurement are proportional
to velocity, and thus the larger the velocity is, the more accurate can be the
determination of the relative distance.

It is sometimes necessary or desirable to use a gas target. In this case, it
is not possible to confine reactions to a distance D from the plunger; instead
the beam enters the gas through a thin window—such as a nickel foil—
traverses a thickness D of gas, is stopped in or beyond the plunger. Assuming that the energy loss in the gas target can be neglected and that the
reaction cross section is constant over the path length, we have

$$I_0/(I_0 + I_s) = \int_0^D e^{-x/v\tau}\, dx \bigg/ \int_0^D dx = (v\tau/D)\,(1 - e^{-D/v\tau}) \qquad (18)$$

and again $v\tau$ is determined by measuring $I_0/(I_0 + I_s)$ as a function of D.

This method was first used by Bizzeti *et al.* (1967). Application of the method to the ^{4}He(^{18}O, n)^{21}Ne reaction (Warburton *et al.*, 1971) is illustrated in Figs. 13 and 14.

Deviations from the zeroth-order description are typically 10% and depend on the degree of sophistication of the apparatus as well as on how well we have achieved a unidirectional, monoenergetic beam.

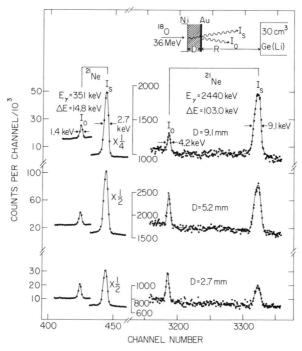

Fig. 13. Partial γ-ray spectra from the ^{4}He(^{18}O, n)^{21}Ne reaction. The full-energy peaks of the 351- and 2440-keV γ rays are shown for three different gas target thicknesses. Both the stopped I_0 and fully Doppler-shifted I_s peaks are evident for both γ rays (Warburton *et al.*, 1971).

The recoil distance apparatus—the plunger assembly—can be a simple gas target chamber with the variable thickness determined by a screw thread for use with lifetimes in the 10^{-8}–10^{-10}-sec range or a sophisticated and expensive example of precision machining for use in the 10^{-10}–10^{-12}-sec range. To get a feeling for the distances involved, note that at $\beta = 5\%$, a nucleus moves 0.6 mils (1 mil $\equiv 10^{-3}$ in.) in 1 psec. In all cases, the crucial factor in design is the

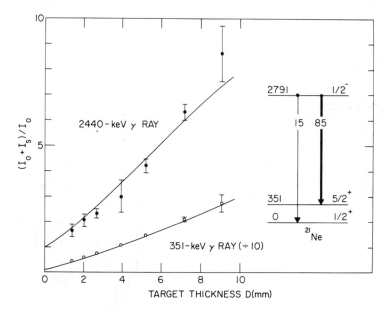

Fig. 14. The recoil distance data of Fig. 13 as a function of the gas target thickness D. The solid curves are the results of nonlinear least-squares fits to the appropriate equations including the corrections discussed in the text (Warburton *et al.*, 1971).

need for an accurate measure of the relative distance between target (or gas target entrance window) and plunger. At various times, an optical interferometer has been suggested as the ultimate method of determining D but as yet no one has found a practical way of harnessing one to an experiment. The almost universal method of varying D is to turn a micrometer screw. Two methods are used for measuring D, the straightforward use of a calibrated micrometer or measurement of the capacitance between the target and plunger. Alexander and Bell[†] (1970) have refined the capacitance method to the point where it is probably the best choice for the shortest lifetimes. One advantage of the capacitance method is that it can either monitor continuously or sample the distance each time an event is recorded, and thus any beam-on, beam-off effects can be observed or eliminated. Otherwise, without careful target design, the change of target position from beam pressure and heating could possibly be important. The preparation of flat targets parallel

[†] It is regretted that these authors did not "ring" in their colleague R. L. Graham as a "silent" coauthor.

to the plunger and under enough tension to eliminate beam-on effects has been considered by Gallant (1970) and Brandolini *et al.* (1971) as well as by Jones *et al.* (1969). Nickel, copper, gold, and carbon foils have been used successfully for target backings. Nickel is preferable mechanically but for some beams gives excessive background, e.g., α particles with $E_\alpha > 10$ MeV. If a coincidence is to be made with backscattered particles, then the plunger can no longer serve as the beam stop for most beams and must itself be a foil thick enough to stop the nuclei M_3^* but thin enough to allow the beam to pass on through. This essentially limits the method to reactions with $m_1 \ll M_2$. The annular counter coincidence arrangement is increasingly popular with Coulomb excitation (McGowan and Stelson, Chapter VII.A; Ward *et al.*, 1971; Diamond *et al.*, 1971).

2. Corrections to the Zeroth-Order Description

If the reaction and kinematics and other experimental conditions are carefully chosen, deviations from the zeroth-order description can be kept below 10% and can be estimated with about 10% accuracy. Thus the recoil distance method is capable of about 1% accuracy, which compares quite favorably with electronic techniques in the range $10^{-6} > \tau > 10^{-10}$ sec. However, most measurements to date have quoted uncertainties in the range ~ 5–20%. Quite simply, nuclear theory has not provided the motivation for more accurate measurements and the corrections summarized here have been often neglected.

Deviations from the zeroth-order description can arise from: (1) effects of velocity and distance distributions; (2) deorientation effects; (3) second-order velocity-dependent effects; (4) the energy dependence of the γ-ray detector efficiency; and (5) effects due to the finite detector–target distance.

We now consider these effects in turn. From now on, it is assumed that the γ-ray detector is at $0°$ to the beam. The axial symmetry of this geometry greatly simplifies the corrections considered here and in fact some have not been considered for other geometries. We consider explicitly only solid targets; the application of these corrections to gas targets is straightforward. For gas targets, there is one further uncertainty, not discussed here; namely, that due to the variation of the reaction cross section along the beam path (Schwalm, 1969).

a. Effects of velocity and distance distributions. We have discussed how to minimize the kinematic spread in velocity. Nevertheless, this remains one of the most important deviations from a zeroth-order description and one which is crucially dependent on the design of the experiment. Thus, we consider it at some length. Other contributions to the spread in velocity arise from the differences in energy loss in the target of the beam and the recoiling nuclei, nonuniformities in target thickness over the beam profile, and the straggling of the beam in the target backing or gas target window.

These various effects will be folded together to produce a velocity distribution. Using the first-order Doppler shift of Eq. (15), we see that there is a one-to-one correspondence between γ-ray energy and the recoil velocity, i.e., $v/c = E/E_o - 1$. Thus the velocity distribution translates into a γ-ray energy distribution which we shall often refer to as the γ-ray line shape. Two essentially different methods for correcting for the spread in velocity have been proposed (Jones et al., 1969). In the first of these, introduced by Ashery et al. (1966, 1967), the velocity distribution $f(v)$ is expressed as a power series in the velocity moments. That is, with $\bar{v}$ now representing the mean velocity, we can rewrite Eq. (17) as (Jones et al., 1969)

$$R = \psi(f) \exp(-D/\bar{v}\tau) \tag{19}$$

where

$$\psi(f) = \sum_{n=0}^{\infty} \alpha_n(\tau) M_n(f) \tag{20}$$

In Eq. (20), $M_n(f)$ is the nth moment of the velocity distribution and the $\alpha_n(\tau)$ are known analytical functions of $D/\bar{v}\tau$. The utility of this method lies in the rapid convergence of Eq. (20) and in the relative ease with which the experimental contribution to the velocity distribution of the detector response can be factored out. A somewhat similar method was used by Schwalm (1969); again the detector response was carefully taken into account.

In the second method, which is equivalent to a numerical integration, R is expressed as a series:

$$R = \sum_{n=1}^{m} a_n e^{-D/v_n \tau} \tag{21}$$

where

$$\sum_{n=1}^{m} a_n = 1 \tag{22}$$

This method necessarily calls for a computer program to perform the least-squares fit to Eq. (21). It is appropriate if the velocity distribution is large enough compared to the detector resolution so that the a_n can be determined directly from the experimental spectrum. Unfortunately, this is seldom possible without introducing nonnegligible uncertainties, so that the effect of the finite detector resolution must be carefully considered. This involves the unfolding of the velocity distribution from the resolution function convolution. Assuming Gaussians, the velocity distribution is simply obtained from the experimental width of the shifted peak and the resolution width. This method has been used by Jones et al. (1969) and McDonald et al. (1971). Neglecting the detector resolution and assuming a multichannel analyzer spectrum, a_n would be the normalized number of counts in the nth channel $v_n/c = E_n/E_0 - 1$, with E_n the γ-ray energy corresponding to the nth channel.

The most likely contributions to a spread in distance D are target non-uniformities and nonparallel alignment of target and plunger. These contributions are expected to be independent of D. In as far as they are, the spread in distance will not affect the slope of the decay curve in first order but only its intercept. The effect on the slope comes in second order, i.e., in a moment expansion, the leading term is proportional to the product of the second moments of the velocity and distance distributions (Jones et al., 1969). The effect of the distance distribution can be estimated through a modification of the a_n of Eq. (21).

b. Angular correlation deorientation effects. This phenomenon is not well understood and so it is important to minimize any uncertainties due to it rather than to attempt to correct for them.

The nuclear spin alignment which may be produced in the formation of a nuclear state by a nuclear reaction can be considerably perturbed when the atom is allowed to recoil into vacuum. The effective fields at the nucleus due to the excited and possibly stripped atomic structure in which it resides in vacuum can be very large and produce considerable loss of alignment even for states with lifetimes in the psec range (Goldring, 1968). Since the perturbations produce a time-dependent variation in the probability of emission at a given detector angle, it is clear that some modification of the intensity of Eq. (16) will result. These effects were first considered by Ashery et al. (1966, 1967). The intensity of γ radiation observed at angle θ with respect to the incident beam direction may be written as follows:

$$I(\theta, t) = N\lambda e^{-\lambda t}[1 + \sum_{k=2, 4, \ldots} A_k G_k(t) Q_k P_k(\cos \theta)] \qquad (23)$$

The quantities A_k are determined by the reaction mechanism exciting the state, the spins of the initial and final states, and the multipole order of the γ rays. Q_k is a geometric attenuation coefficient determined by the detector geometry. The $G_k(t)$ includes all the deorientation effects and are bounded by $0 \leqslant G_k(t) \leqslant 1$.

Recent studies on *light* nuclei by Faessler *et al.* (1971) and Berant *et al.* (1971) indicate that the hyperfine interaction responsible for the attenuation is static. That is, the atomic configuration reaches its ground state in a time short compared to ~ 1 psec. Furthermore, the experiments indicate that the interaction is due mainly to the unpaired 1s electron, with some effects due to 2s and higher electronic configurations which become more important for longer lifetimes.

For heavier nuclei, the hyperfine interaction involves a large number of excited atomic states which decay in successive steps during the nuclear lifetime. Thus the perturbation process is due to rapidly fluctuating magnetic fields composed of multicontributions from a complex atomic environment and the situation can only be described in statistical terms. For either light or heavy nuclei, the magnitude of the effect on a recoil distance measurement can only be determined well by measuring the angular distribution of the γ-ray emission relative to the beam axis; the angular distribution needs to be measured twice. First, with the excited nuclei recoiling into and stopping in a nonperturbing medium (cubic metal is the best choice) and second, with the excited nuclei recoiling into vacuum. If the first distribution is isotropic, there is no deorientation correction. If it is anisotropic, then the ratio of the Legendre polynominal expansion coefficients will give the time-averaged $G_k(t)$, i.e.,

$$\bar{G}_k(\infty) = \int_0^\infty \lambda e^{-\lambda t} G_k(t) \, dt \tag{24}$$

and both the maximum possible effect and an estimate of the effect on the lifetime measurement can be determined from various models for the time dependence of the $G_k(t)$ and a knowledge of the $\bar{G}_k(\infty)$. Recent measurements by Ward *et al.* (1972) on the $2^+ \rightarrow 0^+$, $\tau = 68$ psec ^{150}Sm γ ray have shown a powerful way to investigate deorientation in vacuum utilizing the RDM. These authors observed the attenuation coefficients $G_2(t)$ and $G_4(t)$ as a function of time by measuring the γ-ray angular distribution as a function of plunger distance D. They found that the $G_k(t)$ decay exponentially as the model outlined by Ashery *et al.* (1966, 1967) and others predicts. For most nuclei and especially near the shorter-lifetime limit of the recoil distance method, corrections for deorientation probably cannot be made with reliable accuracy and it is best to design the experiment so such effects are small. Unfortunately, deorientation effects were not considered in many of the pre-1969 measurements and so some of them are liable to an unknown systematic uncertainty.

c.   *Relativistic and detector efficiency effects.*   We include here deviations from the first-order Doppler shift expression. Extending Eq. (15) to second order gives us Eq. (11). If the terms in β^2 are not negligible, then the one-to-one correspondence between γ-ray energy and velocity is distorted and a little extra effort is necessary to extract the velocity distribution from the energy spectrum.

The most important relativistic correction is for the solid angle aberration. That is, since the angular distribution of γ rays becomes forward-peaked in the laboratory system, the solid angle of the detector is slightly bigger for γ rays emitted by moving nuclei relative to that for γ rays emitted by the stopped nuclei. The simple, relativistic transformation of solid angles results in the following expression for the ratio of solid angles of the moving to stopped peaks:

$$\frac{\Omega_s}{\Omega_0} = \frac{(1 + v/c)}{1 - (v/c)\cos\theta_c} \tag{25}$$

where θ_c is the maximum half-angle subtended by the detector. (This assumes the γ-ray emission is isotropic in the center of mass of the emitting nucleus, that the recoil velocity is along the beam direction, and that the counter distance is large compared to D.) As a result of this effect, Eq. (17) must be modified as follows:

$$R = \frac{e^{-D/v\tau}}{e^{-D/v\tau} + (1 - e^{-D/v\tau})(1 + \chi)} \tag{26}$$

where

$$1 + \chi \equiv \frac{(1 + v/c)}{[1 - (v/c)\cos\theta_c]} \tag{27}$$

Since in the usual experimental setup, $\cos\theta_c \simeq 1$ and $v/c \ll 1$, we have

$$1 + \chi \simeq 1 + 2(v/c) \tag{28}$$

as sufficient for all but the most accurate measurements. It is convenient to include the detector efficiency correction with the solid angle correction. A good functional form for the Ge(Li) full-energy peak efficiency is

$$\varepsilon_n = \varepsilon_0 E_0^{\ k}/E_n^{\ k} \tag{29}$$

where k will be close to unity. For the series representation of Eq. (21), the combined solid angle and efficiency corrections can be made through modification of the a_n (McDonald et al., 1971):

$$a_n = E_n^k N_n / E_0^k (1 + \chi_n) I_s \qquad (30)$$

where N_n is the number of counts in the nth channel and I_s is the corrected total yield of γ rays from nuclei decaying in flight,

$$I_s = \sum_n E_n^k N_n / E_0^k (1 + \chi) \qquad (31)$$

If v is not small, one must question the assumptions used to derive Eqs. (25).

   d. *Effects due to the finite detector–target distance.* With a detector of finite size, the shifted peak is lowered in energy slightly because the averaged projection of the recoil velocity on the various γ-ray directions allowed by the detector solid angle is less. Assuming $D \ll R_d$, where R_d is the detector–plunger distance, and an isotropic γ-ray angular distribution over the detector solid angle, the correction for this effect becomes (Jones et al., 1969)

$$E = E_0 [1 + (v/2c)(1 + \cos\theta_c)]$$

where θ_c is the half-angle subtended by the detector.

   In the measurement of very long lifetimes or when using gas targets, the variation of the detector–plunger distance must be taken into account. A sensible approximation is to parameterize the detector efficiency as $e^{-\alpha r}$, which gives a good fit in a useful range of r. The ratio $I_0/(I_0 + I_s) = e^{-D/v\tau}$ should then be replaced (Goosman and Kavanagh, 1967) by

$$e^{-D/v\tau}(1 - v\tau\alpha)/[e^{-D\alpha} - (v\tau\alpha)e^{-D/v\tau}]$$

A formalism suitable for use with a gas target has been given by Warburton et al. (1971). It is straightforward but complicated and so a nonlinear least-squares computer fitting routine becomes the easiest way out.

3. *Applications*

   The RDM is so widely and frequently used that only a few representative examples will be given. The group at SUNY at Stony Brook have measured a number of lifetimes using endothermic (α, n) and (α, p) reactions with α energies of ~ 7–15 MeV and no coincidence conditions. This is just about the simplest method possible. The distance D was measured with a micrometer screw. Figure 12 illustrates some data from the ^{49}Ti$(\alpha, n)^{52}$Cr reaction. This group used the ^{31}P$(\alpha, n\gamma)^{34}$Cl reaction (Snover et al., 1971) and the ^{40}Ca$(\alpha, p)^{43}$Sc, ^{39}K$(\alpha, n)^{42}$Sc, ^{41}K$(\alpha, p)^{44}$Ca, ^{46}Ti$(\alpha, p)^{49}$V, ^{45}Sc$(\alpha, p)^{48}$Ti, ^{45}Sc$(\alpha, n)^{48}$V, and ^{49}Ti$(\alpha, n)^{52}$Cr reactions (Brown et al., 1974) to measure lifetimes from 4.8 to 77 psec. These reactions are listed to illustrate the wide applicability of the use of (α, n) and (α, p) reactions for RDM measurements. Results obtained using the $^{45}Sc(\alpha, p)^{48}Ti$ reaction are illustrated in Fig. 15.

The RDM is still more widely used at the Chalk River Laboratories, where the plunger technique was developed, than anywhere else. Examples are use of the exothermic $^{35}Cl(\alpha, p)^{38}Ar$ reaction with p–γ coincidences to measure mean lives for the 3377- and 4585-keV levels of ^{38}Ar (Ball *et al.*, 1972) and ^{36}Cl and ^{36}Ar lifetimes from singles measurements using "inverse" (d, p) and (d, n) reactions with ^{35}Cl beams incident on titanium deuteride targets

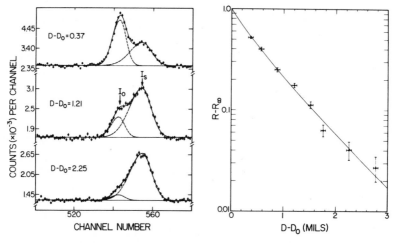

Fig. 15. The lifetime measurement of the ^{48}Ti 6+ 3335-keV level (3335 → 2297 keV). The $^{45}Sc(\alpha, p)^{48}Ti$ reaction ($Q_0 = 2.558$ MeV) was used with $E_\alpha = 10$ MeV. The 1038-keV 6+ → 4+ γ ray was detected at 0° to the beam. The line shape of this γ ray for three different plunger distances (in mils) is shown on the left. The decay on the right yields $\tau = 12.8 \pm 1.2$ psec (Brown *et al.*, 1974).

(Alexander *et al.*, 1971). A recent report (McDonald *et al.*, 1971) illustrates the simultaneous extraction of several lifetimes as well as the scope for ingenuity allowed by the method. From ^{19}F ions incident on self-supporting stretched carbon targets, the $^{12}C(^{19}F, np)^{29}Si$ and $^{12}C(^{19}F, \alpha p)^{26}Mg$ reactions were used in singles to measure mean lives for the ^{29}Si 3626-keV level $(4.2 \pm 0.5$ psec) and the ^{26}Mg 1809-keV level $(0.7 \pm 0.3$ psec), the latter being the shortest lifetime measured by the RDM to date. Ingenuity was shown in extracting a lifetime from the 0° line shape of the decay of the ^{27}Al 843-keV level to the ground state following Coulomb excitation by ^{35}Cl ions.

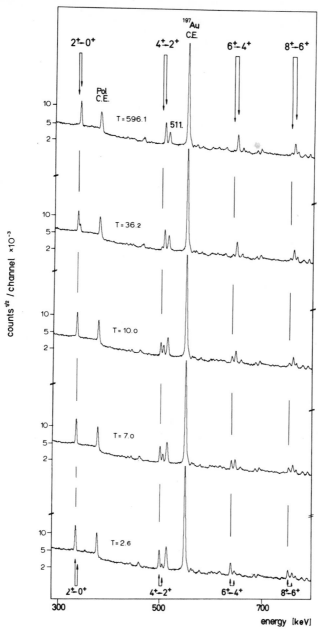

Fig. 16. Gamma-ray spectra observed at 0° to the beam direction following the reaction $^{110}Pd(^{16}O, 4n)^{122}Xe$ for different settings of the target to plunger distance D. The spectra are labeled in terms of the recoil ion flight time (in psec) $T = D/v$, where v is the average recoil velocity. The positions of the unshifted and shifted lines of the ground-state band transitions are indicated; $E(^{16}O) = 75$ MeV. (Kutschera *et al.*, 1972.)

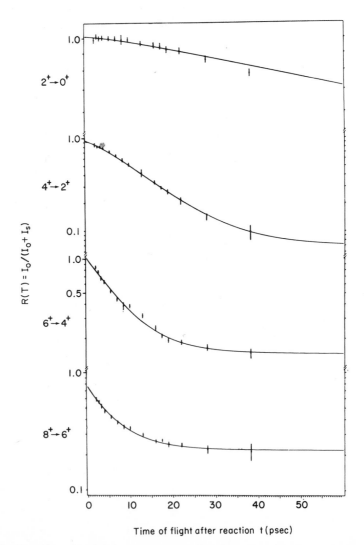

Time of flight after reaction t (psec)

Fig. 17. The ratio of the unshifted to total intensity for several transitions in ^{122}Xe plotted versus the time after the reaction ^{110}Pd(^{16}O, 4n)^{122}Xe. The solid lines are the results of the fitting procedure to complex decays described in the text. The indicated errors are due to statistics (Kutschera *et al.*, 1972). Mean lives of <3.5, 3.9 ± 0.7, 8.2 ± 1.2, and 89.3 ± 8.1 psec for the $8^+ \rightarrow 6^+$, $6^+ \rightarrow 4^+$, $4^+ \rightarrow 2^+$, and $2^+ \rightarrow 0^+$ transitions, respectively, were obtained.

The shifted and stopped peaks were not clearly resolved because of the large variation in recoil velocities allowed by the kinematics. However, the decay curve for an energy region including all of the stopped peak I_0 and a small portion of the unresolved shifted peak I_s has the form of a constant plus an exponential term, from which the lifetime was determined using a representation like the series of Eq. (21).

Coulomb excitation (discussed by McGowan and Stelson, Chapter VII.A) is one example of a reaction suitable for applications to heavier nuclei. Another is provided by reactions of the type (HI, xn) (discussed by Newton, Chapter VII.E), such as the (^{16}O, 4n) results of Kutschera et al. (1972) on states in $^{120, 122}$Xe and $^{126, 128}$Ba. These are illustrated in Figs. 16 and 17. This work is an example of cascade decays that involve more than one lifetime. Thus when a given level of mean life τ_2 is fed both directly (with fraction f_2) and from cascade (with fraction f_1) through a state with lifetime τ_1, we have

$$R = f_2 e^{-D/v\tau_2} + f_1\{[1/(\tau_1 - \tau_2)] (\tau_1 e^{-D/v\tau_1} - \tau_2 e^{-D/v\tau_2})\} \qquad (32)$$

Generalization to more than two lifetimes is straightforward. The fractions f_i can be determined independently from the γ-ray intensities, so there are only as many unknowns as lifetimes in Eq. (32).

B. The Doppler Shift Attenuation Method

1. Basic Technique

The DSAM was developed by Devons and his collaborators (Devons et al., 1955; 1956; Bunbury et al., 1956) in the days when the NaI(Tl) crystal was just coming into its own as the best all-around γ-ray detector. The historic heavy-ion work of Litherland et al. (1963) and the developments of Warburton et al. (1963) are other significant pre-Ge(Li)-detector results. The reviews of Warburton (1967), Schwarzschild and Warburton (1968), and Broude (1969) emphasize the use of the Ge(Li) detector, as we shall do here. Again we start with a zeroth-order description.

Consider a nuclear reaction which produces a monoenergetic, unidirectional beam of excited nuclei $M_3{}^*$ with initial velocity $v(0)$. Consider further that the nuclei are moving in a medium in which they will ultimately slow down and stop. As they slow down, they decay at the rate

$$dN(t)/dt = -(N_0/\tau) e^{-t/\tau}$$

In general there will be a continuous distribution of γ-ray energies between

E_0 and $E_0[1 + \beta(0) \cos \theta]$. To see how this line shape can be used to obtain the mean life τ, we take a very simple form for the energy loss in the stopping material, namely

$$dE/dx = -K_e(v_z/v_0) \tag{33}$$

where we have introduced the Bohr orbital velocity $v_0 \equiv c/137$ as a convenient unit of velocity. Actually, Eq. (33) is a good approximation for the electronic part of dE/dx in the region $\beta \lesssim 0.02$, and the constant K_e may be readily obtained from available experimental data. For ions of mass M, we then have

$$dE/dx = Ma = M\, dv_z/dt = -K_e(v_z/v_0) \tag{34}$$

and solving this differential equation gives

$$v_z(t) = v(0)\, e^{-t/\alpha} \tag{35}$$

where α $(\equiv Mv_0/K_e)$ is called the characteristic slowing-down time of the material. It turns out that even in regions where Eq. (33) is not a good approximation, α gives a good rule-of-thumb value for the time for $v(t)$ to slow down to $1/e$ of its original value. α is characteristically of the order of 1 psec for solids (within a factor of 2–5) and scales as the density for gases. From Eq. (35), we can easily derive the line shape and the average Doppler shift, these being the two experimental quantities from which τ is customarily extracted. The ratio of average shift to maximum shift is termed the attenuation factor and is given by

$$F(\tau) = [1/v(0)\, \tau] \int_0^\infty v_z(t)\, e^{-t/\tau}\, dt \tag{36}$$

and for the energy loss dependence of Eq. (33), we have

$$F(\tau) = \alpha/(\alpha + \tau) \tag{37}$$

One very important attribute of Eq. (37) is that $F(\tau)$ is independent of $v_z(0)$, so that an initial velocity spread, due to a spread in either energy or direction, will not introduce any uncertainty in τ for the energy loss dependence of Eq. (33). For the energy loss actually encountered in practice, this is not true, but since Eq. (33) is often a good approximation, the mean life is often insensitive to any spread in $v_z(0)$.

To derive the line shape, we define $V \equiv v_z(t)/v_z(0) \equiv \beta_z(t)/\beta_z(0)$ so that V varies between zero and one and the line shape is $dN(V)/dV$. From the radioactive decay law, we have

$$N(V) = N(t) = N_0 e^{-t/\tau} = N_0(e^{-t/\alpha})^{\alpha/\tau} \tag{38}$$

and from Eq. (35), we have

$$N(V) = N_0 V^{\alpha/\tau} \tag{39}$$

for the number of excited ions left at velocity V (time t). The velocity spectrum (line shape) is then

$$dN(V)/dV = (\alpha/\tau) N_0 V^{(\alpha/\tau)-1} \tag{40}$$

and the line shape is a function of one variable, α/τ, so that fitting the line shape observed for a given stopping material will give τ. It should perhaps be emphasized that because of the one-to-one correspondence between γ-ray energy and recoil velocity [Eq. (15)], the velocity line shape is also the γ-ray-energy line shape. Once the line shape is obtained, we can use it to derive other quantities, e.g.,

$$F(\tau) = \int_0^\infty \frac{dN(V)}{dV} V \, dV \bigg/ \int_0^\infty \frac{dN(V)}{dV} \, dV \tag{41}$$

A fundamental relation exists between $- dE/dx$ and the range:

$$R(E) = R(E_1) + \int_{E_1}^E \frac{dE}{dE/dx} = R(v_{z1}) + \int_{v_{z1}}^{v_z} \frac{M v_z \, dv_z}{- dE/dx} = \int_0^{v_z} \frac{M v_z \, dv_z}{- dE/dx} \tag{42}$$

For $- dE/dx = K_e(v_z/v_o)$ and $V \equiv v_z(t)/v_z(0)$, we have

$$R(V) = \int_0^{v_z} \frac{M v_z \, dv_z}{K_e(v_z/v_0)} = \alpha \int_0^{v_z} dv_z = \alpha v_z(0) \, V \tag{43}$$

This illustrates that α can be extracted from range data as easily as from energy loss data.

This development contains all the necessary points to illustrate the DSAM. We have already discussed methods of producing beams of excited ions. As already mentioned, departures from unidirectional, monoenergetic beams are not as serious for the DSAM as for the RDM. In the more sophisticated computer programs currently used for evaluating DSAM data, spreads in direction and velocity are taken into account by numerical integration and averaging. With care, this can be done with considerably less uncertainty than for some of the other contributions. Likewise, the measuring error of the attenuation factor $F(\tau)$ or the line shape $dN(V)/dV$ can be kept relatively small so that, at the present time, the accuracy of the DSAM is limited primarily by uncertainties in our knowledge in the stopping cross sections for heavy ions in matter. This subject is a complicated one which is in a state of rapid development. We attempt here only to give a feeling for our present state of knowledge and how it relates to lifetime measurements.

2. *Energy Loss and Scattering of Ions in Matter*

At low velocities, $\beta \lesssim 0.35\%$, the energy loss is dominated by nuclear scattering. The most thorough treatment of this process and the one used as a guide in most DSAM work is that of Lindhard *et al.* (1963). In the low-velocity region, the energy loss takes place in rather large discrete steps accompanied by large-angle scattering. Thus what is needed is not dE/dx but $M\,dv_z/dt$; these quantities are identical only when there is no change of direction in the stopping process. Because the stopping of a given ion involves many discrete collisions with each change in $M\,dv_z/dt$ dependent on the past history of that particular ion, an accurate evaluation of the process requires a statistical treatment such as the Monte Carlo method, which has been quite elegantly applied by Currie (1969). Unfortunately, the Monte Carlo method requires relatively large amounts of computer time and so has not been used routinely but only to check results obtained using computer programs based on the treatment of Blaugrund (1966). The essential point of Blaugrund's treatment is the approximation $\langle v_z(t) \rangle \equiv \langle v(t) \cos\Phi(t) \rangle \simeq \langle v(t) \rangle \langle \cos\Phi(t) \rangle$. This approximation appears to introduce uncertainties in τ of $\lesssim 10\%$ (Broude, 1969; Currie, 1969), which can be partially corrected for, so the method is quite acceptable. Thus the theory of Lindhard *et al.* (1963), which is based on the Fermi–Thomas screened potential, can be applied with the Blaugrund approximation to DSAM data with reasonable accuracy. The theory itself has not been subjected to very extensive experimental checks but what checks have been made have indicated the possibility of large deviations (Blaugrund *et al.*, 1967; Currie *et al.*, 1969). Thus lifetime measurements made with $\beta \lesssim 0.3\%$ (this includes almost all proton capture results) must be viewed as possibly susceptible to large but unknown systematic errors.

As the velocity of the recoiling nucleus increases, the relative importance of nuclear stopping decreases and the electronic stopping process becomes more important. The velocity region $\beta = 0.5$–1.5% is a common one for DSAM measurements initiated by p, d, t, ^{3}He, and ^{4}He beams. In this velocity region, both energy loss processes are important. The electronic energy loss is, to a good approximation, continuous and not accompanied by any change of direction. It is thus much easier to handle analytically than the nuclear process. The theoretical guide is again that of Lindhard *et al.* (1963) but in essence all that is used is the prediction that the electronic energy loss in this energy region is proportional to velocity, i.e., Eq. (33) pertains for the electronic part of dE/dx. Deviations from this prediction have been noted in extensive experimental work, for instance, that of Hvelplund (1971), and, although not serious from the present point of view, these deviations should be kept in mind. At the present time, the magnitude of dE/dx is taken from experiment, either directly or by interpolation; and the theoretical prediction is used only as a reference value. It has been known for some time that the ratio of the measured to predicted energy loss fluctuates up to 30% for differing Z of the moving ion (Omrod *et al.*, 1965; Hvelplund and Fastrup, 1968) and recently it was established that there are similar fluctuations with the Z of the stopping material (Hvelplund, 1971). This means that unless a measurement of dE/dx exists for the particular combination of moving ion and stopping material being used, there can be rather large uncertainties in dE/dx and thus in the mean life. Two recent reports bear on this. Broude *et al.* (1972) obtained the mean life of the 3340-keV level of ^{22}Ne from γ-ray line shapes observed with 39 different solid elemental stopping materials and found a range of lifetimes of 1.9 (but covered by a 15% standard deviation) strongly correlated with the atomic number of the stopping material. Hvelplund (1971) compared the energy loss of ^{16}O ions in various materials and found deviations up to $\sim 30\%$ from the theory and a strong correlation with the atomic number of the stopping material. Except for the results obtained using a carbon stopper (carbon is a notoriously difficult material from which to make uniform thin films of known density), the results of Broude *et al.* (1972) seem in good agreement with those of Hvelplund (1971) in the region of overlap. It would seem that unless data exist to explicitly evaluate dE/dx for the problem at hand, interpolation for $v/v_0 \lesssim 2.5$ can only be trusted to about 15% accuracy and the accuracy can be considerably less if care is not taken in the interpolation process.

At velocities above $v/v_0 = 2.5$ ($\beta = 1.5\%$), the electronic energy loss rises more slowly with velocity, reaches a maximum, and slowly decreases and the nuclear energy loss becomes almost negligible. The total dE/dx (really $M \, dv_z/dt$) is illustrated in Fig. 18 for ^{24}Mg stopping in Mg (Schwalm *et al.*, 1974). This illustrates how energy loss data are obtained for $\beta \simeq 1.5$–7% ($v/v_0 \simeq 2$–9), which is the region commonly reached in heavy-ion-induced reactions. In Fig. 18, the solid points were obtained by interpolation for both Z_1 (moving ion) and Z_2 (stopping material) using the effective-charge

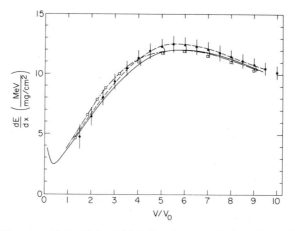

Fig. 18. The energy loss dE/dx of ^{24}Mg ions in natural magnesium. The data points are from interpolation procedures: the γ^2_{eff} (calculated) method (●) is explained in the text; the "universal" curves of Northcliffe and Schilling (1970) (□) were obtained in a similar manner. The three curves are three different possible forms for the analytical expression for dE/dx, which is discussed in the text. The differences in the curves are comparable to the uncertainty in dE/dx (Schwalm et al., 1974).

concept in which the effective charge $Z\gamma_{\text{eff}}$ is defined by

$$dE/dx = (Z^2\gamma^2_{\text{eff}}/\gamma_{\text{p}}{}^2)\,(dE/dx)_{\text{p}} \tag{44}$$

where γ_{p} and dE/dx_{p} are the effective charge and energy loss for protons moving with the same velocity and in the same stopping medium as the heavy ion in question (Booth and Grant, 1965; Northcliffe, 1963; Häusser et al., 1969). This method makes use of the empirical fact that a plot of $(\gamma_{\text{eff}})^2$ versus $EA/Z^{4/3}$ is almost a universal curve, implying among other things that the fluctuations of dE/dx with Z_1 and Z_2 observed in the region $v/v_0 \lesssim 2.5$ are appreciably smoothed out for higher velocities. It is believed that dE/dx for Z_1, $Z_2 < 30$ can be obtained by this method with an accuracy varying from $\pm 5\%$ at $v/v_0 = 10$ to $\pm 10\%$ at $v/v_0 = 2.5$ (Schwalm et al., 1974).

A good phenomenological fit to dE/dx in this region is given by a quadratic in v/v_0, i.e.,

$$-\,dE/dx = a + b(v/v_0) + c(v/v_0)^2, \qquad 2 \leqslant v/v_0 \leqslant 10 \tag{45}$$

The procedure used by Häusser et al. (1969), Warburton et al. (1973), and

Schwalm *et al.* (1974) is to match the analytical form of Eq. (45) to

$$- dE/dx = K_n(v/v_0)^{-1} + K_e(v/v_0) - K_3(v/v_0)^3, \qquad v/v_0 \leqslant v_c/v_0 \qquad (46)$$

at $v/v_0 = v_c/v_0$ by appropriate choice of v_c and K_3. In Eq. (46), $K_e(v/v_0)$ is the electronic energy loss of Lindhard *et al.* (1963) and K_e is chosen accordingly as discussed. The nuclear energy loss and scattering are represented by $K_n(v/v_0)^{-1}$ and so K_n is chosen to best represent $M dv_z/dt$ rather than dE/dx (Warburton *et al.*, 1967). This is accomplished by fitting to projected range data, i.e., using Eq. (42) to relate projected range and dE/dx, or in some other way simulating the effect of nuclear scattering. For initial velocities in the region $v/v_0 \gtrsim 1.5$, *proper* use of the representation of Eq. (46) is believed to introduce an uncertainty small compared to that in the basic form and magnitude of dE/dx.

3. *Applications*

Although spreads in velocity are not as serious as was the case for RDM measurements, we still desire a nearly unidirectional, monoenergetic ensemble of recoiling nuclei for the DSAM measurements, in part because of the greater ease of analysis. The radiative capture of protons or α particles is the prototype of the reaction with a small Δv. These reactions are most useful in the light nuclei, $A \lesssim 40$, where the capture cross sections and the recoil velocities are large enough for routine DSAM measurements. For these nuclei, advantages besides the sharpness of v are the possibility of many simultaneous lifetime determinations and the ease of selection of levels for study. That is, levels are selected by the choice of resonance and a given resonance will, in general, decay by several cascades as will the levels it feeds and thus the decays of many states are observed simultaneously. An additional advantage for a sharp resonance is that the beam energy can be chosen so that the resonance is at the front of the target and the recoils stop in the target. This avoids the complication, often present in particle-in, particle-out reactions, of partial slowing down in the target and the target backing. The main disadvantage is the small value of v/c, which means both small Doppler shifts [and thus larger uncertainties in $F(\tau)$] and also larger uncertainties in $M dv_z/dt$ as has been discussed. Many lifetime determinations have been made by this method. A good example is the work at the University of Utrecht on the $^{37}\text{Cl}(p, \gamma)^{38}\text{Ar}$ reaction (Engelbertink, *et al.*, 1968). Another is the $^{13}\text{C}(p, \gamma)^{14}\text{N}$ results of Bister *et al.* (1971) for three levels of ^{14}N. This latter work is notable for its careful study of the effects of target preparation—which can be sizable—and of the reliability of the theory of Lindhard *et al.* (1963) in the region of velocity where nuclear scattering is dominant.

As in the RDM particle–γ coincidence conditions can be used to select a unidirectional beam of recoiling ions. The low counting rate is often compensated for by the use of two-dimensional analysis with particle energy selection so that the lifetimes of many levels can be obtained simultaneously and/or two or more Ge(Li) detectors or particle detectors can be used at one time. Figure 19 shows the experimental line shapes obtained by Currie *et al.* (1969) for the first excited state of ^{30}Si following the ^{27}Al$(\alpha, \mathrm{p})^{30}$Si reaction with p–$\gamma$ coincidences utilizing an annular particle detector centered at 180° to the target. The recoil velocity was $v/c = 0.93\%$ and the theoretical line shapes shown in Fig. 19 were calculated (Currie, 1969) by the Monte Carlo

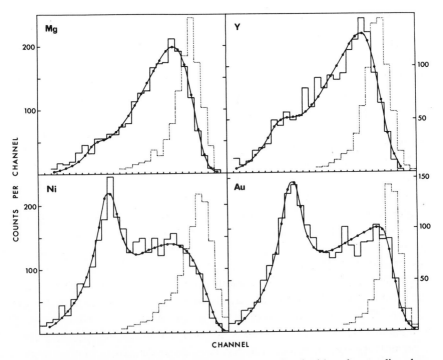

Fig. 19. Computed line shapes (solid curves), convoluted with a detector line shape which was assumed to be the same as that of the fully shifted peak (dotted histogram), compared with the experimentally observed line shapes (solid histograms) for four different backing materials. The lifetimes used in the computation were (in psec) 0.42 (Mg), 0.38 (Y), 0.325 (Ni) and 0.29 (Au) (Currie, 1969, used with permission of North-Holland Publishing Co., Amsterdam).

method as discussed earlier. The Al target was thin compared to the range of the recoils so, aside from a small correction, the recoils slowed down and stopped in the various backing materials. As can be seen from Fig. 19, the Monte Carlo method of calculation gives excellent agreement with experiment. In contrast to this, fits (Currie *et al.*, 1969) with the Blaugrund representation showed noticeable deviations from experiment, especially for the heavier stopping materials, but even so, the detailed comparison made by Currie (1969) showed that the Blaugrund approximation, if used properly, introduced errors small compared to those in dE/dx. By comparing results for different stopping materials, Currie was able to reach conclusions about the relative values of dE/dx for these different materials.

The ^{37}Cl(d, pγ)^{38}Cl results of Engelbertink and Olness (1972) offer an example of the utilization of the particle-γ method to determine several lifetimes simultaneously. The recoiling ^{38}Cl ions were selected in a forward cone of half-angle 9° by means of a proton coincidence requirement in the backward direction, this coincidence also selecting the state to be studied. The ^{38}Cl velocity was $v/c = 0.55\%$. The lifetimes of seven states were determined simultaneously. Results for four γ transitions from three different levels are shown in Fig. 20. In this experiment, as in most others, the main source of uncertainty is in the stopping power for the target and backing. The mean lives of some of the same ^{38}Cl levels were also measured following the bombardment of deuterium targets by a ^{37}Cl beam at nearly the same center-of-

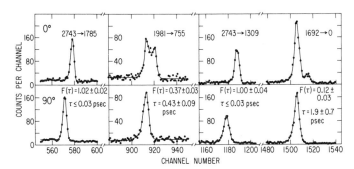

Fig. 20. Doppler shift measurements in the ^{37}Cl(d, pγ)^{38}Cl reaction for primary γ rays measured in coincidence with proton groups leading to the 1692-, 1981-, and 2743-keV levels in ^{38}Cl. The proton detector, centered at 180°, restricted the ^{38}Cl ions to move in a forward cone with a half-angle of 9° and with a velocity of about $v/c = 0.55\%$. The 40-cm^3 Ge(Li) detector was positioned at 0° (upper spectra) or at 90° (lower spectra) and the dispersion is 0.79 keV/channel. A target of 0.52 mg/cm^2 Ba^{37}Cl$_2$ on 42 mg/cm^2 Ta was used (Engelbertink and Olness, 1972).

mass energy (Warburton *et al.*, 1973). A comparison of the $^{37}Cl(d, p\gamma)^{38}Cl$
and $^2H(^{37}Cl, p\gamma)^{38}Cl$ results shows the advantages and disadvantages of
both methods. With the ^{37}Cl beam, the kinematics forced the recoiling ^{38}Cl
ions into a cone of half-angle 2.8° for a 60-MeV beam, which for the purpose
of defining the recoil ion direction, is better than the 9° half-angle in the
$^{37}Cl(d, p\gamma)^{38}Cl$ coincidence results. Furthermore, the ^{38}Cl velocity is
$v/c = 5.4\%$—a gain of a factor of ten in the magnitude of the Doppler shift.
The larger velocity means that most of the slowing down is in a velocity
region where dE/dx is better known and in an interior region of the target
where surface effects are less important. It also permits the study of γ-ray
transitions of lower energy since the Doppler shift effect is correspondingly
larger. In addition, the much greater efficiency of singles measurements allows
better statistics and more time to vary experimental conditions such as the
target backing. Line shapes for two ^{38}Cl transitions from recoils stopping in
Mg, Al, and Cu backings are shown in Fig. 21. The data are from a survey

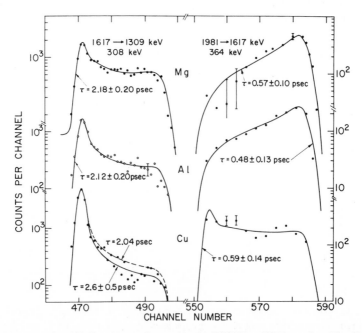

Fig. 21. Doppler shift line shapes observed in the $^2H(^{37}Cl, p\gamma)^{38}Cl$ reaction. The Ge(Li)
detector was at 0°. The kinematics are such that the ^{38}Cl ions recoiled within a cone of
half-angle 2.8° and with a mean velocity of $v/c = 5.4\%$. The background has been subtracted. The three line shapes correspond to the ^{38}Cl ions slowing down in Mg, Al, and
Cu backings with the indicated mean lives (Warburton *et al.*, 1973).

of ~ 2 hr per spectrum as opposed to ~ 2 days for the data on Fig. 20. There are two serious disadvantages to the singles heavy-ion method. First, the background is often very high, and when a line shape covers 50 keV (5% at 1 MeV), the chance of structure in the background, which cannot be determined, is great. The second serious disadvantage is that in singles measurements, there is no control of feeding from higher states and so the observed line shapes may be influenced by preceding nonnegligible lifetimes. This is the case for the $1617 \rightarrow 1309$ keV transition of Fig. 21, for instance, since the 1617-keV level is fed partially by cascade via the 1981-keV level and the fit for the 308-keV line shape is a function of the lifetimes of both the 1617- and 1981-keV levels.

For the general case with heavy-ion bombardment, we do not have a narrow forward cone of recoiling ions such as is forced by $M_{\text{Target}} \ll M_{\text{Projectile}}$ and either a coincidence condition is needed or the angular distribution of the reaction must be known and taken into account. An example of the former is provided by work at Yale on the Coulomb excitation of ^{150}Nd and ^{152}Sm using ^{16}O and ^{32}S beams with coincidence detection of the

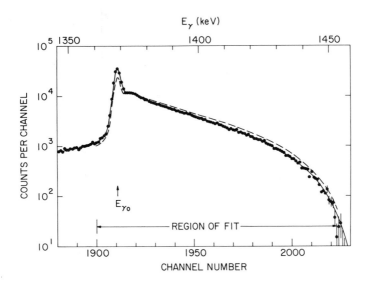

Fig. 22. Doppler shift line shape observed at 0° for the ground-state decay of the 2+ ^{24}Mg first excited state formed via Coulomb excitation by a 53-MeV ^{37}Cl beam. The solid and dashed curves are fits to the experimental line shape for mean lives $\tau = 2.00$ and 1.50 psec, respectively. $E_{\gamma_0} = 1368$-53 keV. The least-squares solution is $\tau = 2.0 \pm 0.3$ psec, where the uncertainty is due almost entirely to the uncertainty in dE/dx (Schwalm et al., 1974).

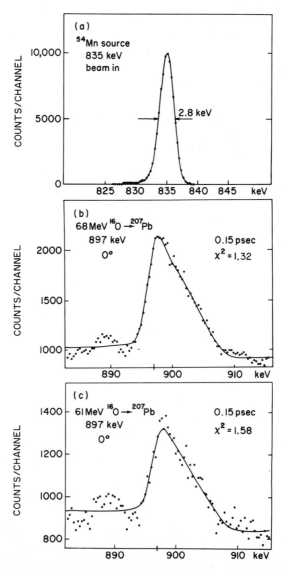

Fig. 23. Curves (b) and (c) are partial γ-ray spectra following excitation of ^{207}Pb with ^{16}O ions of the indicated energies. The curves show the line shape of the $\frac{3}{2}^- \rightarrow \frac{1}{2}^-$ line as observed at $0°$ to the beam. The solid curves are the best fits obtained from the DSAM calculations for the indicated lifetimes (Dost and Rogers, 1972). Curve (a) illustrates the experimental detector response measured with a radioactive source.

inelastically scattered projectile in the backward direction (Stokstad *et al.*, 1970). The Doppler shifts were 1.7 and 3.0% for 60-MeV ^{16}O and 110-MeV ^{32}S, respectively. In this work, nine lifetimes were measured by the DSAM in these two nuclei. An example of a DSAM lifetime obtained from Coulomb excitation without a coincidence condition is given in Fig. 22. Here, the line shape of the ^{24}Mg $2^+ \rightarrow 0^+$ transition observed at 0° to the 53-MeV ^{37}Cl beam is fitted using the known theoretical angular distribution of the reaction to yield a mean life of 2.0 ± 0.3 psec for the first excited state of ^{24}Mg (Schwalm *et al.*, 1974). The maximum Doppler shift was $v/c = 6.8\%$ and the natural Mg target was thick enough to stop the beam and the ^{24}Mg recoils. Input to the theoretical line shape includes the theoretical Coulomb excitation cross section and particle–gamma correlation as a function of beam energy, quadrupole reorientation effects, and the various relativistic effects, all integrated over the Coulomb cross section as the ^{37}Cl ions slow down and stop in the target. The finite integration over the detector solid angle and the detector response function to monoenergetic γ rays were also carefully evaluated and folded in. The dE/dx results of Fig. 18 were used for the ^{24}Mg ions. This is an example of how well line shapes can be determined if care is taken.

A similar example but for the heavy nucleus ^{207}Pb is shown in Fig. 23. Here, the line shape of the ^{207}Pb 897-keV $\frac{3}{2}^- \rightarrow \frac{1}{2}^-$ ground state transition observed in singles was fit (Dost and Rogers, 1972) to a theoretical line shape based on the Monte Carlo program of Currie (1969). The result is a mean life of 0.15 psec with an uncertainty determined by the dE/dx of ^{207}Pb in lead. These authors also observed the line shape at 45, 90, and 130°. From the fits, they concluded, as might be expected, that the most accurate results are obtained at 0°. Similar results for the 897-keV level of ^{207}Pb were obtained by Grosse *et al.* (1971) in an almost identical experiment.

V. Miscellaneous Timing Techniques

In addition to the three main methods of direct timing covered in Sections III and IV, there are several other direct techniques that are important but specialized. Of these, brief discussions will be given of the two newest, namely the channeling and microwave methods.

The channeling method makes use of the planar and axial alignment of atoms in crystals, with nuclei at the various lattice sites serving as targets for a nuclear bombardment. Because of the momentum brought in by the bombarding particles, the recoil nuclei will move (in time) out of the lattice site into one of the open channels between crystal planes. The intensity pattern of reaction flux will therefore be a function of the nuclear lifetime and the crystal orientation relative to the detection direction. As in RDM measurements, the velocity of the recoil nucleus defines a time scale in terms of the displacement distance from the original lattice site. This is discussed in detail by Richter in Chapter IV.D.1.

Because the interplane distances involved are orders of magnitude smaller than those practical in the RDM (for comparison), the range of lifetimes which can be studied are correspondingly orders of magnitude shorter (10^{-16}–10^{-18} sec, approximately). Maruyama et al. (1970) used planar blocking with inelastic proton reactions on ^{70}Ge and ^{72}Ge, and more recently, Clark et al. (1971) have used axial blocking together with two-dimensional detectors for the same reactions. Mean lifetimes of a few times 10^{-17} sec were obtained for compound nuclear states in ^{71}As and ^{73}As. Additional measurements in Ge single crystals (Gibson et al., 1972) have shown variations in the inelastic reaction times over analog resonances. Similar experiments have also recently been made in ^{24}Mg with the ^{27}Al (p, α)^{24}Mg reaction (Komaki et al., 1972). These techniques have also been used for states in heavy compound nuclei (Brown et al., 1968; Gibson and Nielsen, 1970; Melikov et al., 1972) by observing blocking effects for fission fragments; these lifetimes are dominated by neutron and fission branches.

The microwave method is basically an electronic technique with the addition of a microwave shutter; it is intended for lifetimes of low-lying states in heavy nuclei where internal conversion coefficients are large. The method has been successfully used with pulsed accelerator beams (Blaugrund et al., 1960; Gorodetsky et al., 1967) and more recently Ben-Zvi et al. (1968) have employed silicon counters for the electron detection.

The remaining methods for lifetime determination that were mentioned in Section I are indirect in nature and are thus not the main focus of this chapter. These methods involve cross-section measurements from which lifetimes can be extracted. Capture measurements yield lifetime information on levels in the compound system of the target plus bombarding projectile, while Coulomb excitation, resonance fluorescence, inelastic electron scattering, and inelastic particle scattering give information only on excited states of stable nuclei that can be used as targets in the various reactions.

Capture cross-section measurements can yield the total width and the partial γ-ray widths of virtual levels in the compound nucleus. A discussion of capture reactions is given in Chapter VII.D by Rolfs and Litherland.

Coulomb excitation is a very useful technique in obtaining information on $E(L)$ matrix elements connecting to the ground state. The employment of heavy ions has increased the importance of this approach. Coulomb excitation is discussed in detail by McGowan and Stelson in Chapter VII.A.

The resonance fluorescence technique has been reviewed by Metzger (1959), Malmfors (1965), and Booth (1967) (also see Rolfs and Litherland, Chapter VII.D). It involves the resonance absorption of a γ ray which excites the nucleus to a given state, and the subsequent γ-ray emission. The width of the excited state can be obtained from the (γ, γ) cross section. The initial γ-ray beam is often obtained from the same nuclear transition; the increased energy needed to compensate for the nuclear recoil is provided by the Doppler shift. Several techniques have been used to provide the Doppler shift: mechanical methods, thermal methods, and the use of recoil velocities from a nuclear reaction or previous radioactive decay. Compton and bremsstrahlung spectra are also used for the γ-ray beam as well as crystal monochromators for low energies. The bremsstrahlung technique is being applied to heavier nuclei and smaller radiation widths with the use of Ge(Li) detectors, which improve the signal-to-noise ratio (Metzger, 1972). The resonance fluorescence technique is applicable for lifetimes of $\tau < 10^{-10}$ sec and is capable of high accuracy; it is independent of the specific multipolarity, which in certain cases is an advantage. Recoil-free Mössbauer absorption (Mössbauer, 1965) is a special case of this technique; it is usable only for γ-ray energies of < 150 keV.

Inelastic electron scattering has been reviewed by Barber (1962), Bishop (1965), deForest and Walecka (1966), and Theissen (1972). Lifetimes and multipolarities can be obtained from a theoretical analysis of the cross sections. This technique is applicable to short lifetimes generally; however, its accuracy is limited because of energy resolution difficulties and uncertainties in the analysis.

Inelastic particle scattering is used in a manner similar to that for inelastic electron scattering except that the theoretical analysis is model dependent and thus more uncertain. The subject of inelastic particle scattering is discussed by Perey in Chapter IV.B.1 and Madsen in Chapter IX.D.

ACKNOWLEDGMENTS

One of us (DBF) would like to express appreciation for the hospitality extended by the Sektion Physik and Prof. J. de Boer of the University of Munich during the time that a portion of this chapter was being written and to acknowledge several helpful discussions with Drs. C. V. K. Baba, M. R. Maier, and K. E. G. Löbner.

References

Alaga, G., Alder, K., Bohr, A., and Mottelson, B. R. (1955). *Mat. Fys. Medd. Dan. Vid. Selsk.* **29**, no. 9.

Alder, K., Bohr, A., Huus, T., Mottelson, B., and Winther, A. (1956). *Rev. Mod. Phys.* **28**, 432.

Alexander, T. K., and Allen, K. W. (1965). *Can. J. Phys.* **43**, 1563.

Alexander, T. K., and Bell, A. (1970). *Nucl. Instrum. Methods* **81**, 22.

Alexander, T. K., Allen, K. W., and Healey, D. C. (1966). *Phys. Lett.* **20**, 402.

Alexander, T. K., Häusser, O., McDonald, A. B., O'Donnell, J., and Ewan, G. T. (1971). AECL-3742, PR-P-87.

Allen, K. W. (1966). *In* "Lithium-Drifted Geramium Detectors," p. 142 IAEA, Vienna.

Allen, K. W., Alexander, T. K., and Healey, D. C. (1966). *Phys. Lett.* **22**, 193.

Allen, K. W., Alexander, T. K., and Healey, D. C. (1968). *Can. J. Phys.* **46**, 1575.

Arima, A., and Horie, H. (1954). *Prog. Theor. Phys.* **12**, 623.

Ashery, D., Bahcall, N., Goldring, G., Sprinzak, A., and Wolfson, Y. (1966). *Nucl. Phys.* **77**, 650.

Ashery, D., Bahcall, N., Goldring, G., Sprinzak, A., and Wolfson, Y. (1967). *Nucl. Phys.* **A101**, 51.

Astner, G., Bergström, I., Blomqvist, J., Fant, B., and Wikström, K. (1972). *Nucl. Phys.* **A182**, 219.

Baba, C. V. K., Fossan, D. B., Faestermann, T., Feilitzsch, F., Maier, M. R., Raghavan, P., Raghavan, R. S., and Signorini, C. (1973). *Supp. J. Phys. Soc. Japan* **34**, 260.

Ball, G. C., Davies, W. G., Forster, J. S., James, A. N., and Ward, D. (1972). *Nucl. Phys.* **A182**, 529.

Barber, W. C. (1962). *Ann. Rev. Nucl. Sci.* **12**, 1.

Bartl, W., and Weinzierl, P. (1963). *Rev. Sci. Instrum.* **34**, 252.

Barton, R. D., Wadden, J. S., Carter, A. L., and Pai, H. L. (1971). *Can. J. Phys.* **49**, 971.

Bay, Z. (1950). *Phys. Rev.* **77**, 419.

Becker, J. A., and Wilkinson, D. H. (1964). *Phys. Rev.* **134**, B1200.

Bell, R. E. (1955). *In* "β- and γ-ray Spectroscopy" (K. Siegbahn, ed.), North-Holland Publ., Amsterdam.

Bell, R. E. (1965). *In* "α-, β-, and γ-Ray Spectroscopy" (K. Siegbahn, ed.), Vol. 2, pp. 905–30. North-Holland Publ., Amsterdam.

Bell, R. E. (1966). *Nucl. Instrum. Methods* **42**, 211.

Bengtson, B., and Moszynski, M. (1970). *Nucl. Instrum. Methods* **81**, 109.

Ben-Zvi, I., Blaugrund, A. E., Dar, Y., Goldring, G., Hess, J., Sachs, M. W., Skurnik, E. Z., and Wolfson, Y. (1968). *Nucl. Phys.* **117**, 625.

Berant, Z., Goldberg, M. B., Goldring, G., Hanna., S. S., Loebenstein, H. M., Plesser, I., Pepp, M., Sokolowski, J. S., Tandon, P. N., and Wolfson, Y. (1971). *Nucl. Phys.* **A178**, 155.

Bergström, I., Fant, B., Herrlander, C. J., Thieberger, P., Wikström, K., and Astner, G. (1970). *Phys. Lett.* **32B**, 476.

Bertolini, G., Mandl, V., Rota, A., and Cocchi, M. (1966). *Nucl. Instrum. Methods* **42**, 109.

Birk, M., Blaugrund, A. E., Goldring, G., Skurnik, E. Z., and Sokolowski, J. S. (1962). *Phys. Rev.* **126**, 726.

Bishop, G. R. (1965). *In* "Nuclear Structure and Electromagnetic Interactions" (N. MacDonald, ed.), p. 211. Plenum Press, New York.

Bister, M., Anttila, A., Piiparinen, M., and Viitasalo, M. (1971). *Phys. Rev. C* **3**, 1972.

Bizzeti, P. G., Bizzeti-Sona, A. M., Kalbitzer, S., and Povh, B. (1967). *Z. Phys.* **201**, 295.

Blatt, J. M., and Weisskopf, V. F. (1952). "Theoretical Nuclear Physics." Wiley, New York.

Blaugrund, A. E. (1966). *Nucl. Phys.* **88**, 501.

Blaugrund, A. E., Dar, Y., and Goldring, G. (1960). *Phys. Rev.* **120**, 1328.

Blaugrund, A. E., Youngblood, D. H., Morrison, G. C., and Segel, R. E. (1967). *Phys. Rev.* **158**, 893.

Blin-Stoyle, R. J., and Perks, M. A. (1954). *Proc. Phys. Soc. (London)* **67A**, 885.

Blomqvist, J., Freed, N., and Zetterstrom, H. O. (1965). *Phys. Lett.* **18**, 47.

Bohr, A., and Mottelson, B. R. (1953). *Mat. Fys. Medd. Dan. Vid. Selsk.* **27**, No. 16.

Bohr, A., and Mottelson, B. R. (1969). "Nuclear Structure," Vol. 1. Benjamin, New York.

Bonitz, M. (1963). *Nucl. Instrum. Methods* **22**, 238.

Booth, E. C. (1967). *In* "Nuclear Research with Low Energy Accelerators" (J. B. Marion and D. M. Van Patter, eds.). p. 75. Academic Press, New York.

Booth, W., and Grant, I. S. (1965). *Nucl. Phys.* **63**, 481.

Boström, L., Olsen, B., Schneider, W., and Matthias, E. (1966). *Nucl. Instrum. Methods* **44**, 61.

Brandolini, F., Signorini, C., and Kusstatcher, P. (1971). *Nucl. Instrum. Methods* **91**, 341.

Braunsfurth, J., and Körner, H. J. (1965). *Nucl. Instrum. Methods* **34**, 202.

Broude, C. (1969). *In Proc. Int. Conf. Properties Nucl. States, Montreal* (M. Harvey, R. Y. Cusson, J. S. Geiger, and J. M. Pearson, eds.), pp. 221–244. Univ. of Montreal Press, Montreal.

Broude, C., Engelstein, P., Papp, M., and Tandon, P. N. (1972). *Phys. Lett.* **39B**, 185.

Brown, B. A., Fossan, D. B., McDonald, J. M., and Snover, K. A., (1974). *Phys. Rev. C* **9** (in press).

Brown, F., Marsden, D. A., and Werner, R. D. (1968). *Phys. Rev. Lett.* **20**, 1449.

Bunbury, D. St. P., Devons, S., Manning, G., and Towle, J. W. (1956). *Proc. Phys. Soc. (London)* **A69**, 165.

Burde, J., and Cohen, S. G. (1956). *Phys. Rev.* **104**, 1093.

Chase, R. L. (1968). *Rev. Sci. Instrum.* **39**, 1318.

Chemtob, M. (1969). *Nucl. Phys.* **A123**, 449.

Cho, Z. H., and Chase, R. L. (1972). *Nucl. Instrum. Methods* **98**, 335.

Cho, Z. H., and Llacer, J. (1972). *Nucl. Instrum. Methods.* **98**, 461.

Clark, G. J., Poate, J. M., Fuschini, E., Maroni, C., Massa, I. G., Uguzzoni, A., and Verondini, E. (1971). *Nucl. Phys.* **A173**, 73.

Cocchi, M., and Rota, A. (1967). *Nucl. Instrum. Methods* **46**, 136.

Cochavi, S., and Fossan, D. B. (1971). *Phys. Rev. C* **3**, 275.

Cochavi, S., Fossan, D. B., Henson, S. H., Alburger, D. E., and Warburton, E. K. (1970a). *Phys. Rev. C* **2**, 2241.

Cochavi, S., Cue, N., and Fossan, D. B. (1970b). *Phys. Rev. C* **1**, 1821.

Cochavi, S., McDonald, J. M., and Fossan, D. B. (1971). *Phys. Rev. C* **3**, 1352.

Currie, W. M. (1969). *Nucl. Instrum. Methods* **73**, 173.

Currie, W. M., Martin, J., and Earwaker, L. G. (1969). *Nucl. Phys.* **A135**, 325.

deForest, T., and Walecka, J. D. (1966). *Advan. Phys.* **15**, 1.

de-Shalit, A. and Talmi, I. (1963). "Nuclear Shell Theory." Academic Press, New York.

Devons, S., Hereward, H. G., and Lindsey, G. R. (1949). *Nature (London)* **164**, 586.

Devons, S., Goldring, G., and Lindsey, G. R. (1954). *Proc. Phys. Soc. (London)* **A67**, 134.

Devons, S., Manning, G., and Bunbury, D. St. P. (1955). *Proc. Phys. Soc. (London)* **A68**, 18.

Devons, S., Manning, G., and Towle, J. W. (1956). *Proc. Phys. Soc. (London)* **A69**, 173.

Diamond, R. M., Stephens, F. S., Nakai, K., and Nordhagen, R. (1971). *Phys. Rev. C* **3**, 344.

Dost, M., and Rogers, J. D. (1973). *Rev. Brasil. Fis.* **3**, 217.

Ellegaard, C., Barnes, P. D., Eisenstein, R., and Canada, T. R. (1971). *Phys. Lett.* **35B**, 145.

Engelbertink, G. A. P., and Olness, J. W. (1972). *Phys. Rev. C* **5**, 431.

Engelbertink, G. A. P., Lindeman, H., and Jacobs, M. J. H. (1968). *Nucl. Phys.* **A107**, 305.

Ewan, G. T., and Graham, R. L. (1965). *In* "α-, β-, γ-Ray Spectroscopy" (K. Siegbahn, ed.), p. 951. North-Holland Publ., Amsterdam.

Faessler, M. A., Povh, B., and Schwalm, D. (1971). *Ann. Phys.* **63**, 577.

Fossan, D. B., and Poletti, A. R. (1966). *Phys. Rev.* **152**, 984.

Fossan, D. B., McDonald, R. E., and Chase, L. F. (1966). *Phys. Rev.* **141**, 1018.

Fouan, J. P. and Passerieux, J. P. (1968). *Nucl. Instrum. Methods* **62**, 327.

Gallant, J. L. (1970). *Nucl. Instrum. Methods* **81**, 27.

Gatti, E., and Svelto, V. (1966). *Nucl. Instrum. Methods* **43**, 248.

Gedcke, D. A., and McDonald, W. J. (1968). *Nucl. Instrum. Methods* **58**, 253.

Gibson, W. M., and Nielsen, K. O. (1970). *Phys. Rev. Lett.* **24**, 114.

Gibson, W. M., Hashimoto, Y., Keddy, R. J., Maruyama, M., and Temmer, G. M. (1972). *Phys. Rev. Lett.* **29**, 74.

Goldring, G. (1968). *In* "Hyperfine Structure and Nuclear Reactions" (E. Matthias and D. A. Shirley, eds.), p. 640. North-Holland Publ., Amsterdam.

Goosman, D. R., and Kavanagh, R. W. (1967). *Phys. Lett.* **24B**, 507.

Gorni, S., Hochner, G., Nadav, E., and Zmora, H. (1967). *Nucl. Instrum. Methods* **53**, 349.

Gorodetsky, S., Gerber, J., Vivien, J. P., Macher, A., and Armbruster, R. (1967). *Nucl. Instrum. Methods* **57**, 152.

Grosse, E., Dost, M., Haberkaut, K., Hertel, J. W., Klapdor, H. V., Körner, H. J., Proetel, D., and von Brentano, P. (1971). *Nucl. Phys.* **A174**, 525.

Hageman, U., Neubert, W., Schulze, W., and Stary, F. (1971). *Nucl. Instrum. Methods* **96**, 415.

Häusser, O., Pelte, D., Alexander, T. K., and Evans, H. C. (1969). *Can. J. Phys.* **47**, 1065.

Houdayer, A., Mark, S. K., and Bell, R. E. (1968). *Nucl. Instrum. Methods* **59**, 319.

Hvelplund, P. (1971). *Mat. Fys. Medd., Dan. Vid. Selsk.* **38**, no. 4.

Hvelplund, P., and Fastrup, B. (1968). *Phys. Rev.* **165**, 408.

Hyman, L. G. (1965). *Rev. Sci. Instrum.* **36**, 193.

Jain, H. C., Bhattacherjee, S. K., and Baba, C. V. K. (1972). *Nucl. Phys.* **A178**, 437.

Jones, K. W., Schwarzschild, A. Z., Warburton, E. K., and Fossan, D. B. (1969). *Phys. Rev.* **178**, 1773.

Kim, H. J., and Milner, W. T. (1971). *Nucl. Instrum. Methods* **95**, 429.

Koechlin, Y., and Raviart, A. (1964). *Nucl. Instrum. Methods* **29**, 45.

Komaki, K., Fujimoto, F., Nakayama, H., Ishii, M., and Hisatake, K. (1972). *Phys. Lett.* **38B**, 218.

Kutschera, W., Dehnhadt, W., Kistner, O. C., Kump, P., Povh, B., and Sann, H. J. (1972). *Phys. Rev. C* **5**, 1658.

Lederer, C. M., Jaklevic, J. M., and Hollander, J. M. (1971). *Nucl. Phys.* **A169**, 449.

Lefevre, H. W., and Russell, J. T. (1959). *Rev. Sci. Instrum.* **30**, 159.

Lindhard, J., Scharff, M., and Schiøtt, H. E. (1963). *Math. Fys. Medd., Dan. Vid. Selsk.* **36**, no. 14.

Litherland, A. E., Yates, M. J. L., Hinds, B. M., and Ecceleshall, D. (1963). *Nucl. Phys.* **44**, 220.

Lowe, J., McClelland, C. L., and Kane, J. V. (1962). *Phys. Rev.* **126**, 1811.

Lynch, F. J. (1966). *IEEE Trans. Nucl. Sci.* NS-13, No. 2, 140.

Lynch, F. J. (1968). *IEEE Trans. Nucl. Sci.* NS-15, No. 3, 102.

MacDonald, J. R., Benczer-Koller, N., Tape, J., Guthman, L., and Goode, P. (1969). *Phys. Rev. Lett.* **23**, 594.

Maier, M. R. (1970). Ph. D. Thesis, Tech. Univ. München.

Maier, M. R., and Sperr, P. (1970). *Nucl. Instrum. Methods* **87**, 13.

Maier, K. H., Leigh, J. R., Pühlhofer, F., and Diamond, R. M. (1971). *J. Phys.* **32**, Coll. C-6, suppl. 11–12, 221.

Malmfors, K. G. (1965). *In* "α-, β-, and γ-Ray Spectroscopy" (K. Siegbahn, ed.), Vol. II, p. 128. North-Holland Publ., Amsterdam.

Marelius, A., Sparrman, P., and Sundström, T. (1968). *In* "Hyperfine Structure and Nuclear Radiations" (E. Matthias and D. A. Shirley, eds.), p. 1043. North-Holland Publ., Amsterdam.

Maruyama, M., Tsukada, K., Ozawra, K., Fujimoto, F., Komaki, K., Mannami, M., and Sakurai, T. (1970). *Nucl. Phys.* **A145**, 581.

McDonald, R. E., Fossan, D. B., Chase, L. F., and Becker, J. A. (1965). *Phys. Rev.* **140**, B1198.

McDonald, A. B., Alexander, T. K., Häusser, O., and Ewan, G. T. (1971). *Can. J. Phys.* **49**, 2886.

Meiling, W., and Stary, F. (1968). "Nanosecond Pulse Technique." Gordon and Breach, New York.

Melikov, Y. V., Otstavnov, Y. D., Tulinov, A., and Chetchenin, N. G. (1972). *Nucl. Phys.* **A180**, 241.

Metzger, F. R. (1959). *Progr. Nucl. Phys.* **7**, 53.

Metzger, F. R. (1972). *Nucl. Phys.* **A182**, 213.

Mössbauer, R. L. (1965). *In* "α-, β-, and γ-ray Spectroscopy" (K. Siegbahn, ed.), Vol. II, pp. 1293-1312. North-Holland Publ., Amsterdam.

Moszynski, M., and Bengtson, B. (1970). *Nucl. Instrum. Methods* **80**, 233.

Moszynski, M., and Bengtson, B. (1971). *Nucl. Instrum. Methods* **91**, 73.

Neal, W. R. and Kraner, H. W. (1965). *Phys. Rev.* **137**, B1164.

Newton, T. D. (1950). *Phys. Rev.* **78**, 490.

Nilsson, S. G. (1955). *Mat. Fys. Medd. Dan. Vid. Selsk.* **29**, no. 16.

Northcliffe, L. C. (1963). *Ann. Rev. Nucl. Sci.* **13**, 67.

Northcliffe, L. C., and Schilling, R. F. (1970). *Nucl. Data Tables* **A7**, 233.

Nutt, R. (1968). *Rev. Sci. Instrum.* **39**, 1342.

Nutt, R., Gedcke, D. A., and Williams, C. W. (1970). *IEEE Trans. Nucl. Sci.* NS-17, No. 1, 299.

Ogata, A., Tao, S. J., and Green, J. H. (1968). *Nucl. Instrum. Methods* **60**, 141.

Omrod, J. H., MacDonald, J. R., and Duckworth, H. E. (1965). *Can. J. Phys.* **43**, 275.

Present, G., Schwarzschild, A. Z., Spirn, I., and Wotherspoon, N. (1964). *Nucl. Instrum. Methods* **31**, 71.

Quaranta, A. A. (1965). *Nucl. Instrum. Methods* **35**, 93.

Ryge, P., and Borchers, R. R. (1970). *Nucl. Instrum. Methods* **95**, 137.

Schmidt-Whitley, R. D. (1969). *Nucl. Instrum. Methods* **70**, 227.

Schwalm, D. (1969). Ph. D. dissertation, Univ. of Heidelberg.

Schwalm, D., Engelbertink, G. A. P., Olness, J. W., and Warburton, E. K. (1974). (To be published).

Schwarzschild, A. Z. (1963). *Nucl. Instrum. Methods* **21**, 1.

Schwarzschild, A. Z., and Warburton, E. K. (1968). *Ann. Rev. Nucl. Sci.* **18**, 265.

Severiens, J. C., and Hanna, S. S. (1956). *Phys. Rev.* **104**, 1612.

Sherman, I. S., Roddick, R. G., and Metz, A. J. (1968). *IEEE Trans. Nucl. Sci.* **NS-15**, No. 3, 500.

Shipley, E. N., Holland, R. E., and Lynch, F. J. (1969). *Phys. Rev.* **182**, 1165.

Sims, P. C., and Kuhlman, H. (1972). (To be published).

Skorka, S. J., Hertel, J., and Retz-Schmidt, T. W. (1966). *Nucl. Data Sect. A* **2**, 347.

Snover, K. A., McDonald, J. M., Fossan, D. B., and Warburton, E. K. (1971). *Phys. Rev. C* **4**, 398.

Stokstad, R. G., Fraser, I. A., Greenberg, J. S., Sie, S. H., and Bromley, D. A. (1970). *Nucl. Phys.* **A156**, 145.

Talmi, I., and Unna, I. (1960). *Ann. Rev. Nucl. Sci.* **10**, 353.

Tape, J. W., Benczer-Koller, N., Hensler, R., and MacDonald, J. R. (1972). *Phys. Lett.* **40B**, 635.

Theissen, H. (1972). (To be published).

Thirion, J. and Telegdi, V. L. (1953). *Phys. Rev.* **92**, 1253.

Thirion, J., Barnes, C. A., and Lauritsen, C. C. (1954). *Phys. Rev.* **94**, 1076.

Tove, P. A., and Falk, K. (1964). *Nucl. Instrum. Methods* **29**, 66.

Warburton, E. K. (1967). *In* "Nuclear Research with Low Energy Accelerators" (J. B. Marion and D. M. Van Patter, eds.), p. 43. Academic Press, New York.

Warburton, E. K., and Weneser, J. (1969). *In* "Isospin in Nuclear Physics" (D. H. Wilkinson, ed.), p. 175. North-Holland Publ., Amsterdam.

Warburton, E. K., Alburger, D. E., and Wilkinson, D. H. (1963). *Phys. Rev.* **129**, 2180.

Warburton, E. K., Olness, J. W., and Poletti, A. R. (1967). *Phys. Rev.* **160**, 938.

Warburton, E. K., Olness, J. W., Engelbertink, G. A. P., and Jones, K. W. (1971). *Phys. Rev. C* **3**, 2344.

Warburton, E. K., Olness, J. W., Engelbertink, G. A. P., and Alexander, T. K. (1973). *Phys. Rev. C* **7**, 1120.

Ward, D., Rud, N., Ewan, G. T., Graham, R. L. and Geiger, J. S. (1971). AECL-4068 (PR-P-90).

Ward, D., Graham, R. L., Geiger, J. S., Andrews, H. R., and Sie, S. H. (1972). *Nucl. Phys.* **A193**, 479.

White, D. E. S., and McDonald, W. J. (1973). UAE-NPL-1056.

Wieber, D. L., and Lefevre, H. W. (1966). *IEEE Trans. Nucl. Sci.* **NS-13**, no. 1, 406.

Wilkinson, D. H. (1960). *In* "Nuclear Spectroscopy" (F. Ajzenberg-Selove, ed.), Part B, p. 852. Academic Press, New York.

Williams, C. W. (1971a). *In* "Methods of Experimental Physics" (L. Marton, ed.), Vol. 2, Academic Press, New York.

Williams, C. W. (1971b). Private communication.

Yamazaki, T., and Ewan, G. P. (1969). *Nucl. Phys.* **A134**, 81.

Yamazaki, T., Nomura, T., Nagamiya, S., and Katou, T. (1970). *Phys. Rev. Lett.* **25**, 547.

Zamick, L. and McCullen, J. (1965). *Bull. Amer. Phys. Soc.* **10**, 485.

VIII
Other Topics

VIII.A PHOTONUCLEAR REACTIONS

B. L. Berman

LAWRENCE LIVERMORE LABORATORY
UNIVERSITY OF CALIFORNIA, LIVERMORE, CALIFORNIA

I. Photonuclear Techniques

Photonuclear reaction experiments at last have come of age as a means for studying both the spectroscopy of nuclear levels near the particle separation energies and nuclear reaction mechanisms in the continuum. Such studies, which have been hampered in the past by the lack of monoenergetic photon beams and of suitable neutron detectors, now are producing spectroscopic data of high quality and resolution as well as data on reaction mechanisms which are intimately related to many other fields of nuclear research.

This chapter will concentrate on those methods, mostly of recent origin, which are peculiar to photonuclear measurements and which have been most fruitful or show great promise. Since photon production and neutron detection techniques best fit these criteria, several techniques, such as con-

ventional charged-particle and photon detection and bremsstrahlung unfolding, are discussed only briefly or omitted entirely. Also, it is premature
to review photonuclear measurements above the giant resonance, since very
little data exist at present, and the advent of the powerful new machines at
Livermore, Saclay, the Massachusetts Institute of Technology (MIT), and
Amsterdam can be expected in the near future to render anything discussed
here completely out of date. Finally, in the present, experimentalist-oriented
context, no great effort is made to give a comprehensive theoretical treatment
of the data. For these and other topics not discussed here because of space
limitations, the reader is referred to review articles on the subject (Levinger,
1960; Hayward, 1965; Danos and Fuller, 1965; Spicer, 1969; Firk, 1970)
and to Paul, Chapter III.B, and Rolfs and Litherland, Chapter VII.D.
Levinger (1960) gives a summary of early work in the field, together with a
fine theoretical introduction to the subject based on the analogy with the
atomic photoeffect; Hayward (1965) gives a particularly good treatment of
photon scattering and of angular distributions of photonucleons; Danos
and Fuller (1965) give a good theoretical exposition of the shell and collective model treatments of the giant resonance; Spicer (1969) gives a strong
treatment of the dynamic collective model, particularly as applied to vibrational nuclei; and Firk (1970) gives an able treatment of giant-resonance
properties in heavy nuclei, the angular distribution and polarization of
photonucleons from light nuclei, and isospin effects in photonuclear reactions. The reader also is referred to the proceedings of two recent conferences
in the field (Shoda and Ui, 1972; Berman, 1973a).

A. PHOTON PRODUCTION

The central experimental problem in photonuclear reaction studies has
been the production of monoenergetic photon beams with good resolution
and sufficient intensity. In the past, this problem could not be solved—the
required technology, such as high-intensity electron accelerators, simply
did not exist. Therefore one had to use a continuous bremsstrahlung beam
as the source of radiation. This imposed frightful difficulties upon the measurement of photonuclear cross sections. In general, one had to obtain a
yield curve by stepping the bremsstrahlung end-point energy in small
increments, and then differentiate it to determine the cross section. This
required great stability in the accelerator parameters, enormous counting
statistics, a knowledge of the bremsstrahlung flux and spectrum, especially
near the end point (both exceedingly hard either to calculate or to measure),
and tedious data reduction procedures. That any worthwhile nuclear physics

data at all were obtained is a testimony to the diligence and patience of the early workers in the field. In any case, the great value of the photon as a probe of nuclear structure not only stimulated these workers, but also spurred on the technological developments that made possible alternative

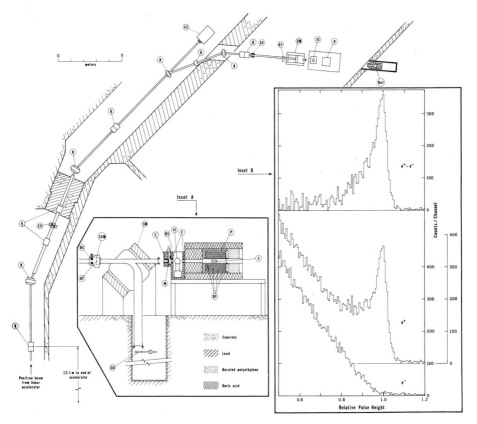

Fig. 1. Experimental layout for annihilation-photon experiments at the new Livermore electron–positron linear accelerator: B—bending magnets; Q—quadrupole magnets; ES—energy-analyzing slit; FC—Faraday cup; SC—steering coils; AT—annihilation target; SM—sweeping magnet; IC—photon ion chamber; D—detector; NaI—photon spectrometer. Inset A: Detail of a (γ, n) experiment: RC—removable collimator; SEM—secondary-emission monitor; SD—swept-beam detector; C—collimators; M—beam-tuning monitor; BS—beam shutter; S—nuclear sample; P—paraffin moderator; BF—$^{10}BF_3$ neutron detectors. Inset B: Photon spectra produced by beams of 16.5-MeV positrons and electrons striking an 0.13-mm beryllium target. The annihilation-photon peak is broadened greatly by the resolution of the 20×20 cm NaI spectrometer. The difference spectrum, in fact, is a measure of the response function of the NaI crystal (see also Chapter III.B).

methods of performing photonuclear measurements. In particular, the problem of the production of a monoenergetic photon beam has been attacked in a variety of ways: five are presented here.

1. *Annihilation in Flight of Fast Positrons*

This technique, developed by Fultz and collaborators at Livermore (Fultz *et al.*, 1962) and Tzara and collaborators at Saclay (Schuhl and Tzara, 1961), consists in producing positrons by pair production when the electron beam from the first part of a high-current linac is allowed to strike a thick, high-Z converter, capturing and accelerating them in the rest of the linac to the desired energy, and allowing them to strike a thin, low-Z target, thus producing, in the forward direction, a monoenergetic beam of annihilation photons. The effect of the positron bremsstrahlung contaminant is subtracted out by repeating the measurements with electrons. The annihilation-photon intensity is about 10^4 sec^{-1} for older and 10^7 sec^{-1} for newer linacs in a 1% energy bin. With the newer machines, 0.1% energy resolution is obtainable (with a corresponding loss in intensity). This technique has proven most useful for photoneutron measurements in the giant resonance region (up to about 30 MeV). An example of the type of photon spectra obtained is shown in Fig. 1, along with the general laboratory layout used at Livermore.

A handy atlas of photoneutron cross sections obtained with annihilation photons now is available (Berman, 1973b).

2. *Tagged Bremsstrahlung*

This technique, developed by Axel and collaborators at Illinois (O'Connell *et al.*, 1962), consists in bending a high-resolution electron beam from a betatron (or synchrotron) in a beta-ray spectrometer after it has passed through a thin, high-Z bremsstrahlung target, and demanding a fast coincidence between a reaction product of the photonuclear event induced by a bremsstrahlung photon and the scattered electron that produced the photon. The intensity and resolution of the quasi-monoenergetic photon beam thus produced are comparable to those of the annihilation-photon technique for older linacs. The chief advantage of this technique stems from the long duty cycle of circular accelerators. This technique has proven most useful for photon scattering and fast photoneutron measurements below 16 MeV.

3. *Fixed-Energy Bremsstrahlung Subtraction*

This technique, developed by Matthews and Owens (1971) at Glasgow, consists in alternating bremsstrahlung targets of markedly different Z in a photonuclear experiment at a fixed end-point energy and taking advantage of the difference in the shape of the bremsstrahlung spectrum near its end point for the two targets. Subtracting one yield spectrum from the other nearly cancels the effect of photons having energy well below the tip region, so that a quasi-monoenergetic beam is effected. Photon intensities can be very high using this technique, and resolution is moderate ($\sim 4\%$). This technique is still in its infancy; the first experiments have been photoproton spectrum measurements above the giant resonance.

4. Capture γ Rays

Capture γ-ray photons from the ^{7}Li(p, γ) reaction have long been used for photonuclear measurements at 17 MeV, but such measurements are now essentially obsolete. More recent measurements using the ^{3}H(p, γ) reaction (Lochstet and Stephens, 1966) have proven more useful in the energy region just above 20 MeV, where 150 keV resolution has been achieved. Most recently, γ rays from resonance capture of reactor neutrons have been used by Ben-David, Moreh, and collaborators in Israel (Arad *et al.*, 1964; Moreh and Nof, 1969) to study properties of single levels just below the particle thresholds. This technique brings to bear high intensities and very high resolution, but suffers from the fact that the energy of the monoenergetic photons is not easily varied; the method depends on the chance overlap of the sharp-line source with the nuclear level under study. This technique, too, is still relatively young, although a sizable number of nuclear states have been studied to date. Experiments using this technique have been summarized by Arad and Ben-David (1973).

5. Virtual Photons

The use of virtual photons, through inelastic electron scattering measurements, is sure to produce a large amount of photonuclear data in the near future. It is not necessary to discuss the experimental techniques here, since they are well represented in the literature. Rather, it should be emphasized that not only can the quantum numbers and electromagnetic transition rates of individual nuclear states be studied in this way, but so can the various kinds of collective excitations that are the core of photonuclear physics. Such studies have been carried on for a number of years at Stanford, Darmstadt, Amsterdam, Orsay, Tohoku, and the Naval Research Laboratory (NRL), and more recently and with better resolution (at times of the order of 0.1% of the energy of the incident electron beam) at the National Bureau of

Standards (NBS). Soon, one hopes, the new MIT facility will make possible experiments requiring 0.01% resolution. In addition, the absorption of virtual photons via inelastic electron scattering can be used to acquire information on such quantities as form factors and transition radii which depend upon the momentum transferred to the nucleus by the scattered electron, and therefore are not obtainable from reactions induced by real photons.

B. Neutron Detection

The development of neutron detection techniques (see Finckh, Chapter VI.A), which has paralleled that for monoenergetic photon beams, also has proven to be essential for photonuclear reaction studies, for several reasons:

(a) Owing to the height of the Coulomb barrier, virtually all the photon absorption strength in medium and heavy nuclei goes into the neutron-producing partial cross sections: (γ, n), $(\gamma, 2n)$, etc. This gave rise to a need for highly efficient 4π neutron detectors (since the efficiency for detecting two neutrons is the square of that for one) and for neutron multiplicity counting techniques [in order, for instance, to distinguish a $(\gamma, 2n)$ event from two (γ, n) events].

(b) Very high resolution ($\ll 0.1\%$) cross-section studies would have been very difficult without photoneutron time-of-flight techniques.

(c) Studies of isospin mixing in light self-conjugate nuclei using photonuclear reactions likewise are very difficult without (γ, n) measurements.

(d) Studies of the interference of other multipoles with the dominant El photoabsorption mode have been facilitated greatly by photoneutron angular distribution and polarization measurements.

1. *Neutron Multiplicity Counting*

A number of highly efficient 4π neutron detectors capable of multiplicity counting have been developed. This is accomplished by employing a slowing-down type of detector in which the neutrons produced during the short beam burst of a pulsed accelerator are moderated before being detected between beam bursts. Both large arrays of $^{10}BF_3$ tubes embedded in a paraffin or polyethylene matrix (Fultz *et al.*, 1962; Berman *et al.*, 1967; Kelly, 1968; Kelly *et al.*, 1969) and large liquid scintillators (Beil *et al.*, 1969) have been used. In order to be able to differentiate between a $(\gamma, 2n)$ event and two (γ, n) events, say, as well as to be able to measure absolute cross sections well, one must know the neutron detector efficiency (which might be 40–60% typically) rather precisely. The Livermore group has attacked the problem by developing the ring-ratio technique (Caldwell *et al.*, 1965; Caldwell, 1967; Berman *et al.*, 1967, 1969a) for measuring the average neutron energy for every data run. This is done by measuring (and calibrating) the ratio of counting rates of various detectors (or subsections of the detector) which have different thicknesses of moderating material between the detector and the source of photoneutrons. Then the ratio of the counting rate in the outer ring of $^{10}BF_3$ detectors to that in the inner ring is a strong, monotonically increasing function of the energy of the photoneutrons. At photon energies above the giant resonance region, where $(\gamma, 3n)$, $(\gamma, 4n)$, and higher neutron multiplicities become important, neutron detectors with even higher efficiencies become necessary. Detectors with 80–90% efficiency, based on gadolinium- or cadmium-loaded plastic scintillators, are currently under development at Livermore (Czirr, 1972).

2. Neutron Time of Flight

The time-of-flight technique is a well-known tool of neutron spectroscopy and has been employed extensively at many laboratories (see Finckh, Chapter VI.A). Nevertheless, certain features of photoneutron reactions have spurred the development of this technique in two important respects:

(a) Photoneutron spectrum measurements of giant resonance states in light nuclei require very fast timing in order to achieve sufficiently high resolution for neutron energies of several MeV. Fast time-of-flight systems have been developed by Firk and collaborators at Harwell and at Yale (Firk, 1964, 1966) and by several other groups, at Rensselaer Polytechnic Institute, at Toronto, and elsewhere.

(b) Threshold photoneutron cross-section measurements likewise require high resolution, but also high efficiency and background rejection in the keV-to-MeV energy region. A unique detector which has all these attributes has been developed and used extensively at Livermore (Van Hemert, 1968; Van Hemert *et al.*, 1970; Baglan, 1970). It consists of a large disk of virgin ^{235}U surrounded by a large cylindrical shell of plastic scintillator, cut into quadrants. Triple coincidences of various kinds are demanded of the prompt γ rays and neutrons emitted in a fission event induced by the incident neutron. The detector efficiency then follows the (large) neutron fission cross section for ^{235}U, the background is reduced to a very low rate by the coincidence requirement, and the scintillators are outside the neutron flight tube and hence can be shielded from the gamma flash from the beam burst. The result is a new kind of keV neutron detector, useful in conventional neutron physics experiments as well.

3. *Photon Detection using* D(γ, n)

In principle, the two most fundamental kinds of photonuclear experiments are measurements of total photon absorption and photon scattering cross sections. In practice, however, these are the most difficult, since conventional electron linear accelerators, by far the most powerful machines for producing intense photon beams, are low-duty-cycle devices. [The electron prototype accelerator (EPA) at Los Alamos, the new high-energy linacs at Saclay, MIT, and Amsterdam, and the superconducting accelerators being built at Stanford and Illinois are exceptions. Of these, only the EPA has produced any significant photonuclear data to date.] This implies that photon transmission and scattering experiments, which involve photon detection during the intense beam burst, are very difficult and have high backgrounds.

A novel technique fashioned in response to this challenge makes use of the D(γ, n) reaction and neutron time of flight, as shown in Fig. 2. A short burst of bremsstrahlung, having passed through the photon transmission sample, is allowed to strike a deuterium photon-to-neutron converter, viewed from afar by a fast neutron detector. The one-to-one correspondence between the energy of the neutron (detected between beam bursts) and the

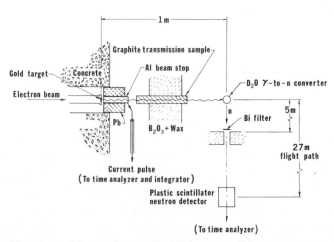

Fig. 2. Experimental layout for photon total cross-section measurements using the D(γ, n) reaction and a photoneutron time-of-flight facility as a photon spectrometer (from Wu *et al.*, 1970a, with permission of North-Holland Publishing Co., Amsterdam).

incident photon permits one to measure the total photon absorption cross section at all energies simultaneously from 2.2 MeV [the D(γ, n) threshold] to the bremsstrahlung end-point energy (which might be 100 MeV or more). The high resolution obtained by use of the time-of-flight technique is particularly valuable at the lower energies, especially below the particle thresholds. Photon scattering measurements can be done in a similar way. The D(γ, n) reaction had been used as a photon spectrometer before (Nielsen, 1963; O'Dell et al., 1968). The feasibility of this method for photonuclear measurements was first demonstrated at Yale (Wu et al., 1970a); it is being pursued at Ottawa and Livermore.

4. *Photoneutron Angular Distributions and Polarizations*

Photoneutron angular distributions, using bremsstrahlung and neutron time of flight, have been measured in two ways. One is by use of multiple flight paths, which has been the approach taken at MIT, Saskatchewan, and Argonne. In many ways, this is the most satisfactory, because all angles are measured simultaneously, with the resulting saving in beam time, but it is expensive and involves calibrating the efficiencies of several neutron detectors. Figure 3 shows the more novel approach of McNeill and collaborators at Toronto (Hewitt et al., 1970; Jury, 1970), where the angle of the photon beam is varied by bending the incident electron beam before it strikes the movable bremsstrahlung target, so that the same 50-m evacuated flight tube and neutron detector can be used for all angles. The angular distribution facility at the new Livermore linac (see Fultz et al., 1971a) contains multiple 10-m flight paths and neutron detectors (nine angles, including 0° and 180°), but uses a monoenergetic annihilation-photon beam rather than bremsstrahlung. An even more novel technique makes use of sandwiches of fission foils and plastic fragment-track detectors (but no time of flight) to measure neutron angular distributions. This technique, first tried at Harwell (Patrick and Bowey, 1971), is still at a rudimentary stage of development, but shows promise.

Photoneutron polarizations have been measured in two ways; both have been used by Firk and collaborators at Yale (Cole, 1970; Nath, 1971; Nath et al., 1971a). One employs a left–right scattering-asymmetry measurement of photoneutrons from ^{12}C or (liquid) ^{4}He, the latter being nearly background-free, since a fast coincidence is demanded between the scattered photoneutron and the recoil α particle. The other method employs a solenoid along the neutron flight path to precess the neutron spin en route to the scatterer; this eliminates the need for the second detector (except for doubling

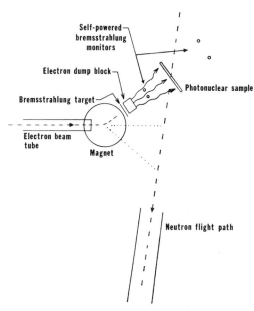

Fig. 3. Schematic diagram showing the Toronto method for performing photoneutron angular distribution measurements with a single flight path and neutron detector (from Hewitt *et al.*, 1970, with permission of North-Holland Publishing Co., Amsterdam).

the counting rate), and thus avoids the annoying problem of intercalibrating the two detectors by interchanging them and doing the experiment a second time. (Also see Walter, Chapter VI.C.)

C. Charged-Particle Detection

Although there exists in the photonuclear literature a large body of (γ, charged particle) reaction data, nearly all of it has been acquired using bremsstrahlung and conventional charged-particle detectors (emulsions, scintillators, semiconductor detectors), and merits no special attention here. Total photoproton cross sections are relatively small for medium and heavy nuclei because of the Coulomb barrier, and (γ,p) measurements which have employed the photon difference technique suffer from even greater uncertainties than their (γ, n) counterparts. Photoproton cross sections for light nuclei are absolutely small, and require either intense beams or large samples. Except for certain favorable cases (1s-shell nuclei, ^{16}O, and perhaps a few others), there is always the ambiguity when one measures, say, the proton energy spectrum at a fixed bremsstrahlung end point, resulting from the contamination of the ground-state spectrum with protons emitted in transitions to excited states of the residual nucleus. In addition, (γ, p_0) cross sections frequently can be obtained from the inverse (p, γ_0) experiments, and the energy resolution which can be achieved in this way (from electrostatic accelerators) is generally far better, so that most of our present knowledge of (γ, p) cross sections comes from (p, γ) work. Similar considerations apply to the (γ, d), $(\gamma, {}^3He)$, and (γ, α) reactions.

The advantage of the inverse-reaction technique in resolution recently has been enhanced greatly by the development of polarized charged-particle beams of reasonable intensity. Indeed, since it is exceedingly difficult to measure photoproton polarizations directly, such measurements, which complement $(\gamma, \vec{n})$ measurements nicely, ultimately might prove to be even more important for understanding photonuclear reaction mechanisms than high resolution. The first such experiment, studying the giant resonance of ^{12}C via the $^{11}B(\vec{p}, \gamma)$ reaction, has been performed at Stanford by Hanna and collaborators (Glavish et al., 1972).

Photofission measurements constitute a special case, but there has not yet been a great deal of work done in this field, although experiments of note have been carried out in the giant resonance region with annihilation photons (Bowman et al., 1964; Bergère et al., 1972) and on several even-even nuclei in the threshold energy region with bremsstrahlung (Rabotnov et al., 1970) (also see Section III.B.2). Finally, two charged-particle detection methods of special interest will be discussed.

1. Magnetic Spectrometers

One of the principal experimental problems in measuring (γ, p) cross sections at high-intensity linacs is the difficulty of detecting the photoprotons in the field of scattered radiation from the sample during the beam burst. Since the slowing-down and time-of-flight techniques are not practical for charged particles, so that one cannot separate the photonuclear event and its detection in time, then one can try to separate them in space. This is conveniently accomplished with a magnetic spectrometer (see Hendrie, Chapter III.C) and an array (a "ladder") of charged-particle detectors, in much the same way as one does electron scattering experiments. Shoda and collaborators at Tohoku have made extensive use of this technique to measure (γ, p) and (γ, p_0) cross sections on many nuclei, employing a virtual photon beam as the source of radiation (Shoda et al., 1968, 1969, 1971a, b). This technique also has been used to measure $(e, e'd)$, $(e, e't)$, and (e, e'^3He) cross sections on 1s-shell nuclei at Saskatchewan (Wait et al., 1970; Kundu *et al.*, 1971) and at NBS (Skopik, 1970; Dodge *et al.*, 1972; Murphy and Dodge, 1972), and (γ, d) and $(e, e'd)$ cross sections on ^{3}H and ^{3}He at Glasgow (Owens and Matthews, 1971).

2. Charged-Particle Detector As the Nuclear Sample

In performing (γ, p) or (γ, α) experiments with monoenergetic photons, one faces the problem that if the nuclear sample is massive enough to use with the low photon beam intensities available, then the charged particles cannot escape from the sample. Therefore one must use the detector itself as the nuclear sample. This has been accomplished by Matsumoto *et al.* (1964, 1965) and Nagel (1970) using silicon detectors, the former using ^{7}Li(p, γ_0) and ^{3}H(p, γ) photons and the latter tagged bremsstrahlung. A more general approach, being pursued at Livermore, is to use a large, high-pressure ionization chamber for total (γ, p) cross-section measurements, in which either the filling gas or the electrodes can serve as the nuclear sample under study.

II. Giant Resonance Studies

The giant dipole resonance, the dominant mode of nuclear photoabsorption, always has been of central interest in photonuclear reaction studies, both experimentally and theoretically. It corresponds to the fundamental frequency for absorption of electric dipole radiation by the nucleus acting as a whole, and is most simply understood as the oscillation of the neutrons against the protons in the nucleus [the semiclassical hydrodynamic model, formulated over 20 years ago (Goldhaber and Teller, 1948; Steinwedel and Jensen, 1950)]. Alternatively, one can construct the giant resonance from a superposition of particle–hole states based upon the shell model. [Indeed, the particle–hole (p–h) theory originally was developed to explain the giant resonance (Elliott and Flowers, 1957; Brown and Bolsterli, 1959).] This latter approach is particularly suited to calculating the decay modes (branching ratios, angular distributions, polarizations, and the like) of the giant resonance states, especially for light nuclei, where the number of states involved is small enough to be manageable. The experimental emphasis seems to be changing gradually from the exploration of the absorption process to the attempt to understand the giant resonance states in detail by studying their decay products. Yet, both these aspects have been scrutinized in ever finer detail in recent years; the former by means of systematic studies in medium and heavy nuclei and the latter through key experiments on specific light nuclei. We shall discuss these two topics in reverse order.

A. LIGHT NUCLEI

The giant dipole resonance is fragmented into considerable structure for nuclei lighter than ^{60}Ni; it is composed of one broad structure (two for heavy deformed nuclei) for nuclei heavier than ^{75}As (although there are suggestions of fine structure superposed for some heavy nuclei, particularly at the closed neutron shells $N=82$ and 126); intermediate cases like the copper isotopes are not so clear cut. In any case, this circumstance serves as an operational definition for "light" and "heavy" nuclei for purposes of discussing photonuclear reactions. Here, then, we discuss light nuclei, for which one might read "special cases." (Nuclear models for light nuclei are discussed by Harvey and Khanna, Chapter IX.A.)

1. Very Light Nuclei and Nuclear Forces

After a number of years of slight theoretical activity, owing chiefly to the lack of detailed experimental data, it appears now that there is a serious lack of understanding of the photonuclear cross section for every 1s-shell nucleus studied.

a. The deuteron. Three recent experiments have contradicted flatly the predictions of the "standard" theoretical calculations of the photodisintegration process for this nucleus (Partovi, 1964). First, Weissman and Schultz (1971) measured the (γ, p) cross section and angular distribution from 25 to 55 MeV, which stands as perhaps the most painstaking and comprehensive experiment of the conventional type (bremsstrahlung source, silicon detectors) ever done. Second, Baglin *et al.* (1973) did a similar measurement from 17 to 25 MeV, this time using the long-duty-cycle EPA. The total cross-section results from these experiments are shown in Fig. 4, together with the theoretical prediction of Partovi. Below 35 MeV, the measured cross section falls considerably below the theoretical prediction (by as much as 20%, at 22 MeV); but almost any change one might make in the theory, it seems (for example, enhancing the role played by E2 transitions), would *raise* the theoretical values. Clearly, one would like an independent measurement of this cross section done with monoenergetic photons or in some other way which does not depend critically on the shape of the bremsstrahlung spectrum or on the calibration of a bremsstrahlung-flux-measuring ionization chamber. Third, Firk and collaborators (Nath, 1971; Nath *et al.*,

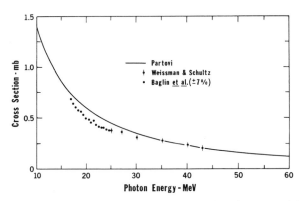

Fig. 4. The ^{2}H(γ, p)n cross-section measurements of Weissman and Schultz (1971) (⬥) and Baglin *et al.* (1973) (●), showing the deviation from the theoretical calculation of Partovi (1964) (curve) (from Baglin *et al.*, 1973).

1971a, b) measured the polarization of photoneutrons from deuterium at two angles (45° and 90°) from 10 to 30 MeV. The 45° results, at least, are at variance with the theory (Partovi, 1964; Feshbach and Lomon, 1968), but here, unlike the case of the cross section, the trouble probably lies in the incorrect treatment of multipolarities other than E1. This is corroborated by the fact that the angular distribution data both of Weissman and Schultz and of Baglin *et al.* indicate that only the value for A_3 (the coefficient of the E1– E2 mixing term) differs substantially from the Partovi prediction.

b. ^{3}He. The first good-resolution measurement of the giant resonance of a three-body nucleus [^{3}He(γ, n)2p] (Berman *et al.*, 1970a) indicates the existence of sharp structures (1–2 MeV wide) at several energies between 11 and 23 MeV. Although four states can be generated from the splitting of $J = \frac{1}{2}$ and $\frac{3}{2}$ and $T = \frac{1}{2}$ and $\frac{3}{2}$ configurations, there has been no serious theoretical attempt to explain this phenomenon to date. The theoretical calculation of Barbour and Phillips (1970) has, however, by introducing final-state interactions, succeeded in reducing the theoretical estimate of the absolute magnitude of the three-body breakup cross section almost to within a reasonable range of the experimental data. Several other experiments on ^{3}He (and ^{3}H) have been performed (among them those by Kundu *et al.*, 1971; Owens and Matthews, 1971; and Bösch *et al.*, 1965), but a detailed discussion of their results would require a separate review.

c. ^{4}He. Photoneutron measurements at Livermore with annihilation

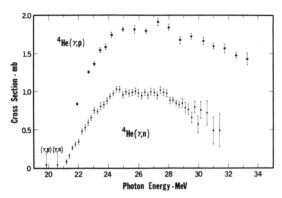

Fig. 5. The ^{4}He$(\gamma, n)^3$He total cross section and the ^{4}He$(\gamma, p)^3$H total cross section derived from the ^{3}H(p, γ) data of Meyerhof *et al.* (1970) [which in turn were normalized to the earlier (p, γ) work of Perry and Bame (1955)] (from Berman *et al.*, 1971).

photons (Berman *et al.*, 1970b, 1971), and at Yale (Berman *et al.*, 1972a) with bremsstrahlung and neutron time of flight, have indicated that the (γ, n) cross section is scarcely more than half the size of the (γ, p) cross section. The ^{4}He (γ, n) cross section is shown in Fig. 5, together with the ^{4}He(γ, p) cross section obtained from the ^{3}H(p, γ) measurements of Meyerhof *et al.* (1970) and Perry and Bame (1955). This implies an anomalously high degree of iso-spin mixing in the giant resonace of this nucleus: $\gtrsim 15\%$ from 23 to 28 MeV (see Barker and Mann, 1957). This result is confirmed by the total photon absorption cross section obtained from the inelastic electron scattering data of Walcher (1970). If this large amount of isospin mixing cannot be explained entirely by *Coulomb* mixing, then the conservation of charge symmetry by nuclear forces is called into question. However, this result is still controver-sial, and might be incorrect (Dodge *et al.*, 1972; Irish *et al.*, 1971; Irish, 1973).

2. *The* 1p *Shell*; *Particle–Hole States*

The 1p-shell nuclei, particularly the self-conjugate nuclei ^{12}C, ^{14}N, and ^{16}O, have continued to receive major attention from both theoretical and experimental photonuclear physicists. This is because there are typically only a few prominent and well-defined giant resonance states in these nuclei, whose strength and location have been reasonably well accounted for by particle–hole calculations, and, unlike the situation for the hydrogen and helium isotopes, these nuclei are plentiful and easy to work with in the laboratory. The situation up to early 1970 has been summarized in the review by Firk (1970), which includes discussion of the open-shell calculation for ^{12}C by Rowe and Wong (1969) (also Wong and Rowe, 1969), the coupled-channel continuum calculations for ^{16}O by Buck and Hill (1967) and Saruis and Marangoni (1969), and the measurements on ^{16}O of the photoproton angular distribution (Frederick et al., 1969; Baglin and Thompson, 1969), the photoneutron polarization (Cole et al., 1969; Cole, 1970), and the deexcitation γ-ray spectrum [from $(\gamma, n\gamma')$ and $(\gamma, p\gamma')$ reactions] (Caldwell, 1967; Caldwell et al., 1967; Murray and Ritter, 1969). New developments include the following:

(a)   The theoretical calculation of Shakin and Wang (1971) demonstrates the potential importance of 3p–3h configurations in the ^{16}O giant resonance states.

(b)   Detailed (γ, p) and $(\gamma, x\gamma')$ measurements by Baglin and collaborators (Baglin et al., 1971; Carr et al., 1971) have been performed at the EPA on ^{16}O and ^{14}N. These, together with earlier work on ^{16}O (Caldwell, 1967; Jury, 1970) and more recent work on the ^{14}N$[(\gamma, n) + (\gamma, pn)]$ total cross section (Berman et al., 1970c), on ^{14}N photoneutron spectra by Sherman et al. (1970), and on the ^{13}C(p, γ_0) reaction by Riess et al. (1971), have helped to unravel the complex details of the decay of the giant resonance states in these nuclei. The branching ratios, for example, integrated over all the giant resonance states for these nuclei, are given in Fig. 6.

(c)   The ^{11}B(p, γ)^{12}C work of Brassard (1970) and the later work of Shay (1972), which contradicts the absolute cross-section results of the earlier (p, γ) results of Allas et al. (1964), is itself contradicted by the absolute photon absorption measurement on ^{12}C by Ziegler and collaborators (Ahrens et al., 1971), in combination with the rather well-established absolute ^{12}C(γ, n) cross section (Lochstet and Stephens, 1966; Fultz et al., 1966): Now it appears once again that the isospin mixing ratio in the ^{12}C giant resonance is large (Barker and Mann, 1957). This view is strengthened by the recent ^{12}C(γ, γ') measurements of Medicus et al. (1970). [However, even the ^{15}N(p, γ_0)^{16}O measurements disagree with each other by as much as 50%, so that one must be cautious in ascribing any given amount of isospin mixing for these nuclei.] The results of the Mainz group for the total photon absorption cross section for ^{12}C are shown in Fig. 7. Shay (1972) also has studied 2p–2h and 3p–3h configurations in the giant resonance of ^{12}C and ^{16}O via (d, γ) and (^{3}He, γ) measurements, and has found important structures at several energies, which are strongly correlated with the predictions of Shakin and Wang (1971).

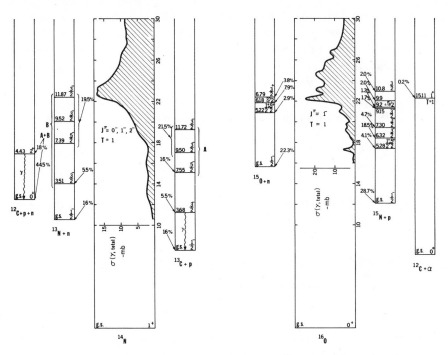

Fig. 6. Decay of the giant resonances of ^{14}N and ^{16}O. The branching ratios given are for the total photoabsorption strength integrated up to about 28 MeV. [The ^{14}N data are from Carr *et al.* (1971); the ^{16}O data are from Caldwell (1967).]

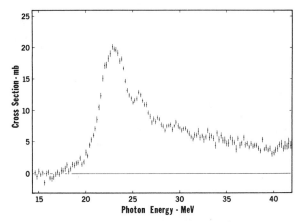

Fig. 7. The ^{12}C(γ, total) photon absorption cross section, measured with bremsstrahlung and Compton spectrometers (from Ahrens *et al.*, 1971).

(d)   A detailed study of giant resonance states of ^{12}C by inelastic electron scattering has been carried out at Tohoku (Yamaguchi *et al.*, 1971). The momentum transfer dependence of a 1$^-$ state at 22.7 MeV is consistent with its excitation by the dipole spin–isospin mode (Lewis and Walecka, 1964; Kelly and Uberall, 1968).

(e)   Inelastic electron scattering studies also have been carried out for ^{13}C at NBS (Bergstrom *et al.*, 1971). The relatively high resolution of these measurements (~ 200 keV) reveals considerable structure, with the giant resonance being split into three major clumps of strength centered at 14, 20.5, and 24.5 MeV, each being several MeV wide. The lower clumps might very well be composed of $T = \frac{1}{2}$ states and the upper of $T = \frac{3}{2}$ states, in keeping with the early calculation of Easlea (1962) [also see Measday *et al.* (1965)]. These results are supported by the recent ^{13}C($\gamma, x\gamma'$) measurements of Winhold *et al.* (1971).

3.   *The 2s–1d Shell*; *Extreme Fragmentation of the Giant Resonance*

A series of measurements on the $A = 23$–27 nuclei (Fultz *et al.*, 1966, 1971b; Alvarez *et al.*, 1971) has established that the even-A nuclei have much more prominent structure than the odd-A nuclei (also see Wu *et al.*, 1970b, for a discussion of recent results for $A = 28$, 32, and 40 nuclei). In fact, below 28 MeV, there are at least 14 distinct peaks in the ^{24}Mg cross section and 32 in the ^{26}Mg cross section. The total photoneutron cross section for ^{26}Mg is shown in Fig. 8. This structure has been confirmed by (e, e') experiments performed at Darmstadt (Titze *et al.*, 1967, 1970). While there probably is considerable fine structure which has not been resolved in the Livermore work (done with $\sim 1\%$ resolution), these intermediate structures were analyzed statistically and found to correspond to a Porter–

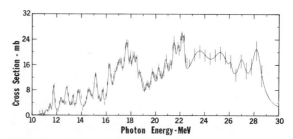

Fig. 8. The ^{26}Mg total photoneutron cross section, fitted with 32 Lorentz lines, illustrating extreme fragmentation of the giant resonance for even-even nuclei in the s-d shell (from Fultz *et al.*, 1971b).

Thomas distribution with one degree of freedom (Fultz *et al.*, 1971b). Equating the number of degrees of freedom with the number of spin states accessible with E1 transitions, Fultz *et al.* conclude that the reason very little structure is seen in the odd-*A* nuclei is because the three such spin states available make for a density of states that could not have been resolved in the measurements performed to date.

While the question of isospin mixing in self-conjugate nuclei has been treated by Firk (1970), the question of isospin splitting of non-self-conjugate nuclei in the s-d shell has achieved prominence more recently. Building upon the apparent splitting of the giant resonance of ^{26}Mg into two large clumps of strength (Titze *et al.*, 1970; Fultz *et al.*, 1971b) (see Fig. 8), and the location of the two low-lying analogs in ^{25}Mg of the ground and first excited states of ^{25}Na by a threshold photoneutron experiment (Berman *et al.*, 1970d), a ^{26}Mg(γ, n) experiment using bremsstrahlung and neutron time of flight was done at Yale (Wu *et al.*, 1970c) which established the isospin splitting of the ^{26}Mg giant resonance. This measurement showed that the group of states between 19 and 23 MeV has isospin $T = 2$, since they decay to the $T = \frac{3}{2}$ states but not to the $(T = \frac{1}{2})$ ground state of ^{25}Mg. (Isospin is treated by Temmer, Chapter IV.A.2.)

B. Medium and Heavy Nuclei

The extensive series of measurements on the (γ, n), $(\gamma, \text{2n})$, and $(\gamma, \text{3n})$ cross sections of medium and heavy nuclei carried out at Livermore (Fultz *et al.*, 1962, 1969; Harvey *et al.*, 1964; Bramblett *et al.*, 1966; Berman *et al.*, 1967, 1969a, b, c) and, more recently, at Saclay (Bergère *et al.*, 1968, 1969; Veyssière *et al.*, 1970; Beil *et al.*, 1971; Carlos *et al.*, 1971; Leprêtre *et al.*, 1971) have made possible the delineation of many of the systematics of the giant resonance, since, because of the Coulomb barrier, the sum of the partial photoneutron reactions is virtually identical to the total photon absorption cross section. An example of the partial photoneutron cross sections and their sum is shown in Fig. 9 for the case of a spherical nucleus, namely ^{120}Sn. Note that it is absolutely essential to measure all the partial cross sections (e.g., with the neutron multiplicity counting technique employed at Livermore and Saclay) in order to obtain the total cross section. An example of a deformed nucleus, ^{165}Ho, is shown in Fig. 10, while the total cross sections for a group of transitional nuclei, the neodymium isotopes, are shown in Fig. 11. A detailed treatment of the giant resonance for vibrational nuclei is given in the review by Spicer (1969). (Nuclear models for heavy nuclei are discussed by Rasmussen, Chapter IX.B.)

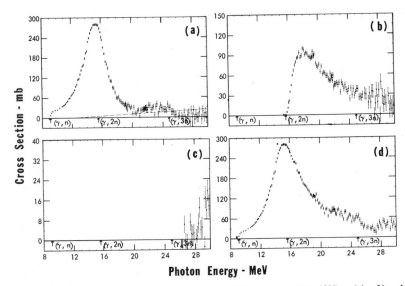

Fig. 9. Photoneutron cross sections for the spherical nucleus ^{120}Sn: (a) $\sigma[(\gamma, n) + (\gamma, pn)]$; (b) $\sigma(\gamma, 2n)$; (c) $\sigma(\gamma, 3n)$; (d) total photoneutron cross section $\sigma[(\gamma, n) + (\gamma, pn) + (\gamma, 2n) + (\gamma, 3n)]$ (from Fultz *et al.*, 1969).

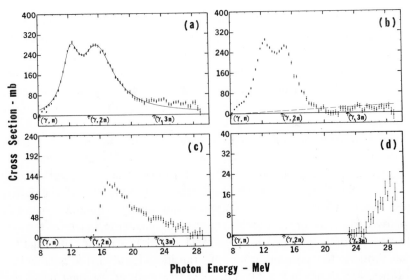

Fig. 10. Photoneutron cross sections for the statically deformed nucleus ^{165}Ho: (a) The total photoneutron cross section $\sigma[(\gamma, n) + (\gamma, pn) + (\gamma, 2n) + (\gamma, p2n) + (\gamma, 3n)]$, fitted with a two-component Lorentz curve; (b) $\sigma[(\gamma, n) + (\gamma, pn)]$; (c) $\sigma[(\gamma, 2n) + (\gamma, p2n)]$; (d) $\sigma(\gamma, 3n)$ (from Berman *et al.*, 1969b).

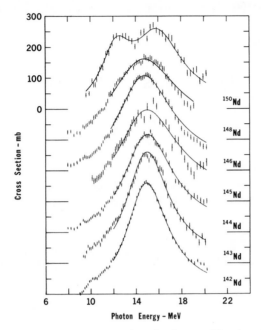

Fig. 11. Total photoneutron cross sections for the transitional nuclei, the neodymium isotopes, fitted with Lorentz curves (a two-component fit for the case of ^{150}Nd) (from Carlos *et al.*, 1971, with permission of North-Holland Publishing Co., Amsterdam).

1. Spherical Nuclei and Shell Effects

The study of the systematics of single-nucleon effects on the giant resonance in the zirconium, tin, and lead isotopes (Berman *et al.*, 1967; Fultz *et al.*, 1969; Harvey *et al.*, 1964) as well as in other nuclei has revealed the not unreasonable fact that it is the neutron number that determines primarily the shape of the giant resonance and the ratios of the partial photoneutron cross sections. In particular, the following results have been found:

(a) The giant resonance has a narrow width (~ 4 MeV) only for closed-neutron-shell nuclei, or nuclei one or two neutrons away from a closed shell, but not for others.

(b) The shell-and-pairing-effect parameter Δ in an expression for the nuclear density of states, which is proportional to $\exp 2[a(U - \Delta)]^{1/2}$, where a is the level-density parameter and U the excitation energy, and which is obtained from the energy dependence of the ratio $\sigma(\gamma, 2n)/\sigma(\gamma, \text{total})$, clearly is largest for closed-neutron-shell nuclei (see also Richter, Chapter IV.D.1).

(c)   The ratio of integrated cross sections

$$\int \sigma[(\gamma, 2n) + (\gamma, 3n)] \, dE \Big/ \int \sigma(\gamma, \text{total}) \, dE$$

up to 30 MeV, shown in Fig. 12, shows dramatic minima at closed neutron shells and subshells (from Fultz *et al.*, 1969).

The isospin splitting of the giant resonance into $T_<$ and $T_>$ giant resonances was predicted by Fallieros and collaborators (Fallieros *et al.*, 1965; Fallieros and Goulard, 1970; Akyüz and Fallieros, 1971) from shell

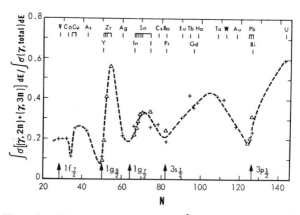

Fig. 12.   The ratio of integrated cross sections $\int \sigma[(\gamma, 2n) + (\gamma, 3n)] \, dE / \int \sigma(\gamma, \text{total}) \, dE$, showing pronounced minima at closed neutron shells and subshells (from Fultz *et al.*, 1969).

model considerations. This splitting, first observed in the photoneutron cross section for ^{90}Zr (Berman *et al.*, 1967), has been discussed by Firk (1970) and more recently confirmed in detail by Paul and collaborators at Stony Brook (Paul *et al.*, 1971) in an extensive series of (p, γ) experiments, as well as by Shoda and collaborators at Tohoku (Shoda and Sugawara, 1971) in a comprehensive series of (e, e'p) experiments.

2.   *Deformed Nuclei and Collective Effects*

It has been known for many years that the giant dipole resonance of statically deformed nuclei is split into two broad components (Fuller *et al.*, 1958). This behavior has been explained both by the hydrodynamic model of proton- and neutron-fluid oscillations (Okamoto, 1956; Danos, 1958) and by the newer dynamic collective model of the nucleus (Danos and Greiner, 1964). The essential difference between the two theories is that the dynamic collective treatment specifically includes the coupling of the initial dipole vibrational mode to the other collective modes of the nucleus. In particular, the coupling to the surface quadrupole oscillations proves to be important, and results in a sharing of the dipole strength with several "satellite" peaks in the giant resonance. The model fails, however, to give a fundamental explanation for the widths (or damping) of the various peaks; consequently, the question of gross structure in the giant resonance of medium and heavy vibrationally or statically deformed nuclei is ambiguous theoretically, and considerable controversy surrounds the topic (see Huber, 1967, and Spicer, 1969, for further details). In any case, for deformed nuclei, the hydrodynamic theory predicts (correctly) that the nuclear eccentricity, and hence the intrinsic quadrupole moment, is simply related to the energy splitting between the two giant resonance humps, since the two humps arise from dipole vibrations along the major and minor axes of the (spheroidal) nucleus. Moreover, the sign of the quadrupole moment is immediately apparent from the relative strengths of the two components.

However, two recent experiments have shown at the same time that these theories are both qualitatively correct and quantitatively inadequate:

(a) Measurements of the photoneutron cross sections for polarized ^{165}Ho, for the target nuclei aligned both parallel and perpendicular to the direction of the photon beam (Kelly *et al.*, 1969), have confirmed the existence of the predicted asymmetry

$$A_t = \int |\sigma_\parallel(\gamma, \text{total}) - \sigma_\perp(\gamma, \text{total})| \, dE$$

but have shown that this asymmetry is only three-fourths as large as either theory predicts.

(b) The ratio of areas under the two humps of the giant resonance $R_A = \sigma\Gamma(\text{lower})/\sigma\Gamma(\text{higher})$, predicted by the hydrodynamic model to be one-half for prolate nuclei [where there are two short (high-frequency) axes of vibration to one long] has been shown (Berman *et al.*, 1969b) to approximate this value only for the most deformed nuclei, and to fall off either

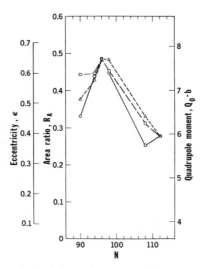

Fig. 13. The area ratio $R_A(\bigcirc)$, nuclear eccentricity $\varepsilon(\square)$, and intrinsic quadrupole moment $Q_0(\triangle)$ of various nuclei plotted versus neutron number. The lines merely connect the three sets of data points, showing that the three quantities vary in a similar fashion (from Berman *et al.*, 1969b).

for heavier or lighter nuclei in a way proportional to the deformation, as illustrated in Fig. 13.

3. Systematic Properties of the Giant Resonance

The 34 medium and heavy nuclei studied at Livermore have yielded information on several systematic properties of the giant resonance. Three of these are presented here:

(a) The peak energy E_m has been predicted by collective models (Goldhaber and Teller, 1948; Steinwedel and Jensen, 1950) to be proportional either to $A^{-1/3}$ or $A^{-1/6}$. The data, shown in Fig. 14, indicate that the answer. lies midway between the two predictions. When the $A^{-1/4.2}$ dependence is decomposed, the result is E_m (in MeV) $= 39A^{-1/3} + 17A^{-1/6}$. Werntz and collaborators (Bach and Werntz, 1968; Brennan and Werntz, 1970) have modified the Steinwedel–Jensen picture to include a more realistic nuclear surface, and obtain an A dependence which changes from $A^{-1/3}$ at low A values to $A^{-1/6}$ at high ones; this appears to be a very promising approach (also see Akhiezer *et al.*, (1971).

(b) A quantity related to the peak energy and width of the giant res-

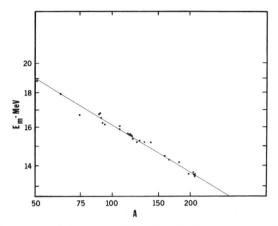

Fig. 14. The energy of the peak of the giant dipole resonance plotted versus mass number on a log-log scale, showing a dependence midway between the $A^{-1/3}$ and $A^{-1/6}$ predictions: $E_m = 47.95A^{-1/4.24}$ (from Livermore data, 1969).

onance is the nuclear symmetry energy K, which is obtained from the expression (Danos, 1958; Bramblett *et al.*, 1966)

$$K = 9.935 \times 10^{-4} \frac{A^{8/3}}{NZ} \frac{E_m^2}{1 - (\Gamma/2E_m)^2}$$

for spherical nuclei [and suitably modified (Berman *et al.*, 1969a) for deformed nuclei], where K and E_m are in MeV. This quantity, which is the coefficient multiplying the $(N - Z)^2/A$ term in the semiempirical mass formula, is shown in Fig. 15 as a function of A. The dot-dashed curve in Fig. 15, which represents the modification of the (volume) symmetry energy by the nuclear surface (Myers and Swiatecki, 1966), is an acceptable fit to the data.

(c) The Thomas–Reiche–Kuhn (TRK) sum rule for photonuclear absorption (Levinger, 1960) predicts that, in the absence of exchange effects, the integrated cross section is given by

$$\sigma_{int} = \int \sigma(\gamma, \text{total}) \, dE = 0.060NZ/A \quad \text{(in MeV-b)}$$

This quantity, integrated up to ~ 30 MeV for each nucleus and divided by the sum-rule prediction, is shown in Fig. 16. It can be seen that the sum rule is exhausted at approximately twice the giant resonance energy.

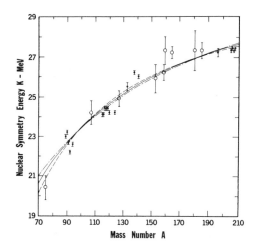

Fig. 15. The nuclear symmetry energy derived from giant resonance parameters plotted versus A, fitted with functions of the form $K = K_0(1 - cA^{-n})$, where $n = 1$ (dashed curve), $\frac{1}{3}$ (dot-dashed curve), and 0.62 (best fit; solid curve); dots for spherical nuclei, open circles for deformed nuclei (from Livermore data, 1971).

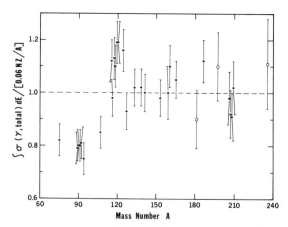

Fig. 16. The measured integrated total photoneutron cross section $\int \sigma(\gamma, \text{total})\, dE$ divided by the TRK sum-rule prediction, plotted versus A (see text) (from Livermore data, 1971).

III. Photonuclear Studies near the Particle Thresholds

In principle, photonuclear reactions constitute the ideal way to study the properties of highly excited nuclear states, since the electromagnetic selection rules can be made to serve as a spectroscopic tool. The difficulty has been, of course, in obtaining data of high quality and resolution. Now, two methods are available which are capable not only of high intensities and counting rates, but of high resolution as well. The threshold photoneutron technique is applicable for studying nuclear states lying just above the neutron separation energy, and the scattering of photons, real or virtual, for states lying below it.

A. THRESHOLD PHOTONEUTRON MEASUREMENTS

The threshold photoneutron technique makes use of a low-energy pulsed bremsstrahlung beam and neutron time of flight to determine with high precision the energy of the photoneutron ejected in a (γ, n) event. When the end-point energy of the bremsstrahlung spectrum is limited to an MeV or so above the (γ, n) threshold, so that the residual nucleus is known to have been left in its ground state, then there is a one-to-one correspondence between the energy of the photoneutron and that of the incident photon. Thus, since it is easy to measure neutron energies with resolution better than 1%, one can achieve a photon-energy resolution, at photon energies near 8 MeV, of 0.01% routinely, and in some cases as good as 1 ppm. Furthermore, since one can study the same compound levels in some nuclei through the (n, γ) reaction on the target-minus-one-neutron nucleus, the combined results can yield more information than either experiment could separately. A typical example of the kind of data achieved to date is shown in Fig. 17, for $^{56}\text{Fe}(\gamma, n)$. The technique was demonstrated initially at MIT (Bertozzi

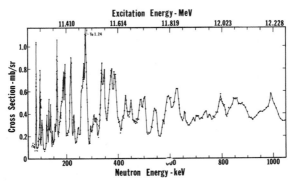

Fig. 17. The 135° differential $^{56}\text{Fe}(\gamma, n)$ cross section near threshold. At the highest energies, the deterioration in the energy resolution resulting from the finite resolving time of the experimental apparatus is evident. From these data, values for $g\Gamma_{\gamma 0}\Gamma_n/\Gamma$ have been extracted for 28 resonances (also see Chapter VII.D) and all but five have been identified with ground-state transitions (from Baglan *et al.*, 1971a).

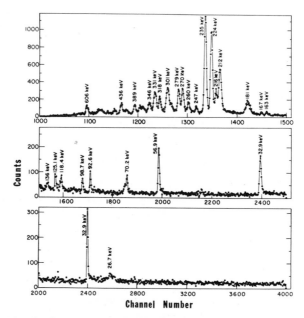

Fig. 18. Threshold photoneutron time-of-flight spectrum for ^{57}Fe$(\gamma, n)^{56}$Fe; $\theta_n = 90°$, $E_e - E_{thr} = 0.81$ MeV (from Jackson and Strait, 1971a).

et al., 1963) and developed to its present state at Livermore (Berman *et al.*, 1966; Bowman *et al.*, 1967; Baglan *et al.*, 1971a). In the Livermore experiments alone, more than 250 resonances in 16 nuclei have been identified and in most cases parameterized, yielding information on several diverse topics in nuclear physics, some of which will be treated here. In the last year or two, several other linac laboratories have begun work in the field, including those at Argonne, Toronto, Japan Atomic Energy Research Institute, Harwell, Ottawa, and Gulf Radiation Technology, so that much additional data soon can be expected. An example of the data from the new Argonne linac is shown in Fig. 18.

1. *Electromagnetic Strength Functions*

Not enough nuclei have been studied yet to enable complete understanding of the systematics of electromagnetic strength functions in the threshold energy region. Values for the ground-state strength function for E1 transitions for ^{26}Mg, ^{53}Cr, ^{57}Fe, ^{207}Pb, and ^{208}Pb have been given by Baglan *et al.* (1971a); the results seem to favor the $A^{2/3}$ dependence predicted by the single-particle theory over the $A^{8/3}$ dependence which results from an extrapolation of a Lorentz curve fitted to the giant resonance down to the threshold energy region (Axel, 1962).

For the case of M1 transitions, it appears that the M1 strength typically is enhanced over the single-particle estimate by about a factor of 15 (Jackson and Strait, 1971a; also see Bollinger, 1969).

2. Analog States: Location and Purity

The threshold photoneutron technique can provide precise measurements of Coulomb displacement energies, and, when the proton channel is closed, a direct measurement of the isospin purity of an analog state in a non-self-conjugate nucleus (since neutron decay is forbidden from a $T_>$ state). An illustration of both these features is given by Berman et al. (1970d) for the case of ^{25}Mg. The location of the analogs of the ground and first excited states of ^{25}Na were located to within ± 4 keV, their isospin impurity was determined to be $\lesssim 0.3$ and 1.3%, respectively, and the mass of ^{25}Si was predicted to within ± 32 keV. (Also see Berman and Phillips, 1972.)

3. The Source of Neutrons in Stars

The elements heavier than iron are synthesized in stars (or supernovae) by neutron capture (Burbidge et al., 1957). The most likely source of neutrons for nucleosynthesis has been found to be the ^{22}Ne(α, n) reaction, which proceeds through a resonance in the ^{26}Mg compound system. This 1^- state, which lies 54 keV above the ^{26}Mg(γ, n) threshold, was identified in a two-angle threshold photoneutron measurement (Berman et al., 1969c). Measurements were made at two angles in order to obtain information on the spin and parity of the level, since α-particle-induced reactions can proceed only through $J^\pi = 0^+, 1^-, 2^+, \ldots$ states.

4. Doorway States and Their Fine Structure

The threshold photoneutron technique has been used to study many nuclei for the presence of intermediate structure in the photon channel. The fine-structure resonances measured with high resolution were examined for clusters of ground-state γ-ray transition strength associated with excited states all having the same spin and parity. Evidence for doorway states in the form of envelopes of γ-ray widths was observed for several nuclei (Baglan et al., 1971b; Berman, 1971; Berman et al., 1972b).

A doorway state common to both the neutron and photon channels has been found in ^{207}Pb (Baglan et al., 1971b; Farrell et al., 1965). This was demonstrated by (a) observing that envelopes of $\frac{1}{2}^+$ strength exist in both channels at the same excitation energy, and (b) finding that the neutron and photon widths for the fine-structure resonances are highly correlated. The ground-state γ-ray width for the doorway was found to be 36.5 eV, about one-tenth of a single-particle unit for E1 transitions. The nature of the doorway is a $\frac{1}{2}^-$ neutron hole coupled to a 1^- collective or p–h excitation; the fine-structure states are 3p–2h states or other more complicated configurations. A comprehensive theoretical analysis of this subject has been given by Lane (1971).

5. The M1 Giant Resonance

A special kind of photon doorway, comprised exclusively of 1p–1h states, is the long-sought M1 giant resonance. Like its counterpart, the E1 giant resonance, it takes nearly all the ground-state transition strength from other states having the same spin and parity and concentrates this strength in a relatively small number of states in a narrow energy band. There is an increasing body of experimental evidence that the characteristic energy for the M1 giant resonance is near the particle separation energy: the 15.11-MeV level in ^{12}C; the strong ground-state M1 transitions to levels near 10 MeV for several s-d-shell nuclei found by Axel and collaborators at Illinois by scattering tagged photons (Kuehne et al., 1967) and by Fagg and collaborators at NRL with inelastic electron scattering at 180° (Fagg et al., 1969, 1970, 1971, 1972; Bendel et al., 1971), including the remarkable level at 11.2 MeV in ^{20}Ne; and now, the evidence for ^{208}Pb and other nuclei from threshold-photoneutron measurements.

Figure 19 shows the values for Γ_{γ_0} for 1^+ resonances in ^{208}Pb, which deexcite by p-wave neutrons (Bowman et al., 1970; Berman, 1971; Berman et al., 1972b). The values for J^π for these resonances were obtained from the ratio of the (γ, n) cross sections obtained at two angles (90 and 135°) combined with ^{207}Pb(n, n) data (Bilpuch, 1970). The M1 strength in ^{208}Pb, concentrated in an energy region 600 keV wide centered at 7.9 MeV, arises from spin-flip transitions from the $i_{13/2}$ neutron shell and the $h_{11/2}$ proton shell; these p–h states form the collective doorway state through which the compound nucleus states observed here are reached. The total γ-ray strength for these states is 51.6 eV, more than five Weisskopf units, and constitutes at least half and perhaps all of the total M1 strength calculated for this nucleus (Weiss, 1970).

Other evidence for the M1 giant resonance comes from the threshold-photoneutron measurements of Jackson and Strait (1971a, b) for ^{53}Cr and

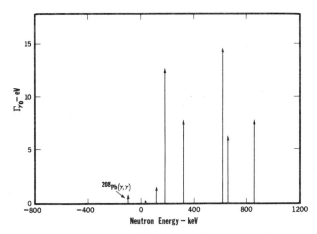

Fig. 19. The M1 giant resonance in [208]Pb. Values for the ground-state transition width for 1+ levels are plotted as a function of neutron energy for the (γ, n) reaction (from Berman *et al.*, 1972b).

[57]Fe, Patrick and collaborators for [117, 119]Sn (Winhold *et al.*, 1970) and other nuclei (Patrick, 1971); the (n, γ) work of Bollinger and Thomas (1971) for [118, 120]Sn; and, confirming the Livermore work on [208]Pb, the (γ, n) work of Fuketa and collaborators (Mizumoto *et al.*, 1971) and McNeill and collaborators (Haacke *et al.*, 1971) and the (n, γ) measurements of Allen and Macklin (1971). Since this topic currently is being pursued actively at several laboratories, one can hope to be able to delineate the systematics of the M1 giant resonance shortly.

B. PHOTON SCATTERING

In order to study the electromagnetic properties of levels below the particle thresholds, one must resort to photon scattering ("resonance fluorescence") or absorption measurements. Photon scattering using bremsstrahlung is very difficult but Ge(Li) detectors have revived the field somewhat, and some good data, e.g., those of Swann (1971a, b) and Shikazono and Kawarasaki (1972), are being acquired. Photon scattering and absorption measurements using $D(\gamma, n)$ (Section I.B.3) have not yet borne fruit. Therefore, we shall confine our attention to those methods mentioned in Sections I.A.2, 4, and 5.

1. *Tagged Photons*

Axel and collaborators also have performed (γ, γ) measurements on Zr,

Sn, and Pb isotopes below the (γ, n) thresholds with an energy resolution of ~ 100 keV (Axel *et al.*, 1963, 1970). The case of tin is particularly interesting, since for the odd tin isotopes, the Harwell group has seen evidence for the M1 giant resonance at ~ 7.8 MeV (Winhold *et al.*, 1970), and this energy lies below the (γ, n) thresholds for the even tin isotopes. The Illinois data are shown in Fig. 20. Clearly, there is an envelope of strength in the

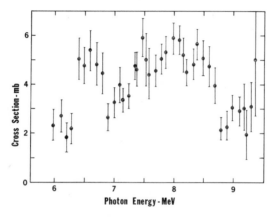

Fig. 20. The photon scattering cross section for natural tin. The broad peak in the cross section in the energy region near 8 MeV might constitute the M1 giant resonance in the even tin isotopes (see text) (from Axel *et al.*, 1970).

(γ, γ) cross section at about the same energy. This, combined with the ^{208}Pb data (Bowman *et al.*, 1970) and the recent Darmstadt results for $N = 82$ nuclei, could lead one to believe that the M1 giant resonance might be located at about 8 MeV for many (if not all) medium and heavy nuclei.

2. *Spectroscopy with Neutron-Capture γ Rays*

The Israelis have taken the lead in utilizing the high intensity and very sharp monochromaticity of neutron-capture γ rays from a reactor to study the properties of individual levels in many nuclei (Arad *et al.*, 1964; Ben-David *et al.*, 1966; Moreh *et al.*, 1970), although previous work on (γ, n) reactions was carried out by Donahue and collaborators at Pennsylvania State University (Welsh and Donahue, 1961; Green and Donahue, 1964; Hurst and Donahue, 1967). Examples are the M1 transitions found in ^{208}Pb (see Fig. 19) and ^{141}Pr (Moreh *et al.*, 1970), the parity of which were determined from the measurements of the polarization of the scattered photons.

Other groups have used (n, γ) photons to measure the ^{3}H(γ, n) cross section just above threshold (Bösch *et al.*, 1965), and photofission cross sections (Manfredini *et al.*, 1966, 1969; Mafra *et al.*, 1970; Anderl *et al.*, 1971; Khan and Knowles, 1972) and angular distributions (Dowdy and Krysinski, 1971; Manfredini *et al.*, 1971).

Another interesting use of the scattering of neutron-capture γ rays is the elegant demonstrations of Delbrück scattering (elastic scattering of photons by a Coulomb field, observed in the forward direction from a high-Z nucleus) on lead and uranium by Jackson and Wetzel (1969) and Moreh *et al.* (1971), and of nuclear Raman scattering (inelastic scattering of photons, populating excited states of the nucleus) on ^{232}Th and ^{238}U by Hass *et al.* (1971) and on ^{159}Tb, ^{232}Th, and ^{238}U by Jackson and Wetzel (1972).

3. *Inelastic Electron Scattering*

Penner and co-workers at NBS (including collaborators from MIT) have measured transition strengths and form factors for a number of bound states in several nuclei. The resolution achieved to date has been as good as 60 keV at 6 MeV excitation energy, which was good enough to attack a long-standing problem in ^{16}O where the 0^+ and 3^- first and second excited states were resolved; the background-subtracted experimental data are shown in Fig. 21 (Bergstrom *et al.*, 1970). Other nuclei studied at the NBS electron

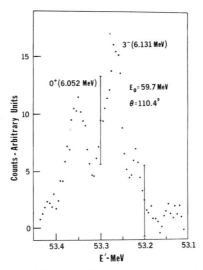

Fig. 21. The spectrum of inelastically scattered electrons from ^{16}O, showing the resolution of the first two excited states at 6.05 and 6.13 MeV (from Bergstrom *et al.*, 1970).

scattering facility include, among others, ^{14}N, ^{19}F, ^{40}Ca, ^{90}Zr, and the even zinc isotopes (Ensslin *et al.*, 1971; Hallowell *et al.*, 1970; Mills *et al.*, 1970; Neuhausen *et al.*, 1971, 1972).

Torizuka and collaborators at Tohoku recently have measured deformation parameters for ^{12}C, ^{20}Ne, ^{24}Mg, and ^{28}Si by combining their elastic and inelastic electron scattering results from the 0^+, 2^+, and 4^+ states in these nuclei (Nakada *et al.*, 1971; Horikawa *et al.*, 1971).

A recent result from Darmstadt indicates the existence of an energy region of large ground-state M1 transition strength around 8.7 MeV in cerium as well as in ^{139}La and ^{141}Pr, and some prominent E2 states atop the giant resonances of these nuclei (Pitthan and Walcher, 1971).

Finally, some of the most interesting recent results from inelastic electron scattering work pertain to the discovery of prominent E2 strength both below and above the E1 giant resonance. These experiments were performed at Tohoku (Torizuka *et al.*, 1973) on ^{44}Ca, ^{56}Fe, ^{90}Zr, and ^{208}Pb and at Darmstadt (Buskirk *et al.*, 1972) on ^{208}Pb. The E2 assignment for the low-energy strength has been confirmed by (p, p') measurements done at Oak Ridge (Lewis, 1973).

IV. Summary

Recent results on a number of aspects of photonuclear reactions have been presented in this chapter, concentrating on those experiments performed with fruitful or promising techniques, especially those peculiar to photonuclear physics. The techniques mentioned include those for producing beams of monoenergetic photons and for detection of photoneutrons and photoprotons. The experimental data pertain largely to (i) the delineation of the properties of the giant dipole resonance and, in light nuclei, the states comprising it, and (ii) the specification of the properties of individual nuclear states or groups of states near the particle thresholds. Considerable progress has been made in these fields in the last few years, and, since many of these techniques are just beginning to bear fruit, considerably more can be expected in the next few years to come.

References

Ahrens, J., Borchert, H., Gimm, G., Gundrum, H., Sita Ram, G., Zieger, A., and Ziegler, B. (1971). Private communication.

Akhiezer, I. A., Barts, B. I., and Lazurik-El'tsufin, V. T. (1971).*Zh. Eksp. Teor. Fiz. Pis'ma Red.* **14**, 535 [*English transl.: JETP Lett.* **14**, 368].

Akyüz, R. Ö., and Fallieros, S. (1971). *Phys. Rev. Lett.* **27**, 1016.

Allas, R. G., Hanna, S. S., Meyer-Schützmeister, L., and Segel, R. E. (1964). *Nucl. Phys.* **58**, 122.

Allen, B. J., and Macklin, R. L. (1971). *In* "Neutron Cross Sections and Technology" (R. L. Macklin, ed.), pp. 764–770. USAEC Div. of Tech. Information Extension, Oak Ridge, Tennessee.

Alvarez, R. A., Berman, B. L., Lasher, D. R., Phillips, T. W., and Fultz, S. C. (1971). *Phys. Rev. C* **4**, 1673.

Anderl, R. A., Yester, M. V., and Morrison, R. C. (1971). Private communication.

Arad, B., Ben-David, G., Pelah, I., and Schlesinger, Y. (1964). *Phys. Rev.* **133**, B684.

Arad, B., and Ben-David, G. (1973). *Rev. Modern Phys.* **45**, 230.

Axel, P. (1962). *Phys. Rev.* **126**, 671.

Axel, P., Min, K., Stein, N., and Sutton, D. C. (1963). *Phys. Rev. Lett.* **10**, 299.

Axel, P., Min, K., and Sutton, D. C. (1970). *Phys. Rev. C* **2**, 689.

Bach, R., and Werntz, C. (1968). *Phys. Rev.* **173**, 958.

Baglan, R. J. (1970). Ph. D. Thesis, Univ. of California, Lawrence Radiat. Lab. Rep. No. UCRL-50902 (unpublished).

Baglan, R. J., Bowman, C. D., and Berman, B. L. (1971a). *Phys. Rev. C* **3**, 672.

Baglan, R. J., Bowman, C. D., and Berman, B. L. (1971b). *Phys. Rev. C* **3**, 2475.

Baglin, J. E. E. and Thompson, M. N. (1969). *Nucl. Phys.* **A138**, 73.

Baglin, J. E. E., Bentz, E. J., Carr, R. W., McConnell, D. B., and Thomas, B. J. (1971). Private communication.

Baglin, J. E. E., Carr, R. W., Bentz, E. J., and Wu, C.-P. (1973). *Nucl. Phys.* **A201**, 593.

Barbour, I. M., and Phillips, A. C. (1970). *Phys. Rev. C* **1**, 165.

Barker, F. C., and Mann, A. K. (1957). *Phil. Mag.* **2**, 5.

Beil, H., Bergère, R., and Veyssière, A. (1969). *Nucl. Instrum. Methods* **67**, 293.

Beil, H., Bergère, R., Carlos, P., Leprêtre, A., Veyssière, A., and Parlag, A. (1971). *Nucl. Phys.* **A172**, 426.

Ben-David, G., Arad, B., Balderman, J., and Schlesinger, Y. (1966). *Phys. Rev.* **146**, 852.

Bendel, W. L., Fagg, L. W., Numrich, S. K., Jones, Jr., E. C., and Kaiser, H. F. (1971). *Phys. Rev. C* **3**, 1821.

Bergère, R., Beil, H., and Veyssière, A. (1968). *Nucl. Phys.* **A121**, 463.

Bergère, R., Beil, H., Carlos, P., and Veyssière, A. (1969). *Nucl. Phys.* **A133**, 417.

Bergère, R., Beil, H., Carlos, P., Veyssière, A., and Leprêtre, A. (1972). *In Proc. Int. Conf. Nuclear Structure Studies Using Electron Scattering and Photoreaction* (K. Skoda and H. Ui, eds.). Sendai, Tohoku, p. 273.

Bergstrom, J. C., Bertozzi, W., Kowalski, S., Maruyama, X. K., Lightbody, Jr., J. W., Fivozinsky, S. P., and Penner, S. (1970). *Phys. Rev. Lett.* **24**, 152.

Bergstrom, J. C., Crannell, H., Kline, F. J., O'Brien, J. T., Lightbody, Jr., J. W., and Fivozinsky, S. P. (1971). *Phys. Rev. C* **4**, 1514.

Berman, B. L. (1971). University of California, Lawrence Radiat. Lab. Rep. No. UCRL-73003 (unpublished).

Berman, B. L. (ed.) (1973a). *Proc. Int. Conf. Photonuclear Reactions and Applications*, Lawrence Livermore Laboratory, Univ. of California.

Berman, B. L. (1973b). Atlas of Photoneutron Cross Sections Obtained with Monoenergetic Photons, Univ. of California Lawrence Livermore Lab. Rep. No. UCRL-74622 (unpublished).

Berman, B. L., and Phillips, T. W. (1972). *Phys. Rev. C* **6**, 2295.

Berman, B. L., Sidhu, G. S., and Bowman, C. D. (1966). *Phys. Rev. Lett.* **17**, 761.

Berman, B. L., Caldwell, J. T., Harvey, R. R., Kelly, M. A., Bramblett, R. L., and Fultz, S. C. (1967). *Phys. Rev.* **162**, 1098.

Berman, B. L., Bramblett, R. L., Caldwell, J. T., Davis, H. S., Kelly, M. A., and Fultz, S. C. (1969a). *Phys. Rev.* **177**, 1745.

Berman, B. L., Kelly, M. A., Bramblett, R. L., Caldwell, J. T., Davis, H. S., and Fultz, S. C. (1969b). *Phys. Rev.* **185**, 1576.

Berman, B. L., Van Hemert, R. L., and Bowman, C. D. (1969c). *Phys. Rev. Lett.* **23**, 386.

Berman, B. L., Fultz, S. C., and Yergin, P. F. (1970a). *Phys. Rev. Lett.* **24**, 1494.

Berman, B. L., Fultz, S. C., and Kelly, M. A. (1970b). *Phys. Rev. Lett.* **25**, 938.

Berman, B. L., Fultz, S. C., Caldwell, J. T., Kelly, M. A., and Dietrich, S. S. (1970c). *Phys. Rev. C* **2**, 2318.

Berman, B. L., Baglan, R. J., and Bowman, C. D. (1970d). *Phys. Rev. Lett.* **24**, 319.

Berman, B. L., Fultz, S. C., and Kelly, M. A. (1971). *Phys. Rev. C* **4**, 723.

Berman, B. L., Firk, F. W. K., and Wu, C.-P. (1972a). *Nucl. Phys.* **A179**, 791.

Berman, B. L., Bowman, C. D., and Baglan, R. J. (1972b). *In* "Statistical Properties of Nuclei" (J. B. Garg, ed.), pp. 611–617, Plenum Press, New York.

Bertozzi, W., Sargent, C. P., and Turchinetz, W. (1963). *Phys. Lett.* **6**, 108.

Bilpuch, E. G. (1970). Private communication.

Bollinger, L. M. (1969). *In* "International Symposium on Nuclear Structure" pp. 317–340. Int. At. Energy Agency, Vienna.

Bollinger, L. M., and Thomas, G. E. (1971). *Bull. Amer. Phys. Soc.* **16**, 1181.

Bösch, R., Lang, J., Müller, R., and Wölfli, W. (1965). *Helv. Phys. Acta* **38**, 8.

Bowman, C. D., Auchampaugh, G. F., and Fultz, S. C. (1964). *Phys. Rev.* **133**, B676.

Bowman, C. D., Sidhu, G. S., and Berman, B. L. (1967). *Phys. Rev.* **163**, 941.

Bowman, C. D., Baglan, R. J., Berman, B. L., and Phillips, T. W. (1970). *Phys. Rev. Lett.* **25**, 1302.

Bramblett, R. L., Caldwell, J. T., Berman, B. L., Harvey, R. R., and Fultz, S. C. (1966). *Phys. Rev.* **148**, 1198.

Brassard, C. (1970). Ph. D. Thesis, Yale Univ. (unpublished).

Brennan, J. G., and Werntz, C. (1970). *Phys. Rev. C* **1**, 1679.

Brown, G. E., and Bolsterli, M. (1959). *Phys. Rev. Lett.* **3**, 472.

Buck, B., and Hill, A. D. (1967). *Nucl. Phys.* **A95**, 271.

Burbidge, E. M., Burbidge, G. R., Fowler, W. A., and Hoyle, F. (1957). *Rev. Mod. Phys.* **29**, 547.

Buskirk, F. R., Gräf, H.-D., Pitthan, R., Theissen, H., Titze, O., and Walcher, Th. (1972). *Phys. Lett.* **42B**, 194.

Caldwell, J. T. (1967). Ph. D. Thesis, Univ. of California, Lawrence Radiat. Lab. Rep. No. UCRL-50287 (unpublished).

Caldwell, J. T., Bramblett, R. L., Berman, B. L., Harvey, R. R., and Fultz, S. C. (1965). *Phys. Rev. Lett.* **15**, 976.

Caldwell, J. T., Fultz, S. C., and Bramblett, R. L. (1967). *Phys. Rev. Lett.* **19**, 447.

Carlos, P., Beil, H., Bergère, R., Leprêtre, A., and Veyssière, A. (1971). *Nucl. Phys.* **A172**, 437.

Carr, R. W., Bentz, E. J., Wu, C.-P., and Baglin, J. E. E. (1971). Private communication.

Cole, G. W., Jr. (1970). Ph. D. Thesis, Yale Univ. (unpublished).

Cole, G. W., Jr., Firk, F. W. K., and Phillips, T. W. (1969). *Phys. Lett.* **30B**, 91.

Czirr, J. B. (1972). Univ. of California Lawrence Livermore Lab. Rep. No. UCRL-73762 (unpublished), submitted to *Nucl. Instrum. Methods*.

Danos, M. (1958). *Nucl. Phys.* **5**, 23.

Danos, M., and Fuller, E. G. (1965). *Ann. Rev. Nucl. Sci.* **15**, 29.

Danos, M., and Greiner, W. (1964). *Phys. Rev.* **134**, B284.

Dodge, W. R., Murphy, II, J. J., and Wyckoff, J. M. (1972). *Bull. Amer. Phys. Soc.* **17**, 152, 153.

Dowdy, E. J., and Krysinski, T. L. (1971). *Nucl. Phys.* **A175**, 501.

Easlea, B. R. (1962). *Phys. Lett.* **1**, 163.

Elliott, J. P., and Flowers, B. H. (1957). *Proc. Roy. Soc. London* **A242**, 57.

Ensslin, N., Bertozzi, W., Bergstrom, J., Hallowell, P., Kowalski, S., Maruyama, X., Sargent, C. P., Turchinetz, W., Williamson, C., Fivozinsky, S., Lightbody, J., and Penner, S. (1971). *Bull. Amer. Phys. Soc.* **16**, 561.

Fagg, L. W., Bendel, W. L., Jones, Jr., E. C., and Numrich, S. (1969). *Phys. Rev.* **187**, 1378.

Fagg, L. W., Bendel, W. L., Numrich, S., and Chertok, B. T. (1970). *Phys. Rev. C* **1**, 1137.

Fagg, L. W., Bendel, W. L., Cohen, L., Jones, Jr., E. C., Kaiser, H. F., and Uberall, H. (1971). *Phys. Rev. C* **4**, 2089.

Fagg, L. W., Bendel, W. L., Jones, Jr., E. C., Cohen, L., and Kaiser, H. F. (1972). *Phys. Rev. C* **5**, 120.

Fallieros, S., and Goulard, B. (1970). *Nucl. Phys.* **A147**, 593.

Fallieros, S., Goulard, B., and Venter, R. H. (1965). *Phys. Lett.* **19**, 398.

Farrell, J. A., Kyker, Jr., G. C., Bilpuch, E. G., and Newson, H. W. (1965). *Phys. Lett.* **17**, 286.

Feshbach, H., and Lomon, E. (1968). *Ann. Phys. (N.Y.)* **48**, 94.

Firk, F. W. K. (1964). *Nucl. Instrum. Methods* **28**, 205.

Firk, F. W. K. (1966). *Nucl. Instrum. Methods* **43**, 312.

Firk, F. W. K. (1970). *Ann. Rev. Nucl. Sci.* **20**, 39.

Frederick, D. E., Stewart, R. J. J., and Morrison, R. C. (1969). *Phys. Rev.* **186**, 992.

Fuller, E. G., Petree, B., and Weiss, M. S. (1958). *Phys. Rev.* **112**, 554.

Fultz, S. C., Bramblett, R. L., Caldwell, J. T., and Kerr, N. A. (1962). *Phys. Rev.* **127**, 1273.

Fultz, S. C., Caldwell, J. T., Berman, B. L., Bramblett, R. L., and Harvey, R. R. (1966). *Phys. Rev.* **143**, 790.

Fultz, S. C., Berman, B. L., Caldwell, J. T., Bramblett, R. L., and Kelly, M. A. (1969). *Phys. Rev.* **186**, 1255.

Fultz, S. C., Whitten, C. L., and Gallagher, W. J. (1971a). *IEEE Trans. Nucl. Sci.* **NS-18**, 533.

Fultz, S. C., Alvarez, R. A., Berman, B. L., Kelly, M. A., Lasher, D. R., Phillips, T. W., and McElhinney, J. (1971b). *Phys. Rev. C* **4**, 149.

Glavish, H. F., Hanna, S. S., Avida, R., Boyd, R. N., Chang, C. C., and Diener, E. (1972). *Phys. Rev. Lett.* **28**, 766.

Green, L., and Donahue, D. J. (1964). *Phys. Rev.* **135**, B701.

Goldhaber, M., and Teller, E. (1948). *Phys. Rev.* **74**, 1046.

Haacke, L. C., Thomas, B. J., and McNeill, K. G. (1971). *Bull. Amer. Phys. Soc.* **16**, 651.

Hallowell, P., Bertozzi, W., Bergstrom, J. C., Kowalski, S., Maruyama, X., Sargent, C.P., Turchinetz, W., Williamson, C., Fivozinsky, S. P., Lightbody, Jr., J. W., and Penner, S. (1970). *Bull. Amer. Phys. Soc.* **15**, 501.

Harvey, R. R., Caldwell, J. T., Bramblett, R. L., and Fultz, S. C. (1964). *Phys. Rev.* **136**, B126.

Hass, M., Moreh, R., and Salzmann, D. (1971). *Phys. Lett.* **36B**, 68.

Hayward, E. (1965). *In* "Nuclear Structure and Electromagnetic Interactions" (N. MacDonald, ed.), pp. 141–209. Plenum Press, New York.

Hewitt, J. S., McNeill, K. G., and Jury, J. W. (1970). *Nucl. Instrum. Methods* **80**, 77.

Horikawa, Y., Torizuka, Y., Nakada, A., Mitsunobu, S., Kojima, Y., and Kimura, M. (1971). *Phys. Lett.* **36B**, 9.

Huber, M. G. (1967). *Amer. J. Phys.* **35**, 685.

Hurst, R. R., and Donahue, D. J. (1967). *Nucl. Phys.* **A91**, 365.

Irish, D. (1973). Ph. D. Thesis, Univ. of Toronto (unpublished).

Irish, D., Berman, B. L., Johnson, R. G., Thomas, B. J., McNeill, K. G., and Jury, J. W. (1971). *Bull. Amer. Phys. Soc.* **16**, 498.

Jackson, H. E., and Strait, E. N. (1971a). *Phys. Rev. C* **4**, 1314.

Jackson, H. E., and Strait, E. N. (1971b). *Phys. Rev. Lett.* **27**, 1654.

Jackson, H. E., and Wetzel, K. J. (1969). *Phys. Rev. Lett.* **22**, 1008.

Jackson, H. E., and Wetzel, K. J. (1972). *Phys. Rev. Lett.* **28**, 513.

Jury, J. W. (1970). Ph. D. Thesis, Univ. of Toronto (unpublished).

Kelly, F. J., and Uberall, H. (1968). *Phys. Rev.* **175**, 1235.

Kelly, M. A. (1968). Ph. D. Thesis, Univ. of California, Lawrence Radiat. Lab. Rep. No. UCRL-50421 (unpublished).

Kelly, M. A., Berman, B. L., Bramblett, R. L., and Fultz, S. C. (1969). *Phys. Rev.* **179**, 1194.

Khan, A. M., and Knowles, J. W. (1972). *Nucl. Phys.* **A179**, 333.

Kuehne, H. W., Axel, P., and Sutton, D. C. (1967). *Phys. Rev.* **163**, 1278.

Kundu, S. K., Shin, Y. M., and Wait, G. D. (1971). *Nucl. Phys.* **A171**, 384.

Lane, A. M. (1971). *Ann. Phys. (N.Y.)* **63**, 171.

Leprêtre, A., Beil, H., Bergère, R., Carlos, P., Veyssière, A., and Sugawara, M. (1971). *Nucl. Phys.* **A175**, 609.

Levinger, J. S. (1960). "Nuclear Photo-Disintegration." Oxford Univ. Press, London and New York.

Lewis, F. H., and Walecka, J. D. (1964). *Phys. Rev.* **133**, B849.

Lewis, M. B. (1973). *In Proc. Int. Conf. Photonuclear Reactions and Applications* (B. L. Berman, ed.), p. 685. Lawrence Livermore Laboratory, Univ. of California.

Lochstet, W. A., and Stephens, W. E. (1966). *Phys. Rev.* **141**, 1002.

Mafra, O. Y., Kuniyoshi, S., and Bianchini, F. G. (1970). Instituto de Energia Atomica (Sao Paulo, Brazil), Rep. No. IEA-211 (unpublished).

Manfredini, A., Muchnik, M., Fiore, L., Ramorino, C., de Carvalho, H. G., Bösch, R., and Wölfli, W. (1966). *Nuovo Cimento* **44B**, 218.

Manfredini, A., Fiore, L., Ramorino, C., de Carvalho, H. G., and Wölfli, W. (1969). *Nucl. Phys.* **A127**, 687.

Manfredini, A., Fiore, L., Ramorino, C., and Wölfli, W. (1971). *Nuovo Cimento* **4A**, 421.

Matsumoto, S., Yamashita, H., Kamae, T., and Nogami, Y. (1964). *Phys. Lett.* **12**, 49.

Matsumoto, S., Yamashita, H., Kamae, T., and Nogami, Y. (1965). *J. Phys. Soc. Japan* **20**, 1321.

Matthews, J. L., and Owens, R. O. (1971). *Nucl. Instrum. Methods* **91**, 37.

Measday, D. F., Clegg, A. B., and Fisher, P. S. (1965). *Nucl. Phys.* **61**, 269.

Medicus, H. A., Bowey, E. M., Gayther, D. B., Patrick, B. H., and Winhold, E. J. (1970). *Nucl. Phys.* **A156**, 257.

Meyerhof, W. E., Suffert, M., and Feldman, W. (1970). *Nucl. Phys.* **A148**, 211.

Mills, M., Bertozzi, W., Bergstrom, J. C., Hallowell, P., Kowalski, S., Maruyama, X. K., Sargent, C. P., Turchinetz, W., Williamson, C., Fivozinsky, S. P., and Penner, S. (1970). *Bull. Amer. Phys. Soc.* **15**, 501.

Mizumoto, M., Nakajima, Y., Bergère, R., and Fuketa, T. (1971). Japan At. Energy Res. Inst. Rep. No. JAERI-M 4520 (unpublished).

Moreh, R., and Nof, A. (1969). *Phys. Rev.* **178**, 1961.

Moreh, R., Shlomo, S., and Wolf, A. (1970). *Phys. Rev.* C **2**, 1144.

Moreh, R., Salzmann, D., and Ben-David, G. (1971). *Phys. Lett.* **34B**, 494.

Murphy, II, J. J., and Dodge, W. R. (1972). *Bull. Amer. Phys. Soc.* **17**, 153.

Murray, K. M., and Ritter, J. (1969). *Phys. Rev.* **182**, 1097.

Myers, W. D., and Swiatecki, W. J. (1966). *Nucl. Phys.* **81**, 1.

Nagel, M. Z. (1970). Ph. D. Thesis, Univ. of Illinois (unpublished).

Nakada, A., Torizuka, Y., and Horikawa, Y. (1971). *Phys. Rev. Lett.* **27**, 745.

Nath, R. (1971). Ph. D. Thesis, Yale Univ. (unpublished).

Nath, R., Firk, F. W. K., Schultz, H. L., and Brooks, F. D. (1971a). Private communication.

Nath, R., Cole, Jr., G. W., Firk, F. W. K., Wu, C.-P., and Berman, B. L. (1971b). *In* "Polarization Phenomena in Nuclear Reactions" (H. H. Barschall and W. Haeberli, eds.), pp. 437–438. Univ. of Wisconsin Press, Madison, Wisconsin.

Neuhausen, R., Lightbody, J. W., Jr., Fivozinsky, S. P., and Penner, S. (1971). *Bull. Amer. Phys. Soc.* **16**, 561.

Neuhausen, R., Lightbody, J. W., Jr., Fivozinsky, S. P., and Penner, S. (1972). *Phys. Rev.* C **5**, 124.

Nielsen, L. C. (1963). Private communication from W. C. Dickinson, 1971.

O'Connell, J. S., Tipler, P. A., and Axel, P. (1962). *Phys. Rev.* **126**, 228.

O'Dell, Jr., A. A., Sandifer, C. W., Knowlen, R. B., and George, W. D. (1968). *Nucl. Instrum. Methods* **61**, 340.

Okamoto, K. (1956). *Progr. Theoret. Phys.* **15**, 75.

Owens, R. O., and Matthews, J. L. (1971). Private communication.

Partovi, F. (1964). *Ann. Phys. (N.Y.)* **27**, 79.

Patrick, B. H. (1971). Private communication.

Patrick, B. H., and Bowey, E. M. (1971). Private communication.

Paul, P., Amann, J. F., and Snover, K. A. (1971). *Phys. Rev. Lett.* **27**, 1013.

Perry, J. E., Jr., and Bame, S. J., Jr. (1955). *Phys. Rev.* **99**, 1368.

Pitthan, R., and Walcher, Th. (1971). *Phys. Lett.* **36B**, 563.

Rabotnov, N. S., Smirenkin, G. N., Soldatov, A. S., Usachev, L. N., Kapitza, S. P., and Tsipenyuk, Yu. M. (1970). *Yad. Fiz.* **11**, 508 [*English transl.*: *Sov. J. Nucl. Phys.* **11**, 285].

Riess, F., O'Connell, W. J., and Paul, P. (1971). *Nucl. Phys.* **A175**, 462.

Rowe, D. J., and Wong, S. S. M. (1969). *Phys. Lett.* **30B**, 147.

Saruis, A. M., and Marangoni, M. (1969). *Nucl. Phys.* **A132**, 433.

Schuhl, C., and Tzara, C. (1961). *Nucl. Instrum. Methods* **10**, 217.

Shakin, C. M., and Wang, W. L. (1971). *Phys. Rev. Lett.* **26**, 902.

Shay, H. D. (1972). Ph. D. Thesis, Yale Univ. (unpublished).

Sherman, N. K., Lokan, K. H., Gellie, R. W., and Johnson, R. G. (1970). *Phys. Rev. Lett.* **25**, 114.

Shikazono, N., and Kawarasaki, Y. (1972). *Nucl. Phys.* **A188**, 461.

Shoda, K., and Sugawara, M. (1971). Private communication.

Shoda, K., and Ui, H. (eds.) (1972). *Proc. Int. Conf. Nuclear Structure Studies Using Electron Scattering and Photoreaction.* Sendai, Tohoku.

Shoda, K., Sugawara, M., Saito, T., and Miyase, H. (1968). *Phys. Lett.* **28B**, 30.

Shoda, K., Sugawara, M., Saito, T., and Miyase, H. (1969). *Phys. Rev. Lett.* **23**, 800.

Shoda, K., Suzuki, A., Sugawara, M., Saito, T., Miyase, H., and Oikawa, S. (1971a). *Phys. Rev. C* **3**, 1999.

Shoda, K., Suzuki, A., Sugawara, M., Saito, T., Miyase, H., Oikawa, S., and Sung, B. N. (1971b). *Phys. Rev. C* **3**, 2006.

Skopik, D. M. (1970). Ph. D. Thesis, American Univ. (unpublished).

Spicer, B. M. (1969). *Advan. Nucl. Phys.* **1**, 1–78.

Steinwedel, H. and Jensen, J. H. D. (1950). *Z. Naturforsch.* **5a**, 413.

Swann, C. P. (1971a). *Nucl. Phys.* **A172**, 569.

Swann, C. P. (1971b). *Bull. Amer. Phys. Soc.* **16**, 651.

Titze, O.. Spamer, E., and Goldmann, A. (1967). *Phys. Lett.* **24B**, 169.

Titze, O., Goldmann, A., and Spamer, E. (1970). *Phys. Lett.* **31B**, 565.

Torizuka, Y., Kojima, Y., Saito, T., Itoh, K., Nakada, A., Mitsunobu, S., Nagao, M., Hosoyama, K., Fukuda, S., and Miura, H. (1973). *In Proc. Int. Conf. Photonuclear Reactions and Applications* (B. L. Berman, ed.) p. 675. Lawrence Livermore Laboratory, Univ. of California.

Van Hemert, R. L. (1968). Ph. D. Thesis, Univ. of California, Lawrence Radiat. Lab. Rep. No. UCRL-50501 (unpublished).

Van Hemert, R. L., Bowman, C. D., Baglan, R. J., and Berman, B. L. (1970). *Nucl. Instrum. Methods* **89**, 263.

Veyssière, A., Beil, H., Bergère, R., Carlos, P., and Leprêtre, A. (1970). *Nucl. Phys.* **A159**, 561.

Wait, G. D., Kundu, S. K., Shin, Y. M., and Stubbins, W. F. (1970). *Phys. Lett.* **33B**, 163.

Walcher, Th. (1970). *Phys. Lett.* **31B**, 442.

Weiss, M. S. (1970). Private communication.

Weissman, B., and Schultz, H. L. (1971). *Nucl. Phys.* **A174**, 129.

Welsh, R. E., and Donahue, D. J. (1961). *Phys. Rev.* **121**, 880.

Winhold, E. J., Bowey, E. M., Gayther, D. B., and Patrick, B. H. (1970). *Phys. Lett.* **32B**, 607.

Winhold, E. J., Bowey, E. M., Patrick, B. H., and Reid, J. M. (1971). *Bull. Amer. Phys. Soc.* **16**, 1417.

Wong, S. S. M., and Rowe, D. J. (1969). *Phys. Lett.* **30B**, 150.

Wu, C.-P., Firk, F. W. K., and Berman, B. L. (1970a). *Nucl. Instrum. Methods* **79**, 346.

Wu, C.-P., Firk, F. W. K., and Phillips, T. W. (1970b). *Nucl. Phys.* **A147**, 19.

Wu, C.-P., Firk, F. W. K., and Berman, B. L. (1970c). *Phys. Lett.* **32B**, 675.

Yamaguchi, A., Terasawa, T., Nakahara, K., and Torizuka, Y. (1971). *Phys. Rev. C* **3**, 1750.

VIII.B NUCLEAR SPECTROSCOPY FROM DELAYED PARTICLE EMISSION

J. C. Hardy

ATOMIC ENERGY OF CANADA LIMITED
PHYSICS DIVISION, CHALK RIVER NUCLEAR LABORATORIES
CHALK RIVER, ONTARIO, CANADA

I. Introduction

Although they did not realize it at the time, the discoverers of β-delayed particle emission were Rutherford and Wood (1916). In examining the scintillations caused by a strong thorium source, they detected, in addition to the familiar groups of α particles from thorium C, small numbers of high-energy α's which were noted to be "of greater velocity than any previously known." This was of no little significance at the time since it marked the first departure from the early belief that each nucleus which α-decays emits particles of one definite energy only. Nevertheless, the experimenters managed to restrain their enthusiasm and introduced a report of the work by stating, "There are still a number of points that require further examination, but as neither of the authors is likely to have time to continue the experiments in the near future, it has been thought desirable to give a brief account of the preliminary results."

The experiments were, in fact, continued and improved; and not only were the puzzling results confirmed, but Rutherford (1919) found high-energy alphas accompanying the decay of radium C as well. More than ten years passed before the true explanation of these radiations was understood. Thorium C (^{212}Bi) can β-decay to thorium C' (^{212}Po); if the latter nucleus "is left in an excited state ... either the α-particle will cross the potential barrier surrounding the nucleus and will fly away with the total energy of the excited level (long range α-particle), or it will fall down to the lowest level, emitting ... γ rays, and will later fly away as an ordinary α-particle" (Gamow, 1930). The same explanation held for radium C (^{214}Bi), and the first two β-delayed-alpha precursors had been identified.

For the next 30 years after that explanation was given, the number of known β-delayed-particle emitters and their importance to nuclear spectroscopy grew but erratically. In 1937, the first artificially produced precursor of β-delayed alphas, ^{8}Li, was identified (Lewis et al., 1937), and in 1939, delayed neutrons were first detected following neutron bombardment of uranium[†] (Roberts et al., 1939); but more than 20 years passed before a β-delayed-proton precursor was observed and identified (Karnaukhov et al., 1963; Barton et al., 1963). However, in the last ten years, the number of known precursors of all three types has increased enormously from fewer than 20 to nearly 75 nuclei. Predictions to be discussed later indicate that this number could eventually be increased to more than 1000. From a rare mode of decay for a few nuclei, β-delayed particle emission has become one of the important classes of nuclear radioactivity.

[†] Actually, β-delayed neutrons may have been seen earlier from a radium C source (Pollard et al., 1938). Radium C″ (^{210}Tl) has a 0.02 % β branch to a neutron-unstable state in ^{210}Pb, but whether this branch was really observed in 1938 is impossible to determine.

Its spectroscopic significance has grown apace. The emission of β-delayed particles can be used simply to identify and measure the half-life of precursors that are far from the valley of β stability or it can provide a means for measuring the precursor's mass. In heavy nuclei, it yields data on the average properties of level widths and densities as well as β-decay strengths to regions of high excitation; while among lighter nuclei, the analysis of delayed-particle spectra can be the most powerful single method for studying the details of complex β decays, from branching ratios to beta–neutrino angular correlations. Delayed neutron emission even has practical importance in the control of nuclear reactors.

In this chapter, the phenomenon of β-delayed particle decay will be described. The physical principles governing its occurrence will be generalized and the specific information that has been derived from it will be discussed in some detail. The related phenomenon of self-delayed particle decay—or particle radioactivity—will also be mentioned, but for the extensive subject of α decay, the reader will be referred elsewhere. Generally speaking, the discussions here will be restricted to nuclei with $A \lesssim 200$, and where a survey of the literature is indicated, it will include work reported before March 1972.

II. The Phenomena

If a nucleus can β-decay to energy levels that are unbound to nucleon emission, its radioactivity will be characterized by the presence of energetic nucleons with the same half-life as the initial β decay. This phenomenon is called β-delayed particle decay. In principle, it should be possible to observe nuclei that decay by the emission of β-delayed neutrons, protons, tritons, ^{3}He, alpha particles and fission fragments, although only neutron, proton, and α emitters have actually been identified so far.[†] As an illustration, the symbolic decay scheme for a β-delayed-proton precursor is shown in Fig. 1; similar schemes apply to any other β-delayed particle emission.

Evidently the particle energy spectrum is determined by two factors: (i) the intensity of β-decay branches from the "precursor" to energy levels in the "emitter" and (ii) the branching ratios for subsequent particle emission to states in the "daughter." Frequently there are only a few of these states available and the particle spectrum corresponding to the population of an individual final state can be obtained experimentally. With this information,

---

[†] Very recently, ^{228}Np, ^{232}Am, and ^{234}Am have been tentatively identified by Skobelev (1972) as β-delayed fission precursors.

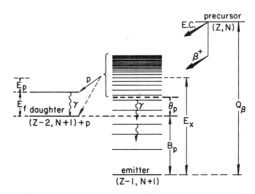

Fig. 1. Decay scheme for a typical delayed-proton precursor, illustrating some symbols and terms used in the text.

it is possible to reconstruct in detail the β decay of the precursor by determining from the energies and relative intensities of peaks in the particle spectrum which states in the emitter are populated by β decay and with what intensities. From additional experimental information, the decay energy can sometimes be determined, and often it is possible to establish the spins, parities, and isospin structure of the states involved. Generally this is information which cannot be determined in any other way.

The observed properties of β-delayed particle decay, and the information learned from them, differ significantly from the other mode of particle decay —self-delayed or direct particle emission. Here the radioactive nuclei, unlike the precursors already described, are themselves energetically unbound to the emission of particles, and all that prolongs their existence is the presence of the centrifugal and/or Coulomb barrier. The probability for penetrating this barrier depends strongly on the energy available and it is this which largely determines the nuclear lifetime as well as the relative intensities of groups in the energy spectrum of emitted particles. As a result, relatively few states at low excitation in the daughter are populated detectably and, although this permits the measurement of relative masses, it does restrict spectroscopic studies. Furthermore, since the lifetime of a self-delayed-particle emitter decreases rapidly with increasing decay energy, a reasonably long experimentally accessible lifetime will usually be associated with lower energy particles than those characteristic of a β-delayed emitter in the same region of nuclei. This presents acute experimental problems in observing the ground-state proton decay of light nuclei where particle energies of at most a few hundred kilovolts are expected; proton energies up to 11 MeV have been observed in the same mass region following β decay. No such difficulties have been encountered, of course, for the α decay of heavy nuclei, where the barrier is much higher.

A. ENERGY REQUIREMENTS FOR BETA-DELAYED PARTICLE DECAY

Many of the symbols that will be used here are defined in Fig. 1, where the subscript p has been used to specify proton decay; alternative subscripts n and α will be used where necessary to denote decay involving neutrons or alphas, while μ will describe the emission of any allowed but unspecified particle.

The minimum requirement for the existence of β-delayed particle emission is that the β-decay energy of the precursor must exceed the particle separation energy of the emitter, i.e.,

$$Q_\beta > B_\mu \tag{1}$$

This condition is not sufficient, though, to ensure that the particle-decay branch will be strong enough to be observed. Even those states above energy B_μ that are populated in the emitter nucleus can de-excite both by γ and particle decay. The probability that a level will decay by either channel is expressed in terms of its partial width, Γ_γ or Γ_μ, respectively; since no other channels are allowed, these partial widths add directly to comprise the total width Γ of each level. To know the conditions required for observable particle emission, it will be necessary to examine the average behavior of Γ_μ/Γ as a function of excitation energy.

For the moment, let us consider only particle decay to the ground state of the daughter nucleus. Following the treatment of Blatt and Weisskopf (1952), the partial width for this channel is given by

$$\Gamma_\mu = T(\mu)/2\pi\varrho(E_x) \tag{2}$$

where $T(\mu)$ is the transmission coefficient that describes the probability of decay through the specific channel μ, and $\varrho(E_x)$ is the density of states in the emitter at excitation energy E_x. It follows that if only one particle channel is open, the probability for any level to particle decay can be written as

$$\Gamma_\mu/\Gamma = T(\mu)/[T(\mu) + 2\pi\varrho(E_x)\,\Gamma_\gamma] \tag{3}$$

The transmission coefficient $T(\mu)$ must be zero for $E_x < B_\mu$; at some energy above B_μ it increases, first rapidly, then more slowly, until it reaches unity above the barrier. [For optical model calculations of $T(\mu)$, see Mani et al.,

1963.] On the other hand, both ϱ and Γ_γ are known (Cameron, 1956) to show a steady increase with E_x but the product $2\pi\varrho\Gamma_\gamma$ rarely exceeds unity over the region of interest here. Thus Γ_μ/Γ is zero below B_μ and remains small at energies immediately above that, then increases in the space of a few hundred keV to nearly its maximum value, rises more slowly, and finally decreases again at much higher excitations. The energy above B_μ at which the effective threshold for particle decay occurs will be called θ_μ. Obviously θ_μ depends on the charge and mass of the particle being emitted as well as on its allowed angular momentum, but it can be determined quite precisely for each individual precursor.

It is of value to obtain a more general, if approximate, prescription for θ_μ throughout the periodic table. To do this, an alternative definition of the particle width is used:

$$\Gamma_\mu = 2P\gamma_\mu{}^2 \tag{4}$$

Here P is the barrier penetrability and $\gamma_\mu{}^2$ is called the reduced width. In contrast with Eq. (2), the internal nuclear structure effects are all contained in one term, $\gamma_\mu{}^2$, and are separated from the external kinematic effects. The penetrability P has similar behavior to $T(\mu)$ but is readily calculable and independent of nuclear properties. It does depend, though, on the orbital angular momentum l_μ of the emitted particle, but in our case only small l values will usually be involved since the emitter state must be of rather low spin, having been populated by β decay from the precursor ground state. As a simple prescription, θ_μ has been calculated to be the energy at which $P(l=0) = 10^{-4}$ since this corresponds roughly to $\Gamma_\mu \sim \Gamma_\gamma$, and is at a point where Γ_μ is rapidly increasing. The values of θ_μ are shown in Fig. 2 for protons and α's as a function of Z; for neutrons, $\theta_n = 0$.

Although the probability for particle decay increases with increasing excitation energy of the emitter, the rate at which states are populated by β decay decreases, and the total intensity of particle emission is governed by the interplay of these two opposing factors. The intensity i_β of an allowed β transition to any individual level (providing it is not the analog of the precursor ground state) may be written as

$$i_\beta \propto f\langle\sigma\rangle^2 \tag{5}$$

where f is the statistical rate function, which takes account of the dependence on total energy release and nuclear charge; the symbol $\langle\sigma\rangle$ stands for the Gamow–Teller matrix element and depends upon the wave functions of the initial and final nuclear states. For the moment, let us not consider

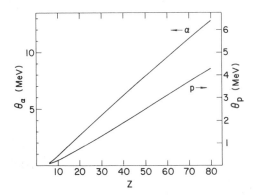

Fig. 2. The approximate threshold energy for particle emission θ_μ plotted as a function of Z of the precursor.

individual states in the emitter but rather the gross features of β decay per unit energy interval. If $I_\beta(E_x)$ is the β-transition intensity to the region of excitation E_x, then it can be expressed analogously to Eq. (5):

$$I_\beta(E_x) \propto f(Z, Q_\beta - E_x) S_\beta(E_x) \qquad (6)$$

The factor $S_\beta(E_x)$ has units MeV^{-1} and is called the β strength function (see, for example, Duke *et al.*, 1970); it is simply the average beta matrix element squared (allowed and forbidden) per unit excitation energy near E_x in the emitter. As will be discussed later, its value is expected to peak sharply near the analog state but otherwise should vary quite slowly with energy. Thus, except for the lightest nuclei, the behavior of I_β over the energy range of particle emission may be approximated by that of the familiar statistical rate function.

The results of these considerations are illustrated in Fig. 3. The calculated gross proton spectrum is shown for the decay of the precursor ^{117}Xe, but for simplicity only $l_p = 0$ proton emission to the daughter ground state has been included in evaluating Γ_p/Γ. It is typical of β-delayed-proton spectra observed experimentally from precursors with $A > 70$. For lighter emitters, the level density decreases and individual rather than average features become salient; the general outline, however, changes little. Except for differences in the behavior of Γ_μ/Γ at low energies, more or less the same holds true for β-delayed neutron and α emission as well.

It is evident from Fig. 3 that corresponding to θ_μ, the effective threshold for particle emission, there is a β energy θ_β below which only some small

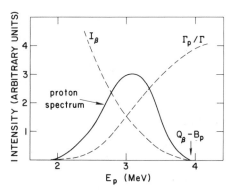

Fig. 3. The partial proton spectrum corresponding to $l_p = 0$ emission calculated for ^{117}Xe assuming $B_p = 2.77$ MeV, $Q_{\beta^+} - B_p = 3.94$ MeV.

proportion (say, one in 10^4) of the precursor decays proceed. As such, the energy θ_β is defined by

$$\int_{Q_\beta - \theta_\beta}^{Q_\beta} I_\beta(E_x)\, dE_x \bigg/ \int_0^{Q_\beta} I_\beta(E_x)\, dE_x = 10^{-4} \qquad (7)$$

Its value for positron decay and electron capture varies from more than 2 MeV for light nuclei (and $Q_\beta \sim 10$ MeV) to a few hundred keV for $Z > 50$, but for electron decay, it remains at about 1 MeV. Thus, the condition that must be met in order that particle emission occur more than once for roughly 10^4 precursor disintegrations is more stringent than that laid down in Eq. (1); it is

$$Q_\beta - \theta_\beta > B_\mu + \theta_\mu \qquad (8)$$

Of course, particle emission can be and has been observed in cases where it occurs as an even weaker branch.

B.   THE OCCURRENCE OF BETA-DELAYED PARTICLE DECAY

With the availability of good semiempirical calculations for nuclear masses, it is possible to use them, together with Eq. (8), to predict the occurrence of β-delayed particle decay. Results are shown for β^+ decay in Fig. 4 and for β^- decay in Fig. 5. Nuclei in the unshaded region between the solid and dashed lines in each figure should decay predominantly by β emission (or electron capture) followed by particle emission with an intensity greater than $\sim 10^{-4}$. The shaded areas in Fig. 4 mark the occurrence of self-delayed alpha decay, which will there compete with β-delayed proton decay, while at and beyond the dashed line in the same figure, proton and double-proton

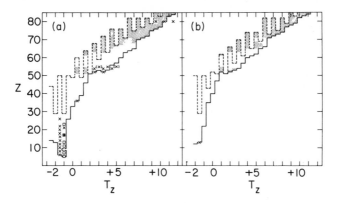

Fig. 4.   Calculated limits for β^+-delayed particle emission: (a) odd N, (b) even N. For a given T_Z, the solid line delineates the lightest precursor for which Eq. (8) is satisfied (i.e., delayed particle emission should occur more than once for every 10^4 disintegrations); the dashed line gives the heaviest nucleus which should be bound in its ground state against self-delayed emission of protons. The shaded regions denote the occurrence of competition with self-delayed α emission. Where masses are unknown, predictions were taken from: Kelson and Garvey (1966), $Z \leqslant 22$; Jänecke (1968), $23 \leqslant Z \leqslant 28$; Zeldes *et al.* (1967), $29 \leqslant Z \leqslant 50$; Myers and Swiatecki (1965), $Z \geqslant 51$. The $\times$ marks the location of observed β^+-delayed-proton emitters, $\square$ marks β^+-delayed-α emitters. For all precursors with $T_Z \geqslant -\frac{1}{2}$, proton emission is expected to predominate.

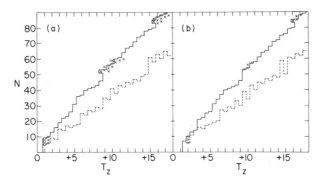

Fig. 5.   Calculated limits for β^--delayed particle emission: (a) odd Z, (b) even Z. For a given T_Z ($\geqslant \frac{1}{2}$), the solid line indicates the heaviest precursor for which delayed neutron emission should occur more than once for every 10^4 disintegrations; the dashed line delineates the neutron drip-line. Where masses are unknown, predictions were taken from Garvey *et al.* (1969). (For $Z \gtrsim 20$, Myers and Swiatecki's (1965) predictions give very nearly the same results.) The symbol $\times$ marks the location of observed β^--delayed-neutron emitters, $\square$ marks β^--delayed-α emitters.

TABLE 1 Observed β^+-Delayed-Proton Precursors with $T_z = -\frac{3}{2}$

Precursor	$t_{1/2}$ (msec)	$Q_{\beta^+} - B_p{}^a$ (MeV)	Proton branch (per disintegration)	Major peaks (relative intensities) (laboratory energy, E_L, in MeV)	References
N odd					
$^{9}_{6}C_{3}$	126.5	16.68	1	8.24(100), 10.92(75); rising continuum at lower energies	Esterl et al. (1972)
$^{13}_{8}O_{5}$	8.9	15.82	1.2×10^{-1}	1.44(100), 6.44(4), 7.00(1)	Esterl et al. (1970)
$^{17}_{10}Ne_{7}$	108.5	13.93	9.9×10^{-1}	1.68(10), 3.77(30), 4.59(100), 5.12(20), 7.04(8)	Hardy et al. (1971)
$^{21}_{12}Mg_{9}$	121	10.66	$\sim 2 \times 10^{-1}$	0.89(17), 1.26(23), 1.77(51), 1.94(100), 2.04(8), 2.49(8), 3.87(10), 4.67(15), 6.23(6)	Hardy and Bell (1965), Sextro et al. (1972)
$^{25}_{14}Si_{11}$	220	10.47	$\sim 3.6 \times 10^{-1}$	0.93(23), 1.87(40), 2.09(24), 2.22(25), 3.33(39), 4.08(100), 4.63(17), 5.40(20)	Reeder et al. (1966)
$^{29}_{16}S_{13}$	189	11.07	$\sim 3.9 \times 10^{-1}$	2.14(65), 2.46(8)^b, 3.44(5), 3.72(16), 5.44(100)	Hardy and Verrall (1964), Verrall (1968)^b
$^{33}_{18}Ar_{15}$	174	9.33	3.4×10^{-1}	1.65(2), 1.79(2), 2.10(9), 3.06(2), 3.17(100), 3.85(2)	Hardy et al. (1971)
$^{37}_{20}Ca_{17}$	175	9.71	$\sim 6 \times 10^{-1}$	1.72(8), 1.93(6), 2.21(3), 2.49(6), 2.73(7), 3.10(100), 4.01(2)	Poskanzer et al. (1966)
$^{41}_{22}Ti_{19}$	88	11.63	$\sim 9 \times 10^{-1}$	1.00(40), 1.09(23), 1.54(21), 2.27(26), 3.09(66), 3.77(26), 4.65(23), 4.75(100)	Poskanzer et al. (1966), Sextro et al. (1972)
$^{49}_{26}Fe_{23}$	75	9.87^c	—	1.92(100)	Cerny et al. (1970a)
N even					
$^{23}_{13}Al_{10}$	470	4.66	—	0.83(100)	Gough et al. (1972)

^{a} Unless otherwise noted, values are experimental measurements or the results of applying the isobaric multiplet mass equation where three members of an isospin quartet are known.

^{b} Because of discrepancies with other data at $E_L > 3$ MeV, the intensities measured by Verrall (1968) have only been used for proton groups below that energy, and have been renormalized to the higher-energy data. Nevertheless, they should be used with caution.

^{c} Calculated using mass values from Jänecke (1968).

TABLE 2

OBSERVED β^+-DELAYED CHARGED-PARTICLE PRECURSORS WITH $T_Z = -1$

Precursor	$t_{1/2}$ (msec)	Particle	$Q_{\beta^+} - B_\mu$ (MeV)	Particle branch (per disintegration)	Major peaks (relative intensities) (E_L in MeV)	References
$^{8}_{5}B_{3}$	769	α	18.07	1	1.50(100) broad, 8.36(8)	Matt et al. (1964)
		p	0.72	—	None	
$^{12}_{7}N_{5}$	10.97	α	9.98	3.5×10^{-2}	Continuum	Schwalm and Povh (1966)
		p	1.39	—	None	
$^{20}_{11}Na_{9}$	446	α	9.16	1.9×10^{-1}	2.15(100), 2.48(5), 3.80(2), 4.44(17), 4.90(1)	Polichar et al. (1967)
		p	1.04	—	None	
$^{24}_{13}Al_{11}$	2066	α	4.57	7.7×10^{-5}	1.57(49), 1.98(100), 2.27(1), 2.33(4), 2.36(3)	Torgerson et al. (1971)
		p	2.19	—	None observed	
$^{24}_{13}Al_{11}{}^{m}$	130	α	5.01	$\lesssim 1 \times 10^{-2}$	1.40(100), 1.77(58), 1.83(12)	Torgerson et al. (1971)
		p	2.63	—	None observed	
$^{32}_{17}Cl_{15}$	298	α	5.74	$\sim 5 \times 10^{-4}$	1.61(65), 2.20(100)	Steigerwalt et al. (1969)
		p	3.83		0.76(32), 1.02(52), 1.35(29)	
$^{40}_{21}Sc_{19}$	182.4	α	7.29	—	None observed	Verrall and Bell (1969)
		p	6.00	$\sim 5 \times 10^{-3}$	1.05(100), 1.24(10), 1.44(3), 1.83(4), 2.10(7), 2.38(5)	
$^{44}_{23}V_{21}$	90	α	8.59^a	—	2.78(100)	Cerny et al. (1971)
		p	5.06^a	—	None observed	

[a] Calculated using mass values from Jänecke (1968).

TABLE 3

OBSERVED β^+-DELAYED-PROTON PRECURSORS WITH $A > 70$

Precursor	$t_{1/2}$ (sec)	$Q_{\beta^+} - B_p$ (MeV)	Proton branch (per disintegration)	Major peaks (relative intensities) (E_L in MeV)	References
$^{73}_{36}Kr_{37}$	34	4.85	6.8×10^{-3}	Many between 1.5 and 3 MeV	Hornshøj et al. (1972c)
$^{109}_{52}Te_{57}$	4.4	7.14	—	2.5(50), 2.8(40), 3.1(20), 3.3(100), 3.7(80)	Siivola (1965)[a] Karnaukhov et al. (1970)
$^{111}_{52}Te_{59}$	19.3	5.07	—	2.42(60), 2.66(70), 2.82(100), 3.25(60), 3.48(40)	Karnaukhov and Ter-Akopian (1967)
$^{113}_{54}Xe_{59}$	~3	8.54[b]	—	—	Hornshøj et al. (1971b)
$^{115}_{54}Xe_{61}$	18.4	6.20	3.4×10^{-3}	Broad hump centered at ~3.1 MeV; some fine structure (line width 15 keV)	Hornshøj et al. (1971b)
$^{117}_{54}Xe_{63}$	65	4.10	2.9×10^{-5}	Broad hump centered at ~2.8 MeV	Hornshøj et al. (1971a)
$^{116}_{55}Cs_{61}$ [c]	16	7.62[b]	—	—	Ravn et al. (1972)
$^{118}_{55}Cs_{63}$		5.40[b]	—	Broad hump centered at ~3.1 MeV	Ravn et al. (1972)
$^{120}_{55}Cs_{65}$ [c]	60.2	3.26[b]	$\leq 10^{-7}$	—	Ravn et al. (1972)
$^{119}_{56}Ba_{63}$	5.0	6.90[b]	—	Broad hump centered at ~3.1 MeV	Karnaukhov (1972)
$^{179}_{80}Hg_{99}$	1.09	7.45[b]	$\sim 2.8 \times 10^{-3}$	—	Hornshøj et al. (1971a)
$^{181}_{80}Hg_{101}$	3.6	5.99[b]	1.8×10^{-4}	Broad hump centered at ~4.5 MeV	Hornshøj et al. (1971a)
$^{183}_{80}Hg_{103}$	8.8	4.54[b]	3.1×10^{-6}	Broad hump centered at ~4.0 MeV	Hornshøj et al. (1971a)

[a] Activity attributed, apparently incorrectly, to ^{108}Te.

[b] Calculated using mass predictions of Myers and Swiatecki (1965).

[c] Identification is tentative.

TABLE 4

OBSERVED β^--DELAYED ALPHA PRECURSORS

Precursor	$t_{1/2}$	$Q_{\beta^-} - B_\alpha$ (MeV)	Alpha branch (per disintegration)	Major peaks (relative intensities) (E_L in MeV)	References
$^{8}_{3}\mathrm{Li}_5$	844 msec	16.10	1	Broad hump centered at ~ 1.5 MeV	Alburger et al. (1963)
$^{11}_{4}\mathrm{Be}_7$	13.78 sec	2.84	3.0×10^{-2}	Continuum up to 1 MeV	Alburger and Wilkinson (1971)
$^{12}_{5}\mathrm{B}_7$	20.41 msec	6.00	1.6×10^{-2}	Continuum, rising toward lower energy	Schwalm and Povh (1966)
$^{16}_{7}\mathrm{N}_9$	7.11 sec	3.26	1.2×10^{-5}	1.84(100)	Hättig et al. (1969)
$^{212}_{83}\mathrm{Bi}_{129}$	60.6 min	11.20	1.4×10^{-4a}	9.50(21), 10.43(12), 10.55(100)[b]	Rytz (1953)
$^{214}_{83}\mathrm{Bi}_{131}$	19.9 min	11.11	3.0×10^{-5a}	9.08(100), 9.50(5), 9.80(5), 10.08(7), 10.33(5)[b]	Leang (1965)

[a] All β branches result eventually in α's; the intensity shown is for β-delayed α's from excited states in the emitter.

[b] Only β-delayed α's are shown; there are many other stronger α peaks in the spectrum.

TABLE 5

OBSERVED SPECTROSCOPICALLY USEFUL β^--DELAYED-NEUTRON PRECURSORS

Precursor	$t_{1/2}$ (msec)	$Q_{\beta^-} - B_n$ (MeV)	Neutron branch (per disintegration)	Major peaks (relative intensities) (E_L in MeV)	References
${}^{9}_{3}\text{Li}_{6}$	172	11.95	7.5×10^{-1}	0.30(100), 0.65(6), 1.00(9) (also a broad α peak at ~ 0.65 MeV)	Chen et al. (1970)
${}^{13}_{5}\text{B}_{8}$	17.3	8.49	2.5×10^{-3}	2.40(59), 3.61(100)	Jones et al. (1969)
${}^{17}_{7}\text{N}_{10}$	4140	4.54	9×10^{-1}	0.39(80), 1.16(100), 1.69(12)	Gilat et al. (1963)
${}^{87}_{35}\text{Br}_{52}$	55,700	1.51[a]	2.6×10^{-2}	0.4, rising continuum toward lower energy	Batchelor and Hyder (1956)
${}^{137}_{53}\text{I}_{84}$	24,400	2.54[a]	3×10^{-2}	0.27(50), 0.38(100), 0.48(55), 0.57(45), 0.76(47), 0.86(40)	Shalev and Rudstam (1972)

[a] Calculated using mass predictions of Myers and Swiatecki (1965).

radioactivity will occur. Even excepting these regions, there is obviously an enormous number of nuclei with β-delayed particle emission contributing significantly to their radioactivity.

All observed β^+-delayed-particle emitters are listed together with a brief outline of their properties in Tables 1–3. A more complete outline and reference list can be found elsewhere (Hardy, 1972). Similarly, all known β^--delayed-α emitters appear in Table 4. Tabulations of all delayed-neutron precursors so far observed have appeared in the literature (del Marmol, 1969; and additions from Talbert et al., 1969 and Talbert, 1970) so only those for which energy spectra have been measured are listed in Table 5. In all tables, the references were chosen to illustrate best the particle energy spectra. The known precursors of all categories have also been indicated in Figs. 4 and 5, where they can be compared with the predictions already described. Without exception, the only precursors known to lie outside the predicted region of nuclei have particle-branch intensities of the order of or less than the chosen limit of 10^{-4}.

C. Specific Properties Governing the Emission of Charged Particles

Particularly for many light-mass precursors, both β-delayed α and proton emission are allowed by the energy conditions set in Eq. (8), yet, except for the decay of ^{32}Cl, only one type of particle has been observed in each case. The reasons for this will best be understood by considering the precursors with $T_z = \frac{1}{2}(N - Z) = -1$, specifically ^{32}Cl and ^{40}Sc. The spin and parity of ^{32}Cl are given as 1^+, so its allowed β decay will populate 0^+, 1^+, and 2^+ states in ^{32}S which can emit protons to the ground state of ^{31}P$(\frac{1}{2}^+)$ or alpha particles to ^{28}Si(0^+). The penetrabilities for protons and alphas with $l = 0$ and 2 are shown in Fig. 6 as a function of excitation energy in ^{32}S. Assuming the reduced widths are approximately the same for both channels, several conclusions can be drawn immediately: 2^+ states will emit alphas; 1^+ states must emit protons (or γ's); and 0^+ states above ~ 10 MeV will emit protons while those below this will emit α's. The energies of observed states and whether they decay by protons or α's are indicated in the figure; it is not difficult to see why both decay modes can occur.

Contrast this case with that of ^{40}Sc, which has spin 4^-. The 3^-, 4^-, and 5^- states it populates in ^{40}Ca can proton-decay to ^{39}K$(\frac{3}{2}^+)$ or α-decay to ^{36}Ar(0^+). The relevant penetrabilities for both channels, which are also plotted in Fig. 6, show a clear predominance for proton emission at all energies, and this is borne out by experiment.

The crucial feature demonstrated by these examples is that penetrability

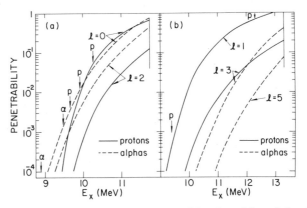

Fig. 6. Calculated penetrabilities P relevant to delayed particle emission following the β^+ decay of (a) ^{32}Cl and (b) ^{40}Sc plotted as a function of excitation energy in the emitter. Arrows in (a) mark individual levels whose decays have been observed; in (b), many levels have been seen in the energy range bracketed by the arrows. The particle emitted is also indicated.

curves for both protons and α's vary rapidly and are nearly parallel to one another over the whole energy range of interest. Thus, a very slight decrease—actually about 600 keV—in the separation energy for protons relative to α's has made the difference between the mixed decay of ^{32}Cl and the pure proton emission of ^{40}Sc. In fact, considering the small region of possible overlap between the two decay modes, it is not surprising that the occurrence of both for a single precursor is so rare.

We can now obtain a general prescription for light nuclei to determine which charged-particle decay mode will occur. It must ignore the specific angular momenta involved in each case and rely upon the fact that the two decay modes available to each state will not differ very greatly in their l values. The symbol ΔE_θ is defined using quantities already discussed:

$$\Delta E_\theta = [B_\alpha + \theta_\alpha] - [B_p + \theta_p] \qquad (9)$$

In terms of the graphs in Fig. 6, ΔE_θ is the energy difference between the $l = 0$ proton and alpha curves for a penetrability of 10^{-4}, but it gives the approximate difference between proton and α curves for similar l values at any penetrability value. Depending on whether ΔE_θ is positive or negative, either proton or α decay, respectively, will dominate. Naturally the approximate nature of this treatment means that if $|\Delta E_\theta| \lesssim 300$ keV, either or both decay modes can occur.

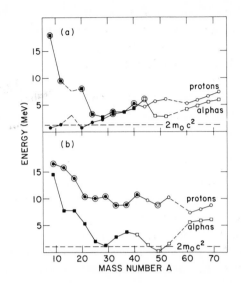

Fig. 7. The total effective decay energy $Q_\beta - B_\mu - \theta_\mu$ for β-delayed proton ($\bullet$, $\circ$) and alpha ($\blacksquare$, $\square$) emission plotted for two families of precursors: (a) $T_Z = -1$, $A = 4n$; (b) $T_Z = -\frac{3}{2}$, $A = 4n + 1$. Black symbols indicate measured values, open ones are values calculated as in Fig. 5. The large open circles mark observed decay modes.

The remarkable success of this prescription is demonstrated in Fig. 7. Here it is not ΔE_θ which has been plotted, but the quantity $Q_\beta - B_\mu - \theta_\mu$ calculated for both protons and alphas; two families of precursors are included: those with $T_Z = -1$ and $A = 4n$, and those with $T_Z = -\frac{3}{2}$ and $A = 4n + 1$. For each allowed mass A, the difference between the plotted points yields ΔE_θ, so the highest point always indicates which decay mode will predominate. Assuming the value of the point exceeds θ_β [see Eq. (8)], that mode should be readily observed. The large open circles in the figure show which modes have been observed and this agrees in all cases with expectation—a considerable success for the $T_Z = -1$ family particularly. These plots, and more like them for other precursor families, suggest that ^{36}K and probably ^{28}P will prove to be precursors of both proton and α decay but, with those exceptions, no other strong β-delayed-α precursors are anticipated.

D.   SELF-DELAYED PARTICLE EMISSION

Alpha radioactivity has a venerable history and, having been described frequently already (see, for example, Rasmussen, 1965), it will not be dealt with here. On the other hand, proton and double-proton decay have attracted attention only quite recently. Those nuclei whose ground states are expected to decay in this manner will be found at and above the dashed line in Fig. 4. Their lifetimes will decrease strongly with increasing available decay energy, and relatively few are anticipated to have half-lives greater than 1 sec. Questions of lifetimes and competition among α, proton, and double-proton decay have been considered in detail by Goldanskii (1966), to which reference the reader is referred for more details. No emitter of this type has been positively identified, although ^{121}Pr has been advanced as a promising possibility (Bogdanov et al., 1971).

The recent observation of proton emission from a high-spin isomer of ^{53}Co (Jackson et al., 1970; Cerny et al., 1970b, 1972) has prompted a theoretical investigation of other possible many-particle isomeric states (Peker et al., 1971); it appears that many such states may occur as proton emitters and since these should lie nearer the β-stable nuclei than do the ground-state emitters, they may well be more amenable to experimental study. They should have the further advantage of a relatively long half-life for particle decay. This arises for two reasons: Their high spin requires the proton to carry away much angular momentum, which increases the barrier it must penetrate; and the structure of these states has little overlap with that of the ground state of their daughters, so the reduced width is correspondingly small ($\gamma_p^2 \sim 4$ meV for ^{53}Co^m). Of course, β decay of these isomers can also occur and will dominate if the partial half-life for proton emission is too long. While the search for other proton-emitting isomers will likely be fruitful, it is always possible that ^{53}Co^m is simply a fortuitous case in which the particle decay is inhibited sufficiently that the total half-life is measurable yet not inhibited so much as to make the particle-decay branch too weak to observe.

Such many-particle isomers should also exist among neutron-rich nuclei and Peker et al. (1971) suggest that partial half-lives and decay energies comparable with ^{53}Co^m might be expected. If this is so, these states would certainly provide the most experimentally accessible self-delayed-neutron emitters. Nuclei whose ground states are predicted to undergo neutron or double-neutron decay (Berlovich, 1970) all lie near the boundary of nucleon stability—the dashed line in Fig. 5. Thus, they are more difficult to produce and because of the absence of a Coulomb barrier or large centrifugal barrier, their decays will be characterized by either a very short lifetime or very low-energy neutrons.

Self-delayed neutron and proton emission are as yet unexplored phenomena.

However, since most emitters will be difficult to study and will exhibit few decay branches, it seems unlikely that their study will be as rich in nuclear spectroscopy as the study of β-delayed emission has already proven to be.

III.   The Experiments

Since the first observation of delayed α's following the reaction $^7Li(d, p)^8Li$ (Lewis et al., 1937), all delayed-particle precursors except RaC" (^{210}Tl) have been artificially produced. Most frequently, this has involved accelerator bombardments but many delayed-neutron precursors have appeared as products of fission induced by reactor neutrons. Techniques used to observe delayed-particle emission have varied considerably but can be generalized into two categories: (i) those techniques in which detectors are used directly to view the target in which an activity has been produced; or (ii) those in which radioactive nuclei produced by bombardment are transported rapidly to a shielded counting location. While the first detection category allows very high counting efficiency, it also suffers in many cases from high background—both prompt and delayed—which can be severe enough to cause radiation damage to some types of detectors even though they may be shielded from the direct accelerator beam; in addition, the thick targets ($\gtrsim 1$ mg/cm^2) which are required to stop the nuclear recoils can cause a spread in the charged-particle energies being observed. Usually with protons or α's this limits the FWHM (full-width at half-maximum) energy resolution to ~ 100 keV.

By transporting the activity to be studied, detection techniques of the second category obviate these difficulties by providing the desired activity in isolation from others which are unwanted. This may be accomplished by introducing into the transport system chemical reagents, traps, filters, an isotope separator, or any combination of these devices. A general discussion of many such techniques has already been given by Klapisch, Chapter II.B and Macfarlane and McHarris, Chapter II.C, so in the following section, experimental spectra will be shown with only brief descriptions of the apparatus used.

A.   BETA-DELAYED ALPHA AND PROTON DECAY

There is no difference in the techniques needed to investigate β-delayed α and proton emission. Accordingly, for illustration, the production and detection of three proton precursors—^{33}Ar, ^{111}Te, and ^{181}Hg—will be described. Experimentally, all involve rapid transport systems, but quite different methods have been employed in each case. The results also serve to illustrate the spectral differences between light and heavy nuclei.

1. ^{33}Ar

The precursor ^{33}Ar has been produced and identified by the reactions ^{35}Cl(p, 3n)^{33}Ar (Hardy and Verrall, 1965) and ^{32}S(^{3}He, 2n)^{33}Ar (Poskanzer et al., 1966; Hardy et al., 1971). The proton spectrum shown in Fig. 8 is

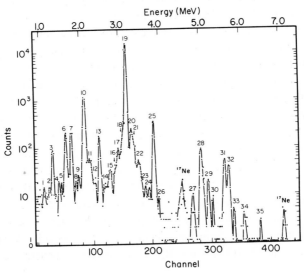

Fig. 8. Spectrum of β^{+}-delayed protons from the decay of ^{33}Ar as observed in a counter telescope with a 14-μm ΔE counter. The proton laboratory energy is indicated at the top. $E(^{3}\text{He}) = 35$ MeV (from Hardy et al., 1971).

taken from the most recent of these references, in which a CS_2 vapor target was bombarded by 35-MeV ^{3}He particles. During bombardment, the vapor was contained in a 20-cm^{3} cell which was periodically swept by helium gas so that the vapor, together with the recoil atoms produced, passed from the target cell, through $\frac{1}{4}$ in. (inside diameter) Teflon tubing to a counting chamber. En route, the vapor, gas, and various activities (including ^{33}Ar) passed through a dry-ice–trichloroethylene trap which removed most of the CS_2 and other condensible impurities. Particulate impurities were then removed by passage through a small glass-wool filter. When the remaining gas reached the counting chamber, it was valved off and counting was begun. The transit time for the 5 m from target to counting chamber was less than 100 msec.

While counting was in progress, the target chamber was replenished with vapor, the whole sequence being controlled electronically with the use of solenoid valves (Esterl et al., 1971).

Decay particles were detected and identified as protons by a cooled, silicon counter telescope mounted inside the counting chamber, with an energy resolution of less than 45 keV (FWHM) being obtained. A Ge(Li) or NaI(Tl) counter could be mounted externally to measure singles or coincident γ-ray spectra, and relative intensities of proton and γ transitions were measured (Hardy et al., 1971). Note should be taken that in the spectrum of Fig. 8 there are more than 30 resolved ^{33}Ar peaks with intensities ranging over more than four orders of magnitude.

2. ^{111}Te

Heavy-ion bombardments have been used to produce ^{111}Te from the reactions ^{92}Mo(^{22}Ne, 3n)^{111}Te, ^{94}Mo(^{20}Ne, 3n)^{111}Te, and ^{102}Pd(^{12}C, 3n) ^{111}Te (Karnaukhov and Ter-Akopian, 1967). A proton spectrum obtained following the last of these reactions is shown in Fig. 9. Unfortunately, experimental excitation functions and cross bombardments with such heavy projectiles do not necessarily lead, as they do with light projectiles, to an unambiguous identification of the nucleus produced. Although ^{111}Te was initially identified in this way, a conflicting assignment was made for the same activity when it was produced by bombarding ^{98}Ru with ^{16}O (Macfarlane, 1967). It was only by observing ^{110}Sn and ^{110}In as products of the delayed proton decay that the identification of ^{111}Te as the precursor could be confirmed (Bogdanov et al., 1968b).

The experiment that produced the spectrum of Fig. 9 employed a technique

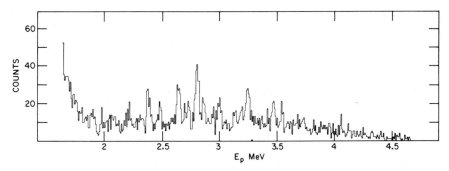

Fig. 9. Spectrum of β^+-delayed protons from the decay of ^{111}Te observed following 63-MeV ^{12}C bombardment of ^{102}Pd (from Karnaukhov and Ter-Akopian, 1967).

originally developed by Macfarlane and Griffioen (1963). Bombarded nuclei recoiled from the target into a small chamber containing atmospheric-pressure helium, where they were thermalized. Contiguous to this chamber was a larger, continuously evacuated one with a tiny nozzle connecting the two. A gas jet expanding from the nozzle contained the recoil nuclei, and these impinged on an aluminum collector, where they were absorbed. Periodically, the collector was moved to bring the active recoils near a counter telescope. Although the experimental energy resolution was comparable to that of the ^{33}Ar work previously described, the proton spectrum from ^{111}Te shows fewer resolved peaks, and these cover a much narrower range of intensities. Evidently, an increased level density for the heavier nucleus and the concomitant averaging of structural properties are beginning to have their effect.

3. ^{181}Hg

Many isotopes of mercury, including ^{181}Hg, have been formed by spallation reactions initiated with 600-MeV protons on a molten lead target (Hansen et al., 1970). The mercury products diffused out of the target and were pumped through a glass transfer line to the ion source of the Isolde electromagnetic isotope separator. An ion beam corresponding to $A = 181$ was extracted through the end plate of the isotope separator onto a thin collector foil behind which a counter telescope was placed to detect and identify protons with a solid angle of 14.5% of 4π (Hornshøj et al., 1971a).

The energy spectrum corresponding to the decay of ^{181}Hg is shown in Fig. 10.[†] All trace of individual level structure is absent from the spectrum, leaving a shape characteristic of the gross features of level widths and β decay as displayed previously in Fig. 3.

B. BETA-DELAYED NEUTRON DECAY

Experiments relating to two delayed-neutron precursors, ^{13}B and ^{137}I, will be described to illustrate two diverse experimental techniques. Once again, the differences between light and heavy nuclei will be apparent, as will be the similarities with delayed proton decay in corresponding mass regions.

[†] The small number of counts in the spectrum of Fig. 10 should not be taken as typical of the experimental method but rather as the result of ^{181}Hg being 15 neutrons removed from the lightest stable isotope of mercury and its proton-decay branches accounting for only one part in 10^4 disintegrations. An example of an Isolde study with good energy resolution and statistics is given for ^{115}Xe in Fig. 16 of Chapter II.B.

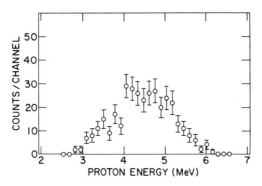

Fig. 10. Spectrum of β^+-delayed protons from the decay of ^{181}Hg (from Hornshøj *et al.*, 1971a, with permission of North-Holland Publishing Co., Amsterdam).

1. ^{13}B

The decay of ^{13}B was studied following its production by a mechanically chopped beam of 3-MeV tritons in the reaction ^{11}B(t, p)^{13}B (Jones *et al.*, 1969). Two plastic scintillators, one to detect electrons and the other neutrons, were placed opposite one another, 50 cm apart and at 90° to the beam direction. Neutron energies were obtained from their time of flight as measured relative to the observation of the preceding β ray. Two neutron peaks at 2.40 and 3.61 MeV were observed; their absolute intensities were determined from the ratio of delayed-neutron yield to the total number of β counts, taking account of detector efficiencies. Energy resolution of $\sim$400 keV (FWHM) was obtained for the observed peaks.

2. ^{137}I

A target of ^{235}U as U$_3$O$_8$ was contained in the discharge chamber of an ion source and placed near the core of a 1-MW reactor. The ionized fission products were extracted and mass separated by the Osiris isotope separator (Borg *et al.*, 1971). To observe the decay of ^{137}I, Shalev and Rudstam (1972) collected ions of the appropriate mass on magnetic tape and transferred them every 60 sec to a remote location where their decays were studied with a neutron spectrometer. The spectrometer consisted of ^{3}He gas contained at up to 10 atm pressure in a gridded ionization chamber (Cuttler *et al.*, 1969) which measured the total energy of products from the reaction ^{3}He(n, p)^{3}H.

Figure 11 shows the observed spectrum. The energy resolution of less

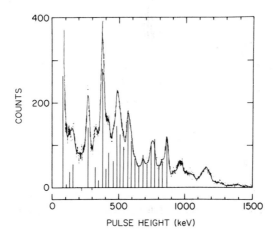

Fig. 11. Spectrum of β^--delayed neutrons from the decay of ^{137}I. The curve through the data points is the sum of energy-dependent response functions, whose energies and relative intensities are shown as vertical lines (from Shalev and Rudstam, 1972).

than 30 keV FWHM was sufficient to allow a number of individual peaks to be observed although the level density is obviously too high to resolve them all. There is really a striking similarity between this delayed neutron spectrum and the proton spectrum from ^{111}Te which was shown in Fig. 9, although the energy ranges covered are somewhat different.

C. Self-Delayed Proton Decay

Except for the lower particle energies involved, the experimental requirements for observing self-delayed emission are similar to those for β-delayed emission, and the same techniques can be used in observing both. However, experiments designed to study the decay of ^{53}Co^m will be described because they illustrate the value of using a counter telescope to view a target directly when the production cross section is low.

The reactions ^{40}Ca(^{16}O, 2np)^{53}Co^m (Jackson *et al.*, 1970) and ^{54}Fe(p, 2n)^{53}Co^m (Cerny *et al.*, 1970b) have been used to produce the proton-emitting isomer. In the latter experiment, a slotted rotating tantalum wheel was placed between the target and a counter telescope. The wheel was used to control the accelerator beam so that it shielded the counters while the target was being bombarded but permitted observation when the beam had been interrupted. The fact that protons from ^{53}Co^m are self-delayed was established by introducing a plastic scintillator to detect electrons; no β rays were observed to be in coincidence with protons recorded by the telescope. In a later experiment (Cerny et al., 1972), a telescope consisting of a 4-μm ΔE and a 50-μm E detector was used, thus permitting the observation and identification of protons ranging from 440 keV to 2.1 MeV. Only a single peak at a laboratory energy of 1.56 MeV (90 keV, FWHM) was observed within that range. This simple result for a self-delayed emitter should be compared with the complex proton spectrum shown in Fig. 8 for a β-delayed emitter of comparable mass.

IV. The Consequences

Certain generalizations can be made about the experimental data for β-delayed particle emitters as outlined in the preceding sections. For low-mass precursors ($A < 50$), the particle energy spectra are characterized by a large number of individual peaks, each corresponding to a specific particle–beta-decay channel. With increasing precursor mass, individual peaks decrease in number and merge into a general continuum of the form described by Fig. 3; at what mass the smearing begins depends upon the density of states in the emitter and on the experimental energy resolution. The qualitatively different spectra for light and heavy nuclei have led to quite different approaches being developed for their analysis, the first emphasizing the spectroscopy of specific levels in the emitter, the second depending upon the gross features of statistically large numbers of levels. Both approaches will be discussed here, together with an indication of where each is applicable. But before examining the specific consequences of delayed particle emission, it will be necessary to describe some aspects of β-decay theory.

A. SOME RELEVANT BETA-DECAY THEORY

The intensity of an allowed β transition is usually expressed in terms of its ft value, where t is the partial half-life of the transition; t equals the total half-life divided by i_β. Expressed in this way (with suitable electromagnetic radiative corrections) the strength can be directly related to nuclear matrix elements

$$ft = \frac{2\pi^3 (ln\,2)\,(\hbar^7/m_0^5 c^4)}{G_V'^2 \langle 1 \rangle^2 + G_A'^2 \langle \sigma \rangle^2} = \frac{6.15 \times 10^3}{\langle 1 \rangle^2 + 1.50 \langle \sigma \rangle^2} \quad \text{sec} \qquad (10)$$

Here G_V' and G_A' are the effective vector and axial-vector coupling constants; the numerical form of the equation results from using the best current values (Blin-Stoyle and Freeman, 1970) for $G_V'^2$ and $(G_A'/G_V')^2$. Equation (10) is a general and more explicit version of Eq. (5) in which the Fermi matrix element $\langle 1 \rangle$ has been included together with $\langle \sigma \rangle$.

Let us consider a β transition between two specified states whose wave functions are $|\Psi_i\rangle$ for the initial state and $|\Psi_f\rangle$ for the final state. The $\beta^\pm$-decay matrix elements are then written as

$$\langle 1 \rangle = \langle \Psi_f | \sum_n \tau_\pm(n) | \Psi_i \rangle, \qquad \langle \sigma \rangle = \langle \Psi_f | \sum_n \sigma(n) \tau_\pm(n) | \Psi_i \rangle \qquad (11)$$

where τ_+ (τ_-) is the isospin ladder operator which changes the nth proton (neutron) into a neutron (proton), σ is the Pauli spin operator, and the summations over n extend to all nucleons. Evidently $\langle 1 \rangle$ vanishes unless $\Delta J = \Delta T = 0$, while $\langle \sigma \rangle$ vanishes unless $\Delta J = 0, \pm 1$ $(0 \nrightarrow 0)$ and $\Delta T = 0, \pm 1$; for both, there can be no parity change.

1. Fermi Transitions and Analog States

One of the implications of these selection rules is that the Fermi matrix element will be significant in only a few cases. If the final-state wave function can be expressed as

$$|\Psi_f\rangle = A[\sum_n \tau_\pm(n) |\Psi_i\rangle] + B |\chi\rangle \qquad (12)$$

then the matrix element will be given by

$$\langle 1 \rangle^2 = [T_i(T_i + 1) - T_{z_i} T_{z_f}] A^2 \qquad (13)$$

Since the isospin of the initial state is T_i (with Z component T_{z_i}) (a discussion of isospin is given by Temmer, Chapter IV.A.2), then the configuration $[\sum_n \tau_\pm(n) |\Psi_i\rangle]$ must also have isospin T_i. The symbol $|\chi\rangle$ is intended to stand for all components of $|\Psi_f\rangle$ which are not so simply related to $|\Psi_i\rangle$; it can in general involve several isospin values including T_i. We must recognize, though, that states with isospin T_i exist in the final nucleus only if $|T_{z_f}| \leqslant |T_{z_i}|$. When this condition is met, there is, in the absence of charge-dependent forces, one state in that nucleus which has the wave function of Eq. (12) with components $A = 1$, $B = 0$. The actual presence of the Coulomb force, and charge dependence in the nuclear force, causes mixing of this pure state with others, but the resultant value for B is expected generally to be small. For those cases where it is, the state involved is said to be the analog of the parent ground state.

Since $A \approx 1$ for the analog state, the β transition feeding it will be characterized by a Fermi matrix element very nearly at its maximum value [see Eq. (13)]. This will mean a strong transition with low ft value ($\log ft <$ 3.8), which is usually referred to as superallowed. Although there will be other states containing some portion of the analog configuration, their coefficient A [see Eq. (12)] will be quite small. Thus, the transition strength of the Fermi matrix element will be restricted mostly to a single state with only weak contributions to other (usually nearby) states. It must be emphasized that, apart from the magnitude of A, the Fermi matrix element is independent of the details of the nuclear wave functions involved, so that if $\langle 1 \rangle^2$ can be measured in a β-decay transition, the result relates directly to the effects of charge-dependent nuclear mixing. However, as illustrated in Fig. 12, superallowed transitions are energetically permitted only between those relatively few nuclei with $N < Z$ that decay by positron emission or electron capture. In all other cases, the allowed β spectrum will be exclusively determined by the Gamow–Teller transition strength.

2. Gamow–Teller Transitions and Strength Functions

In contrast with $\langle 1 \rangle$, the Gamow-Teller matrix element $\langle \sigma \rangle$ can be seen from Eq. (11) to depend very much on details of the nuclear wave functions, and its behavior is much more difficult to generalize. If nuclear forces were spin and isospin independent, though, it would be concentrated in one transition to a state whose wave function was given by $[\sum_n \sigma(n)\tau_{\pm}(n)|\Psi_i\rangle]$; but this level and the analog state would be one and the same so both matrix elements would be concentrated in a single transition. Such a simple supermultiplet scheme (Wigner, 1939) is certainly violated by strong spin-dependent nuclear forces, but much of the Gamow–Teller strength remains in a "giant resonance" of collective transitions to states within a few MeV of the analog state (Fujita and Ikeda, 1965; Fujita et al., 1968). The width of the resonance will of course be greater than that for Fermi transitions since the latter reflects only the relatively smaller deviations from charge independence. In discussing "resonances" in transition intensities, it is convenient to return to the strength function S_β as defined in Eq. (6). Properly normalized, the β intensity per MeV at excitation E_x can be rewritten as

$$I_\beta(E_x) = f(Z, Q_\beta - E_x) S_\beta(E_x) \left[\int_0^{Q_\beta} f(Z, Q_\beta - \varepsilon) S_\beta(\varepsilon) \, d\varepsilon \right]^{-1} \quad (14)$$

For β^+ decay, f stands for the sum of the statistical rate functions corresponding to positron decay and electron capture. It is $S_\beta(E_x)$, the average reduced transition probability per unit energy, which will exhibit a broad peak or giant (Gamow–Teller) resonance near the energy of the analog state. It will also show smaller peaks or "pygmy resonances" at other energies corresponding to more nearly single-particle transitions. The general behavior of the strength function can be inferred from the three schemes illustrated in Fig. 12. For β^+ decay from a nucleus with $N < Z$, both the

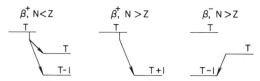

Fig. 12. Schematic illustration of the β decays of various classes of nuclei showing the lowest-energy states with the indicated isospin.

superallowed transition and the giant resonance will appear, while if $N > Z$, only the tail of the giant resonance will contribute to the spectrum of β^+ or β^- decays. For all cases, pygmy resonances will appear depending upon the shell model states that are available. While this description lends itself more to heavy nuclei, where the density of final states is large enough to justify the concept of a smoothly varying strength function, it also applies as a general picture of lighter nuclei, providing an average is taken over the visible microstructure.

To gain more insight into the behavior of the strength function for nuclei with $N > Z$, let us assume that the giant resonance energy coincides exactly with that of the analog state. The magnitude of the strength function can then be expressed as a function of E_s, the final-state excitation energy relative to that of the analog state; $E_s = 0$ marks the position of the resonance. If ΔE_C is the Coulomb energy difference between adjacent isobars and Δm is the neutron–hydrogen mass difference $(= 0.782 \text{ MeV})$, then β transitions to levels at excitation energy E_x are characterized by

$$
\begin{aligned}
|E_s| &= (\Delta E_C - \Delta m) - (Q_\beta - E_x) \qquad \text{for } \beta^+ \text{ decay} \\
|E_s| &= (\Delta E_C - \Delta m) + (Q_\beta - E_x) \qquad \text{for } \beta^- \text{ decay}
\end{aligned}
\qquad (15)
$$

The strength function decreases as E_s increases in magnitude. Therefore Eq. (15) indicates that for β^+ decay, S_β will decrease with increasing excitation energy while the opposite will be true for β^- decay. Furthermore, if ΔE_C is the same for both β^+ and β^- decay, it is the former which will sample the strength function nearest its resonance region, and presumably therefore it will show the greatest variation with energy. However, the value of E_s for both is of the order of 10 MeV or more (if $Z \gtrsim 30$), so the strength functions should not vary rapidly with excitation energy in either case.

Consider first β^+-delayed proton precursors with $N > Z$. We have seen that such nuclei should exhibit transitions which result in a slowly varying strength function. However, most of these transitions must involve nucleons that are initially in paired configurations, so they cannot populate states in the final nucleus which are below a certain minimum excitation energy C. That energy is simply related to the pairing energy, and the actual values used for it (Duke *et al.*, 1970; Hornshøj *et al.*, 1972b) are: $C = 0$ for odd-odd, $C = 2\Delta$ for odd-even, and $C = 4\Delta$ for even-even nuclei, where in the case of delayed particle decay the nucleus being described is the emitter. The gap parameter is $\Delta = 12A^{-1/2}$ (MeV). Thus, while the strength function should be nearly constant for excitation energy $E_x \gtrsim C$, transitions to states below that energy will in general be considerably weaker. All this is consistent with experiment (Duke *et al.*, 1970).

For the β^- decay of nuclei with $N > Z$, i.e., delayed-neutron emitters, the strength function should be slowly varying throughout. Yet its rate of change with energy has often in the past been assumed to follow the condition $S_\beta(E_x) \propto \varrho(E_x)$, where ϱ is the density of those states with appropriate spin at energy E_x. This implies a constant transition strength per level rather than per energy interval. In light of the preceding discussion of Eq. (15), it appears that such a rate of change has the correct sign but is too high for the region of excitation explored by delayed neutron emission. Preliminary experimental results (Nielsen *et al.*, 1971) do confirm the expectation of a lower rate of change, but whether it is as low as that observed for β^+ decay is not yet clear.

3. *Beta–Neutrino Angular Correlations*

When a positron (or electron) is emitted in nuclear β decay, its direction of motion is correlated with that of the neutrino (or antineutrino) emitted simultaneously in a way that depends upon the nature of the transition. If the angle between them is θ, the correlation function $\omega(\theta)$ takes the form (Schopper, 1966)

$$\omega(\theta) = 1 + a(v/c) \cos \theta$$

with

$$a = \tfrac{1}{3}(3\langle 1 \rangle^2 - 1.50\langle \sigma \rangle^2)/(\langle 1 \rangle^2 + 1.50\langle \sigma \rangle^2) \tag{16}$$

where v is the electron velocity, c is the speed of light, and, as in Eq. (10), the number 1.50 is the best current value for $G_A'^2/G_V'^2$. For pure Fermi transitions, $a = +1$, with the result that the lepton and neutrino are emitted preferentially in the same direction; the opposite is true for Gamow–Teller transitions, for which $a = -\frac{1}{3}$. Mixed transitions will have intermediate a values.

4. *Electron Capture and Positron Decay*

Positron emission will occur if the end-point decay energy $W_0 > 2m_0c^2$, where m_0c^2 is the rest mass of the electron (i.e., 511 keV). Under these conditions, the same transition can also occur by the capture of an orbital electron, and the relative contribution of this mode of decay increases both with increasing Z and with decreasing decay energy. For energies below $2m_0c^2$, electron capture remains as the only allowed decay mode providing $W_0 > |E_B|$, where E_B is the electron binding energy ($\ll 1$ keV).

The theoretical ratio of electron capture to positron decay has been plotted as a function of β^+ end-point energy for various Z values by Lederer *et al.* (1967). The ratio is a strong function of W_0 and suggests that if the end-point energy of a particular transition is unknown, it can be determined with reasonable accuracy simply by measuring the ratio. For a complex β^+ decay, such as that of a heavy delayed-proton precursor which feeds many unresolved final states in the emitter, a measurement of the electron capture to positron ratio as a function of the final-state excitation energy can yield an accurate value for the relative mass of the precursor itself.

B. ANALOG STATES AND COULOMB MIXING

The first β-delayed proton precursors to be identified and studied were those light nuclei with $T_z = -\frac{3}{2}$ and odd N; they have been listed in Table 1. When the energies and intensities of peaks in the proton spectra were used to reconstruct the precursors' β decays, a striking feature appeared: all except ^{9}C, ^{13}O, and (at first) ^{17}Ne exhibited one branch with an ft value more than an order of magnitude lower than most of the others observed. In each case, though, some β transitions fed states in the emitter that were below its proton separation energy B_p, and since the intensities of these were unmeasured, absolute ft values could not be obtained for any of the decay branches. Nevertheless, the anomalously strong transitions fed states whose energies agreed with the expected positions of analog states and they were assigned as superallowed on that basis (e.g. Hardy and Verrall, 1964).

The precursor ^{17}Ne was somewhat unusual; it has $J^\pi = \frac{1}{2}^-$ while the only two states below the proton separation energy in ^{17}F—its ground and first excited state—are $\frac{5}{2}^+$ and $\frac{1}{2}^+$, respectively. Being first-forbidden, the β transitions feeding these states absorb less than 2% of the ^{17}Ne decay and furthermore this proportion can be estimated with good accuracy from the known mirror decay of ^{17}N. Thus the *absolute* intensities of the other branches can be determined from the *relative* intensities of peaks in the proton spectrum. By comparing the intensities of peaks in the spectra of other precursors with those of ^{17}Ne and assuming a simple relationship between the production cross sections, Hardy *et al.* (1966) were able to measure approximate absolute *ft* values for transitions from ^{21}Mg, ^{25}Si, ^{29}S, ^{33}Ar, and ^{37}Ca. The transitions believed to be superallowed all had measured log *ft* values between 2.9 and 3.6 (with error bars of ± 0.3) thus confirming their assignment. For ^{17}F, ^{21}Na, ^{25}Al, ^{29}P, ^{33}Cl, ^{37}K, ^{41}Sc, and ^{49}Mn, such delayed proton measurements provided the first identification of their lowest $T = \frac{3}{2}$ states. Those in the nuclei with $A \leqslant 37$ have since been produced and studied in nuclear reactions (Cerny, 1968; Temmer, Chapter IV.A.2); for ^{41}Sc, ^{49}Mn, and the others as yet undiscovered, delayed proton measurements will probably be the only means of observing their lowest $T = \frac{3}{2}$ states for some time to come.

Data and predictions are shown in Fig. 13 for superallowed decay branches

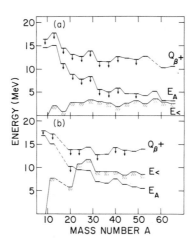

Fig. 13. Data and predictions for superallowed decay branches from two families of precursors with odd N: (a) $T_Z = -\frac{3}{2}$, (b) $T_Z = -1$. An arrow on the β-decay energy $Q_{\beta+}$ signifies that the precursor has been observed, while one on the analog-state energy E_A indicates the superallowed branch is known. The energy $E_<$ is the minimum energy of a state from which particle emission could be observed. All energies are relative to the emitter ground state.

from precursors with an odd number of neutrons and $T_Z = -1$ or $T_Z = -\frac{3}{2}$. For each A value, the energy scale is referred to the ground state of the emitter and shows the appropriate Q_β, the excitation of the analog state E_A, and the minimum excitation energy $E_<$ from which appreciable particle emission could be expected ($\equiv B_\mu + \theta_\mu$). For the $T_Z = -1$ series, where both proton and alpha emission can be involved, $E_<$ refers to whichever process has the lowest threshold. A small arrow below the Q_β line indicates that the precursor has been observed; below the E_A line, it indicates that particle emission from the analog state has been identified. The richness of the $T_Z = -\frac{3}{2}$ series in the study of analog states is evident, as is the fact that it

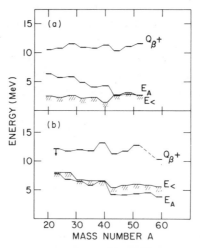

Fig. 14. Data and predictions for superallowed decay branches from two families of precursors with even N: (a) $Tz = -2$, (b) $Tz = -\frac{3}{2}$. The notation is the same as in Fig. 13.

will probably yield a number more. However, delayed particle emission will clearly not yield more analog states for the $T_Z = -1$ series. Similar data and predictions are contained in Fig. 14 for even-N precursors with $Tz = -\frac{3}{2}$ and -2. It can also be noted that for all nucleon-stable precursors with T_Z more negative than those illustrated in the figures, the observed delayed-proton spectra will include contributions from superallowed β decay between analog states. Consideration of Fig. 12 shows that apart from these, no other families of β-delayed particle emitters will be useful in this regard. No superallowed branch can contribute, for example, to delayed-neutron spectra.

Once superallowed decay branches between $T = \frac{3}{2}$ states (and the $T = 1$ states in mass 20) had been observed, improving experimental techniques invited more accurate measurements of their transition intensities with a view to extracting the analog "purity," A^2 [see Eqs. (12) and (13)], of the final states. This is of considerable interest since these $T = \frac{3}{2}$ states lie quite high in excitation and are surrounded by a relatively large density of $T = \frac{1}{2}$ states with some of which—those of the same J^π—they might be expected to mix. After all, $(T = \frac{1}{2})$ proton emission to $(T = 0)$ states in the daughter nuclei can only occur through $T = \frac{1}{2}$ impurities in the predominantly $T = \frac{3}{2}$ analog states. These admixtures can be very small ($\sim 0.1\%$) and still preserve the condition that $\Gamma_p \gg \Gamma_\gamma$, so the total width Γ of such analog states is usually considerably less than that of neighboring $T = \frac{1}{2}$ states.

Recently, accurate measurements of ft values for the decays of ^{17}Ne and ^{33}Ar have been reported (Hardy et al., 1971). The experimental proton spectrum for ^{33}Ar from that work has already appeared in Fig. 8; the corresponding decay scheme and some of the results are shown in Fig. 15. The lifetime was determined by recording proton spectra sequentially and determining the decay rate of individual peaks; this is an accurate method which all but removes any possible contribution from contaminant activities. Absolute transition strengths were obtained by measuring the intensity of the 810-keV γ ray relative to proton emission, and calculating the ground-state transition from that of the known mirror decay of ^{33}P. The procedure for ^{17}Ne has already been described.

For the superallowed transitions from ^{17}Ne and ^{33}Ar, Eq. (13) gives $\langle 1 \rangle^2 = 3A^2$, while model calculations for the Gamow–Teller contribution (Hardy and Margolis, 1965) yield values for $\langle \sigma \rangle^2$ of 0.11 and 0.28, respectively. Although some uncertainty must be attached to the model-dependent results, the relative magnitudes of $\langle 1 \rangle^2$ and $\langle \sigma \rangle^2$ are such that even reasonably large uncertainties in $\langle \sigma \rangle^2$ will only affect the calculated ft values in Eq. (10) by a few percent. Consequently, the measured ft values for ^{17}Ne and ^{33}Ar were used directly to obtain a measurement of A^2—the higher the ft value, the lower the purity. The results show that for the $T = \frac{3}{2}$ state in ^{17}F, $A^2 > 0.95$, while for ^{33}Cl, $A^2 = 0.81 \pm 0.09$; even in the limit of $\langle \sigma \rangle^2 = 0$ for ^{33}Cl, $A^2 = 0.95^{+0.05}_{-0.10}$. This strongly suggests an impurity, perhaps of the order of 10%, in the analog state in ^{33}Cl, and the close proximity of β-fed $T = \frac{1}{2}$ states, with log ft values progressively decreasing the nearer they are in energy to the analog level, is compelling circumstantial evidence that these are the sources of mixing; furthermore, the magnitudes of ft values for transitions to these states are consistent with

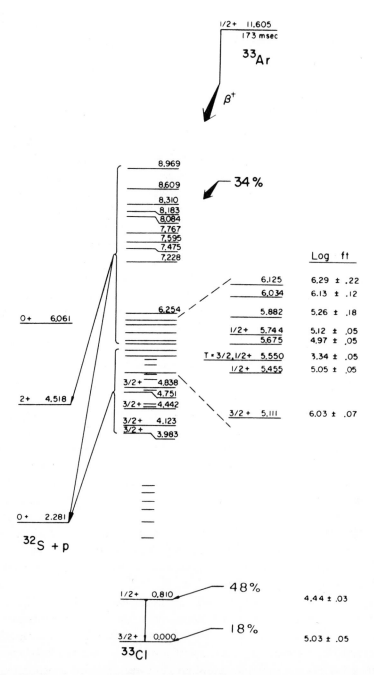

Fig. 15. Decay scheme of ^{33}Ar. Levels whose energies are marked were observed to be populated by β decay and deexcited by proton emission; log ft values were determined for all these transitions, though only a few are shown (from Hardy *et al.*, 1971).

this explanation (Hardy et al., 1971). Equivalent results, obtained from the delayed α decay of ^{20}Na (Torgerson and Macfarlane, 1971), also indicate mixing in the lowest $T = 1$ state of ^{20}Ne; these will be discussed in Section IV.D.

C. GAMOW–TELLER TRANSITIONS AND STRENGTH FUNCTIONS

For every excited state of ^{33}Cl marked in Fig. 15 with its excitation energy, the absolute intensity of the β-decay branch feeding it from ^{33}Ar was determined from delayed-proton studies. There are 25 observed transitions in all, with log ft values ranging from 3.34 for the superallowed branch to 6.39. Except for the few near the analog state, which probably share some strength from the Fermi matrix element, most are pure Gamow–Teller transitions. Similarly, nine Gamow–Teller transitions have been observed from ^{17}Ne, and large numbers from all other light delayed-particle precursors. Although frequently only relative intensities have been measured, in many of the cases, intensities could be normalized to the calculated superallowed branch (Hardy and Margolis, 1965; Reeder et al., 1966; etc.).

The most obvious application of such large numbers of experimental transition intensities is to use them as tests of model wave functions for the levels involved. For example, an examination of the results for ^{17}Ne decay shows reasonable qualitative agreement with calculations for levels in ^{17}F which include up to two particles in the (2s-1d) shell and one hole in the 1p shell (Margolis and de Takacsy, 1966). One interesting conclusion regards the so-called antianalog configuration. In the calculations, the $\frac{1}{2}^-$ $(T = \frac{3}{2})$ analog state is well represented by a $(1p\frac{1}{2})$ hole coupled to the lowest 0^+ $(T = 1)$ state in mass 18 (e.g., the ground state of ^{18}Ne). The same components coupled to $T = \frac{1}{2}$ comprise the antianalog configuration which, unlike the analog, may be shared extensively among many levels. In general, the antianalog configuration is favored by β decay and these levels would be marked by stronger transitions—a pygmy resonance in the strength function. Despite the fact that in ^{17}F there is a specific cancellation which significantly reduces some of this strength, the experimental results have been used to locate components of the antianalog configuration (Hardy et al., 1971).

Another interesting application of the measured transition rates is in the comparison of mirror decays. The mirror of the $(T_Z = -\frac{3}{2})$ delayed-proton precursor ^{13}O is the $(T_Z = +\frac{3}{2})$ delayed-neutron precursor ^{13}B. Based on the ^{13}B transitions observed by Jones et al. (1969) and assuming perfect mirror symmetry, the half-life of ^{13}O can be predicted providing corrections

are made for additional ^{13}O branches observed in delayed-proton studies
(Esterl *et al.*, 1970). The result differs by $15 \pm 3\%$ from the measured half-
life. There are a number of other such discrepancies among light nuclei
with large β-decay end-point energies and this has led to the suggestion by
Wilkinson (1970) that second-class currents in the β interaction might be
the cause. By comparing delayed α spectra from the mirror ^{8}Li and ^{8}B
decays, Wilkinson and Alburger (1971) showed that the induced tensor interaction originally proposed cannot be responsible, but whether some other
interaction is involved or simply T_Z-dependent mixing in the nuclear wave
functions is not yet understood (Wilkinson, 1971).

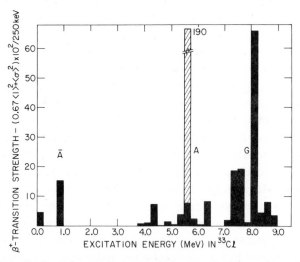

Fig. 16. Distribution of transition strength for the β^{+}-decay scheme ^{33}Ar $\overset{\beta^{+}}{\rightarrow}$ ^{33}Cl.
Decay to the analog state is denoted by diagonal lines.

The decay of ^{33}Ar provides a striking link to the study of strength functions which is usually associated with heavier nuclei. Figure 16 shows a
histogram of the β-decay transition strength as a function of excitation
energy in ^{33}Cl. The quantity plotted is $4.10 \times 10^{5}/ft$ summed over all decays
in 250-keV intervals and is proportional to the strength function used in
Eqs. (6) and (14). Distinct concentrations of decay strength are to be seen.
The Fermi decay to the analog state is marked by diagonal lines and the
letter A; the antianalog configuration, $\bar{\text{A}}$, is at much lower excitation
and very likely is concentrated in the first excited state (Hardy, 1971a),

while the small increase in intensity near the analog state has been noted in Section IV.B as the probable result of Coulomb mixing. However, it is the concentration at about 8 MeV (marked by G) which must be the giant resonance predicted by Fujita *et al.* (1968) and already discussed in Section IV.A.2. This is the first example where the gross structure of the peak itself has been observed in β decay, and here there is the added advantage that the fine structure can be examined as well.

It has also been shown in Section IV.A.2 that for heavier nuclei where $N > Z$, β^+ decay populates states that reflect the tail of the giant resonance. Karnaukhov (1970) was the first to attempt a detailed examination of the strength function in this region through an analysis of the delayed-proton spectrum from the precursor ^{111}Te. In general, the intensity of the particle spectrum $I_\mu(E_\mu)$ at particle energy E_μ can be written

$$I_\mu(E_\mu) = \sum_{\text{if}} \frac{\Gamma_\mu^{\text{if}}(E_\mu, B_\mu)}{\Gamma^{\text{i}}(E_\mu, B_\mu)} I_\beta(E_x) \, \omega(J_i, J) \tag{17}$$

$\Gamma_\mu^{\text{if}}(E_\mu, B_\mu)$ is the partial width of a level at $E_x (= B_\mu + E_f + E_\mu)$ with spin J_i for particle decay to the state at E_f in the daughter (see Fig. 1); $\Gamma^{\text{i}}(E_\mu, B_\mu)$ is the corresponding total width; and $\omega(J_i, J)$ is the proportion of all β decays from the spin-J precursor that populate levels of spin J_i. The β^+ intensity I_β has already been defined in Eq. (14).

For ^{111}Te, Eq. (17) can be simplified considerably. The precursor's spin is believed to be $\frac{5}{2}^+$, so the proton spectrum should be dominated by the $l_p = 2$ decay of $\frac{3}{2}^+$ and $\frac{5}{2}^+$ states in the emitter ^{111}Sb to the 0^+ ground state of ^{110}Sn. Since $E_p \lesssim 4$ MeV (see Fig. 9), the centrifugal barrier inhibits the $l_p = 4$ proton decay of $\frac{7}{2}^+$ states by at least a further two orders of magnitude (Mani *et al.*, 1963) and this is sufficient to make γ emission their predominant mode of decay. In addition, the Coulomb barrier ensures that only a few percent of the observed spectrum can be associated with proton decay to the first excited 2^+ state of ^{110}Sn at 1.2 MeV. Consequently, in evaluating the shape of the proton spectrum, both the summation and the factor ω can be omitted from Eq. (17), although they would still be necessary to determine the absolute intensity. The ratio of partial to total widths can be evaluated from Eq. (3) using $l_p = 2$ transmission coefficients together with Cameron's (1956) expressions for ϱ and Γ_γ, providing the proton separation energy B_p is supplied either from experimental measurements or from mass predictions. Then, if Q_β is taken from a similar source, I_β can be determined through Eq. (14) by postulating a particular form for the strength function S_β, and the result used in Eq. (17) to compare with the experimental proton-spectrum shape. Since $Q_\beta - B_p$ had been measured previously, Karnaukhov (1970) tested various strength functions against the ^{111}Te proton spectrum with B_p as the only parameter. He demonstrated convincingly that over the observed energy range, $S_\beta(E_x)$ must either be constant or very slowly varying.

More complex decay schemes for the precursors ^{115}Xe, ^{117}Xe, ^{181}Hg, and ^{183}Hg have been analyzed with the same techniques by Hornshøj et al. (1972b). In each case, they assumed several different spin-parities J^π for the precursor. Generally, more than one l_p value was involved in each decay scheme, so it was necessary to know the statistical weight factor $\omega(J_i, J)$; this they took to be equal to $(2J_i + 1)[3(2J + 1)]^{-1}$, which is valid for transitions of tensor rank one from paired configurations. In the absence of mass measurements, they used the predictions of Myers and Swiatecki (1965) for Q_β and B_p and, assuming the strength function to be

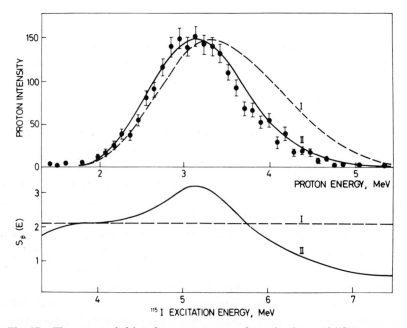

Fig. 17. The measured delayed-proton spectrum from the decay of ^{115}Xe compared with two calculated spectra both assuming $J^\pi = \frac{5}{2}^+$. Strength functions used for the calculations are shown at the bottom of the figure. Masses were taken from Myers and Swiatecki (1965) since the experimental $Q_{\beta^+} - B_p$ coincides with these predictions (from Hornshøj et al., 1972b).

constant, calculated the proton spectra to have approximately the right shape but to be shifted slightly in energy from the experimental results.

This discrepancy could be corrected for the mercury isotopes by a modest increase in the assumed Q_β but the authors proposed a different explanation for ^{115}Xe and ^{117}Xe; they showed that a considerable improvement in the agreement with experiment could be effected by postulating a varying strength function. The results are illustrated for ^{115}Xe in Fig. 17 and suggest that the strength function goes through a pygmy resonance at about 5 MeV. The excitation energy of the resonance corresponds approximately with that expected for states involving a single proton hole in the $1g^9_2$ shell and it may well be caused by β transitions of the type $(1g^9_2)_p \rightarrow (1g^7_2)_n$. A similar interpretation also applies to ^{117}Xe, but Hornshøj et al. (1972b) pointed out that in view of the uncertainties inherent in the calculations for both nuclei, their conclusions must simply be regarded as qualitative indications.

The intensity of proton decay relative to the total disintegration rate is calculated by integrating Eq. (17) over all values of E_p. Since the answer depends upon the assumed spin of the precursor, Hornshøj et al. (1972b) also attempted to determine spins by comparing such calculations with measured intensities. Their results for ^{115}Xe are shown in Table 6, where the relative intensities corresponding to population of individual states in the daughter, ^{114}Te, are also tabulated. The indicated spin for ^{115}Xe is $\frac{5}{2}^+$

TABLE 6

COMPARISON OF EXPERIMENTAL AND CALCULATED PROTON INTENSITIES FOR THE DECAY OF ^{115}Xe

^{115}Xe assumed J^π	Relative intensities to states in ^{114}Te (%)			Total absolute intensity
	0.0 MeV (0+)	0.71 MeV (2+)	1.48 MeV (4+)	
$\frac{1}{2}^-$	92.3	7.7	0.0	9.6×10^{-3}
$\frac{3}{2}^-$	79.7	20.0	0.3	5.6×10^{-3}
$\frac{5}{2}^-$	62.4	36.0	1.6	2.8×10^{-3}
$\frac{7}{2}^-$	2.1	87.8	9.8	7.0×10^{-4}
$\frac{1}{2}^+$	83.5	16.4	0.0	9.6×10^{-3}
$\frac{3}{2}^+$	72.8	27.1	0.1	6.4×10^{-3}
$\frac{5}{2}^+$	41.1	55.7	3.2	2.3×10^{-3}
$\frac{7}{2}^+$	33.2	56.3	10.5	1.1×10^{-3}
Experiment[a]	~ 40	58 ± 7	< 2	$(3.6 \pm 0.6) \times 10^{-3}$

[a] Hornshøj et al. (1972b).

and, although certainly tentative, it is consistent with shell model expectations. However, the reliability of such spin assignments depends very strongly upon the parameters used and even upon the prescribed variation of level densities, so the technique cannot be applied indiscriminately to all precursors.

In principle, delayed-neutron spectra can also be analyzed by using Eq. (17) to yield β^--decay strength functions and possibly the spins and parities of precursors. At present, only the recently measured spectrum of ^{137}I (Shalev and Rudstam, 1972) is well enough known for practicable analysis, though the powerful new experimental techniques introduced for its measurement will undoubtedly be used in the future to study other precursors in equivalent detail. There will be considerable interest in determining whether the strength function for electron emission in fact behaves as predicted in Section IV.A.2 or shows the more rapid energy dependence frequently assumed.

D. BETA–NEUTRINO ANGULAR CORRELATIONS

It has been shown in Section IV.A.3 that an experimental determination of the β–v correlation coefficient a can be used to establish the relative contributions of Gamow–Teller and Fermi matrix elements to a particular transition. Obviously, this can be helpful in examining, among other things, the effects of Coulomb mixing on states in the emitter.

Because of the difficulty in detecting neutrinos, the angle θ between the directions of motion of β particles and neutrinos in a nuclear decay cannot be measured directly. Nevertheless, it can be inferred from a measurement of the energies of both the β and the recoil nucleus, since

$$\cos \theta = (p_R{}^2 - p_v{}^2 - p_\beta{}^2)/2p_v p_\beta \tag{18}$$

where p_R, p_v, and p_β are momenta for the recoil nucleus, neutrino, and β particle, respectively, and p_v can be related to the other two by energy conservation. Experimental difficulties remain, though, in detecting the heavy recoil nucleus since its energy is very low, and a subsequent particle emission can be of great value. The energy of the emitted particle is shifted by an amount which corresponds to the energy and direction of motion of the recoil nucleus itself (i.e., the emitter). A schematic diagram of the geometry involved appears in Fig. 18. Since the distance traveled by the emitter E is small compared to experimental dimensions, a measurement of the energies of decay particles observed in coincidence with betas at fixed angle ϕ is a sensitive probe of $\omega(\theta)$ for the transition involved.

The first application of these principles was a decay study of ^{8}Li. This precursor emits electrons to the first excited 2^+ state in ^{8}Be, which breaks up into two α particles. If it were not for the preceding β decay, the α particles would be emitted in opposite directions with equal energies, but deviations of up to 7° in angle and 20% in energy appear as a result of the ^{8}Be recoil energy. The distribution of angles between the α particles was measured by Lauritsen *et al.* (1958), who showed that it is a strong function not only of the beta interaction involved but also of the spin of ^{8}Li. The former is

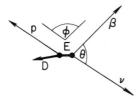

Fig. 18. Schematic diagram of the geometry involved in the decay of a delayed-particle precursor. The path of motion is shown for the electron (β), neutrino (v), emitter (E), daughter (D), and particle (p) which could be either proton or α particle. The definition of ϕ depends on the distance traveled by E being very short.

important because it controls the relative direction of electron and neutrino emission and hence the recoil velocity, while the latter determines the alignment of the recoil nucleus (whose axis of alignment must be perpendicular to the plane of α emission) relative to its direction of motion. The authors concluded from the data that ^{8}Li must be 2^+ and its decay must be more than 90% determined by the Gamow–Teller interaction. In a related experiment, Barnes *et al.* (1958) measured the energy distribution of alphas in coincidence with electrons at $\phi = 90$ and 180°. Their results led to an important general conclusion that the Gamow–Teller interaction is more than 90% axial vector.

Spectroscopic information can also be gained from applying these techniques to heavier β-delayed-particle precursors where the shape and energy of peaks in the β-coincident particle spectrum can be studied as a function of ϕ. In addition to requiring clearly resolvable energy peaks, the success of the method depends upon the precursor having a high Q_β and a light enough mass to ensure a large recoil velocity. Recent calculations (Hardy, 1971b) indicate that for delayed-proton precursors, these conditions are probably satisfied if $A \lesssim 40$, and that an accuracy of $\pm 10\%$ can be achieved in many cases for determining a from the proton line shape providing the energy of the coincident β^+ particles is restricted to 500 keV $\lesssim$ $E_\beta \lesssim 0.7E_{max}$.

A less sensitive approach, but one that is more tractable experimentally, is to study the shapes of peaks in the singles particle spectrum using a system with very high instrumental resolution. Such a study has been reported for the decay of ^{20}Na by Macfarlane *et al.* (1971), and data for two peaks from that work appear in Fig. 19 together with calculated curves from which the best values of *a* were obtained. Taking account of nuclear alignment for the α-emitting states, the authors determined *a* to be 0.80 $\pm$ 0.08 for the superallowed transition to the analog state at 10.28 MeV and -0.30 ± 0.03 for the other illustrated transition. These numbers should be compared with the values $+1$ and $-\frac{1}{3}$ expected for pure Fermi and Gamow–Teller transitions, respectively (see Section IV.A.3). With the help of Eq. (16), the result for the superallowed branch yields a matrix element ratio $\langle\sigma\rangle^2/\langle 1\rangle^2 = 0.11 \pm 0.05$, but the absolute *ft* value is not known accurately enough to determine $\langle 1\rangle^2$ alone. However, the authors also report that other β^+ transitions to states near the analog state have correlation coefficients which show Fermi components are involved there, too (see also Torgerson and Macfarlane, 1971), thus indicating that, through Coulomb mixing, the $T = 1$ analog state in ^{20}Ne has spread some of its strength to nearby levels

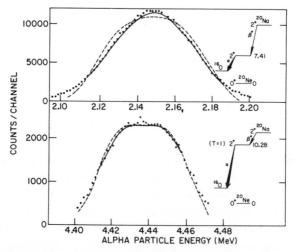

Fig. 19. Comparison of experimental line shapes with fitted theoretical curves assuming the α emission is isotropic in ϕ (dashed curves) and taking account of anisotropy introduced by the nuclear alignment following β decay (solid curves) (from Macfarlane *et al.*, 1971, with permission of North-Holland Publishing Co., Amsterdam).

of the same spin. The result is not unlike that already described (in Section IV.B) for the $T = \frac{3}{2}$ analog state in ^{33}Cl. Detailed studies of this type could, in the future, yield a precise knowledge of the isospin purity and sources of mixing for analog states in a variety of light nuclei.

E. MASS MEASUREMENTS

Delayed particle emission can provide a convenient tag to identify one nucleus among a host of competing activities, and for this reason, it is useful in the measurement of threshold energies or excitation functions. Among light nuclei, this can sometimes yield a highly accurate precursor mass, such as in the case of ^{9}C, whose ground-state energy, accurate to ± 4 keV, was obtained by Mosher et al. (1971) from the threshold of the ^{7}Be(^{3}He, n)^{9}C reaction. Unfortunately, heavier precursors are rarely produced in such one-step reactions. Heavy-ion bombardments, multiple-neutron evaporation, and spallation reactions cannot be similarly analyzed for mass measurements, although in some cases the mass number A of the precursor can be established with the help of excitation functions from these reactions.

The shape of the delayed-particle spectrum itself, though, depends upon the mass of the nuclei involved through the quantities B_μ and Q_β. It is evident from Fig. 3 and the discussion in Section II.A that the low-energy part of the spectrum depends (weakly) upon B_μ since that quantity determines the relationship between Γ_μ and Γ_γ, while the high-energy part depends upon $Q_\beta - B_\mu$, which is, in fact, the maximum allowed particle energy. However, we have seen in Fig. 17 that the high-energy spectrum also depends strongly upon the β-decay strength function, so unless that is known from an independent source, it would be risky to attempt to determine $Q_\beta - B_\mu$ by fitting such data. Nevertheless, Bogdanov et al. (1968a) were successful in using a constant strength function for the decay of ^{111}Te to extract an experimental value of $Q_\beta - B_p$ which was subsequently verified by the following, more reliable technique.

It was noted in Section IV.A.4 that the probability ratio of electron capture to positron emission $\lambda_{EC}/\lambda_{\beta^+}$ for a particular transition can be a sensitive measure of its end-point energy. Bacso et al. (1968) used this principle to study ^{111}Te by measuring the spectrum of protons in coincidence with positrons and comparing it with the singles spectrum. Denoting the counts in each spectrum at proton energy E_p by $N_{p\beta}(E_p)$ and $N_p(E_p)$, then we have

$$N_{p\beta}(E_p)/N_p(E_p) = \sum_f R_f(E_p)\, \varepsilon_\beta(E) \left[1 + \lambda_{EC}(E)/\lambda_{\beta^+}(E)\right]^{-1} \qquad (19)$$

where $E(= Q_\beta - B_p - E_f - E_p)$ is the beta end-point energy, $R_f(E_p)$ is the fraction of protons with energy E_p populating the state at E_f in the daughter, and ε_β is the positron detection efficiency; the summation extends over all states available in the daughter. In Fig. 20, the term $[1 + \lambda_{EC}/\lambda_{\beta^+}]^{-1}$ is plotted as a function of E_p, where $X = Q_\beta - B_p - E_f - 2m_0 c^2$; curves are shown for several representative values of Z. For ^{111}Te, only the ground state of ^{110}Sn is populated significantly, so $E_f = 0$, $R_f = 1$, and the summation in Eq. (19) is omitted. The appropriate curve based on Fig. 20 is then fitted to the data for $N_{p\beta}(E_p)/N_p(E_p)$ by correcting the ordinate to take account of the experimental ε_β and adjusting the abscissa to optimize agreement; the latter adjustment yields directly the value of $Q_\beta - B_p (= X + 2m_0 c^2)$. Bacso et al. (1968) obtained a result accurate to ± 70 keV from data shown in the inset of Fig. 20.

Hornsøj et al. (1972a) have recently analyzed the more complicated decays of ^{115}Xe and ^{117}Xe by the same methods. In these cases, excited states were populated in the daughter nucleus, thus requiring an experimental measurement of $R_f(E_p)$. This was accomplished by observing proton coincidences with γ rays corresponding to deexcitation of states at the various energies E_f in the daughter. The results for $Q_\beta - B_p$ were quoted with accuracy better than ± 200 keV.

In assessing the applicability of this method of mass measurement to other precursors, it is necessary to combine with Fig. 20 a knowledge of the decrease in total β rate with decreasing decay energy since this will determine

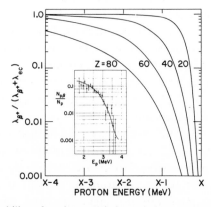

Fig. 20. The probability of positron emission relative to the total decay probability (positron emission plus electron capture) plotted as a function of the subsequent proton energy, where $X = Q_\beta - B_p - E_f - 2m_0 c^2$. Curves are shown for four different Z values. The inset shows the fit to experimental results for ^{111}Te taken from Bacso et al. (1968).

the statistical accuracy with which $N_{p\beta}/N_p$ can be measured at high proton energy. It can be seen from Fig. 20 that the high-energy region becomes of primary importance for nuclei with lower Z, so it seems unlikely that the method can be useful if Z is much less than 30.

It would be inappropriate here to elaborate greatly on the importance of absolute or relative mass measurements. Let it suffice to say that there are many semiempirical prescriptions for calculating masses over a wide region of the periodic table. Since all are based on the masses of nuclei near the valley of beta stability, in that region they tend to agree well with one another and with experiment. It is by comparing the predictions with measurements for remote nuclei, like delayed-particle precursors, that their precepts can be genuinely tested. Among light nuclei, the masses of analog states are related by the "isobaric multiplet mass equation" (see Cerny, 1968; Jänecke, 1969) which is derived by considering the effects of charge-dependent forces in first-order perturbation theory. Deviations from its predictions are very small and reflect the effects of Coulomb mixing, but to detect them, the masses of at least four members of an isobaric multiplet must be accurately known. Light precursors, such as ^9C, provide an essential link in the study of accessible multiplets.

F. SPECIAL APPLICATIONS TO FUNDAMENTAL QUESTIONS

The discussion so far has been concerned mostly with general spectroscopic properties shared by many nuclei. Delayed particle emission can in certain special instances shed light on specific questions of a more fundamental nature, and this chapter will conclude with a brief description of several such instances.

1. CVC Theory

The experiment of Barnes *et al.* (1958) studying the delayed α decay of ^8Li has already been mentioned in Section IV.D as providing evidence that the Gamow–Teller interaction is axial-vector in character. A later experiment by Nordberg *et al.* (1962) measured β–α angular correlations in the decay of ^8Li and its mirror ^8B. Denoting the angle between α and β directions by ϕ (see Fig. 18), the angular correlation is written

$$W(\phi) = 1 + A\cos\phi + B\cos^2\phi \qquad (20)$$

The term in $\cos\phi$ is a kinematic effect of the same type as that which caused the α energy to change with ϕ in Barnes's experiment. The value of B, however, depends upon "weak magnetism," the effects of which can be isolated by measuring the difference $\delta = B(^8\text{Li}) - B(^8\text{B})$. The experimental results provided important confirmation for predictions of the conserved vector current (CVC) theory.

A more recent experiment by Oakey and Macfarlane (1970) measured $W(\phi)$ for the β^+ decay of ^{20}Na to an α-unstable 2^+ state at 7.415 MeV in ^{20}Ne. In this case, the measured value of B was a factor of six larger than the prediction of CVC assuming the single-particle value for the M1 γ-transition rate from the $T = 1$ analog state in ^{20}Ne. The interpretation of these results in terms of contributions from other second-order matrix elements must await measurements of the relevant γ-decay strengths, but correlation measurements for other transitions appear to be warranted.

2. Parity Violation

Beta-delayed particle emission has produced the only experimental example in light nuclei of parity nonconservation occurring through the effects of nonleptonic weak interactions. The 2^- state at 8.87 MeV in ^{16}O is fed by an allowed (1%) β-decay branch from ^{16}N and is 1.71 MeV above the α-particle separation energy. Being an unnatural-parity state, its decay by α emission to the 0^+ ground state of ^{12}C is forbidden.

The precursor ^{16}N was obtained by Hättig et al. (1970) from the reaction $^{15}\text{N}(d, p)^{16}\text{N}$; data were collected during a 600 hr bombardment with 6 μA of 2.8-MeV deuterons. The dominant feature of the α spectrum was a peak from the decay of the 1^- state at 9.61 MeV in ^{16}O which is fed by a $1.2 \times 10^{-3}\%$ β-decay branch, but a detailed statistical analysis indicated the existence of a weak peak from the parity-forbidden decay of the 2^- state. The measured α width of $(1.8 \pm 0.8) \times 10^{-10}$ eV corresponds to an admixed 2^+ component of about $10^{-11}\%$. Since α decay can only occur from $T = 0$ configurations, the measured admixture must be caused by strangeness nonchanging currents, and in fact its magnitude agrees well with calculations involving rho exchange (e.g., Henley et al., 1969). It must be hoped that the inevitable increase in the number of known delayed-particle precursors will lead to other decay schemes which are favorable for this type of measurement. Independent confirmation of such an important result would have great value.

ACKNOWLEDGMENTS

I should like to thank J. Cerny, P. G. Hansen, V. A. Karnaukhov, R. D. Macfarlane, G. Rudstam, and V. I. Goldanskii for their correspondence and the communication of data prior to publication. Comments and suggestions for improvement of the first draft of this chapter were made by P. G. Hansen, R. D. Macfarlane, V. A. Karnaukhov, A. M. Poskanzer, and I. S. Towner; their efforts will be appreciated as much by the reader as they were by the author.

References

Alburger, D. E., Donovan, P. F., and Wilkinson, D. H. (1963). *Phys. Rev.* **132**, 334.

Alburger, D. E., and Wilkinson, D. H. (1971). *Phys. Rev.* C 3, 1492.

Bacso, I., Bogdanov, D. D., Darocsy, S., Karnaukhov, V. A., and Petrov, L. A. (1968). *Sov. J. Nucl. Phys.* **7**, 689.

Barnes, C. A., Fowler, W. A., Greenstein, H. B., Lauritsen, C. C., and Nordberg, M. E. (1958). *Phys. Rev. Lett.* **1**, 328.

Barton, R., McPherson, R., Bell, R. E., Frisken, W. R., Link, W. T., and Moore, R. B. (1963). *Can. J. Phys.* **41**, 2007.

Batchelor, R., and Hyder, H. R. McK. (1956). *J. Nucl. Energy* **3**, 7.

Berlovich, E. Ye. (1970). CERN rep. 70–30, p. 497.

Blatt, J. M., and Weisskopf, V. F. (1952). "Theoretical Nuclear Physics," Chapter 8. Wiley, New York.

Blin-Stoyle, R. J., and Freeman, J. M. (1970). *Nucl. Phys.* **A150**, 369.

Bogdanov, D. D., Darotsi, Sh., Karnaukhov, V. A., Petrov, L. A., and Ter-Akopian, G. M. (1968a). *Sov. J. Nucl. Phys.* **6**, 650.

Bogdanov, D. D., Bacho, I., Karnaukhov, V. A., and Petrov, L. A. (1968b). *Sov. J. Nucl. Phys.* **6**, 807.

Bogdanov, D. D., Bochin, V. P., Karnaukhov, V. A., and Petrov, L. A. (1971). *Proc. Int. Conf. Heavy-Ion Phys.* Dubna rep. D7-5769, p. 299.

Borg, S., Bergström, I., Holm, G. B., Rydberg, B., de Geer, L.-E., Rudstam, G., Grapengiesser, B., Lund, E., and Westgaard, L. (1971). *Nucl. Instrum. Methods* **91**, 109.

Cameron, A. G. W. (1956). *Can. J. Phys.* **35**, 666.

Cerny, J. (1968). *Ann. Rev. Nucl. Sci.* **18**, 27.

Cerny, J., Cardinal, C. U., Evans, H. C., Jackson, K. P., and Jelley, N. A. (1970a). *Phys. Rev. Lett.* **24**, 1128.

Cerny, J., Esterl, J. E., Gough, R. A., and Sextro, R. G. (1970b). *Phys. Lett.* **33B**, 284.

Cerny, J., Goosman, D. R., and Alburger, D. E. (1971). *Phys. Lett.* **37B**, 380.

Cerny, J., Gough, R. A., Sextro, R. G., and Esterl, J. E. (1972). *Nucl. Phys.* **A188**, 666.

Chen, Y. S., Tombrello, T. A., and Kavanagh, R. W. (1970). *Nucl. Phys.* **A146**, 136.

Cuttler, J. M., Shalev, S., and Dagan, Y. (1969). *Trans. Amer. Nucl. Soc.* **12**, 63.

del Marmol, P. (1969). *Nucl. Data* **A6**, 141.

Duke, C. L., Hansen, P. G., Nielsen, O. B., and Rudstam, G. (1970). *Nucl. Phys.* **A151**, 609.

Esterl, J. E., Hardy, J. C., Sextro, R. G., and Cerny, J. (1970). *Phys. Lett.* **33B**, 287.

Esterl, J. E., Sextro, R. G., Hardy, J. C., Ehrhardt, G. J., and Cerny, J. (1971). *Nucl. Instrum. Methods* **97**, 229.

Esterl, J. E., Allred, D., Hardy, J. C., Sextro, R. G., and Cerny, J. (1972). *Phys. Rev.* C **6**, 373.

Fujita, J-I., and Ikeda, K. (1965). *Nucl. Phys.* **67**, 145.

Fujita, J-I., Futami, Y., and Ikeda, K. (1968). *Proc. Int. Conf. Nucl. Structure, Tokyo. Suppl. J. Phys. Soc. Japan* **24**, 437.

Gamow, G. (1930). *Nature (London)* **126**, 397.

Garvey, G. T., Gerace, W. J., Jaffe, R. L., Talmi, I., and Kelson, I. (1969). *Rev. Mod. Phys.* **41**, S1.

Gilat, J., O'Kelley, G. D., and Eichler, E. (1963). Oak Ridge rep. ORNL-3488, p. 4.

Goldanskii, V. I. (1966). *Ann. Rev. Nucl. Sci.* **16**, 1.

Gough, R. A., Sextro, R. G., Cerny, J. (1972). *Phys. Rev. Lett.* **28**. 510.

Hansen, P. G., Nielsen, H. L., Wilsky, K., Alpsten, M., Finger, M., Lindahl, A., Naumann, R. A., and Nielsen, O. B. (1970). *Nucl. Phys.* **A148**, 249.

Hardy, J. C. (1971a) *Proc. Int. Conf. Heavy-Ion Phys.* Dubna Rep. D7-5769, p. 261.

Hardy, J. C. (1971b). At. Energy of Canada Limited rep. AECL-3912, p. 23.

Hardy, J. C. (1972). *Nucl. Data Tables* **11**, 327.

Hardy, J. C., and Bell, R. E. (1965). *Can. J. Phys.* **43**, 1671.

Hardy, J. C., and Margolis, B. (1965). *Phys. Lett.* **15**, 276.

Hardy, J. C., and Verrall, R. I. (1964). *Phys. Lett.* **13**, 148.

Hardy, J. C., and Verrall, R. I. (1965). *Can. J. Phys.* **43**, 418.

Hardy, J. C., Verrall, R. I., and Bell, R. E. (1966). *Nucl. Phys.* **81**, 113.

Hardy, J. C., Esterl, J. E., Sextro, R. G., and Cerny, J. (1971). *Phys. Rev. C* **3**, 700.

Hättig, H., Hünchen, K., Roth, P., and Wäffler, H. (1969). *Nucl. Phys.* **A137**, 144.

Hättig, H., Hünchen, K., and Wäffler, H. (1970). *Phys. Rev. Lett.* **25**, 941.

Henley, E. M., Keliher, T. E., and Yu, D. U. L. (1969). *Phys. Rev. Lett.* **23**, 941.

Hornshøj, P., Wilsky, K., Hansen, P. G., Jonson, B., Alpsten, M., Andersson, G., Appelqvist, Å., Bengtson, B., and Nielsen, O. B. (1971a). *Phys. Lett.* **34B**, 591.

Hornshøj, P., Wilsky, K., Hansen, P. G., Jonson, B., and Nielsen, O. B (1971b). *Proc. Int. Conf. Heavy-Ion Phys.* Dubna rep. D7-5769, p. 249.

Hornshøj, P., Wilsky, K., Hansen, P. G., Jonson, B., and Nielsen, O. B. (1972a). *Nucl. Phys.* **A187**, 599.

Hornshøj, P., Wilsky, K., Hansen, P. G., Jonson, B., and Nielsen, O. B. (1972b). *Nucl. Phys.* **A187**, 609.

Hornshøj, P., Wilsky, K., Hansen, P. G., and Jonson, B. (1972c). *Nucl. Phys.* **A187**, 637.

Jackson, K. P., Cardinal, C. U., Evans, H. C., Jelley, N. A., and Cerny, J. (1970). *Phys. Lett.* **33B**, 281.

Jänecke, J. (1968). *Nucl. Phys.* **A114**, 433.

Jänecke, J. (1969). *In* "Isospin in Nuclear Physics" (D. H. Wilkinson, ed.), Chapter 8. North-Holland Publ., Amsterdam.

Jones, K. W., Harris, W. R., McEllistrem, M. T., and Alburger, D. E. (1969). *Phys. Rev.* **186**, 978.

Karnaukhov, V. A. (1970). *Sov. J. Nucl. Phys.* **10**, 257.

Karnaukhov, V. A. (1972). Private communication.

Karnaukhov, V. A., and Ter-Akopian, G. M. (1967). *Arkiv Fysik* **36**, 419.

Karnaukhov, V. A., Ter-Akopian, G. M., and Subbotin, V. G. (1963). *In Proc. Asilomar Conf. Reactions between Complex Nuclei*, (A. Ghiorso, R. M. Diamond, and H. E. Conzett, eds.), p. 434. Univ. of California Press; see also Dubna rep. JINR P-1072.

Karnaukhov, V. A., Bogdanov, D. D., and Petrov, L. A. (1970). CERN rep. 70-30, p. 457.

Kelson, I., and Garvey, G. T. (1966). *Phys. Lett.* **23**, 689.

Lauritsen, T., Barnes, C. A., Fowler, W. A., and Lauritsen, C. C. (1958). *Phys. Rev. Lett.* **1**, 326.

Leang, C.-F. (1965). *C. R. Acad. Sci. Paris* **260**, 3037.

Lederer, C. M., Hollander, J. M., and Perlman, I. (1967). "Table of Isotopes," pp. 575–576. Wiley, New York.

Lewis, W. B., Burcham, W. E., and Change, W. Y. (1937). *Nature (London)* **139**, 24.

Macfarlane, R. D. (1967). *Archiv Fysik* **36**, 431.

Macfarlane, R. D., and Griffioen, R. D. (1963). *Nucl. Instrum. Methods* **24**, 461.

Macfarlane, R. D., Oakey, N. S., and Nickles, R. J. (1971). *Phys. Lett.* **34B**, 133.

Mani, G. S., Melkanoff, M. A., and Iori, I. (1963). Centre D'Etudes Nucléaires de Saclay rep. C.E.A.-2379 (unpublished).

Margolis, B., and de Takacsy, N. (1966). *Can. J. Phys.* **44**, 1431.

Matt, E., Pfander, H., Rieseberg, H., and Soergel, V. (1964). *Phys. Lett.* **9**, 174.

Mosher, J. M., Kavanagh, R. W., and Tombrello, T. A. (1971). *Phys. Rev. C* **3**, 438.

Myers, W. D., and Swiatecki, W. J. (1965). Lawrence Radiat. Lab. Rep. UCRL-11980 (unpublished).

Nielsen, K. B., Johansen, H., and Rudstam, G. (1971). Private communication referred to by Shalev and Rudstam (1972).

Nordberg, M. E., Morinigo, F. B., and Barnes, C. A. (1962). *Phys. Rev.* **125**, 321.

Oakey, N. S., and Macfarlane, R. D. (1970). *Phys. Rev. Lett.* **25**, 170.

Peker, L. K., Volmyansky, E. I., Bunakov, V. E., and Ogloblin, S. G. (1971). *Phys. Lett.* **36B**, 547.

Polichar, R. M., Steigerwalt, J. E., Sunier, J. W., and Richardson, J. R. (1967). *Phys. Rev.* **163**, 1084.

Pollard, E., Schultz, H. L., and Brubaker, G. (1938). *Phys. Rev.* **53**, 351.

Poskanzer, A. M., McPherson, R., Esterlund, R. A., and Reeder, P. L. (1966). *Phys. Rev.* **152**, 995.

Rasmussen, J. O. (1965). *In* "Alpha-, Beta- and Gamma-Ray Spectroscopy" (K. Siegbahn, ed.), Vol. I, pp. 701–743. North-Holland Publ., Amsterdam.

Ravn, H. L., Sundell, S., and Westgaard, L. (1972). *Phys. Lett.* **39B**, 337.

Reeder, P. L., Poskanzer, A. M., Esterlund, R. A., and McPherson, R. (1966). *Phys. Rev.* **147**, 781.

Roberts, R. B., Meyer, R. C., and Wang, P. (1939). *Phys. Rev.* **55**, 510.

Rutherford, E. (1919). *Phil. Mag.* **37**, 571.

Rutherford, E., and Wood, A. B. (1916). *Phil. Mag.* **31**, 379.

Rytz, A. (1953). *J. Rech. Centre Nat. Rech. Sci., Lab. Bellevue (Paris)* **25**, 254.

Schopper, H. F. (1966). "Weak Interactions and Nuclear Beta Decay" Chapter 5. North-Holland Publ., Amsterdam.

Schwalm, D., and Povh, B. (1966). *Nucl. Phys.* **89**, 401.

Sextro, R. G., Gough, R. A., and Cerny, J. (1972). Private communication.

Shalev, S., and Rudstam, G. (1972). *Phys. Rev. Lett.* **28**, 687.

Siivola, A. T. (1965). *Phys. Rev. Lett.* **14**, 142.

Skobelev, N. K. (1972). *Yadernaya Fiz.* **15**, 444.

Steigerwalt, J. E., Sunier, J. W., and Richardson, J. R. (1969). *Nucl. Phys.* **A137**, 585.

Talbert, W. L. (1970). *Phys. Rev. C* **1**, 1135.

Talbert, W. L., Tucker, A. B., and Day, G. M. (1969). *Phys. Rev.* **177**, 1805.

Torgerson, D. F., and Macfarlane, R. D. (1971). *Proc. Int. Conf. Heavy-Ion Phys.* Dubna Rep. D7-5769, p. 288.

Torgerson, D. F., Oakey, N. S., and Macfarlane, R. D. (1971). *Nucl. Phys.* **A178**, 69.

Verrall, R. I. (1968). New Results in Delayed Proton Emission. McGill Univ. Thesis (unpublished).

Verrall, R. I., and Bell, R. E. (1969). *Nucl. Phys.* **A127**, 635.

Wigner, E. P. (1939). *Phys. Rev.* **56**, 519.

Wilkinson, D. H. (1970). *Phys. Lett.* **31B**, 447.

Wilkinson, D. H. (1971). *Phys. Rev. Lett.* **27**, 1018.

Wilkinson, D. H., and Alburger, D. E. (1971). *Phys. Rev. Lett.* **26**, 1127.

Zeldes, N., Grill, A., and Simievic, A. (1967). *Kgl. Danske Videnskab. Selskab Mat.-fys.* **3**, no. 5.

VIII.C IN-BEAM ATOMIC SPECTROSCOPY

Indrek Martinson

RESEARCH INSTITUTE FOR PHYSICS, STOCKHOLM, SWEDEN

I. INTRODUCTION

Atomic spectroscopy deals with investigations of the energy levels of atoms and ions usually by observations of the photons emitted by these systems.

Such quantities as wavelengths, excited-state lifetimes, radiative transition probabilities, and atomic fine and hyperfine level structure form the experimentally required information. It has been shown in recent years that several of these problems can be very successfully studied with in-beam techniques, i.e., by using accelerated ion beams which are stripped and excited in a thin solid foil or in a gas target. The deexcitation radiation emitted by the ions can be easily observed, and therefore the excited ion beam constitutes a light source with several quite attractive properties, such as very high chemical and isotopic purity and excellent time resolution. Present heavy-ion accelerators cover a wide range of energies and, consequently, atoms in many ionization states can be investigated. The results are not only of importance for atomic theory and quantum electrodynamics but they often have significant applications in astrophysics, plasma physics, and laser physics.

Investigations of the light emitted by fast, excited ions were initiated by Wien (1919), who observed the wavelength spectra of canal rays, i.e., positive ions from discharge experiments which had traversed a tiny hole in the cathode. The spectra showed characteristic lines of different decay lengths, which were correlated to the radiative lifetimes of excited levels in ions and atoms. Due to systematic errors, the lifetime results were unreliable and the technique soon became of "historical interest." Not until the early 1960's were experiments developed that achieved the aims of Wien's work. Kay (1963) and Bashkin (1964) independently discovered that ions from Van de Graaff accelerators were strongly excited after passing a very thin foil. Careful spectroscopic studies made it clear that the emitted light (line spectra) definitely originated from the accelerated ions, whereas no excitation of the rest gas in the target chamber ($< 10^{-5}$ Torr) was found. Furthermore, by accelerating various elements, different characteristic spectra were found. Bashkin (1964) surveyed in detail the possibilities of this experimental method—called beam–foil spectroscopy—for investigation of highly ionized atoms, atomic lifetimes and their astrophysical applications, fine-structure effects, etc.

In the subsequent years, this technique has developed rapidly and produced a wealth of quantitative results. The accelerated ions can also be excited by means of a differentially pumped gas cell, and the method of beam–gas spectroscopy is also useful for studies of atomic spectra and lifetimes, as well as various atomic collision mechanisms. This chapter will be limited to the investigations of free atoms and ions. For detailed information about atomic collision experiments, the review article by de Heer (1966) can be consulted.

A. Survey of the Experimental Methods

The experimental equipment needed for in-beam atomic experiments consists roughly of an accelerator, e.g., a Van de Graaff, an optical spectrograph or monochromator, a photon detector, and some standard equipment in electronics, including amplifiers, discriminators, counters, etc. Figure 1 shows a typical experimental arrangement for studies of atomic spectra and radiative lifetimes of excited states.

Monoenergetic ions from the accelerator are mass-analyzed in a magnet and then directed through a thin exciter foil (self-supporting carbon foils, 5–20 $\mu g/cm^2$, are often used) or a differentially pumped gas cell (typical pressure 10^{-3} Torr). The light emitted by the excited ions is dispersed in a grating spectrometer and detected with a photomultiplier (or a photographic plate). After amplification and discrimination, the pulses are fed into a multiscaler or via a ratemeter displayed on a strip chart recorder. When the grating is rotated, a wavelength spectrum is obtained. Part of such a spectrum is shown in Fig. 2. This particular spectrum was obtained by sending 300-keV Fe^+ ions through a 5 $\mu g/cm^2$ foil.

By setting the spectrometer on a given spectral line and measuring how

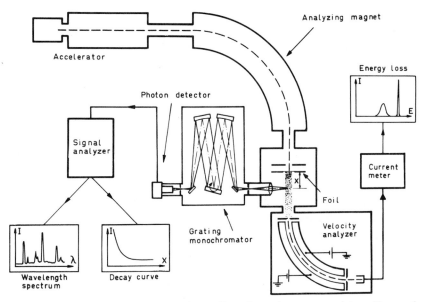

Fig. 1. Experimental arrangement for studies of atomic spectra and transition probabilities with the beam–foil technique.

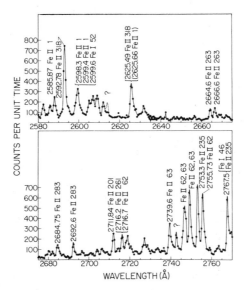

Fig. 2. Beam–foil spectrum of iron beam at 300 keV in the UV, observed with a 0.3 m scanning monochromator (Smith *et al.*, 1970, used with permission of North-Holland Publishing Co., Amsterdam).

the intensity of this line varies with the distance x downstream from the exciter, a decay curve can be obtained, which in the simplest case, follows the relation

$$I(x) = I(0) e^{-x/v\tau} \tag{1}$$

where $I(0)$ is the initial intensity and $I(x)$ the intensity at a distance x from the foil, v is the velocity of the ions after the exciter, and τ is the radiative lifetime of the decaying upper level. Atomic lifetimes are typically of the order of 10^{-7}–10^{-10} sec (allowed transitions) and they are in most cases easily measurable with ion beam techniques. For example, 2-MeV $^{14}N^+$ ions travel 5.3 mm/nsec. One of the shortest decay times measured with in-beam methods, $\tau = 0.084 \pm 0.008$ nsec for the N(V) 2s ^{2}S–3p ^{2}P transition,[†] corresponds to a $1/e$ decay length of 0.45 mm (Heroux, 1967).

The velocity v can be determined in several ways; for example, with an electrostatic analyzer as shown in Fig. 1. (See also Section I.B.3.) By applying

[†] Conventional spectroscopic notation is being used; N(V) means nitrogen ionized four times. The upper level of the transition is 3p ^{2}P.

electric or magnetic fields over the excited beam, ions of various charge states can be separated. It is also easy to perturb the foil-excited ions using external ac or dc fields. With such techniques, finer details of the atomic structure, e.g., fine-structure splittings, can be investigated (Section V).

1. *Accelerators*

Several kinds of accelerators, including Van de Graaffs, Cockcroft–Walton generators, electromagnetic isotope separators, and heavy-ion linear accelerators, have already been applied for in-beam atomic spectroscopy. So far, the energies used range from about 10 keV to more than 400 MeV. At the lowest energies, the energy loss in the foil becomes a large fraction of the incoming particle energy, and therefore gas exciters are much more suitable. A low-energy accelerator equipped with a radio frequency (rf) ion source has been described by Chin-Bing et al. (1970). This accelerator gives mass-analyzed beams of more than 1 μA. In the low-energy region, isotope separators equipped with universal ion sources are also being used. Bickel et al. (1969) initiated beam–foil studies with an 80-kV isotope separator capable of giving μA currents of most elements. At such low energies, the electron capture cross sections are high, and these accelerators are therefore well suited for studies of transitions in neutral atoms. However, higher charge states can also be investigated. In a beam–foil experiment, Curtis et al. (1971a) found intense P(V) transitions when P^{2+} ions were accelerated over 80 kV. A 600-kV heavy-ion accelerator, equipped with a universal ion source has been used by Andersen et al. (1969), who obtained beams of 1-μA Al^+ through a 2.5 mm-diameter foil. With many accelerators, the energy range can be further expanded by accelerating doubly or triply charged ions, but lower intensities must usually be taken into account.

An impressive amount of research in atomic spectroscopy has been done with Van de Graaff accelerators. Most investigators have used machines with terminal voltages of 400 kV (e.g., Chupp et al., 1968), 2 MV (e.g. Bashkin et al., 1966; Lewis et al., 1967), or 5.5 MV (e.g., Bakken et al., 1969). Elements such as H, He, C, N, O, Ne, and Ar, which are easily ionized with rf ion sources have been thoroughly studied in beam–foil experiments with Van de Graaffs. However, thermal ion sources have also been inserted in these accelerators, and spectra of Li (Buchet et al., 1967), Na (Brown et al., 1968; Dufay et al., 1971), and Fe (Whaling et al., 1969) have been investigated in this way.

Tandem Van de Graaffs have also been exploited in atomic physics and Sellin et al. (1968), Donnally et al. (1971), and Hallin et al. (1971) have studied transitions in highly ionized N, O, F, and Cl with beam energies ranging from 6 to 42 MeV. Very high ion energies have also been obtained with heavy-ion linear accelerators. Using the Orsay linear accelerator (1.15 MeV/nucleon), Dufay *et al.* (1970) and Denis *et al.* (1971) obtained μA beams of energetic C, N, O, and Ar ions. The Berkeley linear accelerator HILAC, first employed for atomic lifetime measurements by Berkner *et al.* (1965), has more recently been used by Marrus and Schmieder (1970, 1972), who accelerated beams of Si, S, and Ar to 288, 330, and 412 MeV, respectively (10.3 MeV/nucleon).

2. *Spectrometers*

While most of the observations of the foil-excited ions have been made by studying light in the visible and UV regions, both X rays and electrons originating from the ions have also been detected. In the wavelength region above 2000 Å, comparatively simple optical spectrometers can be used, whereas vacuum instruments are needed below 2000 Å. Often normal-incidence spectrometers equipped with concave gratings are used, but various other types have also been successfully tried. Most investigators have employed relatively fast, 0.25–1.0-m commercial instruments, such as the Mc-Pherson 225 1-m normal-incidence scanning monochromator, which has an inverse linear dispersion of 16.7 Å/mm (600 lines/mm grating) and covers the wavelength region 400–6000 Å. For observations of shorter wavelengths, grazing-incidence spectrometers are usually needed.

Using a 2-m grazing-incidence spectrometer, Heroux (1967) extended the measurements of beam–foil radiation down to 100 Å. More recently, Hallin *et al.* (1971) have reached as far as 15 Å using a commercial 2-m grazing-incidence spectrometer. In most beam–foil and beam–gas experiments, photoelectric detection has been employed; however, the photographic method, i.e., by using a stigmatic spectrograph with the slit parallel to the excited beam, has its merits, especially in combination with image intensifiers (Brown *et al.*, 1968). For quantitative studies of lifetimes, careful considerations of the plate darkening are needed. Also, the Rutherford scattering of the ions in the foil may present some problems (Assousa *et al.*, 1970). Most authors therefore prefer the directness and simplicity of the photoelectric detection techniques. In order to increase the intensity for lifetime measurements, narrowband interference filters have also been tried (cf., e.g., Goodman and Donahue, 1966). This technique demands that the spectral lines be well isolated, however.

In very highly ionized atoms, the de-excitation radiation lies in the X-ray region. In the experiments of Marrus and Schmieder (1970, 1972) and Schmieder and Marrus (1970), solid-state Si(Li)X-ray detectors were used to detect 1–4 keV photons from Si(XIII), S(XV), Ar(XVII), and Ar(XVIII). For their study of autoionization electrons (400–1600 eV) emitted by highly stripped O and F beams, Donnally *et al.* (1971) used an electrostatic, cylindrical electron spectrometer.

3. *Detectors*

When photoelectric methods are used, the photons are generally registered with commercial photomultipliers, e.g., the EMI 6256S. It is usually advisable to cool the photomultiplier with dry ice or liquid N_2 so as to increase the signal-to-noise ratio. This is particularly important in lifetime measurements where the subtraction of a high background may introduce large uncertainties. In the region below 1500 Å, open detectors, such as the Bendix Channeltron, are very useful. The open detectors have very low noise and they can also register electrons and ions. A detailed discussion of various detectors is given by Heroux (1970).

B. DATA ANALYSES

1. *Wavelengths*

It is desirable in all spectroscopic work to use as high resolution as possible. With in-beam methods, certain problems complicate high-resolution experiments. The high particle velocities v introduce wavelength shifts and line broadenings which are difficult to correct for. The relativistic Doppler formula can be approximated as follows:

$$\lambda_0 - \lambda = \lambda_0[(v/c)\cos\theta - \tfrac{1}{2}(v^2/c^2)] \tag{2}$$

where λ and λ_0 are the observed and the rest-frame wavelengths, respectively, and θ is the angle between the direction of the observation and the ion beam. With the side-on geometry of Fig. 1 ($\theta = 90°$), the first-order Doppler shift vanishes. However, besides the second-order wavelength displacement, additional corrections must be considered. Due to the finite acceptance angle α of the spectrometer, there also arises a line broadening given by

$$\Delta\lambda \sim \lambda\alpha v/c \tag{3}$$

In typical beam–foil experiments, this broadening may exceed 5 Å at a wavelength of 4000 Å, which certainly rules out precision spectroscopy. However, by refocusing the spectrometer for a moving light source, Stoner and Leavitt (1971) practically eliminated this uncertainty. Figure 3 displays one of their results. Although the aperture effects can be reduced effectively, there remains a contribution to the line broadening caused by the beam divergence after the foil (Rutherford scattering). This effect should be kept as low as possible, by using very thin foils of low Z. A detailed study of the influence of this effect on the line widths has been presented by Stoner and Radziemski (1970). The results show that gas targets are preferable for high-resolution spectroscopic studies. Recently, Stoner and Radziemski (1972) found line widths as narrow as 0.1 Å in the 5000-Å range.

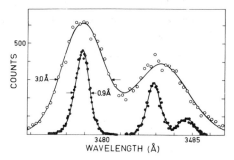

Fig. 3. The effects of refocusing the spectrometer with respect to the movable light source (fast beam). The line width decreases from 3.0 Å to 0.9 Å and all three components in a N(IV) multiplet are clearly resolved (Stoner and Leavitt, 1971).

The end-on configuration ($\theta = 0°$) has also been successfully used (Bakken *et al.*, 1969; Dufay, 1970). The technique consists in placing a lens collinear with the beam axis and collecting all the light emitted in the forward direction. The spectral lines exhibit appreciable blue shifts due to the first-order Doppler effect, but this can be taken into account. The line widths are quite satisfactory, however: Bakken *et al.* report 0.7 Å at 5000 Å. Unfortunately, this geometry cannot be used for decay-time measurements.

The high chemical and isotopic purity of the accelerated and mass-analyzed ion beam facilitates spectral analyses. For example, if in a beam–foil experiment of Be, the bending magnet selects mass 9, all the observed spectral lines should belong to transitions in Be(I)–Be(IV). However, in some cases, spurious coincidences must be considered. Studies of Si spectra are easily obscured by contaminations because the mass-28 beam may contain Si^+, CO^+, and N_2^+, the last originating from impurities in the ion source. Berry *et al.* (1971a) were able to circumvent this particular problem by selecting the less abundant isotope ^{29}Si. In many cases—especially when heavy projectiles of low kinetic energy are used—spectral lines characteristic of the exciter foil or gas can be observed but they can be recognized, e.g., from their much smaller Doppler shifts.

2. Line Identification

Even if a particular beam–foil or beam–gas experiment may be directed toward lifetime determinations and not a complete spectral study, a careful analysis of all the observed lines is always useful. It is generally an easy task to identify those transitions in the spectra that have been reported earlier. For this purpose, the "Atomic Energy Levels" of Moore (1949, 1952, 1958) and other compilations can be consulted. The existing spectroscopic material is certainly not complete, however, and in-beam experiments may produce lines that have not been reported in the literature. Furthermore, the collision processes may strongly excite atomic levels which are insignificantly populated with other spectroscopic light sources (arcs and sparks).

In the process of classifying previously unreported transitions, the wavelength measurement is usually followed by a charge-state assignment. The distribution of charges after the exciter is highly sensitive to the incoming beam energy. For a first check, the data on equilibrium charge distributions are very useful (e.g., Marion and Young, 1968). Such data show, for example, that 2-MeV incoming O^+ ions appear mostly as O^{3+} and O^{4+} after the foil.

Several methods have been tried for making more definite assignments. Kay (1965) noted that known spectral lines of nitrogen belonging to various charge states showed different intensity variations with beam velocity. Such regularities can guide one in deducing charge states for unknown lines. Figure 4 illustrates such intensity–velocity variations for carbon ions.

Another possibility consists in applying a transverse electric field across the foil-excited beam. In this way, the various electric charges can be effectively separated (Malmberg et al., 1965; Fink, 1968). When this method is used, two directions of observation are possible. The spectrometer axis is either (a) perpendicular to both beam and field directions or (b) perpendicular to the beam but parallel to the field (in this case, slotted field plates must be used). When arrangement (a) is applied, the separated charge states can be viewed individually with the optical spectrometer, and this gives clean, charge-analyzed spectra. With (b), the ions receive a velocity component toward the observer or away from him, depending on the field direction. The observed spectral lines therefore exhibit blue or red shifts. Simple calculations show that in this case, the relative shift $\Delta\lambda/\lambda$ is proportional to the ionic charge

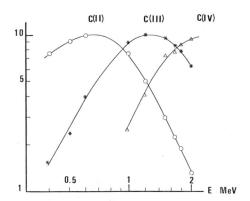

Fig. 4. The intensity variations for C(II), C(III), and C(IV) transitions as a function of beam energy (Poulizac *et al.*, 1971).

after the foil. Using a 4-MeV N_2^+ beam, Bashkin and Leavitt (1971) observed a red shift of 2.5 Å for the N(IV) 1718-Å transition when a field of 60 kV/cm was applied.

The intensity–velocity method for charge-state determinations is not a rigorous one, and examples have been found where transitions belonging to the same ionization state show different variations with the beam energy (Dufay, 1970). Not only the ionization state, but also the excitation energy of the transition within a given ion plays a role. The charge splitting methods appear as more unambiguous but the practical difficulties in applying the fields (typically 50 kV/cm) make observations of rapidly decaying lines [such as the 0.084-nsec N(V) transition mentioned in Section I.A] very difficult because these may have decayed before entering the field region. Furthermore, the strong fields may cause broadening of the spectral lines (Bashkin and Carriveau, 1970) due to the Stark effect. Whatever method one selects, it is important to realize that the charge assignment is only a first step toward final classification.

The analyses of optical spectra are thoroughly discussed by Edlén (1964). Several of the methods described can be directly applied to beam–foil and beam–gas spectra. To facilitate line identifications, comparisons with similar spectra are very useful. We may safely expect, e.g., Be(I), B(II), C(III), N(IV), O(V), etc., to show very similar level structure, since they all have two electrons outside the $1s^2$ core. These spectra form the Be(I) isoelectronic sequence. Several regularities can be found for such isoelectronic systems. For transitions between two given levels, we can form the ratio E/ζ, where E is the transition energy (expressed, e.g., in cm^{-1} by inverting the wavelength) and $\zeta = 1$ for the neutral atom, $\zeta = 2$ for the singly ionized atom, etc. By displaying E/ζ versus ζ, we expect to find relatively smooth curves, by which new spectral lines can be classified. Additional valuable graphical methods for isoelectronic sequences are discussed by Edlén (1964).

Atomic terms can be written as

$$T = R\zeta^2/(n - \mu)^2 \tag{4}$$

where T is the term value (the difference between the excitation energy and the ionization limit), R the Rydberg constant, n the principal quantum number, and μ the quantum defect, which is a measure of the deviation from the purely hydrogenic case. The quantum defects possess regularities which facilitate line identifications. Within a spectral series, such as $1s^2nd\ ^2D$ in Li(I) ($n = 3, 4, 5, ...$), the quantum defects decrease but slightly with increasing n. More irregular trends may indicate that the terms are perturbed. The quantum defects are largest for the S terms ($L = 0$) and they decrease with increasing L because the orbits become less penetrating. Finally, there are isoelectronic regularities for the quantum defects, and for a given level, say $3p\ ^1P$ in the He(I) sequence, μ decreases with increasing atomic number Z.

If lines belonging to the same multiplet can be resolved, the measured relative intensities can be compared to the theoretical LS coupling intensity ratios (White and Eliason, 1934), which may provide additional checks of the correctness of an assignment. The tabulated energy levels (e.g., Moore, 1949) can also be used to identify transitions obeserved in beam–foil or beam–gas spectra. This is often done by searching for all possible allowed combinations between the known terms by using computer programs (Lewis et al., 1967; Fink et al., 1970). However, the results must be interpreted with great care. Especially when the wavelength resolution is moderate, spurious coincidences may easily occur. Decay-time measurements can also be valuable for energy level studies. If a given term is depopulated in several ways, the decay-time measurements should give identical results, within the experimental uncertainties. This fact serves as an extra check of the assignments.

Finally, it should be stressed that theoretical calculations of atomic energy levels may be very accurate, even for relatively complex atoms. In simpler cases, e.g., the He(I) isoelectronic sequence, the calculations can compete favorably with the most precise spectroscopic measurements.

3. *Lifetimes*

As mentioned briefly in Section I.A, measurements of intensity decays for well-isolated spectral lines yield the radiative lifetimes of the corresponding upper levels. By using Eq. (1) and plotting $\ln I(x)$ versus x, a straight line should be obtained, the slope of which $-(1/v\tau)$ gives the lifetime τ if the velocity v is known.

The accelerator energies are usually calibrated with various well-known nuclear reactions, and the incoming ion energies are in general precisely known. However, in many cases, particularly at low energies, the energy loss in the foil must be taken into account. The energy loss of 1-MeV incoming O^+ ions in a 10 $\mu g/cm^2$ C-foil is approximately 80 keV, and neglecting this would thus produce a 4% systematic lifetime error. If the foil thickness is known, corrections can be made by using the tables of Northcliffe and Schilling (1970). Direct determinations of the ion energy distribution after the foil as shown in Fig. 1 are also valuable, particularly when the foil thickness is not accurately known. Using an 18-cm 90° electrostatic analyzer, Bergström et al. (1969) measured the energy losses for light atoms. Boron ions $^{11}B^+$, of 56.5 keV and a full-width at half-maximum of less than 0.4 keV, were found to lose 8.8 keV in a 10 $\mu g/cm^2$ C-foil, accompanied by a width increase to 2.3 keV. Consistent results were also found for He, Li, Be and C. Velocity determinations from Doppler shifts are also possible. By observing the light emitted by a foil-excited 1.2-MeV oxygen beam, Pinnington (1970) found that a spectral line at 4350 Å shifted by 33 Å when the viewing direction changed from $\theta = 90°$ to $\theta = 30°$. Also, resonance phenomena can be used for velocity measurements. Liu et al. (1971) showed that the excited beam velocity can be calibrated from known Landé g factors.

A simple way of obtaining the intensity decay of a spectral line consists in moving the foil in steps along the x direction and recording the number of photons for various distances. In order to obtain meaningful results, the beam fluctuations must be compensated for, e.g., by collecting the beam into a Faraday cup connected to a charge integrator. It is also possible to monitor the emitted light, however. Brand et al. (1970) discuss such an arrangement, consisting of two monochromators (one movable and the other stationary) which pass the same spectral line. An automatic method has been developed by Oona and Bickel (1969). It consists of a fast-scanning foil drive of constant velocity synchronized to a multichannel analyzer. By sweeping the x direction many times, beam fluctuations are averaged out.

The simple equation (1) is valid only when the repopulation of the decaying level under study can be neglected. In most cases, this is not the case. Ion–atom collisions are characterized by rather unselective excitation, and therefore cascading transitions from higher levels into the level under study must be taken into account. If the decaying level i is fed by an upper level j with lifetime τ_j, the intensity function becomes

$$I(x) = Ae^{-x/v\tau_i} + Be^{-x/v\tau_j} \tag{5}$$

where A and B are functions of the lifetimes τ_i and τ_j and the initial intensities. Depending on the relative magnitudes of these constants, the observed decay curve is a sum or a difference of two exponentials. In practice, the cascading situation can be even more complex, however, and additional exponentials must be included (Curtis *et al.*, 1970). Figure 5 shows three typical decay curves from lifetime studies with the beam–foil method.

The decomposition of the complex decay curves into the various exponentials is often quite difficult, particularly when the lifetimes are not too different from each other. Computer programs have been used for this decomposition. If possible, the cascade transitions should be directly measured and then subtracted from the decay curve of interest. A study of the energy levels helps in judging which levels are the main contributors to the cascading. If only one cascade is dominant, a method of undetermined multipliers introduced by Curtis *et al.* (1971b) is very helpful. Numerical differentiations and integrations of the intensity curve have also been used to enhance certain contributions and suppress others (Curtis *et al.*, 1971a). In order to minimize systematic errors introduced by cascading, the decays of the spectral lines can be measured at several beam velocities.

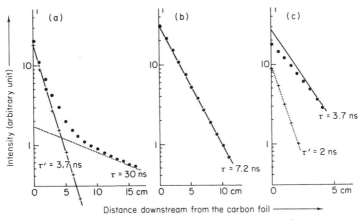

Fig. 5. Three different decay curves. (a) The decay of the 4076 Å line [C(II), $E = 1.2$ MeV] is influenced by a longer-lived cascade. (b) No cascade effects are observable for the 2297-Å line [C(III), $E = 1$ MeV]. (c) A shorter-lived cascade affects the initial decay of the 1335-Å line [C(II), $E = 1.5$ MeV] (Poulizac *et al.*, 1971).

Additional uncertainties in lifetime measurements are introduced by the scattering of the ions in the foil, which is particularly serious for low-velocity ions of heavy mass. Detailed discussions of various factors influencing the lifetime results can be found in a review article by Bickel (1968).

II. Results: Atomic Energy Levels

In spite of the imperfections concerning line shapes and resolution (Section I.B.1), in-beam experiments have already yielded much new information about wavelengths and atomic energy levels. Experience has shown that spectroscopy with accelerated ions is particularly well suited for studies of multiply excited atomic levels which are formed when more than one electron is excited. Often these states lie embedded in the continua above the ionization limit.

Ion beams can also be used for studies of hydrogenlike states in highly ionized atoms. Hydrogenlike states are characterized by high values of the quantum numbers n and l. In very highly stripped ions, forbidden transitions can be observed (e.g., Section III.C).

However, in addition to such results, new information about lower lying, singly excited levels has also been obtained. The lists of new lines observed with in-beam methods are quite impressive, but in several cases, much work was necessary before these lines were assigned to appropriate level schemes.

So far, spectra of more than 50 elements have been investigated with in-beam methods, and the charge states observed range from neutral atoms to atoms ionized 17 times. Transitions in Ar(XVIII) were observed by Marrus and Schmieder (1970). A few specific examples of the results are given in this section.

A. Doubly Excited States

The simplest atom in which doubly excited states can be observed is He(I). It may appear to be surprising that new transitions can be found in the He spectrum. The singly excited levels, denoted by $1snl$, are indeed thoroughly known, but much less experimental material can be found for the doubly excited states. The lowest of the latter, $2s^2$ ^1S, lies 57.9 eV above the He(I) $1s^2$ ^1S ground state and 33.3 eV above the He(I) ionization limit. Such levels usually decay by autoionization into He$^+$ and a free electron. However, autoionization via the electrostatic interaction, with typical probabilities of 10^{13}–10^{15} sec^{-1}, also has the very restrictive selection rules $\Delta L = 0$, $\Delta S = 0$, $\Delta J = 0$, and no parity change. Therefore certain levels, such as $2pnp$ ^1P, ^3P, $2pnd$ ^1D,

³D are metastable with respect to this process (there are no final states that are allowed by the selection rules). Instead, these levels decay with photon emission to singly excited levels. Two such transitions, at 309 Å (1s3p ¹P–2p3p ¹P) and 320 Å (1s2p ³P–2p² ³P), have been known for a long time (Moore, 1949). The doubly excited He(I) transitions were investigated by Berry *et al.* (1971b) in a beam–foil experiment. In addition to the lines just mentioned, a number of previously unknown He transitions appeared. Lifetime measurements for the doubly excited levels confirmed recent theoretical calculations (Drake and Dalgarno, 1970). Transitions from higher doubly excited He(I) levels, 2p3d ¹D and 2p3d ³D, have also been observed with the beam–foil method (Berry *et al.*, 1972a). The wavelengths of such transitions lie between 2000 and 3000 Å. The majority of these transitions have not been reported in He spectra obtained with other light sources. Their remarkable intensities in the beam–foil spectra show that the collision mechanisms very much favor the production of multiply excited configurations.

Similar doubly excited states can also be found for Li(I). Their excitation energies are more than ten times higher than the ionization energy. The "normal" system of Li(I) consists of doublets, i.e., levels of the type 1s²nl ²L (n = 2, 3, 4, ...), but the doubly excited states can also be quartets (all three electrons with parallel spins) such as 1s2snl ⁴L, 1s2pnl ⁴L, etc. Due to the ΔS = 0 selection rule, these quartets are metastable against autoionization. The first beam–foil studies of Li spectra showed that the doubly excited levels

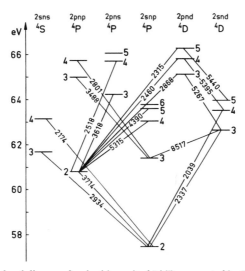

Fig. 6. Energy level diagram for doubly excited Li(I), suggested by Berry *et al.* (1972b).

were very strongly populated. Several experimental investigations of these states have been made in recent years. Figure 6 shows an energy level diagram for the Li(I) doubly excited system, largely based on the work of Berry *et al.* (1972b). Most of the transitions observed in beam–foil spectra have been indicated. Only four of these lines have been reported in the arc and spark spectra of lithium. Similar transitions have also been observed for Be(II), B(III), and C(IV), isoelectronic to Li(I) (cf., e.g., Martinson, 1970).

Figure 6 indicates that all doubly excited levels finally decay to the metastable $1s2s2p$ ^4P term. The latter cannot make radiative transitions; instead, it autoionizes via higher-order mechanisms (spin–spin or spin–orbit interaction). These are very slow processes for Li(I) (of the order of 10^6 sec^{-1}) but the rates increase drastically with Z. Donnally *et al.* (1971) and Sellin *et al.* (1971) have investigated the autoionization processes for O(VI) and F(VII), which belong to the Li(I) sequence. Figure 7 shows two of the elec-

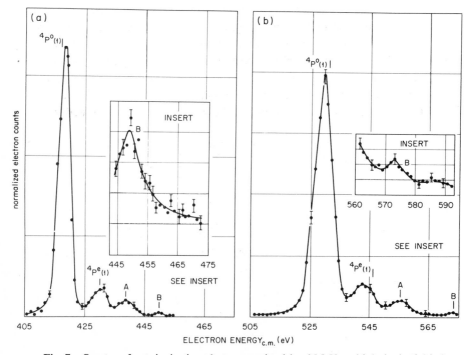

Fig. 7. Spectra of autoionization electrons emitted by 6-MeV multiply ionized (a) O and (b) F beams. The strongest peak $^4P^o(1)$ is due to the autoionization of the doubly excited $1s2s2p$ ^4P level in Li(I)-like O(VI) and F(VII), whereas the smaller peaks are caused by the autoionization of higher quartet terms (Sellin *et al.*, 1971).

tron spectra obtained in these experiments. The strongest peaks, labeled $^4P^0(1)$, represent autoionization electrons from the decay of the 1s2s2p ^4P term, whereas the other peaks can be related to the autoionization of higher quartet levels, e.g., 1s2p^2 ^4P. Lifetime measurements for these levels in highly ionized spectra permit quantitative estimates of the autoionization probability through the magnetic interactions.

B. SINGLY EXCITED STATES

It would be a tedious task to discuss all the information on transitions and energy levels that has been obtained with ion beam methods. Even studies of relatively light atoms, e.g., Be, B, C, and N, at moderate energies (< 2 MeV) have revealed many transitions that were not known earlier. The beam–foil spectra of boron (Martinson *et al.*, 1970) exhibited more than 40 previously unreported transitions, and several new C transitions have been reported by Poulizac *et al.* (1971). It is less surprising that many new transitions are found in the spectra of heavier atoms, which have much more complex structures than B or C.

The beam–foil interaction abundantly populates high-angular-momentum states in multiply ionized atoms. These levels are also excited in spark light sources, but here the interionic fields may cause serious problems in the form of line broadenings, forbidden transitions, etc. With the very low particle density of the in-beam light source (typically 10^5 ions/cm^3), such effects are negligible. The transitions between high n and l states (hydrogenic transitions) can be classified with the formula

$$T = (R\zeta^2/n^2) + \Delta_r + \Delta_p \qquad (6)$$

where Δ_r is the relativistic and Δ_p the polarization correction to the purely hydrogenic term value. Edlén (1964) shows that these corrections become rather insignificant for high n and l values. Indeed, many transitions in beam–foil spectra can be directly identified even if these corrections are neglected.

Even with 2-MV Van de Graaffs, many hydrogenic transitions have been found in multiply ionized spectra. Druetta *et al.* (1971) reported a number of transitions in O(VI), whereas Bashkin and Martinson (1971) classified a number of transitions from G, H, and I terms ($L = 4$, 5, and 6, respectively) for Cl(VII).

Hydrogenic transitions belonging to considerably higher charge states are also prominent at very high beam energies. Using 12–40-MeV ion energies, Denis *et al.* (1971) observed many such transitions in highly ionized C, N, O, Ne, and Ar spectra. Charge states as high as Ar(XIV) were registered. By

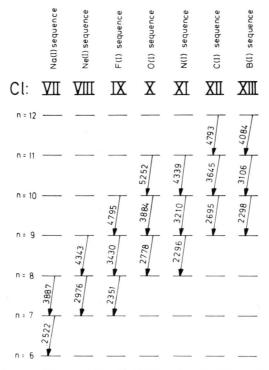

Fig. 8. The hydrogenlike transitions in highly stripped Cl ions, observed by Hallin *et al.* (1971).

accelerating O and Cl beams to 6–42 MeV, Hallin *et al.* (1971) discovered hydrogenlike transitions in O(VI)–O(VIII) and Cl(VII)–Cl(XIII). The Cl transitions so far observed are schematically shown in Fig. 8. None of these lines has appeared in previous spectral studies. Such measurements, preferably at higher wavelength resolution, will facilitate precise determinations of ionization potentials (electron binding energies) in highly ionized atoms.

III.   Results: Atomic Lifetimes and Transition Probabilities

The experimental aspects of lifetime measurements with in-beam techniques were discussed in Section I.B.1. The radiative lifetime τ_i is related to atomic transition probabilities A_{if} according to

$$\tau_i^{-1} = \sum_f A_{if} \qquad (7)$$

The sum is over all lower levels f that are directly populated in the decay of the level i. Equation (7) shows that a lifetime measurement yields the transition probability A_{if} only when the excited level studied decays to a single lower level. In other cases, only the sum of the transition probabilities is obtained. If individual transition probabilities are needed—these are more fundamental quantities than radiative lifetimes—the relative intensities (branching ratios) of the various decay modes must be investigated. The lifetime always gives the upper limit of any transition probability of interest. Furthermore, quite often, one of the branches may be strongly favored, in which case the measured lifetime gives a good approximation for this transition probability.

For allowed transitions in atoms (electric dipole or E1) the probability can be expressed as

$$A_{if} = (4\omega^3/3\hbar c^3) \, |\langle u_i| \, Q \, |u_f\rangle|^2 \quad (\sec^{-1}) \tag{8}$$

where $\hbar\omega$ is the photon energy, c the speed of light, u_i and u_f the wave functions of the upper and lower states, respectively, and Q the dipole operator, $Q = \sum_i er_i$.

It is often convenient to use a dimensionless quantity denoted the absorption oscillator strength, or f value, which is numerically related to A_{if} according to

$$f = 1.499\lambda^2 (g_i/g_f) \, A_{if} \tag{9}$$

where λ is the wavelength of the transition (in cm), and g_i and g_f are the statistical weights of the upper and lower states, respectively.

Measurements of atomic transition probabilities provide information about the accuracy of the atomic wave functions applied in calculating excitation energies and line intensities. The f values are indeed very sensitive to the effects of interaction between the atomic electrons, e.g., configuration mixing, which means that two (or more) electron states of the same parity and angular momentum interfere with each other. In such cases, the various independent-particle models—which are capable of giving good theoretical f values for simple spectra and for transitions between nonpenetrating orbits—may fail completely. The 2s2p ^{1}P–2p^2 ^{1}D transition in B(II) is a very good example of such shortcomings. Here, one-electron theories predict $f = 0.65$, in clear disagreement with the values close to 0.18 which are obtained from beam–foil experiments and very refined theoretical calculations. When such differences occur even in these relatively simple systems, we can easily imagine theoretical complexities in calculating accurate f values in Fe or Ni.

Detailed information about various theoretical models for calculating transition probabilities can be found in the review articles by Layzer and Garstang (1968), Crossley (1969), and Weiss (1970).

Besides their theoretical importance, atomic f values are also crucial for many astrophysical problems, e.g., the determination of element abundances in stellar objects (Section IV).

An impressive amount of empirical data on lifetimes and transition probabilities has been obtained in recent years. Besides in-beam techniques, the following experimental methods have been widely used: (a) anomalous dispersion, the variation of the refractive index in the vicinity of absorption lines; (b) the Hanle effect, or zero-field level crossing, measurements of the intensity decrease of scattered resonance radiation in an external magnetic field; (c) electron excitation, using (1) pulsed beams for delayed coincidence studies or (2) sinusoidally modulated beams for phase-shift measurements; and (d) emission measurements, determination of spectral line intensities in arc light sources. For very thorough discussions of these and other experimental techniques, the review articles by Foster (1964) and Corney (1970) should be consulted.

In-beam experiments have contributed significantly to the present information about atomic f values. Many of the initial studies, e.g., Bickel and Goodman (1966), dealt with comparatively simple atoms [H(I), He(II), etc.] for which the theoretical f values are very accurate. The experimental results showed good agreement with theory, which encouraged studies of more complex atoms and ions.

It is particularly valuable to find systematic trends in the available material. Wiese and Weiss (1968) have pointed out that important regularities for f values are found in (a) isoelectronic and (b) homologous systems. The beam–foil results will be discussed in accordance with these classifications. We will also discuss some very important results for forbidden transitions in highly ionized atoms.

A. Isoelectronic Sequences

According to perturbation theory, the oscillator strength f for a given transition can be expressed as a power series in the inverse nuclear charge $1/Z$ as

$$f = f_0 + f_1(1/Z) + f_2(1/Z^2) + \cdots \tag{10}$$

where f_0 is the hydrogenic f value (e.g., $f_0 = 0$ when $\Delta n = 0$, the degenerate case), which can be easily calculated. By displaying the f value for a certain transition versus $1/Z$ along an isoelectronic sequence, the effects of configuration mixing can be estimated from the shape of the curve.

By measuring the lifetime of the 2p ^{2}P level in C(IV), N(V), O(VI), F(VII), and Ne(VIII), Berkner *et al.* (1965) made the first systematic experimental study of this kind. More recently the lifetimes of the lower members of the sequence have also been determined, and now experimental f values are available for the 2s ^{2}S–2p ^{2}P resonance transition all the way from Li(I) to Ne(VIII). These beam–foil data support the calculated values (Wiese *et al.*, 1966). The theoretical difficulties increase for the Be(I) and B(I) sequences, as mentioned earlier, and for several transitions, the beam–foil experiments yielded f values that were considerably lower than the theoretical estimates. However, the very complex calculations of Weiss (1969) and Westhaus and Sinanoğlu (1969), based on many-electron models, have resulted in very good accord with the experimental values. Figure 9 presents a good example of this fact. It shows the f values for the $2s^2 2p$ ^{2}P–$2s2p^2$ ^{2}D transition in the B(I) isoelectronic sequence. In these spectra, with three electrons outside the closed $1s^2$ shell, the effects of

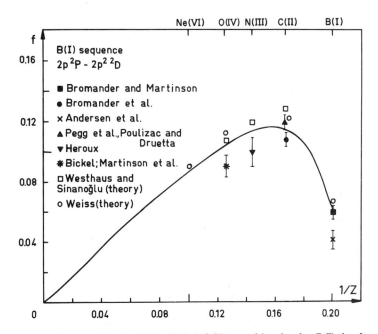

Fig. 9. The f values for the $2s^2 2p$ ^{2}P–$2s2p^2$ ^{2}D transition in the B(I) isoelectronic sequence. The experimental data, which originate from various beam–foil measurements, support the many-electron calculations.

perturbations may be very substantial. The "normal" spectral series $2s^2nd$ ^2D $(n = 3, 4, 5, ...)$ is seriously perturbed by a "foreign" term, $2s2p^2$ ^2D, which has the same angular momentum and parity. From the quantum mechanical point of view, a mixing of eigenfunctions occurs. The f values are very sensitive indicators of these effects. For the B(I) transition, the perturbation is particularly dominant, as can be judged from the curve (Fig. 9). One-electron models in this case are off by a factor of six. For more highly ionized atoms, the energy separation between the perturbing term and the series increases and the interaction weakens considerably.

Another example which also illustrates the importance of configuration interaction is shown in Fig. 10. For this transition, $2s^22p$ ^2P–$2s^23s$ ^2S in the B(I) sequence, several theoretical calculations resulted in a very smooth variation of f values with $1/Z$, whereas the many-electron SOC (super-

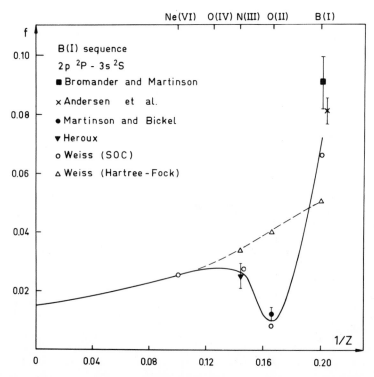

Fig. 10. The f values for the $2s^22p$ ^2P–$2s^23s$ ^2S transition in the B(I) isoelectronic sequence. The many-electron theory of Weiss (1969) is in good agreement with the beam–foil experiments, whereas the independent model calculations deviate from the experimental results.

position of configurations) theory of Weiss (1969) predicted a pronounced minimum around C(II) due to destructive interference between the 3s ^{2}S level and a perturbing 2s2p^2 ^{2}S term. This theoretical result stimulated a beam–foil measurement of the C(II) lifetime (Martinson and Bickel, 1970) which clearly confirmed the new theoretical picture.

These examples show that much can be learned about many-electron effects in atoms and ions by making lifetime measurements. The configuration-mixing effects are often most pronounced for low-lying levels which only decay to the ground state. Branching may sometimes occur, but quite often, lifetime measurements yield the f values directly.

Very many lifetime measurements have been made for noble gas spectra (Wiese *et al.*, 1966, 1969), partly because of their importance for laser physics. In most cases, experiments based on electron excitation or emission studies have been pursued, but there also exist several beam–foil or beam–gas results. The beam–foil method permits lifetime studies even for ions with noble gas structure, e.g., Na(II), Mg(III), K(II), Ca(III), etc., and iso-electronic comparisons may show interesting regularities. One such example is displayed in Fig. 11, which shows the results for the 3p^{5}4p configuration obtained by Andersen *et al.* (1970a). There are ten sublevels, which have

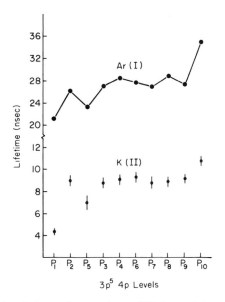

Fig. 11. Comparison between the experimental lifetimes of the ten sublevels belonging to the 3p^{5}4p configuration of Ar(I) and K(II) (Andersen *et al.*, 1970a).

the spectroscopic notation p_1-p_{10}. (Noble gas spectra do not follow the *LS* approximation very well.) One can easily note the similar variations in lifetimes for Ar(I) and K(II).

Beam–foil measurements have also given a wealth of new information about f values for the C(I), N(I), O(I), Na(I), Mg(I), Al(I), and Si(I) iso-electronic sequences. For example, the f values for the 3s ^{2}S–3p ^{2}P resonance transition have been measured from Na(I) to Ar(VIII).

B. HOMOLOGOUS ATOMS

Systematic trends in f values can also be found for homologous atoms, i.e., atoms with the same outer electron structure, such as Li, Na, K, etc. Wiese *et al.* (1969) point out that the f values for the resonance transitions in the alkali metals (2s–2p in Li, 3s–3p in Na, 4s–4p in K) are very close to each other. There are many other examples of such regularities which permit estimates of f values for transitions that have not been measured or calculated explicitly. Using the beam–foil technique, Andersen and Sørensen (1972) systematically examined f values for Al(I), Ga(I), In(I), and Tl(I), all of which have three electrons outside closed shells. In all cases the f values were very close to each other. Andersen and Sørensen also extended the work to ions and studied the np ^{1}P–$(n+1)$ ^{1}S transitions for Al(II) $(n=3)$, Ga(II) $(n=4)$, In(II) $(n=5)$, and Tl(II) $(n=6)$. The results for these alkaline

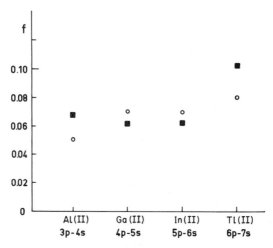

Fig. 12. Comparison between experimental (Andersen and Sørensen, 1972) (■) and theoretical (Helliwell, 1964) (○) oscillator strengths for the np ^{1}P–$(n+1)$ ^{1}S transitions in Al(II) $(n=3)$, Ga(II) $(n=4)$, In(II) $(n=5)$, and Tl(II) $(n=6)$.

earth-type spectra are shown in Fig. 12. The experimental values have estimated uncertainties of about 20%, and they confirm the calculations. Only for Tl(II) can a discrepancy be noted. However, in very heavy atoms, relativistic effects may become important, which might explain the deviation.

Similar systematic studies have also been undertaken by Fink *et al.* (1970), who measured lifetimes for noble gases and found monotonically increasing lifetimes with increasing Z for homologous levels.

C. FORBIDDEN TRANSITIONS

Forbidden transitions can appear in atoms if the LS coupling selection rules are not fulfilled. For example, the selection rule $\Delta S = 0$ is often violated in heavy atoms where spin-forbidden transitions can be quite strong. In certain cases, possibilities arise also for electric quadrupole (E2), magnetic dipole (M1), and even magnetic quadrupole (M2) transitions. Besides their importance for atomic theory, as tests of relativistic or higher-order radiative processes, many forbidden transitions are of great astrophysical interest because they have been observed in the solar corona or the nebulae.

Beam–foil experiments at high energies have facilitated quantitative studies of several important transitions in highly ionized heliumlike and hydrogenlike systems. Sellin *et al.* (1968) accelerated O^{6+} and N^{5+} to energies between 6 and 42 MeV and studied the intercombination line $1s^2\ ^1S_0$–$1s2p\ ^3P_1$ (E1 transition, caused by the spin–orbit interaction) in foil-excited O(VII) (at 21.8 Å) and N(VI) (at 29.1 Å). In neutral He, the $1s2p\ ^3P_1$ state has allowed transitions to $1s2s\ ^3S_1$ and intercombinations have extremely low probability. However, this probability has a very drastic Z-dependence (of the order of Z^{10}), being already ten times more probable than the allowed transition for N(VI) and O(VII). The lifetime measurements of Sellin *et al.* (1968) confirmed the theoretical estimates of the forbidden-decay rate.

More recently, Marrus and Schmieder (1970, 1972) and Schmieder and Marrus (1970) have investigated forbidden transitions in Ar(XVII), which has only two electrons. Figure 13 shows the levels investigated and their observed lifetimes.

The $1s2p\ ^3P_2$ state decays to $1s2s\ ^3S_1$ by allowed transitions (E1). However, decays to the ground state, $1s^2\ ^1S_0$, have also been considered. The latter de-excitation, which, according to theory, has a rate proportional to Z^8, is of the magnetic quadrupole (M2) type, because of the parity change and $\Delta J = 2$. The calculations of Drake (1969) implied that these two processes had approximately equal probabilities for Ar(XVII). This was confirmed by

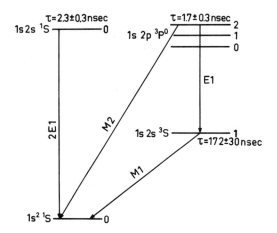

Fig. 13. Partial energy level diagram for Ar(XVII) [He(I) sequence]. The lifetime results of Marrus and Schmieder (1970) have been indicated.

Marrus and Schmieder (1970), who obtained a 2p $^3P_2$ lifetime of 1.7 ± 0.3 nsec, in good agreement with the theoretical value of 1.49 nsec (Drake, 1969). Competitions between magnetic and electric multipole radiation are well known in nuclear spectroscopy. It is interesting to note that they also occur in the spectra of very highly ionized atoms.

Schmieder and Marrus (1970) also studied the 1s2s $^3S_1$ state, for which most calculations had predicted a decay by two-photon emission (2 E1) to $1s^2 \, ^1S_0$ (Fig. 13). However, astrophysical evidence also supported single-photon decays of the M1 type induced by relativistic effects. The in-beam measurements established that the latter decay mode is much more probable for Ar(XVII). The authors further investigated the decay of the metastable 2s $^2S_{1/2}$ state for hydrogenlike Ar(XVIII). In H(I), this state decays by two-photon emission, with a lifetime of 0.13 sec. For highly ionized one-electron spectra, competing mechanisms in the form of magnetic dipole decay were also considered. The measurements of Schmieder and Marrus (1970) showed that the photons had a continuous energy distribution but their sum equalled the $1s_{1/2}$–$2s_{1/2}$ energy difference. The lifetime measurements further showed that the nonrelativistic 2 E1 decay was much more probable than the relativistic M1 process.

D. MOLECULAR LIFETIMES

The in-beam methods are not limited to atomic and ionic lifetime measurements. Radiative lifetimes in molecules and molecular ions can also be investigated if molecular ions are accelerated and subsequently excited in collisions with a gas target at 10^{-3}–10^{-4} Torr. Foils are less suitable because they cause dissociation of the molecule. With this method, lifetimes for vibrational levels have been determined for N_2^+, N_2, N_2O^+, CO^+, and CO_2^+ (Nichols and Wilson, 1968; Desesquelles et al., 1968; Desesquelles, 1969; Poulizac and Druetta, 1969; Head, 1971). The results are consistent with those obtained by other techniques, such as the phase-shift method, the pulsed-electron excitation technique, etc. The applications of various experimental methods to molecular lifetime measurements are summarized by Corney (1970).

E. COMPARISON OF VARIOUS EXPERIMENTAL METHODS

In Table 1 we compare beam–foil data with the results of other experimental techniques for two frequently studied levels, 3p ^{1}P in He(I) and 4p ^{1}P in Ca(I). In the He(I) case, the lifetime has been measured both by using the time-of-flight technique for foil-excited He and by combining the beam–gas and Hanle effects. The more "usual" Hanle effect level crossing technique and the atomic beam method have also been used. Table 1 shows that the results are internally consistent and in excellent agreement with the theoretical lifetime. However, in this case, the theoretical accuracy is considered to be higher than the present experimental one.

For the Ca(I) lifetime (Table 1), the situation is somewhat different. The

TABLE 1

LIFETIMES OF EXCITED LEVELS IN He(I) AND Ca(I)

Atom	Upper level	Lifetime (nsec)	Method	Reference
He(I)	3p ^{1}P	1.66 ± 0.05	Atomic beam	Korylov and Odintsov (1965)
		1.78 ± 0.10	Beam–Foil	Martinson and Bickel (1969)
		1.73 ± 0.11	Level crossing	Burger and Lurio (1971)
		1.8 ± 0.1	In-beam Hanle effect	Carré et al. (1971)
		1.726	Theory	Schiff and Pekeris (1964)
Ca(I)	4p ^{1}P	5.0 ± 0.3	Atomic beam	Odintsov (1963)
		4.48 ± 0.15	Level crossing	Lurio et al. (1964)
		4.7 ± 0.5	Phase shift	Hulpke et al. (1963)
		5.4 ± 0.3	Anomalous dispersion	Penkin and Shabanova (1969)
		6.2 ± 0.5	Beam–foil	Andersen et al. (1970b)

beam–foil result is considerably longer than the value found from Hanle-effect experiments. Although corrections were made for cascades, it is possible that the level repopulation picture is very complex.

These examples should illustrate that the uncertainties in lifetimes from in-beam measurements are around 5% in favorable cases with good statistics, cascade-free curves, and well-determined velocities after the exciter. When cascades are prominent, it is difficult to obtain less than 10% uncertainty; often much larger uncertainties must be taken into account. For heavy atoms, the velocity dispersion after the foil may introduce uncertainties. In such cases, it is particularly difficult to compete with the accuracy of experiments based on the Hanle effect, threshold excitation with electrons, or photon–photon correlation measurements (Corney, 1970).

Perhaps the most important advantage of the beam–foil source lies in the ease with which lifetimes in multiply ionized atoms can be determined. In addition, the relative simplicity of the in-beam method, the virtual absence of secondary collision or resonance absorption after the excitation, and the possibilities for accelerating practically any element are other merits which must be considered.

IV. Astrophysical Consequences of Lifetime Measurements

Experimental information about atomic and molecular parameters is of great astrophysical importance. For element identifications in the sun and the stars, accurately determined wavelengths from laboratory experiments are needed, whereas transition probabilities are crucial in estimating the cosmic abundances of elements.

Until the late 1940's, the wavelength region below 3000 Å was practically inaccessible to astronomical observations, but with the development of rockets and satellites, observations of the solar and stellar spectra in the UV and soft X-ray regions have become possible, and an impressive amount of information has been gained. In addition to well-known transitions, the UV spectra have shown many previously unknown lines, a great many of which have been classified with the help of theoretical calculations and laboratory experiments (Tousey, 1963). However, several observed transitions have so far remained unidentified.

Accurate information about the abundances of chemical elements is needed for testing various theories of star formation. The abundances are determined from spectroscopic studies of characteristic lines present in solar or stellar spectra. From the observed line intensities or widths, the abundances can be determined, provided the appropriate f values are known. The intensity of a spectral line is proportional to the number of radiating atoms and the transition probability. The latter therefore enters directly into the abundance uncertainties.

Solar abundances are usually given on a logarithmic scale relative to the hydrogen abundance; by definition $\log N_H = 12.00$. More than 70 elements have been observed in the sun's spectrum, but for several of these, abundance estimates are hampered by unsatisfactory f values (Engvold and Hauge, 1970).

The possibilities of investigating astrophysical problems with the beam–foil technique have been discussed by Bashkin (1964). Several of the lifetime measurements performed since then have indeed been motivated by the astrophysical need for accurate transition probabilities. A few such results will be discussed.

A. Solar Abundances of Iron-Group Elements

The solar abundances of the iron-group metals (Sc–Zn) are generally estimated from allowed transitions in neutral and singly ionized atoms, which appear in the photosphere, and forbidden transitions in highly ionized atoms, present in the corona. For many elements, the coronal abundances have been reported to be more than ten times higher than the photospheric ones. For iron, typical abundance values were $\log N_{Fe} = 6.65$ (photosphere) and $\log N_{Fe} = 7.87$ (corona). For the allowed Fe(I) and Fe(II) transitions, the f values originated from various emission measurements, whereas those for the coronal lines were derived from calculations. Although there might be physical reasons (e.g., thermal diffusion) for the abundance difference, the f-value accuracies were also questioned. Whaling et al. (1969) therefore made a direct measurement of Fe(I) lifetimes with the beam–foil method. The results found were 4–21 times longer than the previously established lifetime values. This result, supported by another beam–foil experiment (Andersen and Sørensen, 1971) and a precise emission measurement (Bridges and Wiese, 1970) implied that the estimated photospheric Fe abundance had to be drastically changed. In a subsequent experiment, Martinez-Garcia et al. (1970) made careful studies of the relative intensities for the Fe(I) lines of interest, which, in combination with the lifetimes, yielded individual f values. In certain cases the transition probabilities so obtained were more than 100 times lower than the results of early emission experiments. The new results enabled Martinez-Garcia and Whaling (1971) to derive a new photospheric abundance of $\log N_{Fe} = 7.45$, a value that agrees with the coronal abundance and the meteoritic value. However, the revised photospheric abundance seems to have some very far-ranging consequences. According to Watson (1969), this abundance increases the adopted opacity in the center of the sun, which then would require a higher central temperature and nuclear reaction rate than previously assumed.

Decay measurements for several astrophysically important Fe(II) transitions have also been made with the beam–foil method (Smith *et al.*, 1970; see also Fig. 2 in this chapter). The beam–foil lifetimes are significantly longer (roughly by a factor of three) than previously assumed values.

Precise lifetime data are also needed for other transition metals. A good deal of new information is already available, however. Andersen *et al.* (1971) measured Ti(I)–Ti(III) lifetimes using the beam–foil method, whereas Cocke *et al.* (1971) determined Cr(I) decay times. In both cases, the new f values are lower than previous results (from emission measurements) and they may call for significant revisions of the photospheric abundances of Ti and Cr. For a given line intensity, a reduction of the f value demands an increased number of radiating atoms. Consequently, lower f values for the astrophysical transitions show that the abundances are higher. In all these cases, the gaps between photospheric and coronal abundances were narrowed.

The lightest element in the iron group, scandium, was studied with the beam–foil technique by Buchta *et al.* (1971), who measured several Sc(I) and Sc(II) lifetimes. The new measurements largely confirmed previous Sc(II) f values, whereas they indicated that the emission results for Sc(I) should be reduced by 30–50%. The solar Sc abundance, derived from Sc(I) lines, had been reported as 50% higher than the value obtained from observations of photospheric Sc(II) transitions. The beam–foil experiment showed that this difference was mostly due to inaccuracies in Sc(I) lifetimes.

B. PROCESSES IN THE SOLAR CORONA

The solar and stellar spectra below 2000 Å show many strong emission lines of C, N, O, Ne, Mg, Si, Fe, etc., often in high charge states (Tousey, 1963). The oscillator strengths of many of these lines have been precisely determined with in-beam techniques. The thorough study of Si(II)–Si(IV) lifetimes by Berry *et al.* (1971a) is a representative example of such an experiment. Because of the astrophysical significance of this element, some very extensive calculations of configuration mixing in Si ions have been made, and these results were confirmed by the lifetime measurements of Berry *et al.* There still remains a difference between the photospheric and coronal Si abundances, however, the latter being estimated as three times higher.

Besides the questions of coronal and photospheric element abundances, beam–foil experiments have been used to study other processes observed in the solar corona. These experiments, discussed in Section III.C, involved the forbidden transitions in Ar(XVII) (Marrus and Schmieder, 1970; Schmieder and Marrus, 1970). The measurement of the Ar(XVII) 1s2s $^3S_1$ lifetime is a very good example of the importance of laboratory experiments. The forbidden M1 transition, $1s^2$ ^1S–1s2s ^3S (Fig. 13) has been identified in the solar corona for several ions [C(V)–Mg(XI)] in the He(I) isoelectronic sequence by Gabriel and Jordan (1969, 1970), who also calculated the electron densities in the corona. A crucial parameter in their theory is the Z-dependence of the M1 transition probability. The lifetime measurements for Ar(XVII) (Schmieder and Marrus, 1970) were in accord with theory, thereby supporting the estimated electron densities. Other forbidden transitions in He-like atoms, discussed in Section III.C, have also been observed in the solar corona, and the decay measurements for these lines in highly ionized atoms therefore serve as very valuable tests of the calculations underlying various coronal models.

The transitions from doubly excited levels in He(I) and Li(I) sequences (Section II.A) have also been observed in the solar corona. Walker and Rugge (1971) identified such lines in Mg(X), Al(XI), and Si(XII) [Li(I) sequence] and Mg(XI) and Si(XIII) [He(I) sequence]. Considering the fact that the doubly excited levels are strongly populated in the beam–foil interaction, a search for such transitions using high-energy ion beams would be worthwhile.

C. WOLF–RAYET STARS AND PLANETARY NEBULAE

The spectra of the Wolf–Rayet and O-type stars show strong emission lines of singly and multiply ionized C, N, and O. These stars are comparatively young (typically 10^7 y) and they have very hot atmospheres (up to 10^5 °K). Several beam–foil experiments have been made in order to determine the lifetimes of the levels excited in these stars. Pinnington *et al.* (1970) systematically measured the decays of more than 20 O(II) and O(III) lines that have been observed in these stars. Beam–foil methods have also been used for line identification. One of the diffuse emission lines observed in these stellar spectra, at 5806 Å, could not be classified with certainty; it was believed to belong to N(IV) or C(IV). In a beam-foil study of N spectra, Desesquelles (1970) only found very weak transitions at this wavelength, whereas the C experiment of Poulizac *et al.* (1971) revealed very strong lines at the expected wavelength. These experiments were able to confirm previous suggestions that the observed lines should be ascribed to C(IV) (Swings and Swings, 1968).

Strong O(III) transitions appear also in planetary nebulae, where they are excited because of a remarkable coincidence between the energy of an O(III) level and that of the He(II) 303-Å resonance line. Lifetime measurements for O(III) levels by Pinnington *et al.* (1970), Berry *et al.* (1970a), and Druetta and Poulizac (1969) have helped in eliminating certain discrepancies between the calculated and observed intensities for these important O(III) transitions.

Lifetime measurements are also crucial in estimating the density of interstellar matter. The rocket observations of the UV spectra of several stars (Stone and Morton, 1967) showed several absorption lines between 1300 and 1700 Å, which have been identified as transitions in interstellar C, O, Al, and Si atoms. The mean densities of these atoms in interstellar space (typically $10^{-2}/cm^3$) have been estimated by combining the absorption intensities with the results of precise lifetime measurements.

Only a fraction of the astrophysical applications of in-beam data have been discussed in this section. However, the material presented should be sufficient to show that one of the most important advantages of beam–foil studies is the good possibility of studying astrophysical processes under controlled conditions in the laboratory.

V. Results: Atomic Fine Structure

In Section I.B.2, we discussed how the time resolution of the ion beam light source is used for lifetime measurements. In all such experiments, the excited atoms decay spontaneously in an evacuated, field-free region. However, other important atomic parameters, such as separations between fine-structure levels, can be conveniently investigated when the excited atoms are perturbed by electric or magnetic fields during their decay. In certain cases, resonance effects occur even in the absence of fields (see Section V.B). In all these experiments, the spatial resolution of the excited beam can easily be converted into time intervals and thus frequencies and small energy separations.

Several factors may contribute to atomic fine structure, e.g., the interaction between the spin and orbit of an electron, the interaction between the spins of two electrons, and the interaction of an electron with the radiation field (Lamb shift). The fine structure due to the spin–orbit and spin–spin interactions can be described with the relativistic Dirac theory, whereas the Lamb shift is a quantum electrodynamic effect (see, e.g., Sakurai, 1967).

Measurements of atomic fine-structure separations are needed as tests of the fundamental theories which have been used to explain these effects.

A. STARK MIXING IN WEAK FIELDS

When weak, static electric or magnetic fields are applied on the foil-excited beam, interesting phenomena can often appear. The first experiment of this kind was reported by Bashkin *et al.* (1965), who sent foil-excited hydrogen atoms through weak electric (50 V/cm) or magnetic (8 G) fields. The Balmer series ($n = 2$–$n = 3, 4, 5, \ldots$) was detected photographically with the spectrograph slit parallel to the beam. The presence of the fields caused periodic intensity fluctuations in the spectral lines, as shown in Fig. 14. These fluctuations were explained as a quantum mechanical interference

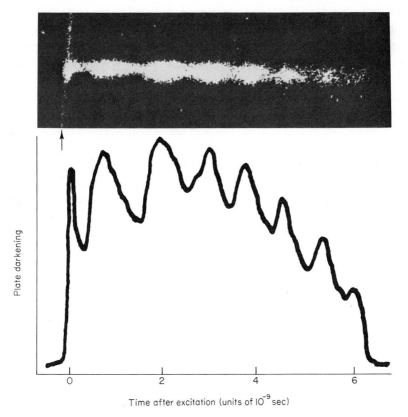

Fig. 14. Intensity oscillations for the H$_\delta$ line ($n = 2$–$n = 6$), observed photographically by Bickel (1968).

between various angular momentum states. For instance, in the case of H_α ($n = 2$–$n = 3$), the upper levels 3s, 3p, and 3d must be considered. In the absence of fields, these decay independently of each other with different mean lifetimes, $\tau_{3s} = 159$ nsec, $\tau_{3p} = 5.40$ nsec, $\tau_{3d} = 15.6$ nsec, but they become coupled under the influence of the electric field (Stark mixing). Since the states have different lifetimes, the coupling manifests itself in oscillating changes in the registered decay rate for the unresolved H_α line. Due to the very small energy differences between adjacent l states in hydrogen, the oscillation frequencies are low enough to be observable as superpositions on the decay curve.

These effects were also found in a study of helium (Bickel and Bashkin, 1967). This experiment confirmed the Stark mixing picture because it established that spectral lines of hydrogenlike He(II) showed field-induced oscillations, whereas He(I) lines were not influenced by the perturbing field.

The agreement between the measured and theoretically expected beat frequencies was initially not very good for the Balmer series. The theoretical problems were substantial, however, because in most cases, several fine-structure levels could take part in the oscillations, and the measured results were superpositions of many frequencies. However, by analyzing the influence of the various components and critically selecting the important contributors, Sellin et al. (1969) showed that satisfactory agreement could be obtained in most cases. They also studied the simplest case, Ly_α ($n = 1$–$n = 2$), for which the measured frequency mainly represents oscillations between the $2s_{1/2}$ ($\tau = 0.13$ sec) and $2p_{1/2}$ ($\tau = 1.6$ nsec). Figure 15 shows the lowest energy levels of hydrogen. Measurements in a field of 197 V/cm yielded a beat frequency of $(1.34 \pm 0.07) \times 10^9$ sec^{-1}, in good agreement with the theoretical value 1.37×10^9 sec^{-1} (Lamb, 1952). Andrä (1970a) reinvestigated the theory for these oscillations and also made very careful

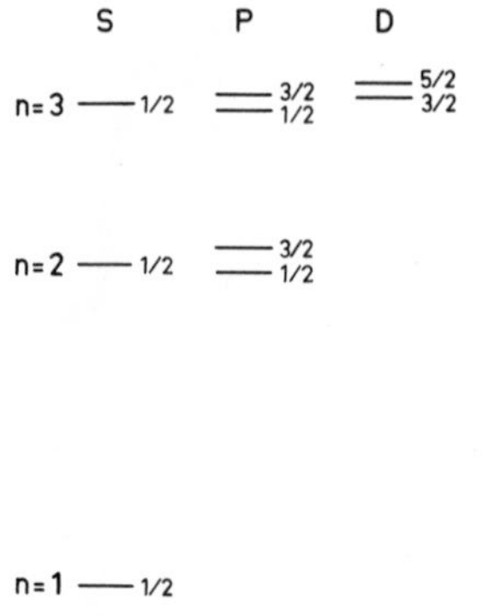

Fig. 15. Partial energy level diagram for hydrogen [H(I)] (not to scale).

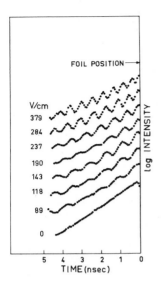

FOIL POSITION⟶

V/cm

379

284

237

190

143

118

89

0

log INTENSITY

5 4 3 2 1 0
TIME (nsec)

Fig. 16. Intensity oscillations for the Ly_α ($n = 1$–$n = 2$) line, showing Stark beats. In the absence of the electrostatic field, the $2p_{1/2}$ level decays exponentially with a lifetime of 1.60 nsec (Andrä, 1970a).

experimental observations. His results for the hydrogen Ly_α are shown in Fig. 16. Note that the beat amplitude and frequency increase with increasing electric field. Figure 17 summarizes the results for different electric field strengths parallel to the beam. The excellent agreement between the experimental and theoretical frequencies is evident. Note that the intercept of the curve for $F = 0$ represents the energy separation between $2s_{1/2}$ and $2p_{1/2}$, which is the $n = 2$ Lamb shift.

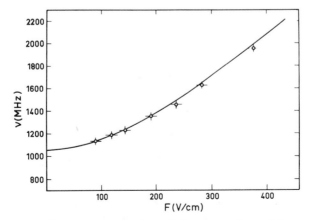

2200

2000

1800

1600

ν(MHz)

1400

1200

1000

800

100   200   300   400
F (V/cm)

Fig. 17. Comparison between the theoretical and experimental frequencies for Ly_α oscillations as a function of an electric field applied parallel to the beam (Andrä, 1970a).

B. ZERO-FIELD OSCILLATIONS

The oscillations discussed were caused by the presence of a perturbing electric or magnetic field which coupled atomic states of different parity. However, even in the absence of fields, oscillations may sometimes occur (Macek, 1970). In such a case, fine-structure levels of the same parity, e.g., $2p_{1/2}$ and $2p_{3/2}$ in hydrogen, which decay to the same lower level show interference effects. The frequency of these oscillations is a direct measure of the energy separation of the levels. According to theory (Macek, 1970), these oscillations occur if there is alignment, i.e., the various magnetic sublevels M_L are not uniformly populated at the excitation. This would also mean that the emitted light shows polarization. The amplitude of the oscillations is a measure of the relative initial population of the magnetic sublevels.

Such effects have been known from electron excitation of gases but they can also be studied with the beam–foil technique. The first quantitative measurements of zero-field oscillations with the beam–foil method were made by Andrä (1970b), who observed such intensity fluctuations in the decay of the He(I) 3889-Å triplet ($1s2s\ ^3S_1$–$1s3p\ ^3P_{2,1,0}$) and also in H_β and H_γ. The oscillation frequencies yielded fine-structure separations (Fig. 18). For example, Andrä found the separation between the He(I) $3p\ ^3P_1$ and $3p\ ^3P_2$ levels to be 655 ± 6 MHz, in good accord with previous rf measurements. The amount of alignment could also be estimated. For symmetry reasons, the cross sections for exciting $M_L = -1$ and $M_L = +1$ in P states should be equal, whereas they should differ from the $M_L = 0$ cross section in the aligned case. Using 475-keV He$^+$ beams, Andrä (1970b) obtained the cross-section ratio $\sigma_1/\sigma_0 = 0.8 \pm 0.1$. This ratio also shows that the alignment is not very pronounced, but it is sufficient to permit fine-structure studies.

Burns and Hancock (1971) investigated in detail the zero-field blips for H_β radiation and determined the splittings, whereas Lynch et al. (1971) observed oscillations between $3p_{1/2}$ and $3p_{3/2}$ in hydrogen. They also showed that the cross-section ratio σ_1/σ_0 varied between 1.3 and 2.3 when the beam energy was increased from 230 to 500 keV.

The fine-structure splittings for H and He that have been measured with the beam–foil technique are less accurate than the results of rf measurements (Radford, 1967). However, the latter are mostly limited to neutral atoms, whereas no such limitations exist for the beam–foil source. Berry and Subtil (1971) observed zero-field oscillations for several Li(II) lines and they measured the previously unknown $4p\ ^3P$ and $5p\ ^3P$ fine-structure splittings.

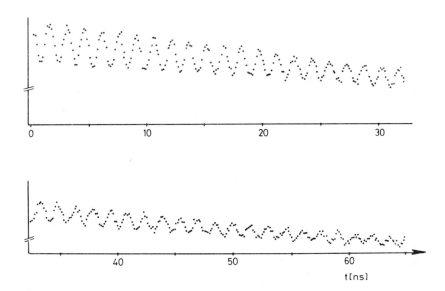

Fig. 18. Zero-field oscillations between the 3p $^3P_1$ and 3p $^3P_2$ states in ^{3}He(I), 658 MHz, observed from the decay of the 3889-Å line (1s2s $^3S_1$–1s3p $^3P_{0,1,2}$). These oscillations, which are superimposed on the exponential decay of the line intensity ($\tau_{3p}{}^3P = 102$ nsec), are observed when the foil-excited light is viewed through a polarizer. The oscillation frequencies yield the energy separations between atomic fine-structure levels. The figure originates from unpublished work of Andrä (1972). See also Andrä (1970b).

C. LANDÉ g FACTORS

The presence of zero-field oscillations shows that the beam–foil excitation creates alignment of the magnetic substates. This phenomenon permits studies of the Hanle effect, various level crossing phenomena, and quantum beats, using accelerated ion beams. These well-known experimental techniques can therefore be extended to singly and multiply ionized atoms.

Liu *et al.* (1971) showed that the alignment condition permits measurements of Landé g factors.[†] Figure 19 shows their experimental arrangement. The excited beam is observed at a fixed distance from the foil, while the

---

[†] When LS coupling is present, the g values can be written as

$$g = 1 + [J(J+1) + S(S+1) - L(L+1)]/2J(J+1)$$

The g values are important quantities in atomic spectra, e.g., they give information about deviations from LS coupling.

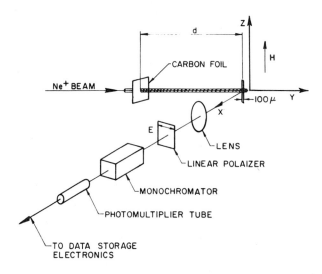

Fig. 19. Experimental arrangement for g-value measurements. The magnetic field H, perpendicular to the traveling beam, varies linearly, which causes sinusoidal variations in the observed spectral line intensities (Liu *et al.*, 1971).

magnetic field H continuously increases. If alignment is present, the intensity of a given spectral line contains an oscillation term, proportional to $(\cos 2\mu_0)\, gHt/\hbar$, where μ_0 is the Bohr magneton and t the time after excitation. Liu *et al.* (1971) used an accurately known Ne(I) g value to calibrate their system and were then able to measure g values for levels in Ne(II) and Ne(III) (Liu and Church, 1971).

D. Lamb Shift Measurements

In-beam techniques can be applied for direct measurements of Lamb shifts. The $n = 2$ Lamb shift in hydrogen is the small energy separation between the $2s_{1/2}$ and $2p_{1/2}$ states; according to relativistic Dirac theory these states should be exactly degenerate, because in this theory, energies in hydrogenlike systems only depend on n and j. The Lamb shift is due to the interaction of the electron with the radiation field and to the effect of vacuum polarization (Sakurai, 1967). Measurements of the Lamb shift serve as important tests of quantum electrodynamics. For H and He$^+$ very high accuracies have been obtained with rf methods (Radford, 1967), but these cannot be applied to other one-electron ions. Experimental information about the Lamb shifts in higher ions is very much needed, however, because the theoretical expressions for this quantity show a very strong Z dependence, and possible inaccuracies of the present theories are most likely to appear for higher values of Z.

The feasibility of in-beam measurements of Lamb shifts was demonstrated by Fan et al. (1967), who studied Li^{2+}. They accelerated a beam of Li ions to 3 MeV and sent it through a charge-exchange chamber. The outgoing ions, part of which were in the metastable Li(III) $2s_{1/2}$ state, were then directed through a strong electric field which mixed the $2s_{1/2}$ state with the short-lived $2p_{1/2}$ state. By varying the electric field, the slope of the $2s_{1/2}$ decay curve could be changed. The $2s_{1/2}$–$2p_{1/2}$ energy separation enters as a parameter into the theoretical expression for the field dependence of the $2s_{1/2}$ lifetime. Measurements of this lifetime at several field strengths yielded the experimental Lamb shift of 63031 ± 327 MHz, in excellent agreement with the theoretical value.

These measurements were extended to hydrogenlike carbon by Murnick et al. (1971), who used C ions of 25–35 MeV. After excitation, part of the ions were in the $2s_{1/2}$ state of C(VI). A magnetic field $\mathbf{B}$, equivalent to an electric field $(1 - v^2/c^2)^{1/2} \mathbf{v} \times \mathbf{B}$, mixed the C(VI) $2s_{1/2}$ and $2p_{1/2}$ states. From decay measurements, a Lamb shift of 780 ± 8.0 GHz was obtained, which is in very good agreement with the theoretical value of 783.7 GHz (Erickson, 1971).

A somewhat different technique was tried by Hadeishi et al. (1969). The excited atoms of velocity v were sent through a region with a spatially varying, periodic potential. If the spatial period of this potential is d, the traveling particles experience an oscillating field with a frequency $v = v/d$. The authors observed the H_α line, and they found a resonance signal at the frequency expected from the $n = 3$ Lamb shift. That experiment and a subsequent refinement by Andrä (1970c) showed that such a technique can be used for Lamb shift studies, although the experimental accuracies are still limited.

Very interesting combinations of in-beam and rf techniques have been introduced by Fabjan and Pipkin (1970, 1971) and Fabjan et al. (1971). In their experiments, 20–40-keV protons were neutralized and excited in a foil or a gas cell, after which they entered an rf chamber of very constant power over a broad bandwidth. [For very precise Lamb shift measurements, the effects of hyperfine splittings must be considered (in all cases with nonzero nuclear spin); in hydrogen, these splittings are as high as a few percent of the Lamb shift.] The authors used two rf chambers in series, which made it possible to select the F states desired (F is the total angular momentum of the atom, i.e., the vector sum of J and I). With this technique, very precise measurements for the $n = 3$, 4, and 5 states in hydrogen were possible. For example, the experiments gave a $3s_{1/2}-3p_{1/2}$ separation of 314.810 ± 0.052 MHz, which confirms the theoretical value of 314.894 MHz.

VI. Foil-Excitation Mechanisms

In most of the work summarized in previous sections, the ion–foil or ion–gas interactions have been merely used as a way of exciting the atomic energy levels of interest. However, several experiments have also been designed which study the physical properties of the beam–foil and the beam–gas light source. The experimental and—particularly—the theoretical problems related to these studies, which enter the domains of ion–atom collisions and the interaction of ions with solids, are quite intricate. No extensive explanations of all the observations can therefore be expected in the near future.

The charge states of the ions after the exciter have been quite thoroughly studied for ions of a number of elements over an energy range of a few keV to several hundred MeV. Such work yields information about charge-changing mechanisms (ionization or electron capture) when fast ions traverse a gas or a solid. These experiments are summarized in a review article by Betz (1972).

The electron capture process for protons traversing a thin C foil has been theoretically investigated by Garcia (1970). According to these calculations, the mean free path for ionization is only about 20 Å inside the foil when 200-keV incoming particles are used. This implies that only those excited states that are created in the last few atomic layers of the foil can enter the downstream region. The calculations showed further that the rate of electron capture is proportional to E_p^{-3}, where E_p is the kinetic energy of the proton. This energy dependence was experimentally found by Bickel (1967), who measured the intensity of Ly_α radiation for several incoming proton energies.

By studying the variations of the light yield with different foil materials, Berry et al. (1970b) concluded that the excitation also depends on the density of the target material, which determines the collision rate.

As already mentioned in Section II.A, multiply excited states are abundantly populated in the beam–foil interaction. N. Andersen et al. (1971) have made explicit studies of the excitation mechanisms which create these states. Beams of Li^+, 10–80 keV, were excited in a foil or a gas and the radiation, consisting of transitions from singly and doubly excited Li(I) and singly excited Li(II), was analyzed spectroscopically. The experiments showed that the Li(I) doubly excited states and the Li(II) states have very similar excitation functions (Fig. 20), whereas the singly excited Li(I) states show a different energy variation.

By replacing the foil with a gaseous target, the intensity of the doubly excited Li(I) lines was very much reduced, which shows that the creation of these states is probably a two-step mechanism.

The light emission of He$^+$ ions after traversing a single crystal of gold has been investigated by Andersen *et al.* (1970c), who observed channeling phenomena for the detected light. The light yield showed distinct minima when the ions traversed the crystal along the 110 and 111 directions. No such

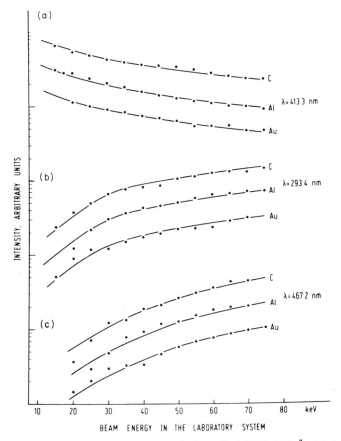

Fig. 20.   Excitation functions for (a) the singly excited Li(I) 4133-Å, (b) the doubly excited Li(I) 2934-Å, and (c) the singly excited Li(II) 4672-Å transitions, measured between 20 and 80 keV using different foils (N. Andersen *et al.*, 1971).

effects could be established in a similar experiment by Poizat *et al.* (1971), however. The experimental conditions (crystal thicknesses, ion energies, etc.) were not identical in these experiments and therefore additional experiments of this type are required.

Additional information about the ion–foil interactions has been provided by the observations of alignment (Section V.B). Continued studies of the relative cross sections for exciting various sublevels may clarify several problems of the ion–foil mechanism.

Besides the experiments sketched here various other possibilities have been investigated, e.g., spectroscopic studies of the light emitted by the particles that are ejected from the foil. Berry *et al.* (1970b) and N. Andersen *et al.* (1971) discuss several useful experimental approaches.

Better understanding of the ion–foil interaction mechanisms will finally permit reductions of the present uncertainties in studies of spectra and lifetimes utilizing the in-beam technique.

Note added in proof: Recent developments in beam–foil spectroscopy are presented in the proceedings of the Third International Conference on Beam–Foil Spectroscopy, held in Tucson, Arizona in October 1972 (S. Bashkin, 1973).

ACKNOWLEDGMENTS

The author wishes to thank Professors J. D. Garcia and L. J. Curtis for enlightening discussions and Professor H. J. Andrä for critical reading of the manuscript.

References

Andersen, N., Bickel, W. S., Boleu, R., Jensen, K., and Veje, E. (1971). *Physica Scripta* **3**, 255.

Andersen, T., and Sørensen, G. (1971). *Astrophys. Letters* **8**, 39.

Andersen, T., and Sørensen, G. (1972), *Phys. Rev. A* **6**, 2447.

Andersen, T., Jessen, K. A., and Sørensen, G. (1969). *J. Opt. Soc. Amer.* **59**, 1197.

Andersen, T., Desesquelles, J., Jessen, K. A., and Sørensen, G. (1970a). *J. Opt. Soc. Amer.* **60**, 1199.

Andersen, T., Desesquelles, J., Jessen, K. A., and Sørensen, G. (1970b). *J. Quant. Spectrosc. Radiat. Transfer* **10**, 1143.

Andersen, T., Datz, S., Hvelplund, P., and Sørensen, G. (1970c). *Phys. Lett.* **33A**, 121.

Andersen, T., Roberts, J. R., and Sørensen, G. (1971). Private communication.

Andrä, H. J. (1970a). *Phys. Rev. A* **2**, 2200.

Andrä, H. J. (1970b). *Phys. Rev. Lett.* **25**, 325.

Andrä, H. J. (1970c). *Phys. Lett.* **32A**, 345.

Andrä, H. J. (1972). Unpublished work.

Assousa, G. E., Brown, L., and Ford, W. K., (1970). *J. Opt. Soc. Amer.* **60**, 1311.

Bakken, G. S., Conrad, A. C., and Jordan, J. A. (1969). *J. Phys. B (At. Mol. Phys.)* **2**, 1378.

Bashkin, S. (1964). *Nucl. Instrum. Methods* **28**, 88.

Bashkin, S. (ed.) (1973). *Proc. Int. Conf. Beam–Foil spectroscopy, 3rd, Tucson, Arizona, Oct. 1972*; *Nucl. Instrum. Methods* **110**, 1–522.

Bashkin, S., and Carriveau, G. W. (1970). *Phys. Rev. A* **1**, 269.

Bashkin, S., and Leavitt, J. A. (1971). Private communication.

Bashkin, S., and Martinson, I. (1971). *J. Opt. Soc. Amer.* **61**, 1686.

Bashkin, S., Bickel, W. S., Fink, D., and Wangsness, R. K. (1965). *Phys. Rev. Lett.* **15**, 284.

Bashkin, S., Fink, D., Malmberg, P. R., Meinel, A. B., and Tilford, S. G. (1966). *J. Opt. Soc. Amer.* **56**, 1064.

Bergström, I., Bromander, J., Buchta, R., Lundin, L., and Martinson, I. (1969). *Phys. Lett.* **28A**, 721.

Berkner, K., Cooper, W. S., Kaplan, S. N., and Pyle, R. V. (1965). *Phys. Lett.* **16**, 35.

Berry, H. G., and Subtil, J. L. (1971). *Phys. Rev. Lett.* **27**, 1103.

Berry, H. G., Bickel, W. S., Martinson, I., Weymann, R. J., and Williams, R. E. (1970a). *Astrophys. Lett.* **5**, 81.

Berry, H. G., Bromander, J, and Buchta, R. (1970b). *Nucl. Instrum. Methods* **90**, 269.

Berry, H. G., Bromander, J., Curtis, L. J., and Buchta, R. (1971a). *Physica Scripta* **3**, 125.

Berry, H. G., Martinson, I., Curtis, L. J., and Lundin, L. (1971b). *Phys. Rev. A* **3**, 1934.

Berry, H. G., Desesquelles, J., and Dufay, M. (1972a). *Phys. Rev. A* **6**, 600.

Berry, H. G., Pinnington, E. H., and Subtil, L. J. (1972b). *J. Opt. Soc. Amer.* **62**, 767.

Betz, H. D. (1972). *Rev. Mod. Phys.* **44**, 465.

Bickel, W. S. (1967). Unpublished work.

Bickel, W. S. (1968). *Appl. Opt.* **7**, 2367.

Bickel, W. S., and Bashkin, S. (1967). *Phys. Rev.* **162**, 12.

Bickel, W. S., and Goodman, A. S. (1966). *Phys. Rev.* **148**, 1.

Bickel, W. S., Martinson, I., Lundin, L., Buchta, R., Bromander, J., and Bergström, I. (1969). *J. Opt. Soc. Amer.* **59**, 830.

Brand, J. H., Cocke, C. L., Curnutte, B., and Swenson, C. (1970). *Nucl. Instrum. Methods* **90**, 63.

Bridges, J. M., and Wiese, W. L. (1970). *Astrophys. J.* **161**, L71.

Brown, L., Ford, K., Rubin, V., Trächslin, W., and Brandt, W. (1968). *In* "Beam-Foil Spectroscopy" (S. Bashkin, ed.), pp. 45–77. Gordon and Breach, New York.

Buchet, J. P., Denis, A., Desesquelles, J., and Dufay, M. (1967). *C. R. Acad. Sci. Paris* **265B**, 471.

Buchta, R., Curtis, L. J., Martinson, I., and Brzozowski, J. (1971). *Physica Scripta* **4**, 55.

Burger, J. M., and Lurio, A. (1971). *Phys. Rev. A* **3**, 64.

Burns, D. J., and Hancock, W. H. (1971). *Phys. Rev. Lett.* **27**, 370.

Carré, M., Desesquelles, J., Dufay, M., and Gaillard, M. L. (1971). *Phys. Rev. Lett.* **27**, 1407.

Chin-Bing, S. A., Head, C. E., and Green, A. E. (1970). *Amer. J. Phys.* **38**, 352.

Chupp, E. L., Dotchin, L. W., and Pegg, D. J. (1968). *Phys. Rev.* **175**, 44.

Cocke, C. L., Curnutte, B., and Brand, J. H. (1971). *Astron. Astrophys.* **15**, 299.

Corney, A. (1970). *Advan. Electron. Electron Phys.* **29**, 116–231.

Curtis, L. J., Schectman, R. M., Kohl, J. L., Chojnacki, D. A., and Shoffstall, D. R. (1970). *Nucl. Instrum. Methods* **90**, 207.

Curtis, L. J., Martinson, I., and Buchta, R. (1971a). *Physica Scripta* **3**, 197.

Curtis, L. J., Berry, H. G., and Bromander, J. (1971b). *Phys. Lett.* **34A**, 169.

Crossley, R. J. S. (1969) *Advan. At. Mol. Physics* **5**, 237–288.

Denis, A., Desesquelles, J., and Dufay, M. (1971). *C. R. Acad. Sci. Paris* **272B**, 789.
de Heer, F. J. (1966). *Advan. At. Mol. Phys.* **2**, 327.
Desesquelles, J. (1969). *C. R. Acad. Sci. Paris* **269B**, 972.
Desesquelles, J. (1970). Thesis, Lyon.
Desesquelles, J., Dufay, M., and Poulizac, M. C. (1968). *Phys. Lett.* **27A**, 96.
Donnally, B., Smith, W. W., Pegg, D. J., Brown, M., and Sellin, I. A. (1971). *Phys. Rev. A* **4**, 122.
Drake, G. W. F. (1969). *Astrophys. J.* **158**, 1199.
Drake, G. W. F., and Dalgarno, A. (1970). *Phys. Rev. A* **1**, 1325.
Druetta, M., and Poulizac, M. C. (1969). *Phys. Lett.* **29A**, 651.
Druetta, M., Ceyzeriat, P., and Poulizac, M. C. (1971). *C. R. Acad. Sci. Paris* **271B**, 846.
Dufay, M. (1970). *Nucl. Instrum. Methods* **90**, 15 (1970).
Dufay, M., Denis, A., and Desesquelles, J. (1970). *Nucl. Instrum. Methods* **90**, 85.
Dufay, M., Gaillard, M., and Carré, M. (1971). *Phys. Rev. A* **3**, 1367.
Edlén, B. (1964) *In* "Handbuch der Physik" (S. Flügge, ed.), Vol. 27, pp. 80–204. Springer-Verlag, Berlin.
Engvold, O., and Hauge, Ø. (1970). *Nucl. Instrum. Methods* **90**, 351.
Erickson, G. W. (1971). *Phys. Rev. Lett.* **27**, 780.
Fabjan, C. W., and Pipkin, F. M. (1970). *Phys. Rev. Lett.* **25**, 421.
Fabjan, C. W., and Pipkin, F. M. (1971). *Phys. Lett.* **36A**, 69.
Fabjan, C. W., Pipkin, F. M., and Silverman, M. (1971). *Phys. Rev. Lett.* **26**, 347.
Fan, C. Y., Garcia-Munoz, M., and Sellin, I. A. (1967). *Phys. Rev.* **161**, 6.
Fink, U. (1968). *Appl. Opt.* **7**, 2373.
Fink, U., Bashkin, S., and Bickel, W. S. (1970). *J. Quant. Spectrosc. Radiat. Transfer* **10**, 1241.
Foster, E. W. (1964). *Rep. Progr. Phys.* **27**, 469.
Gabriel, A. H., and Jordan, C. (1969). *Mon. Not. Roy. Astron. Soc.* **145**, 241.
Gabriel, A. H., and Jordan, C. (1970). *Phys. Lett.* **32A**, 166.
Garcia, J. D. (1970). *Nucl. Instrum. Methods* **90**, 295.
Goodman, A. S., and Donahue, D. J. (1966). *Phys. Rev.* **141**, 1.
Hadeishi, T., Bickel, W. S., Garcia, J. D., and Berry, H. G. (1969). *Phys. Rev. Lett.* **23**, 65.
Hallin, R., Lindskog, J., Marelius, A., Pihl, J. and Sjödin, R. (1971). Private communication.
Head, C. E. (1971). *Phys. Lett.* **34A**, 92.
Helliwell, T. M. (1964). *Phys. Rev.* **135**, A325.
Heroux, L. (1967). *Phys. Rev.* **153**, 156.
Heroux, L. (1970). *Nucl. Instrum. Methods* **90**, 173.
Hulpke, E., Paul, E., and Paul, W. (1963). *Z. Phys.* **177**, 257.
Kay, L. (1963). *Phys. Lett.* **5**, 36.
Kay, L. (1965). *Proc. Phys. Soc.* **85**, 163.
Korylov, F. A., and Odintsov, V. I. (1965). *Opt. Spectrosc. (USSR)* **18**, 547.
Lamb, W. E. (1952). *Phys. Rev.* **85**, 259.
Layzer, D., and Garstang, R. H. (1968). *In* "Ann. Rev. Astron. Astrophysics" pp. 449–494. Annual Reviews, Palo Alto, California.
Lewis, M. R., Marshall, T., Carnevale, E. H., Zimnoch, F. S., and Wares, G. W. (1967). *Phys. Rev.* **164**, 94.
Liu, C. H, Bashkin, S., Bickel, W. S., and Hadeishi, T. (1971). *Phys. Rev. Lett.* **26**, 222.

Liu, C. H., and Church, D. A. (1971). *Phys. Lett.* **35A**, 407.

Lurio, A., deZafra, R. L., Goschen, R. J. (1964). *Phys. Rev.* **134**, A1198.

Lynch, D. J., Drake, C. W., Alguard, M. J., and Fairchild, C. E. (1971). *Phys. Rev. Lett.* **26**, 1211.

Macek, J. (1970). *Phys. Rev. A* **1**, 618.

Malmberg, P. R., Bashkin, S., and Tilford, S. G. (1965). *Phys. Rev. Lett.* **15**, 98.

Marion, J. B., and Young, F. C. (1968). "Nuclear Reaction Analysis." North-Holland Publ., Amsterdam.

Marrus, R., and Schmieder, R. W. (1970). *Phys. Lett.* **32A**, 431; *Phys. Rev. Lett.* **25**, 1689.

Marrus, R., and Schmieder, R. W. (1972). *Phys. Rev. A* **5**, 1160.

Martinez-Garcia, M., and Whaling, W. (1971). *Bull. Amer. Phys. Soc.* **16**, 107.

Martinez-Garcia, M., Whaling, W., Mickey, D. L., and Lawrence, G. M. (1970). *Astrophys. J.* **165**, 213.

Martinson, I. (1970). *Nucl. Instrum. Methods* **90**, 81.

Martinson, I., and Bickel, W. S. (1969). *Phys. Lett.* **30A**, 524.

Martinson, I., and Bickel, W. S. (1970). *Phys. Lett.* **31A**, 25.

Martinson, I., Bickel, W. S., and Ölme, A. (1970). *J. Opt. Soc. Amer.* **60**, 1213.

Moore, C. E. (1949, 1952, 1958). "Atomic Energy Levels," Vols. I–III. Nat. Bur. Std. 467, U.S. Govt. Printing Office, Washington, D.C.

Murnick, D. E., Lewenthal, M., and Kugel, H. W. (1971). *Phys. Rev. Lett.* **27**, 1625.

Nichols, L. L., and Wilson, W. E. (1968). *Appl. Opt.* **7**, 167.

Northcliffe, L. C., and Schilling, R. F. (1970). *In* "Nuclear Data Tables" (K. Way, ed.), Vol 7, p. 233. Academic Press, New York.

Odintsov, V. I. (1963). *Opt. Spectrosc. (USSR)* **14**, 172.

Oona, H., and Bickel, W. S. (1969). Unpublished material.

Penkin, N. P., and Shabanova, L. N. (1969). *Opt. Spectrosc. (USSR)* **26**, 191.

Pinnington, E. H. (1970). *Nucl. Instrum. Methods* **90**, 93.

Pinnington, E. H., Kernahan, J. A., and Lin, C. C. (1970). *Astrophys. J.* **161**, 339.

Poizat, J.-C., Remillieux, J., and Desesquelles, J. (1971). *Phys. Lett.* **37A**, 427.

Poulizac, M. C., and Druetta, M. (1969). *C. R. Acad. Sci. Paris* **269B**, 114.

Poulizac, M. C., Druetta, M., and Ceyzeriat, P. (1971). *J. Quant. Spectrosc. Radiat. Transfer* **11**, 1087.

Radford, H. E. (1967) *In* "Methods of Experimental Physics" (V. W. Hughes and H. L. Schultz, eds.), Vol. 4B. Academic Press, New York.

Sakurai, J. J. (1967). "Advanced Quantum Mechanics." Addison-Wesley, Reading, Massachusetts.

Schiff, B., and Pekeris, C. L. (1964). *Phys. Rev.* **134**, A638.

Schmieder, R. W., and Marrus, R. (1970). *Phys. Rev. Lett.* **25**, 1245, 1692.

Sellin, I. A., Donnally, B. L., and Fan, C. Y. (1968). *Phys. Rev. Lett.* **21**, 717.

Sellin, I. A., Moak, C. D., Griffin, P. M., and Biggerstaff, J. A. (1969). *Phys. Rev.* **184**, 56; **188**, 217.

Sellin, I. A., Pegg, D. J., Brown, M., Smith, W. W., and Donnally, B. (1971). *Phys. Rev. Lett.* **27**, 1108.

Smith, P. L., Whaling, W., and Mickey, D. L. (1970). *Nucl. Instrum. Methods* **90**, 47.

Stone, M. E., and Morton, D. C. (1967). *Astrophys. J.* **149**, 23.

Stoner, J. O., and Leavitt, J. A. (1971). *Appl. Phys. Lett.* **18**, 368; **18**, 477.

Stoner, J. O., and Radziemski, L. J., (1970). *Nucl. Instrum. Methods* **90**, 275.

Stoner, J. O., and Radziemski, L. J. (1972). *Appl. Phys. Lett.* **21**, 165.

Swings, P., and Swings, J. P. (1968). *In* "Beam-Foil Spectroscopy" (S. Bashkin, ed.), pp. 525–538. Gordon and Breach, New York.

Tousey, R. (1963). *Space Sci. Rev.* **2**, 3.

Walker, A. B. C., and Rugge, H. R. (1971). *Astrophys. J.* **164**, 181.

Watson, W. D. (1969). *Astrophys. J.* **158**, L189.

Weiss, A. W. (1969). *Phys. Rev.* **188**, 119.

Weiss, A. W. (1970). *Nucl. Instrum. Methods* **90**, 121.

Westhaus, P., and Sinanoğlu, O. (1969). *Phys. Rev.* **183**, 56.

Whaling, W., King, R. B., and Martinez-Garcia, M. (1969). *Astrophys. J.* **158**, 389.

White, H. E., and Eliason, A. Y. (1933). *Phys. Rev.* **44**, 753.

Wien, W. (1919). *Ann. Phys.* **60**, 597. See also **66**, 229; **73**, 483; **83**, 1.

Wiese, W. L., and Weiss, A. W. (1968). *Phys. Rev.* **175**, 50.

Wiese, W. L., Smith, M. W., and Glennon, B. M. (1966). "Atomic Transition Probabilities," Vol. I, NSRDS-NBS 4. U.S. Govt. Printing Office, Washington, D.C.

Wiese, W. L., Smith, M. W., and Miles, B. M. (1969). "Atomic Transition Probabilities," Vol. II, NSRDS-NBS 22. U.S. Govt. Printing Office, Washington, D.C.

VIII.D EFFECTS OF EXTRANUCLEAR FIELDS
ON NUCLEAR RADIATIONS

D. A. Shirley

DEPARTMENT OF CHEMISTRY AND LAWRENCE BERKELEY LABORATORY
UNIVERSITY OF CALIFORNIA
BERKELEY, CALIFORNIA

I. Introduction

Nuclear physics interacts with a wide variety of other areas of science, and some very profound recent advances in science and technology have arisen through this interaction. There has been a steady flow of scientists trained in nuclear physics across interdisciplinary lines into related areas where nuclear concepts and techniques have been profitably applied. Several chapters would be necessary to cover this subject thoroughly. Instead, one aspect of it—the effects of extranuclear fields on nuclear radiations—is discussed in this chapter. This topic has been important in radioactivity studies for some time, and it has recently attracted interest among accelerator physicists. At this writing, many accelerators have at least one group doing experiments involving extranuclear fields. While the light coverage in this chapter cannot impart much new knowledge to these specialists, other accelerator physicists may find it a useful introduction to the subject.

In this treatment, concepts will be introduced as needed to precede the discussions where they are used. Thus the hyperfine Hamiltonian is introduced at the outset, while statistical tensors do not appear until Section V. The theoretical unity of the subject is stressed, especially in the discussion of oriented nuclear states. Emphasis is given to principles, but a number of applications are cited as examples.

II. The Hyperfine Hamiltonian

The electromagnetic interaction of a nucleus with its environment is most conveniently described in terms of nuclear moments coupling with extranuclear fields. This approach is usually successful because nuclear dimensions are very small compared to the distances over which extranuclear fields arising from electron distributions can vary appreciably. If an electron-nucleon system (e.g., an atom) is described by a Hamiltonian $\mathscr{H}$ which can be broken down into terms involving internucleon, interelectron, and nucleon–electron interactions,

$$\mathscr{H} = \mathscr{H}_{nn} + \mathscr{H}_{ee} + \mathscr{H}_{ne} \tag{1}$$

then the eigenenergy differences that are usually considered relevant to hyperfine structure problems will vary by many orders of magnitude among these three terms. The $\mathscr{H}_{nn}$ term yields energy differences measured in MeV, while electronic energy differences are of the order of electron volts, and the $\mathscr{H}_{ne}$ term gives hyperfine splitting in the micro-electron volt, or μeV, range.

Two *caveats* should be issued at this point. These statements are not applicable to meson–nucleus systems, as discussed in Chapter V.A. They must also be qualified by distinguishing between *electromagnetic* moments and *radial* moments.

Nuclear moments are usually derived from a multipole expansion. Parity conservation allows only even-rank electric or odd-rank magnetic moments to be nonzero. Thus the possible multipole moments are those of electric monopole (E0), magnetic dipole (M1), electric quadrupole (E2), etc., character. In fact, only these three are of sufficient importance to this chapter to be considered further, although both M3 and E4 moments are known.

For point nuclei, only these electromagnetic moments would be observable. Real nuclei are finite, however, and in some cases, their radial density variations must be taken into account. One further criterion must be met before radial moments need be considered: The "extranuclear" field due to the environment must show an appreciable radial variation within the nucleus. To understand this, let us consider the matrix element of an electromagnetic field operator $O(\mathbf{r}, ...)$ with a nuclear state ψ. Assuming for simplicity that the radial and angular coordinates are separable, we can write very schematically $\psi \sim R(r) \, Y(\theta, \phi)$. The matrix element is then separable provided that the radial dependence of $O(\mathbf{r})$ can be factored out,

$$O(\mathbf{r}, ...) = f(r) \, \Theta(\theta, \phi, ...) \tag{2}$$

Thus we have

$$\langle RY | O(\mathbf{r}, ...) | RY \rangle = \langle R | f(r) | R \rangle \langle Y | \Theta(\theta, \phi, ...) | Y \rangle$$
$$\equiv F(r) \langle Y | \Theta(\theta, \phi, ...) | Y \rangle \tag{3}$$

The radial integral $F(r)$ can in practice be accurately represented as a radial moment, and the matrix element of $\Theta(\theta, \phi, ...)$ is an electromagnetic E0, M1, or E2 moment. From (2) and (3), it is clear that if $O(\mathbf{r}, ...)$ did not depend on r, then the matrix element $\langle R | f(r) | R \rangle$ would just be unity. In the discussion to be given, the electromagnetic moments are taken up separately and the $F(r)$ factors are discussed together with the appropriate moments.

A. THE ELECTRIC MONOPOLE INTERACTION

The E0 interaction occupies an anomalous position in our scheme. In this case, the electromagnetic "moment" is just the nuclear charge Ze. Although by far the most important of the nuclear moments, the charge cannot by itself yield observable hyperfine structure (since the E0 moment is a scalar quantity, only shifts in energy, and no level splitting, could be expected anyway). At first, this seems surprising, because the nucleus–electron interaction energy associated with the E0 moment, i.e., the atomic binding energy, is very large, ranging up to hundreds of keV in heavy atoms.

The shift in this energy from one chemical environment to another is also large, ranging up to ~ 10 eV. The problem is that there exists no way to measure such shifts (although they are in fact practically identical to ESCA shifts, as discussed later).

Observable E0 shifts arise entirely through the radial moment effect. They appear as the isomer shift in Mössbauer spectroscopy and as the volume-dependent isomer or isotope shift in atomic spectroscopy. In both cases, four states are formed from two electronic and two nuclear states. The transition energies are studied, and the isomer or isotope shift appears as a change in the nuclear transition energy arising from variation of the electronic state (Mössbauer isomer shift) or as a change in the atomic transition energy arising from variation of the nuclear state (optical isomer or isotope shift). The relationship between these two shifts is illustrated by the energy diagram for ^{197}Hg shown in Fig. 1. It is not feasible to observe the Mössbauer effect in this case, but the optical isomer shift was reported in 1959 by Melissinos and Davis (1959), as Δv(optical) $= 0.021$ cm^{-1}. Conservation of energy gives Δv(nuclear) $= \Delta v$(optical), so the Mössbauer isomer shift would also be 0.021 cm^{-1} if it were observable. The shift arises because of the different interactions of the 6s and 6p electrons with the isomeric states. More specifically, the 6s electron is more strongly attracted to the smaller nuclear

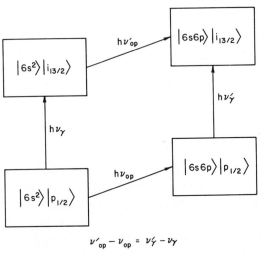

Fig. 1. Relationship between the optical and nuclear transitions in the isomeric states of ^{197}Hg. Energy conservation requires that the two shifts would be equal. Atomic states are denoted by $|6s^2\rangle$ or $|6s6p\rangle$.

$\frac{1}{2}^-$ ground state than to the slightly larger $\frac{13}{2}^+$ isomeric state. The reason for this will be described.

Figure 2 shows a simple illustration of the origin of isomer and isotope shifts. We imagine that the Schrödinger equation has been solved for the atomic system in the point-nucleus approximation, yielding one-electron states. These states are then perturbed by the finite nuclear charge distribution $\varrho(r)$. Outside the nucleus, this charge distribution (assumed spherical)

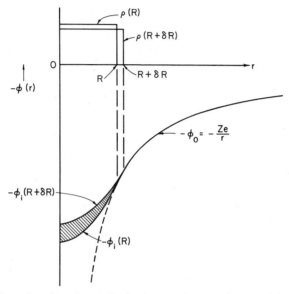

Fig. 2.   Effect of nuclear charge distribution on electrostatic potential experienced by an electron near the nucleus.

creates the same $1/r$ potential that would arise from a nuclear point charge. The potential felt by an electron within the nucleus is modified, however. At a radius r less than the nuclear radius R, the electron is attracted only by the nuclear charge at radii less than r. This charge produces the same potential that it would if concentrated at the origin. Since Coulombic forces vary as $1/r^2$, the nuclear charge "outside" the electron's radial position r exerts zero net force, as in the similar gravitational problem. The potential exerted by the nucleus can easily be shown to have the form

$$\phi_i = (Ze/2R)[3 - (r/R)^2], \qquad r < R$$
$$\phi_o = Ze/r, \qquad\qquad\qquad r > R$$

$$(4)$$

as shown in Fig. 2. When two nuclear states of radii R and $R + \delta R$, respectively, are compared, the external potentials ϕ_0 are terminated at different radii, yielding different ϕ_i's. The difference, $\delta\phi_i = -(3Ze/2R^2) \times [1 - (r/R)^2] \, \delta R$, is shown as the shaded area in Fig. 2.

Integration of the perturbation potential $\delta\phi_i$ over the electron charge density gives the shift in transition energy

$$\Delta E = -\langle \psi_e | \, e \, \delta\phi_i \, | \psi_e \rangle = -e \int_0^R \varrho_e(r) \, \delta\phi_i \, dv \tag{5}$$

Since the potential outside the nucleus is independent of nuclear size, there is no contribution to the integral for $r > R$. Thus only electrons in states with substantial density inside the nucleus, i.e., in s states, can contribute appreciably to ΔE. Equation (5) can be evaluated to give

$$\Delta E = \tfrac{4}{5}\pi Ze^2 |\psi(0)|^2 R \, \delta R \tag{6}$$

In this expression, the electron density within the nucleus has been taken as constant and denoted by $|\psi(0)|^2$. Relativistic electronic wave functions are not constant inside the nucleus. A correction denoted by $S'(Z)$ can be applied to correct the nonrelativistic electron density $|\psi(0)|^2$ for relativity. Finally, an isotope or isomer shift is obtained by comparing the shifts ΔE in two different electronic environments, obtaining (Shirley, 1964)

$$\mathrm{IS} = \tfrac{2}{5}\pi Ze^2 \, S'(Z) [\delta |\psi(0)|^2] [\delta R^2] \tag{7}$$

The penultimate factor is termed the "electronic factor." It is the difference of electron densities at the nucleus between source and absorber in Mössbauer spectroscopy or between the two electronic states in the optical case. The final "nuclear factor" is written as $\delta R^2 = 2R \, \delta R$.

B.   THE MAGNETIC DIPOLE INTERACTION; HYPERFINE MAGNETIC FIELDS

The interaction between a nuclear magnetic moment $\boldsymbol{\mu} = g\mu_N \mathbf{I}$ and an unpaired electron on the same atom can be written

$$\mathscr{H} = 2\mu_B g\mu_N \left[\frac{\mathbf{I} \cdot \mathbf{l}}{r^3} - \frac{\mathbf{I} \cdot \mathbf{s}}{r^3} + 3 \frac{(\mathbf{I} \cdot \mathbf{r})(\mathbf{s} \cdot \mathbf{r})}{r^5} + \frac{8\pi}{3} (\mathbf{I} \cdot \mathbf{s}) \, \delta(\mathbf{r}) \right] \tag{8}$$

Here μ_B and μ_N are the Bohr and nuclear magnetons, g is the nuclear g factor, and $\mathbf{r}$ is the electron's position. The first term describes the interaction between $\boldsymbol{\mu}$ and the magnetic field created at the nucleus by the electron's orbital motion. The next two terms describe dipole–dipole interaction between the nuclear moment and the electron's magnetic moment. The last term is the "Fermi contact term" describing the coupling between μ and the unpaired electron spin density within the nucleus (Fermi, 1930). Only s electrons (and to a lesser extent $p_{1/2}$ electrons in heavy elements) contribute to the Fermi contact term. Explicit evaluation of the delta function gives a "spin Hamiltonian" for this term of the form

$$\mathscr{H}_{\text{spin}} = (16\pi/3)\, g\mu_B\mu_N |\psi(0)|^2\, (\mathbf{I}\cdot\mathbf{s}) \tag{9}$$

In nonrelativistic approximation, the electron density "at" the nucleus is the same as that given in Eq. (6) for isomer and isotope shifts. In an exact relativistic treatment, however, the nonrelativistic densities must be corrected by different factors to give the appropriate relativistic expressions.

In most systems that are studied by Mössbauer spectroscopy or angular distribution methods, the electron operators $\mathbf{l}$ and $\mathbf{s}$ are diagonalized by an external field or by exchange forces. The nucleus in turn experiences an effective magnetic field $\mathbf{H}_{\text{eff}}$. This field consists of several parts, of which the largest is usually the hyperfine field $\mathbf{H}_{\text{hf}}$.

In a free atom, the electrons are vector-coupled to form total angular momentum $\mathbf{J}$, either in the Russell–Saunders limit described by

$$\mathbf{J} = \mathbf{L} + \mathbf{S}, \qquad \mathbf{L} = \sum_i \mathbf{l}_i, \qquad \mathbf{S} = \sum_i \mathbf{s}_i \tag{10}$$

or with varying degrees of spin–orbit coupling. Magnetic hyperfine coupling between the nucleus and the total electronic angular momentum $\mathbf{J}$ can be written as $A\mathbf{I}\cdot\mathbf{J}$, with all the numerical constants that arise from electronic coupling absorbed into the hyperfine structure constant A.

The interaction between a free atom–nucleus system and an external field $\mathbf{H}_0$ is given by

$$\mathscr{H} = g_J\mu_B\mathbf{H}_0\cdot\mathbf{J} - g\mu_N\mathbf{H}_0\cdot\mathbf{I} + A\mathbf{I}\cdot\mathbf{J} \tag{11}$$

Here the electronic g factor is denoted as g_J. For small $\mathbf{H}_0$, M_I is not a "good" quantum number (i.e., the energy eigenstates are not diagonal in an M_I representation). Thus it is not correct to describe the hyperfine interaction in terms of a hyperfine field. As $\mathbf{H}_0$ increases, M_I and M_J tend to become good quantum numbers, and for large $\mathbf{H}_0$ the spin Hamiltonian is given by

$$\mathscr{H} = g_J\mu_B H_0 J_z - g\mu_N H_0 I_z + A J_z I_z + \cdots \tag{12}$$

where off-diagonal terms may be neglected in the high-field limit. For the usual case of $Jg_J\mu_B \gg Ig\mu_N$, the system's eigenstates are resolved into $2J + 1$ "M_J" manifolds of $2I + 1$ states each. In each manifold, the second and third terms in Eq. (12) can be combined, and the resultant expression has the form of an interaction between a nuclear moment and an effective field H_{eff}. The nuclear magnetic substates within each manifold have energies $-g\mu_N H_{eff} M_I$. Thus a hyperfine field may be defined as (Shirley et al., 1968)

$$\mathbf{H}_{hf} = \mathbf{H}_{eff} - \mathbf{H}_0 = -AM_J/g\mu_N \tag{13}$$

Hence in the free-atom case the definition of a hyperfine field is straightforward. The system is isotropic, $\mathbf{H}_0$, $\mathbf{H}_{eff}$, and $\mathbf{H}_{hf}$ are collinear, and each M_J manifold has a unique value of H_{hf}.

An atom or ion in a crystal presents a more complicated problem. In the weak crystal-field case, which applies to the rare earths, J is a good quantum number, and the J manifold is split by the crystal-field potential into crystal-field multiplets each of which is characterized by an effective spin S'. The value of S' is established by setting the *electronic* degeneracy equal to $2S' + 1$. For each crystal-field multiplet, an expression analogous to Eq. (11) can be written (Elliott and Stevens, 1953),

$$\mathscr{H} = -\mu_B \sum_i g_i H_i S_i' - g\mu_N \mathbf{H}_0 \cdot \mathbf{I} + \sum_i A_i S_i' I_i \tag{14}$$

The anisotropy of this Hamiltonian introduces complexities. In the high-field limit, each crystal-field level is resolved into $2S' + 1$ "$M_{S'}$" manifolds of $2I + 1$ states each. Each manifold has an effective field

$$\mathbf{H}_{eff} = \mathbf{H}_0' + \mathbf{H}_{hf}' \tag{15}$$

While this relation is formally similar to the free-atom result, $\mathbf{H}_{hf}'$ and $\mathbf{H}_0'$ are not necessarily collinear. Furthermore, both are angle-dependent. For a crystal with "axial" symmetry (i.e., trigonal, tetragonal, or hexagonal symmetry), $\mathbf{H}_{hf}$ may be related to an effective hfs constant A_f by

$$|\mathbf{H}_{hf}'| = A_f M_{S'}/g\mu_B \tag{16}$$

where

$$A_f = (A_z^2 g_z^2 \cos^2\theta + A_x^2 g_x^2 \sin^2\theta)^{1/2}/g_{S'} \tag{17}$$

Here $g_{S'} = (g_z^2 \cos^2\theta + g_x^2 \sin^2\theta)^{1/2}$, and θ is the angle between $\mathbf{H}_0$ and the z axis. In this case,

$$H_0' = H_0^{corr}(A_z g_z \cos^2\theta + A_x g_x \sin^2\theta)/g_{S'} A_f \tag{18}$$

where H_0^{corr} is the external field at the ionic site, i.e., after correction for demagnetization, etc.

In a ferromagnet, the electronic spin is polarized by spin–spin exchange forces. For the weak crystal-field case, with the spin–orbit coupling $\lambda \mathbf{L} \cdot \mathbf{S}$ large compared to either the exchange energy or the crystal-field energy (as in the rare earths), exchange forces will diagonalize the spin Hamiltonian in the (J, M_J) representation or the $(S', M_{S'})$ representation. It is important to note that the spin $\mathbf{S}$ is oriented by exchange, and that $\mathbf{S}$ in turn orients $\mathbf{J}$ or $\mathbf{S}'$ through spin–orbit coupling. Then $\mathbf{H}_{\text{hf}}$ is determined by $\mathbf{J}$ or $\mathbf{S}'$. For the general case of a nucleus in an atom which is a solute in a ferromagnetic metal, a formal spin Hamiltonian of the form (Shirley *et al.*, 1968)

$$\mathcal{H} = V_{\text{cf}} + 2\mu_B \mathbf{H}_{\text{ex}} \cdot \mathbf{S} + \lambda \mathbf{L} \cdot \mathbf{S} - g\mu_N \mathbf{H}_{\text{c}} \cdot \mathbf{I} + 2g\mu_B\mu_N \langle r^{-3} \rangle$$
$$\times \{\mathbf{L} \cdot \mathbf{I} + [\xi L(L+1) - \kappa] \mathbf{S} \cdot \mathbf{I} - \tfrac{3}{2}\xi[(\mathbf{L} \cdot \mathbf{S})(\mathbf{L} \cdot \mathbf{I}) + (\mathbf{L} \cdot \mathbf{I})(\mathbf{L} \cdot \mathbf{S})]\}$$
(19)

can be useful. In this equation V_{cf} is the crystal-field potential, $\mathbf{H}_{\text{ex}}$ is the "exchange" field between $\mathbf{S}$ and the lattice magnetization, λ is the spin–orbit coupling constant, $\mathbf{H}_{\text{c}}$ is the effective field at the nucleus from conduction-electron polarization, ξ is an angular momentum factor, and κ gives the contribution of unpaired localized s electrons to the hyperfine structure (this last is hard to calculate and cannot be written simply in terms of one-electron operators). Although quite general, this Hamiltonian is essentially symbolic: It is most useful in keeping track of contributions to $\mathbf{H}_{\text{hf}}$. The first three terms, which involve only electronic coordinates, are large. They must be evaluated first to obtain zeroth-order wave functions for calculating matrix elements of the terms linear in $\mathbf{I}$. Both the term in H_{c} and the $\kappa \mathbf{S} \cdot \mathbf{I}$ term (which represents core polarization) involve the Fermi contact interaction.

Nuclei in ferromagnets experience effective magnetic fields arising in part from several rather esoteric effects. In a "demagnetized" ferromagnet, there is no macroscopic magnetic moment $\mathbf{M}$ (or there may be a small one), but randomly oriented magnetic domains exist. Within each domain, the electronic magnetism is saturated and each nucleus experiences an effective magnetic field parallel to that domain's magnetic moment. The domains are separated by domain walls, which are several tens of angstroms thick. On traversing the wall from one domain to another, the direction of magnetization changes gradually from that of the first to that of the second domain. As the sample is magnetized by an externally applied field $\mathbf{H}_0$ the walls move and the domains with favorably oriented magnetization grow at the expense of the rest. During this stage no external field penetrates the domains. The effective hyperfine field in the walls is given by

$$\mathbf{H}_{hf} = \mathbf{H}_{hf}^{corr} + \mathbf{H}_{L'} \qquad (20)$$

Here the corrected hyperfine field $\mathbf{H}_{hf}^{corr}$, which is microscopic in origin, is augmented by $\mathbf{H}_{L'}$, the sum of the usual Lorentz field plus a field arising from dipoles within the Lorentz cavity (Kittel, 1971). Fortunately $\mathbf{H}_{hf}^{corr}$ and $\mathbf{H}_{L'}$ are essentially collinear, and in practice, the observable quantity $\mathbf{H}_{hf}$ is taken as the "hyperfine field." After the ferromagnet is magnetized, the effective field felt by nuclei in the domain is

$$\mathbf{H}_{eff} = \mathbf{H}_{hf}^{corr} + \mathbf{H}_{L'} + \mathbf{H}_0 - D\mathbf{M} = \mathbf{H}_{hf} + \mathbf{H}_0 - D\mathbf{M} \qquad (21)$$

Here D is the demagnetization factor (Kittel, 1971). In most experiments, D cannot be neglected. Even in low-accuracy measurements, D must be considered because the sample cannot be magnetized until $\mathbf{H}_0$ exceeds both $D\mathbf{M}$ and the anisotropy field (Bozorth, 1951).

Many hyperfine structure studies of direct or indirect interest in nuclear physics are based on the hyperfine magnetic fields induced at nuclei of solutes in ferromagnets, especially in iron. Figure 3 shows an updated plot of solute hyperfine fields versus solute atomic number, after Shirley and Westenbarger (1965). These hyperfine fields are interesting for nuclear moment studies because they provide the possibility of obtaining fields in the 10^5–10^7 G range. In the solid-state area, they are of interest both intrinsically and for diagnostic work. The periodic dependence of H_{hf} on Z in Fig. 3 is striking. In the 3d, 4d, and 5d series, H_{hf} appears to arise mainly through "core polarization" of inner s shells by unpaired spins on d electrons. In the p-shell elements, the outer (closed) s shells are probably polarized by d electrons on neighboring atoms. The 4f (and 5f?) series are unique in that $\mathbf{L}$ and $\mathbf{S}$ are coupled to form $\mathbf{J}$. In the first half of the series, $\mathbf{L}$ and $\mathbf{S}$ couple anti-parallelly, producing positive fields. Parallel coupling in the second half of the series gives negative fields. In both cases, $\mathbf{L}$ tends to be the more important contributor to H_{hf} (except in Gd), while $\mathbf{S}$ determines the orientation of the atomic moment.

The discussion of magnetic hyperfine interactions has dealt entirely with the multipole aspects of this phenomenon. There is also a radial moment effect in magnetic interactions. It is called the hyperfine anomaly, denoted Δ. The term anomaly was given because this effect is observed when the ratio of the magnetic hfs constants A of two isotopes of the same element is found to vary from one electronic state to another or to differ from the ratio of

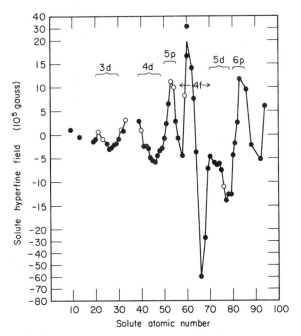

Fig. 3. Hyperfine fields at solutes in iron. Circles are filled where the sign of H_{hf} is known. Scales are linear, but the ordinate scale changes at ± 20.

their g factors. The anomaly arises through the Fermi contact interaction between an unpaired s electron within the nucleus and distributed nuclear magnetism. It is appreciable only in relatively heavy elements, because the electron spin density must vary appreciably throughout the nucleus. Nonrelativistic s-electron wave functions give a constant value of $|\psi(r)|^2$ within the nucleus, while the relativistic density required in heavy elements falls off from the center of the nucleus to the nuclear surface by up to $\sim 20\%$ in the Pb region, and more for the heavy elements.

A hyperfine anomaly observation requires four measurements for the same reason that observation of the E0 radial effect—the isomer shift—involves four states. The hyperfine structure constant for two isomers or isotopes of an element (which we shall denote 1 and 2) must be observed in an electronic state that has nonzero s-electron spin density in the nucleus. In addition, *either* hfs constants must be observed for these two nuclear levels in combination with electronic states in which the hfs constants have a different percentage composition of contact interaction, *or* the two nuclear g factors must be measured directly. Let us consider the latter case for simplicity. An externally applied magnetic field produces a uniform field throughout the nucleus, and magnetic substate splittings yield directly the g factors of the two nuclear states under study, i.e., g_1 and g_2. If both states were point nuclei, then in the electronic s state they would experience the same electronic spin density, and their hyperfine structure constants $A_1{}^0$ and $A_2{}^0$ would stand in the same ratio as the g factors,

$$A_1{}^0/A_2{}^0 = g_1/g_2 \tag{22}$$

In fact, the nuclear moment itself arises from intrinsic spin and orbital contributions. Thus we can write

$$g_1 = \alpha_s{}^1 g_s + \alpha_l{}^1 g_l \tag{23}$$

and similarly for g_2. Here $\alpha_s{}^1$ is the fractional contribution of nuclear intrinsic spin. The relation $\alpha_s + \alpha_l = 1$ always holds, although α_s or α_l may exceed unity. The nuclear orbital motion can be regarded as a current loop that samples an average of the electron spin density at all smaller radii. In contrast, the nuclear intrinsic spin samples electron spin density only at the nucleon's instantaneous position, thus experiencing a smaller average electron spin density (Bohr and Weisskopf, 1950). The fractional reductions in electron spin densities experienced by nuclear intrinsic spin and orbital motion, relative to that at the center of the nucleus, are denoted by $\bar{\kappa}_s$ and $\bar{\kappa}_l$, respectively. These quantities have been tabulated (Eisinger and Jaccarino, 1958). The fractional change in A is thus given by

$$\varepsilon = -\left(\bar{\kappa}_s \alpha_s + \bar{\kappa}_l \alpha_l\right) \tag{24}$$

The real hfs constant A is related to the point-nucleus value by

$$A = A^0(1 + \varepsilon) \tag{25}$$

Thus, from Eqs. (22) and (25), the relation between g ratios and A ratios is

$$g_2/g_1 = (A_2/A_1)(1 + \varepsilon_2)/(1 + \varepsilon_1) \tag{26}$$

The hyperfine anomaly is most conveniently defined as

$$\Delta_{1,2} = (A_1/A_2)(g_2/g_1) - 1 \tag{27}$$

in terms of measurable quantities. The approximation

$$\Delta_{1,2} \cong \varepsilon_1 - \varepsilon_2 \tag{28}$$

is less satisfactory both in principle, because it does not permit comparison with experiment, which allows no means for measuring ε, and in practice, because terms of order ε^2 are not always negligible. There is a widely held misconception that Δ cannot exceed 1% or so. However, values of $\sim 10\%$ are known. Easley *et al.* (1966) pointed out Δ can in principle even be infinite. They showed that for nominal $p_{1/2}$ or $d_{3/2}$ proton states, with l and s coupled to form $j = l - s$, very large values of Δ are possible, because both α_s and α_l are very large in magnitude, with the spin and orbital contributions to the nuclear moment nearly canceling.

C. THE ELECTRIC QUADRUPOLE INTERACTION

The first nonvanishing term in the expansion in Legendre polynomials of the electrostatic interaction between the nuclear charge distribution and the electronic charge distribution in the vicinity of the nucleus,

$$\mathscr{H} = \int\int r_{Ne}^{-1} \varrho(\mathbf{r}_N)\, \varrho_e(\mathbf{r}_e)\, dv_e\, dv_N \tag{29}$$

is the E0 interaction, discussed earlier. The second such term is the E2, or quadrupole, term. If the nuclear charge distribution $\varrho(\mathbf{r}_N)$ is taken as cylindrically symmetric, the quadrupole moment can be uniquely defined as (Townes, 1958)

$$Q = e^{-1} \int \varrho(3z_N^2 - r_N^2)\, dv_N \tag{30}$$

This quadrupole moment interacts with, and is oriented by, the electric field gradient q at the nucleus. In a free atom with one "unbalanced" electron, i.e. one electron not in closed shells, and with angular momentum j, q_j is given by

$$q_j = \langle j, m_j = j | (\partial^2 V/\partial z^2) | j, m_j = j \rangle = -e\langle jj | (3\cos^2\theta - 1)/r | jj \rangle \tag{31}$$

The quadrupole interaction is

$$\mathscr{H} = [eq_jQ/(2j-1)(2I)(2I-1)] [3(\mathbf{I}\cdot\mathbf{j})^2 + \tfrac{3}{2}(\mathbf{I}\cdot\mathbf{j}) - I^2j^2] \tag{32}$$

In the more general case, the electric field gradient tensor can contain several components, including cross terms such as $q_{xy} = \partial^2 V/\partial x\, \partial y$. The latter can be eliminated by a suitable choice of coordinate frame. This leaves the three diagonal components q_{xx}, q_{yy}, and q_{zz}. However, these are related by Laplace's equation, $\nabla^2 V = 0$, leaving only two independent quantities to be specified. These are usually taken to be q_{zz}, which is accorded the name "electric field gradient," and denoted simply by q, and the asymmetry parameter η, defined by

$$\eta = (q_{xx} - q_{yy})/q_{zz} \tag{33}$$

In order that η should be unique, the coordinate axes are chosen such that $q_{zz} \geqslant q_{xx} \geqslant q_{yy}$. The general form of the quadrupole spin Hamiltonian is then

$$\mathscr{H} = (eqQ/4I(2I-1))\,[3I_z^2 - \mathbf{I}^2 + \tfrac{1}{2}\eta(I_+^2 + I_-^2)] \tag{34}$$

where the nuclear spin terms in square brackets are operators. For cases in which the electronic environment has axial symmetry, the η term disappears and the rest of this expression can be evaluated to give energies

$$E(M_I) = [eqQ/4I(2I-1)]\,[3M_I^2 - I(I+1)] \tag{35}$$

Actual evaluation of the field gradient is usually difficult. A single bound electron can be shown to create a field gradient of

$$q_{nl0} = 2l(l+1)\,e\langle r^{-3}\rangle/(2l-1)\,(2l+3) \tag{36}$$

where n and l are the principal and orbital quantum numbers and the third subscript denotes the $m_l = 0$ substate. The evaluation of $\langle r^{-3}\rangle$ requires a self-consistent field calculation if an accurate value is required, but an order-of-magnitude estimate can be obtained from the expression for hydrogenic orbitals (Townes, 1958),

$$q_{nl0} = 4Z^3 e/n^3 a_0^3 (2l-1)\,(2l+1)\,(2l+3) \tag{37}$$

This expression can be corrected in two ways to bring it closer to the real value of q. First, it is nonrelativistic. A relativity correction factor $R(Z)$ can be applied. The factor $R(Z)$ is near unity for light elements and increases with Z. For a $p_{3/2}$ electronic orbital in gold, for example, $R(Z)$ is 1.32. Equation (37) must also be corrected to give an effective value for Z^3. Semiclassical arguments lead to the factor $Z_i Z_o^2$ for Z^3, where Z_i is the effective value of Z for the inner regions of the atom and Z_o is an effective value in the outer regions (thus $Z_i \cong Z$, and $Z_o \cong 1$). The final expression becomes

$$q_{nl0} = 4Z_i Z_o^2 eR(Z)/n^3 a_0^3 (2l-1)\,(2l+1)\,(2l+3) \tag{38}$$

This kind of approach is useful for making quick estimates of field gradients when accurate calculations are not feasible, but self-consistent field calculations are preferable. Even these must be used with care, because knowledge of $\langle r^{-3}\rangle$ for a one-electron state may not yield an adequate estimate of q, especially if configuration interaction is present. An improved value of q can be obtained by considering additional configurations. Often the corrected value is expressed in terms of an atomic shielding factor R_Q and the one-electron value as

$$q_{nl}(\text{corr}) = (1 - R_Q)\, q_{nl}(\text{uncorr}) \tag{39}$$

This factor may be assigned to the $\langle r^{-3} \rangle$ term explicitly. Thus Freeman and Watson (1963) gave the expression

$$\langle r^{-3} \rangle_Q = (1 - R_Q)\, \langle r^{-3} \rangle_{4f} \tag{40}$$

for rare earth 4f orbitals. When this notation is used, it is important to note that the value of $\langle r^{-3} \rangle$ used for quadrupole-coupling constants, namely $\langle r^{-3} \rangle_Q$, is different from that used for magnetic interactions, as in Eq. (8). This precludes using $\langle r^{-3} \rangle$ values deduced from magnetic hfs constants in evaluating field gradients.

In crystal lattices, the evaluation of q may be done in one of several different ways, depending on the nature of the sample. For molecular crystals, the greatest contribution to q usually arises from the molecular orbitals themselves, especially if unbalanced p, d, or f orbitals are present. Ionic crystals are usually treated by expanding the crystal potential in spherical harmonics consistent with the point-group symmetry, and approximating the separate terms by sums over point charges of appropriate magnitude on the lattice sites. For crystals with trigonal or tetragonal symmetry, it is the crystal field term transforming as $Y_2{}^0$ that contributes to q. This term is subject to a correction for the Sternheimer antishielding effect (Sternheimer, 1954) which takes the form

$$A_2{}^0(\text{corr}) = (1 - \gamma_\infty)\, A_2{}^0 \tag{41}$$

where γ_∞ is the "antishielding factor." The quantity $(1 - \gamma_\infty)$ can be as large as 100 or more: hence the name antishielding. The physical picture of how antishielding comes about can be presented in several ways. One approach is to consider configuration interaction, as did Judd et al. (1962). For a filled 5p shell, for example, in an atom subject to the interaction,

$$\mathscr{H} = \mathscr{H}(\text{quadrupole}) + \mathscr{H}(\text{crystal field}) \tag{42}$$

perturbation theory gives cross terms involving the matrix elements

$$\langle 5p^6\,{}^1S_0|\,\mathscr{H}(q)\,|5p^5 np\,{}^1D_2 \rangle \langle 5p^5 np\,{}^1D_2|\,\mathscr{H}(\text{cf})\,|5p^6\,{}^1S_0 \rangle \tag{43}$$

After suitable simplification, it can be shown that these terms contribute to

γ_∞ as

$$\frac{4e^2}{25} \sum_{n>5} \frac{\langle 5p|\, r^2\, |np\rangle \langle np|\, r^{-3}\, |5p\rangle}{E(5p^5 np)} \tag{44}$$

For $n = 6$, both radial integrals are large, and the $n = 6$ term in fact comprises a sizable fraction of γ_∞.

The evaluation of q in a metallic lattice is especially difficult because it is not simple to deal with the conduction electrons. Watson *et al.* (1965) have discussed this problem.

The "radial moment" effect is smaller in the E2 case than in the M1 and E0 interactions, and will not be discussed here. This effect can be large in mesic atoms, however (see Anderson and Jenkins, Chapter V.A).

III.   Mössbauer Spectroscopy

Resonant absorption of radiation between an excited state and a ground state is commonplace in optical spectroscopy. For γ radiation, however, the nucleus ordinarily recoils by an amount sufficient to shift the photon energy off resonance. In emission, energy conservation gives a γ-ray energy

$$E_\gamma = E_0 - E_R \tag{45}$$

where E_0 is the transition energy. Momentum conservation relates the recoil momentum and the γ-photon momentum:

$$p_R = -p_\gamma = -E_\gamma/c \tag{46}$$

Thus the recoil energy is given by

$$E_R = p_R{}^2/2m = E_\gamma{}^2/2mc^2 \tag{47}$$

For a nucleus with $A = 50$ and $E_\gamma = 30$ keV, this gives a recoil energy of 0.1 eV. This is several orders of magnitude greater than the line width of an isomeric state. Since the recoil effect acts to decrease the photon energy in emission but increase the required energy for resonant absorption, the latter process is precluded if recoil is present, as shown in Fig. 4.

A.   The Mössbauer Effect

In 1958, R. Mössbauer showed (Mössbauer, 1958) that nuclei bound in a solid lattice would undergo recoil-free emission and absorption a certain fraction of the time. This effect is represented in Fig. 4 as a peak at E_0 with width given by the excited-state lifetime. The recoil of a single nucleus bound

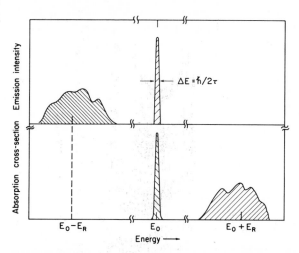

Fig. 4. Energy spectrum for γ-ray emission and absorption near the transition energy E_0, with recoil energy E_R taken into account (not drawn to scale). Mössbauer effect arises from recoil-free events at energy E_0.

in a lattice would correspond to emission or absorption of a phonon. "Recoil-free" events involve no phonon, but rather the entire crystallite can be regarded as recoiling. Thus the recoiling mass in the denominator of E_R in Eq. (47) becomes so large that E_R can be neglected. The probability that emission or absorption will occur without recoil is expressed as the recoil-free fraction, f. This fraction increases with lattice "stiffness," which can be expressed quantitatively in terms of the Debye temperature θ. The recoil-free fraction is given by

$$f = \exp\{-\tfrac{3}{2}(T_R/\theta)\left[1 + F(T/\theta)\right]\} \tag{48}$$

where $F(x)$ approaches zero as x approaches zero. Here the recoil temperature is related to the recoil energy by $E_R = kT_R$. From Eq. (47), we have, after substitution of numerical constants,

$$T_R = \frac{622(E_\gamma/100 \text{ keV})^2}{(A/100)} \tag{49}$$

Most solids are not very well characterized by a Debye temperature, but values of θ in the 100–500 °K range give the best fit. In order to have a recoil-free fraction of $\sim 10^{-2}$ or larger, it turns out that E_γ should be $\sim 10^5$ eV or less, and preferably in the 10^4-eV range. Since the ground state should be stable or very long-lived to provide a suitable absorber, Mössbauer spectroscopy is restricted for the most part to nuclides with low-lying excited states, in the energy region $E \leqslant 100$ keV. Another requirement for a high-quality resonance is that the excited-state lifetime should be in a rather limited range. Here two factors are important. If the lifetime is much longer than 10^{-5} sec, it is difficult to observe the Mössbauer effect at all, in part because of vibrational broadening but more importantly because of broadening arising from microscopic inhomogeneity in the sample. For lifetimes much shorter than 10^{-9} sec, on the other hand, the intrinsic line width is so great as to preclude obtaining very detailed information from the spectra. The ideal region for most solid-state studies is the 10^{-6}–10^{-8}-sec range. This still leaves a number of useful resonances.

B. STATIC SPECTRA

The analysis of Mössbauer spectra to yield hyperfine-structure parameters is a rather subtle problem in the most general case. The simplest case for which all three multipole interactions, E0, M1, and E2, are present is that in which the M1 term has the hyperfine-field form and the quadrupole interaction is axially symmetric ($\eta = 0$) and collinear. The energy levels of each state are then given by the expression

$$E(M_I) = E(I) - g\mu_N H_{hf} M_I + [e^2 Qq/4I(2I - 1)][3M_I{}^2 - I(I + 1)] \qquad (50)$$

There are two states in both source and absorber, so the spectrum is potentially very complicated. Whenever possible, either the source or the absorber is made up of a cubic lattice (where $q = 0$ by Laplace's equation) with no magnetic interactions, thus providing single-line emission or absorption by itself. The observed spectrum is then that of the complementary absorber or source lattice. Even then the spectrum involves all the allowed components of the γ-ray transition between two nuclear levels split as in Eq. (50).

The actual appearance of the spectrum depends on the manner in which the relative energies of source and absorber (or scatterer) are modulated. The most common procedure is to Doppler-shift the absorber at velocities in the cm/sec range and observe transmitted intensity as a function of velocity. In this case, the centroid of the spectrum gives the isomer shift, since the tensor operators describing both the M1 and E2 interactions are traceless. For $\mathscr{H}(M1) \gg \mathscr{H}(E2)$, and provided that Eq. (50) applies, the spectrum will be nearly symmetric and the E2 interaction will show up as a small asymmetry. Taken alone, the E2 spectrum usually appears quite asymmetric, except for the case of a $\frac{3}{2} \to \frac{1}{2}$ transition.

Because there are two nuclear states, a wide variety of spectra are possible. These can be systematized fairly easily, however. Consider, for example, the case of a $\frac{7}{2}$ (M1) $\frac{5}{2}$ transition with $\mu_{ex}/\mu_g > 0$ and only a hyperfine-field interaction present. The only adjustable parameter is the nuclear moment ratio. All cases are covered by a "g-factor diagram" of the kind shown in Fig. 5. In this figure, which was first used to determine the magnetic parameters of the 22 keV ^{151}Eu resonance (Barrett and Shirley, 1963), each line represents a transition component connecting a substate M of the $\frac{7}{2}$ state

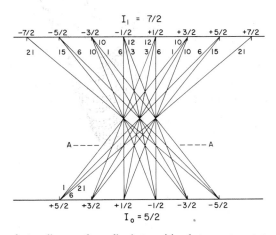

Fig. 5. The g-factor diagram for a dipole transition between two states of spins $\frac{5}{2}$ and $\frac{7}{2}$ having magnetic moments of the same sign. The line A-A fits the spectra of ^{121}Sb and ^{151}Eu. Numbers near lines denote relative intensities (Barrett and Shirley, 1963).

and a substate M' of the $\frac{5}{2}$ state. The intensity of each component is proportional to the square of a Clebsch–Gordan coefficient:

$$\mathscr{I}[\tfrac{7}{2} M \,(M1)\, \tfrac{5}{2} M'] = (\tfrac{5}{2} M' \, 1 \, M - M' | \tfrac{7}{2} M)^2 \qquad (51)$$

These intensities appear as numbers near the component lines in Fig. 5. By comparing an experimental spectrum with the appropriate diagram of this type (similar diagrams are used for quadrupole splitting), it is usually possible to narrow down the assignment of μ_{ex}/μ_g to one, or at most two or three, rough values. Final assignments can then be made by least-squares fitting procedures, using the rough ratios as initial estimates. A recent example of a quadrupole spectrum that was fitted in this way, the quadrupole spectrum of ^{181}Ta, was given by Kaindl *et al.* (1972). These workers used a single crystal of rhenium doped with ^{181}W as a source and a Ta foil as a (single line) absorber. They found $Q(\frac{9}{2})/Q(\frac{7}{2}) = 1.133 \pm 0.010$ and $e^2qQ(\frac{7}{2}) = -2.15 \pm 0.02$ from this work.

C. RELAXATION STUDIES

In addition to determining the "equilibrium" hfs parameters that are found in a stable extranuclear environment, Mössbauer spectra can also be useful in studying relaxation effects. Relaxation studies can yield useful data about the dynamic behavior of the extranuclear environment. A classic case is ^{57}Fe in the high-spin ferric ion, Fe(IV) (3d^5 ^{6}S). In many crystal-field environments, the electronic degeneracy of this ion is partially removed and the $M_s = \pm\frac{5}{2}$ doublet, or a state derived from it, lies lowest in energy. The atomic system can equally well be in the $M_s = +\frac{5}{2}$ state or in the $M_s = -\frac{5}{2}$ state. In any real system, it will in fact "jump" back and forth between these two states because of interactions with the lattice (spin–lattice relaxation). The resulting energy level diagram is shown in Fig. 6a. The ^{57}Fe nucleus

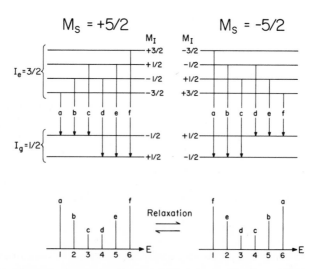

Fig. 6a. The effect of spin–lattice relaxation on line positions in Mössbauer spectra. Energy level diagrams for ^{57}Fe are shown for the two electron-spin orientations $M_s = \pm\frac{5}{2}$ of a ferric ion in an axial crystal field. The corresponding Mössbauer spectra are shown schematically below each case. When M_s flips from $+\frac{5}{2}$ to $-\frac{5}{2}$, the hyperfine field is reversed, and the energy level ordering of the magnetic substates is inverted. Thus transition a (the $-\frac{3}{2} \rightarrow -\frac{1}{2}$ transition) moves from energy position 1 to position 6, etc. For very fast relaxation, all the pairs of transitions collapse to the centroid energy, and a single line results.

experiences a hyperfine field of approximately

$$H_{hf} = -2 \cdot 10^5 M_s \quad \text{G} \tag{52}$$

Thus when the electron spin projection changes sign, so will the hyperfine field. The ^{57}Fe hyperfine-field spectrum consists of six lines. These lines are paired, with the two lines in each pair (i.e., lines 1 and 6, 2 and 5, and 3 and 4) corresponding to opposite signs for H_{hf}. For cases with long spin–lattice relaxation times, the Mössbauer transition takes place while M_s, and thus H_{hf}, has a unique value, and the characteristic six-line spectrum is observed. When the relaxation time is very short, the spin M_s is neither "up" nor "down," and the two components of each doublet coalesce into a single line halfway between them, in complete analogy to the situation for magnetic resonance spectra. Intermediate relaxation rates give more complicated spectra, but these can be calculated using the theoretical treatments that were developed earlier for explaining relaxation via exchange in proton magnetic resonance spectroscopy. The nuclear lifetime affects the spectrum only by establishing a minimum line width that can be observed in either the fast or slow relaxation-time limit. The relevant quantity to which the relaxation rate κ should be compared is the frequency difference $\Delta\omega$ between the two components of a pair that differ in energy by $\Delta E = \hbar \, \Delta\omega$. As the relaxation time is shortened in a system containing ^{57}Fe, the inner doublet (lines 3 and 4) therefore should disappear first, then the doublet made of lines 2 and 5, and finally the outer lines. In many lattices, the relaxation time of the electronic system can be varied through the sensitive range $\kappa \, \Delta\omega \sim 1$ by varying the absolute temperature, with the relaxation rate increasing with temperature. Relaxation effects, and particularly the aforementioned selective relaxation of the pairs of components, has been observed for ^{57}Fe in many lattices. Figure 6b shows a typical set of relaxation curves, reported by Wickman et al. (1966), using the compound ferrichrome a.

D. INTERFERENCE EFFECTS

Two kinds of interference effects have been observed using Mössbauer spectroscopy. The first is a single-atom effect involving the two atomic processes of Mössbauer absorption and photoelectric absorption. The two-step process, in which the excited state of an absorber is resonantly excited by γ absorption, then decays to the nuclear ground state by electron conversion, will yield the same final state as will the photoelectric effect. When a resonant and a nonresonant process both lead to the same final state in this way, interference appears as an asymmetry in the resonance

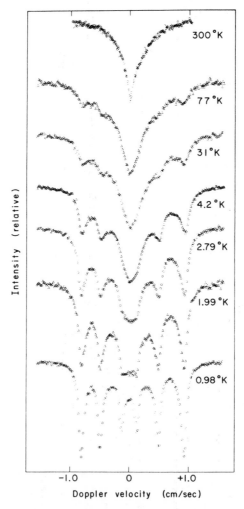

Fig. 6b. Mössbauer spectrum of ^{57}Fe in ferrichrome a, after Wickman *et al.* (1966). Relaxation time decreases with increasing temperature. Note that the inner doublet disappears first, as predicted.

line. The cross section in the resonant region has the form (Hammell and Hannon, 1969)

$$\sigma = \sigma_0(1 - 2\xi x)\,(1 + x^2)^{-1} + \sigma_e \qquad (53)$$

where σ_0 is the Mössbauer-absorption cross section, σ_e the photoelectric cross section, x the energy in units of half-width at half-maximum height, and ξ the asymmetry parameter. Interference effects of this kind are most readily observed in El transitions, although they can be seen in other cases as well. A large asymmetry effect has been reported (Sauer *et al.*, 1968; Kaindl and Salomon, 1970) in ^{181}Ta, from which $2\xi = 0.30 \pm 0.01$ was derived, in excellent agreement with the theoretical value of 0.31.

In the other kind of interference effect, the γ quantum interacts collectively with a periodic crystal lattice. At a Bragg angle, nodes in the electric field $\mathscr{E}$ describing the propagating γ ray can correspond exactly with nuclear positions at lattice sites, and Mössbauer absorption is suppressed. The crystal is therefore transparent to resonant radiation. This effect, predicted by Afanas'ev and Kagan (1967), is the nuclear analog of the Borrmann effect for X rays (Borrmann, 1950; Laue, 1952). A number of experiments have been reported, in which collective effects were observed both in transmission and in reflection, both with ^{57}Fe (Smirnov *et al.*, 1970) and with ^{119}Sn (Voitovetskii *et al.*, 1970).

E. Applications

Mössbauer spectroscopy has been applied to a very wide range of interesting studies. Perhaps the best-known application is the measurement of the gravitational red shift by Pound and Rebka (1960). They confirmed that photons traveling toward the earth are frequency-shifted by gravity. The frequency shift Δv of a photon of frequency v traveling a distance h in a gravitational field g is given by

$$\Delta v / v = gh/c^2 \tag{54}$$

The measurement of this effect requires that the ratio of level width to transition energy

$$\Gamma / E_\gamma = \hbar/\tau E_\gamma \tag{55}$$

be very small, as is the case for Mössbauer radiation.

An important chemical application of Mössbauer spectroscopy was made by Perlow and co-workers (Perlow, 1968), who studied xenon compounds. They studied isomer shifts and quadrupole splittings in halides and oxides of xenon, establishing the p_σ character of the Xe–F bonds in xenon fluorides, and showing in detail the similarity between xenon compounds and the isoelectronic iodine complexes.

Another interesting study was reported by Wertheim and Guggenheim (1968), who used the ^{57}Fe resonance to study critical exponents in antiferromagnetic iron fluorides. They used the hyperfine field to "track" the sublattice magnetization $M(T)$ and measured the parameter β given by the relation

$$M(T) = \text{const} \times (1 - T/T_N)^\beta \qquad (56)$$

which is valid near the Néel point T_N. The derived β values were in good agreement with predictions of the three-dimensional Ising model (i.e., $\beta \cong \frac{1}{3}$).

IV. Comparison of Inner-Shell Spectroscopies

There are now several types of measurements that involve the inner electronic shells of atoms. These have been compared in some detail recently to clarify the kind of extranuclear information that each can yield (Hollander and Shirley, 1970). Since all of these inner-shell methods entail interactions of E0 multipolarity, it is appropriate to consider them here before going on to angular distribution studies.

Four distinct effects will be considered. They are: (1) shifts in nuclear decay rates, (2) isomer shifts, (3) extranuclear effects on conversion-electron spectra, and (4) shifts in core-electron binding energies. They are discussed in this order and compared in Table 1.

Chemical effects can alter $|\psi(0)|^2$, the electron density at the nucleus, enough to change the nuclear decay rate perceptibly for processes that

TABLE 1

COMPARISON OF INNER-SHELL METHODS[a]

Method	Measured property	Derived quantity	Range of application
Shifts in total decay rate	Decay-rate change	$\Delta\|\psi(0)\|^2$	A few isotopes (^{7}Be)
Decay-rate shifts in low-energy transitions	Decay-rate change	$\Delta\|\psi(0)\|^2$	Several isotopes ^{90}Nb, ^{99}Tc, ^{235}U
Isomer shifts	Shifts in peak energies	$N\Delta\|\psi(0)\|^2$	~ 20 heavy elements
Conversion-electron spectroscopy	Peak intensity changes	Valence s-electron population	A few isotopes (^{119}Sn)
Photoelectron spectroscopy	Shifts in peak energies	Shifts in potential	All elements with $Z \geqslant 3$

[a] From Hollander and Shirley (1970).

depend directly on $|\psi(0)|^2$; i.e., electron capture and conversion. The total decay rate may be affected in the case of electron capture in light nuclei, as in ^{7}Be. Larger effects of up to a few percent can be observed in transitions for which only the contributions to $|\psi(0)|^2$ of outer electronic shells are eligible to affect the decay rate. The reason for this is that the decay constant λ varies as $|\psi(0)|^2$. Thus

$$\Delta\lambda/\lambda = \Delta|\psi(0)|^2/|\psi(0)|^2 \tag{57}$$

i.e., the two quantities have equal logarithmic derivatives when the environment is varied. In gross-decay-rate studies, the *total* electron density at the nucleus must go into the denominator. Since the contributions of inner electrons to $|\psi(0)|^2$ are both much larger than those of the valence shell and also much less affected by the chemical environment, the decay rate is usually altered too little for the change to be detectable. When the inner electronic shells are not eligible to participate in the decay process, as in electron-conversion cases for which the transition energies are smaller than inner-shell binding energies, only outer shells contribute to the denominator in Eq. (57), and $\Delta\lambda/\lambda$ can be much larger.

Isomer shifts were discussed earlier. They are more widely applicable and more sensitive than decay-rate shifts. They suffer, however, from the disadvantage that they yield not $\Delta|\psi(0)|^2$ but $N\,\Delta|\psi(0)|^2$, where N is a hard-to-determine nuclear factor [cf. Eq. (7)].

An intriguing improvement on the first method can be made by observing the conversion-electron spectrum, as Bocquet *et al.* (1966) have done for ^{119}Sn. Spectral data allow direct comparisons of peak intensities arising from different electronic orbitals. These intensities are proportional to the individual contributions of the separate atomic orbitals. They therefore yield the atomic orbital composition of the relevant molecular orbitals in each compound. These are quantities of considerable chemical interest.

Shifts in binding energies of core electrons, or ESCA shifts (Siegbahn *et al.*, 1967, 1969), comprise the fourth effect. These shifts are entirely extranuclear. Core electrons are ejected by characteristic (e.g., K_α) X rays of known energy $h\nu$. In the simplest case of a gaseous specimen, where there is no contact potential, the binding energy E_B of a given core level is related to the X-ray energy and the kinetic energy K of the photoelectron by the expression

$$E_B = h\nu - K \tag{58}$$

Energy analysis yields K, and E_B can be deduced. The chemical environment affects E_B, causing variations of 10 eV or more. This chemical shift, δE_B, is closely related to the shift in the electrostatic potential ϕ_e at the nucleus that arises from the electronic environment:

$$\delta E_B = -e\,\delta\phi_e \qquad (59)$$

Binding-energy shifts therefore yield an electronic parameter that is different from those given by the other three methods. Specifically,

$$\delta E_B \cong -e\,\delta \int \left[e\varrho_e(r)/r \right] dv = -e^2\,\delta \langle \Psi | r_i^{-1} | \Psi \rangle$$
$$= -e^2\,\delta \sum_j \langle \psi_j | r^{-1} | \psi_j \rangle \qquad (60)$$

Here Ψ is the total electronic wave function and ψ_j is a one-electron orbital. It is straightforward to write a linear relation between binding-energy shifts and isomer shifts,

$$\delta E_B = A(\text{IS}) - e^2\,\delta \sum_{\substack{\text{non s}\\\text{orbitals}}} \langle \psi | r^{-1} | \psi \rangle \qquad (61)$$

The coefficient A can be deduced from Eqs. (7) and (60),

$$A = -\left[5/2\pi ZS'(Z) \right] (\delta R^2)^{-1}\,\delta \left[\sum_{\text{s orbitals}} \langle \psi_j | r^{-1} | \psi_j \rangle \right] \Big/ \delta \left[\sum_{\text{s orbitals}} |\psi_j(0)|^2 \right] \qquad (62)$$

If the nuclear factor can be determined, this relation can prove useful in some cases. Specifically, for nontransition series elements, in which only the s electrons in the valence shell are affected by the environment, each sum in Eq. (62) reduces to a single term, and the ratio of these sums becomes essentially the ratio

$$\langle \phi | r^{-1} | \phi \rangle / |\phi(0)|^2 \qquad (63)$$

which can be evaluated for the valence s orbital ϕ. Transition series compounds are not so easily treated, because shielding effects change $|\psi(0)|^2$ for s orbitals in closed shells.

V. Radiations from Oriented Nuclear States

Under this heading fall those experimental methods that require the foregoing hyperfine-structure formalism and that also involve the anisotropic angular distribution of radiations from prepared states. These include perturbed angular correlations, nuclear orientation, and the perturbed angular distribution of radiations from nuclei oriented by nuclear reactions (see Recknagel, Chapter VII.C). These methods will be abbreviated PAC, ON, and PAD, respectively. These three methods can be treated by a common theoretical formalism that emphasizes their similarities. Each method involves the preparation of an oriented nuclear state in which hyperfine interactions take place, and in each case, the subsequent evolution of that state under the influence of the hyperfine Hamiltonian is studied through detection of radiation that it emits. The methods differ only in the ways in which the oriented state is prepared; after that, they have a common theoretical description. The description of an oriented nuclear state is first given. Then interactions with static fields are described, followed by resonance experiments and relaxation phenomena.

A. DESCRIPTION OF THE ORIENTED STATE

Figure 7 gives a generalized representation of the geometry that pertains to experiments which involve observing nuclear radiation emitted from oriented states undergoing hyperfine interactions. The unit vector k_1 represents the direction about which the nuclear system is (in the ensemble average) cylindrically symmetric at the initial time $t = 0$ when the oriented state is created. This state evolves under the influence of a hyperfine Hamil-

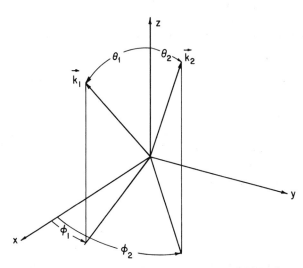

Fig. 7. Laboratory coordinate frame representation of the initial orientation direction (unit vector k_1) and the detector direction (unit vector k_2). (Arrows in figure indicate vector quantities.)

tonian $\mathscr{H}$, and decay quanta are detected by a detector in the direction of the unit vector $\mathbf{k}_2$. The $\mathbf{k}_1$ direction represents the direction of the first detector in a PAC experiment, the beam direction in a PAD measurement, and the orientation axis in the ON case. In any case, it is convenient to describe the oriented state in terms of a density matrix formalism, and in particular to formulate statistical tensors $\varrho_q^\lambda(t)$ defined by

$$\varrho_q^\lambda(t) = \sum_m (-1)^{I+m'} \langle I - m'Im \mid \lambda q \rangle \, \varrho_{m'm}(t) \tag{64}$$

Here $\varrho_{m'm}(t)$ is an element of the density matrix in an m representation. When the $\mathbf{k}_1$ direction is chosen as the axis of quantization, the density matrix is diagonal at $t = 0$ (Matthias et al., 1971), i.e.,

$$\begin{aligned}
\varrho_{m'm}(0)_{\mathbf{k}_1} &= 0 \qquad \text{for} \quad m' \neq m \\
\varrho_q^\lambda(0)_{\mathbf{k}_1} &= 0 \qquad \text{for} \quad q \neq 0
\end{aligned} \tag{65}$$

It is usually straightforward to specify $\varrho_0^\lambda(0)_{\mathbf{k}_1}$, although different approaches must be taken in the three types of experiment. For a PAC experiment, in which the oriented state is produced by detection of γ rays of multipolarity L emitted along $\mathbf{k}_1$ from a randomly oriented state of spin I_0,

$$\varrho_0^\lambda(0)_{\mathbf{k}_1} = (-1)^\lambda (2I+1)^{-1/2} F_\lambda(LLI_0I) \tag{66}$$

where $F_\lambda(LLI_0I)$ is a standard angular correlation coefficient (Frauenfelder and Steffen, 1965). In the ON case, the expression is

$$\varrho_0^\lambda(0)_{\mathbf{k}_1} = \sum_m (-1)^{I+m} \langle I - mIm \mid \lambda 0 \rangle \, (e^{-\mathscr{H}/kT}/\mathrm{Tr}\, e^{-\mathscr{H}/kT}) \tag{67}$$

In PAD studies, no such closed-form expression is available because the details of the nuclear reaction path are not usually very well known. Since momentum is transferred perpendicular to the beam direction, substates with the lowest values of $|m|$ are the most heavily populated. Often a Gaussian distribution is assumed, i.e.,

$$\begin{aligned}
\varrho_{mm}(0)_{\mathbf{k}_1} = A \exp(-m^2/2\sigma^2), \qquad & m = 0, \pm 1, \pm 2, ..., \pm I \\
& \text{or} \quad \pm \tfrac{1}{2}, \pm \tfrac{3}{2}, ..., \pm I
\end{aligned} \tag{68}$$

The constants A and σ are determined by the condition $\mathrm{Tr}\, \varrho = 1$ and by observing the magnitude of an anisotropic radiation distribution. The statistical tensors $\varrho_0^\lambda(0)_{\mathbf{k}_1}$ follow from Eqs. (64) and (68).

B. Evolution under Perturbations

The time evolution of the density matrix under the influence of the hyperfine-interaction Hamiltonian $\mathscr{H}$ is given by the von Neumann equation

$$i\hbar\dot{\varrho} = [\mathscr{H}, \varrho] = \mathscr{H}\varrho - \varrho\mathscr{H} \tag{69}$$

This equation is usually solved by introducing the time evolution operator $\Lambda(t)$, which represents a unitary transformation of ϱ:

$$\varrho(t) = \Lambda(t)\,\varrho(0)\,\Lambda(t)^+ \tag{70}$$

It can be shown that $\varrho(t)$ satisfies Eq. (69) if $\Lambda(t)$ satisfies the Schrödinger equation

$$\partial\Lambda(t)/\partial t = -(i/\hbar)\,\mathscr{H}(t)\,\Lambda(t) \tag{71}$$

If $\mathscr{H}$ is time independent, the solution is

$$\Lambda(t) = \exp(-i\mathscr{H}t/\hbar) \tag{72}$$

and $\varrho(t)$ is given by

$$\varrho(t) = [\exp(-i\mathscr{H}t/\hbar)]\,\varrho(0)\,[\exp(i\mathscr{H}t/\hbar)] \tag{73}$$

It is usually most convenient to calculate $\varrho(t)$ in a frame in which $\mathscr{H}$ is either time independent or at least as simple as possible. Most often $\mathscr{H}$ can be diagonalized by a suitable unitary transformation, and the matrix elements of $\Lambda(t)$ are then easily evaluated. After several steps, the statistical tensors at time t can be written in terms of those at $t = 0$ as

$$\varrho_{\bar{q}}^{\bar{\lambda}} = \sum_{\lambda q} G_{\lambda\bar{\lambda}}^{q\bar{q}}(t)^*\,\varrho_q^{\lambda}(0) \tag{74}$$

where the perturbation coefficient is given by

$$G_{\lambda\bar{\lambda}}^{q\bar{q}}(t) = \sum_{m',\bar{m}} (-1)^{2I+m'+\bar{m}'} \langle I-\bar{m}'I\bar{m}\,|\,\bar{\lambda}\bar{q}\rangle \langle I-m'Im\,|\,\lambda q\rangle$$
$$\times \langle \bar{m}|\,\Lambda(t)\,|m\rangle \langle \bar{m}'|\,\Lambda(t)\,|m'\rangle^* \tag{75}$$

These coefficients contain all the information available about the interaction of the oriented state with its environment. They appear in the angular distribution function as

$$W(\mathbf{k}_1, \mathbf{k}_2, t) = 4\pi \sum_{q,\,\lambda,\,\bar{q},\,\bar{\lambda}} \frac{A_\lambda(x_2)}{[(2\lambda+1)\,(2\bar{\lambda}+1)]^{1/2}}$$
$$\times \varrho_0^{\lambda}(0)_{\mathbf{k}_1}\,G_{\lambda\bar{\lambda}}^{q\bar{q}}(t)_z\,Y_{\lambda q}^*(\theta_1, \phi_1)\,Y_{\bar{\lambda}\bar{q}}(\theta_2, \phi_2) \tag{76}$$

Here A_λ is a radiation parameter: for γ rays it is simply an angular correlation coefficient. Since $\varrho_0^\lambda(0)$ is evaluated in the $\mathbf{k}_1$ frame and $G_{\lambda\lambda}^{q\bar{q}}$ in the laboratory frame, the spherical harmonics arise from transformations between these two frames, i.e.,

$$D_{q0}^{(\lambda)}(\phi, \theta, 0) = [4\pi/(2\lambda + 1)]^{1/2} Y_{\lambda q}^*(\theta, \phi) \tag{77}$$

Experiments are usually designed in such a way as to allow the perturbation coefficients $G_{\lambda\lambda}^{q\bar{q}}$ to be extracted from the data unambiguously and as directly as possible. The interpretation of an experiment then requires the extraction of hyperfine-structure information from the perturbation coefficients. The forms that these coefficients take for various extranuclear situations will be discussed.

C. Time-Independent Interactions

When the hyperfine Hamiltonian is time independent, Eq. (73) is directly usable, and the only remaining problem is to choose the reference frame efficiently. The hyperfine Hamiltonian can be represented as the sum of a magnetic part and a quadrupole part,

$$\mathscr{H} = \mathscr{H}_{\mathrm{M}} + \mathscr{H}_{\mathrm{Q}} \tag{78}$$

A unitary transformation can be made into a representation in which $\mathscr{H}$ is diagonal, and the operator terms $\exp(-i\mathscr{H}t/\hbar)$ are evaluated as energy terms $\exp(-iEt/\hbar)$. In the general case, $G_{\lambda\lambda}^{q\bar{q}}(t)$ is oscillatory but aperiodic. The general expression for $G_{\lambda\lambda}^{q\bar{q}}(t)$ under the restriction that $\mathscr{H}$ has axial symmetry is (Frauenfelder and Steffen, 1965)

$$G_{\lambda\lambda}^{qq}(t) = \sum_m [(2\lambda + 1)(2\bar{\lambda} + 1)]^{1/2} \begin{pmatrix} I & I & \lambda \\ m' & -m & q \end{pmatrix} \begin{pmatrix} I & I & \bar{\lambda} \\ m' & -m & q \end{pmatrix}$$

$$\times \exp[-(i/\hbar)(E_m - E_{m'})t] \tag{79}$$

When either $\mathbf{k}_1$ or $\mathbf{k}_2$ is taken along the direction of axial symmetry in $\mathscr{H}$, it can be shown from Eqs. (76) and (79) that the observed angular correlation is unperturbed. This follows because $Y_{\lambda q}(\theta = 0, \phi) = \delta_{q0}[(2\lambda + 1)/4\pi]^{1/2}$. Thus only terms with $E_m = E_{m'}$ survive and $G_{\lambda\bar{\lambda}}^{qq} = \delta_{\lambda\bar{\lambda}}$. In a nuclear orientation experiment, $\mathscr{H}$ is usually symmetric about $\mathbf{k}_1$, and the "correlation" appears unperturbed. Thus ϱ_0^λ is time independent, and hfs parameters may be deduced from Eqs. (67) and (76). In PAC and PAD studies, it is often convenient to set $\mathbf{k}_1$ or $\mathbf{k}_2$ along the axis of $\mathscr{H}$ to measure the extent of orientation.

In the important case of interaction with an external magnetic field,

$$\mathscr{H} = -\boldsymbol{\mu} \cdot \mathbf{H} = -g\mu_N \mathbf{H} \cdot \mathbf{I} \tag{80}$$

the system exhibits Larmor precession. That is, there exists a symmetry vector $\mathbf{K}(t)$ about which the statistical tensors are cylindrically symmetric and time independent (except for decay). This vector evolves according to the equation (Matthias *et al.*, 1971)

$$d\mathbf{K}(t)/dt = (g\mu_B/\hbar)\, \mathbf{K} \times \mathbf{H} \tag{81}$$

with $\mathbf{K}(0) = \mathbf{k}_1$. This is reminiscent of Larmor precession, and indeed $G_{\lambda\bar{\lambda}}^{qq}$ has the form

$$G_{\lambda\bar{\lambda}}^{qq}(t) = \delta_{\lambda\bar{\lambda}} e^{-iq\omega_L t} \tag{82}$$

where ω_L is the Larmor frequency $\omega_L = -g\mu_B H/\hbar$. It may seem strange that a system precessing at the Larmor frequency should exhibit higher-frequency behavior. In fact, this arises simply because tensors of ranks higher than one are required to describe the angular distribution. A spherical harmonic of rank λ can have components of period down to $2\pi/\lambda$ in the angle of precession. In the most common case of $\lambda = 2$, the frequency $2\omega_L$ is usually dominant, although for the "random fields" configuration, the frequency ω_L is at least equally important (Matthias *et al.*, 1965).

The "quality factor" of a magnetic interaction case is very easily stated. It is $\omega_L \tau$, the product of the nuclear lifetime τ and the Larmor frequency. Experiments have been done for values of $\omega_L \tau$ so low that the mean angle of rotation, as given by

$$\Delta\phi = \tan^{-1}(qg\mu_N H\tau/\hbar) \tag{83}$$

is of order 10^{-2}. In these cases, g factors of nuclei with lifetimes in the psec range were sought, and only time-integrated measurements of $G_{\lambda\bar{\lambda}}^{qq}$ were possible. At the other end of the scale, time-differential studies have been carried out on cases with values of $\omega_L \tau$ so high that hundreds of oscillations could be observed. In these cases, it is useful to Fourier-transform $G_{\lambda\bar{\lambda}}^{qq}(t)$ into the frequency domain rather than fitting it directly. The transform,

$$F(\omega) = \int_0^\infty G_{\lambda\bar{\lambda}}^{qq}(t)\, e^{-t/\tau} \cos \omega t \, dt \tag{84}$$

has the form of a Lorentzian line of width $(\pi\tau)^{-1}$ centered about frequencies of $q\omega_L$. Such lines can be interpreted in the same way as are NMR lines, in terms of line shapes, satellites, etc. They actually correspond to NMR lines in the absence of radiofrequency fields. In the limiting case of very long lifetimes and many cycles of oscillation in $G_{\lambda\lambda}^{qq}(t)$, the statistical scatter is such as to preclude observing any single cycle with certainty. An autocorrelation function may then be formed. It is most convenient to define the normalized autocorrelation function (Matthias and Shirley, 1966) in terms of the coincidence counting rate $W(t)$ and the lag time τ as

$$\frac{C(\tau)}{C(0)} = \frac{\text{Ave}\{[W(t) - \bar{W}(t)][W(t+\tau) - \bar{W}(t+\tau)]\}}{\text{Ave}\{[W(t) - \bar{W}(t)]^2\}} \tag{85}$$

This ratio can be Fourier-transformed in place of $G_{\lambda\lambda}^{qq}(t)$ in Eq. (84). Quite narrow lines (in relation to the resonant frequencies) have been observed in this way. For ^{100}Rh in ferromagnetic nickel, a line of 2.5 MHz FWHM was observed at the Larmor frequency of 339.5 MHz, and the line at $2\omega_L$ was also found (Koicki $et\ al.$, 1970), as shown in Fig. 8.

In accelerator (PAD) experiments, the Fourier transformation can be done as the data are collected by using the "stroboscopic" method (Christiansen $et\ al.$, 1968). In this approach, a resonance is observed when the beam

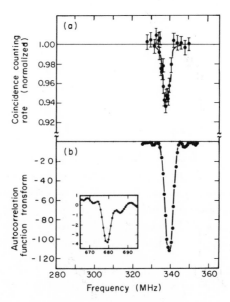

Fig. 8. PAC spectra for ^{100}Rh in nickel (from Koicki $et\ al.$, 1970.). (a) NMR/PAC line; (b) the Fourier transform of a PAC autocorrelation function; the line at $2\omega_L$ is shown in the insert. Frequency discrepancy arises from radio-frequency heating.

pulsing frequency and $\omega_L/2\pi$ are equal or related by simple integers. Usually the applied field H_0 is swept, and the counting rate in the $\mathbf{k}_2$ direction exhibits a resonance when plotted against H_0, at $H_0 = q\hbar\omega_L/g\mu_N$.

For the case of quadrupole splitting with $\eta = 0$, the energy levels are given by Eq. (35). Thus Eq. (79) becomes

$$G_{\lambda\lambda}^{qq}(t) = [(2\lambda + 1)(2\bar{\lambda} + 1)]^{1/2} \sum_m \begin{pmatrix} I & I & \lambda \\ m' & -m & q \end{pmatrix} \begin{pmatrix} I & I & \bar{\lambda} \\ m' & -m & q \end{pmatrix}$$

$$\times \exp[3i(m^2 - m'^2)\omega_Q t] \tag{86}$$

where $\omega_Q = eqQ/4I(2I - 1)\hbar$. Since the level energies vary quadratically with m, there are several frequencies apparent in $G_{\lambda\lambda}^{qq}(t)$. For example, in the spin-$\frac{5}{2}$ case, terms with $\Delta m^2 = 2, 4$, and 6 all contribute, and $G_{22}^{00}(t)$ is periodic in the lowest common multiple.

When $\eta \neq 0$, there is no necessity for the energy level spacings to be related by ratios of integers. Thus $G_{\lambda\lambda}^{qq}(t)$ is oscillatory but aperiodic. A similarly complicated result arises when both magnetic and quadrupole effects are present (Alder *et al.*, 1963).

D.   TIME-DEPENDENT INTERACTIONS: RESONANCE

Hyperfine structure effects on angular correlations may be observed in the resonant mode, by observing a change in the radiation pattern on application of a radio-frequency field

$$H_1(t) = 2H_1 \cos \omega t$$

perpendicular to the static field H_0. This resonant process was predicted for PAC and ON experiments in 1953 (Abragam and Pound, 1953; Bloembergen and Temmer, 1953). It has been observed in PAC, PAD, and ON experiments (Shirley, 1969). Figure 9 indicates how an NMR/PAC experiment works for a hypothetical 0(dipole) 1(dipole) 0 cascade. In an NMR/PAD experiment, the resonance takes place in a reaction-product final state oriented perpendicular to the beam, while in NMR/ON, the resonance is done on a polarized radioactive isotope. In each case, the theory is straightforward. The external field direction $\mathbf{H}_0$ is taken as the z axis and the rf field direction is the $\pm x$ axis. A coordinate transformation into the "rotating frame"—a frame that rotates with angular velocity ω about the z axis—leaves an effective field $\mathbf{H}_e$ with x component H_1 and z component $[(\omega_L - \omega)/\omega_L] H_0$ in this new frame. At the resonance frequency $\omega = \omega_L$, these components have the values H_1 and 0. Resonance can be regarded as the rotation of the

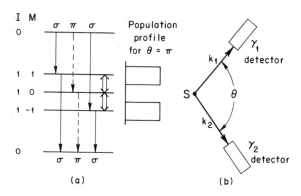

Fig. 9. (a) Typical level scheme and (b) experimental coincidence arrangement for NMR/PAC experiment. A 0 (dipole) 1 (dipole) 0 cascade is depicted. For the detector-source–detector angle $\theta = \pi$ and $H_0 \parallel k_1, k_2$, only σ–σ coincidences are observed and only $M = \pm 1$ components of the intermediate state are populated: symmetry of population is a consequence of parity conservation. Double arrows denote rf transitions (Shirley, 1969, used with permission of North-Holland Publishing Co., Amsterdam).

vector $\mathbf{K}(t)$ around H_1 in this frame. The equation of motion is

$$dK/dt = (g\mu_B/\hbar)\, \mathbf{K} \times \mathbf{H}_1 \qquad (87)$$

analogous to Eq. (81) in the laboratory frame. The resonance phenomenon usually competes with some other process (e.g., nuclear decay, spin–lattice relaxation) having a characteristic time τ. The criterion for an easily observable resonance is that H_1 should be large enough to cause $\mathbf{K}$ to rotate by ~ 1 rad in time τ. Since the rotation frequency is given by $\omega_1 = -g\mu_N H_1/\hbar$, this criterion becomes

$$g\mu_N H_1 \tau/\hbar \geqslant 1$$

In practice, this condition is rather easily met for many cases.

The computation of resonance lines affords a very nice example of the general procedures discussed in Section B. The unitary transformation is in this case simply a product of three rotations in space, and it takes the system into a frame in which the hyperfine Hamiltonian vanishes identically (Matthias *et al.*, 1971).

E. Time-Dependent Interactions: Relaxation

When an oriented state interacts with a randomly fluctuating external field, the orientation is destroyed at a rate that can be characterized by one or more relaxation times. The basic theory for this relaxation (Matthias *et al.*,

1971; Frauenfelder and Steffen, 1965) was given originally by Abragam and Pound (1953). Applications have ranged from "rotational tracer" studies of tumbling macromolecules in solutions (Leipert *et al.*, 1968) to measurements of spin–lattice relaxation in metals at low temperatures (Bacon *et al.*, 1972). Let us close this chapter by using the latter application as an illustrative example.

Nuclear orientation in metals is usually effected through a hyperfine field H_{hf} taken along the z axis. However, conduction electrons produce random instantaneous fields in other directions through the x and y components of the $A\mathbf{I} \cdot \mathbf{S}$ interaction,

$$A[S_x I_x + S_y I_y] = (A/2) [S_+ I_- + S_- I_+] \qquad (88)$$

The static interaction $-\boldsymbol{\mu} \cdot \mathbf{H}_{hf}$ yields $2I + 1$ magnetic substates equally spaced in energy by $\Delta E = g\mu_N H_{hf}$. The relaxation mechanism of Eq. (88) allows transitions between adjacent substates. The "upward" and "downward" transitions have the forms (Bacon *et al.*, 1972)

$$W_{up} = (hv/2kC) [I(I + 1) - M(M + 1)]/(e^{x_L} - 1)$$
$$W_{down} = (hv/2kC) [I(I + 1) - M(M + 1)]/(1 - e^{-x_L}) \qquad (89)$$

for the $M \leftrightarrow M + 1$ transitions. Here C is the Korringa constant that relates the spin–lattice relaxation time T_1 to temperature at high temperatures: $T_1 T = C$. The argument x_L is $g\mu_N H_{hf}/kT$, where T is the lattice temperature. The last factors in each case arise from the Fermi statistics of conduction electrons. They resemble boson distribution functions, however. In fact, it can be shown from Eq. (89) that

$$W_{down} = W_{up} + (hv/2kC) [I(I + 1) - M(M + 1)] \qquad (90)$$

This result is strongly analogous to the problem of a two-level system interacting with the radiation field. Further analysis has shown that W_{up} is equivalent to absorption of energy from the "radiation field" (the lattice, in this case). The constant term in Eq. (90) is equivalent to spontaneous emission, while the rest of W_{down} amounts to induced emission. As $T \to 0$, the temperature-dependent part of W_{down}, as well as that of W_{up}, vanishes, and only spontaneous emission remains. The spin–lattice relaxation time, which varies as T^{-1} at high temperatures, therefore approaches constancy as $T \to 0$ (Brewer *et al.*, 1968). This behavior has been observed in several cases (Bacon *et al.*, 1972) by relaxation studies (Templeton and Shirley, 1967) on oriented nuclei.

References

Abragam, A., and Pound, R. V. (1953). *Phys. Rev.* **92**, 943.

Afanas'ev, A. M., and Kagan, Yu. (1967). *Sov. Phys. JETP* **25**, 124, and references therein.

Alder, K., Matthias, E., Schneider, W., and Steffen, R. M. (1963). *Phys. Rev.* **129**, 1199.

Bacon, F., Barclay, J. A., Brewer, W. D., Shirley, D. A., and Templeton, J. E. (1972). *Phys. Rev. B* **5**, 2397.

Barrett, P. H., and Shirley, D. A. (1963). *Phys. Rev.* **131**, 123.

Bloembergen, N., and Temmer, G. M. (1953). *Phys. Rev.* **89**, 883.

Bocquet, J. P., Chu, Y. Y., Kistner, O. C., Perlman, M. L., and Emery, G. T. (1966). *Phys. Rev. Lett.* **17**, 809.

Bohr, A., and Weisskopf, V. F. (1950). *Phys. Rev.* **77**, 94.

Borrmann, G. (1941). *Phys. Z.* **42**, 157; (1950). *Z. Phys.* **127**, 297.

Bozorth, R. M. (1951). "Ferromagnetism." Van Nostrand Reinhold, Princeton, New Jersey.

Brewer, W. D., Shirley, D. A., and Templeton, J. E. (1968). *Phys. Lett.* **27A**, 81.

Christiansen, J., Mahnke, H.-E., Recknagel, E., Riegel, D., Weyer, G., and Witthuhn, W. (1968). *Phys. Rev. Lett.* **21**, 554.

Easley, W. C., Edelstein, N., Klein, M. P., Shirley, D. A., and Wickman, H. H. (1966). *Phys. Rev.* **141**, 1132.

Eisinger, J., and Jaccarino, V. (1958). *Rev. Mod. Phys.* **30**, 530.

Elliott, R. J., and Stevens, K. W. H. (1953). *Proc. Roy. Soc. (London)* **A218**, 553.

Fermi, E. (1930). *Z. Phys.* **60**, 320.

Frauenfelder, H., and Steffen, R. M. (1965). *In* "Alpha-, Beta-, and Gamma-Ray Spectroscopy" (K. Siegbahn, ed.), Chapter XIX A. North-Holland Publ., Amsterdam.

Freeman, A. J., and Watson, R. E. (1963). *Phys. Rev.* **132**, 706.

Hammell, G. T., and Hannon, J. P. (1969). *Phys. Rev.* **180**, 337.

Hollander, J. M., and Shirley, D. A. (1970). *Ann. Rev. Nucl. Sci.* **20**, 435.

Judd, B. R., Lovejoy, C. A., and Shirley, D. A. (1962). *Phys. Rev.* **128**, 1733.

Kaindl, G., and Salomon, D. (1970). *Phys. Lett.* **32B**, 364.

Kaindl, G., Salomon, D., and Wortmann, G. (1972). *Phys. Rev. Lett.* **28**, 952.

Kittel, C. (1971). "Introduction to Solid-State Physics," 4th ed., Chapter 13. Wiley, New York.

Koicki, S., Koster, T. A., Pollak, R., Quitmann, D., and Shirley, D. A. (1970). *Phys. Lett.* **32B**, 351.

Laue, M. (1949). *Acta. Crystallogr.* **2**, 106; (1952). **5**, 619.

Leipert, T. K., Baldeschwieler, J. D., and Shirley, D. A. (1968). *Nature (London)* **220**, 907.

Matthias, E., and Shirley, D. A. (1966). *Nucl. Instrum. Methods* **45**, 309.

Matthias, E., Rosenblum, S. S., and Shirley, D. A. (1965). *Phys. Rev. Lett.* **14**, 46.

Matthias, E., Olsen, B., Shirley, D. A., Templeton, J. E., and Steffen, R. M. (1971). *Phys. Rev. A* **4**, 1626.

Melissinos, A. C., and Davis, S. P. (1959). *Phys. Rev.* **115**, 130.

Mössbauer, R. (1958). *Z. Phys.* **151**, 124.

Perlow, G. J. (1968). *In* "Chemical Applications of Mössbauer Spectroscopy" (V. I. Goldanskii and R. H. Herber, eds.). Academic Press, New York.

Pound, R. V., and Rebka, G. A., Jr. (1960). *Phys. Rev. Lett.* **4**, 337.

Sauer, C., Matthias, E. and Mössbauer, R. L. (1968). *Phys. Rev. Lett.* **21**, 961.

Shirley, D. A. (1964). *Rev. Mod. Phys.* **36**, 339.

Shirley, D. A. (1969). A review was given in "Colloque Ampere XV," p. 81. North-Holland Publ., Amsterdam.

Shirley, D. A., and Westenbarger, G. A. (1965). *Phys. Rev.* **138**, A170.

Shirley, D. A., Matthias, E., and Rosenblum, S. S. (1968). *Phys. Rev.* **170**, 363.

Siegbahn, K., Nordling, C., Fahlman, A., Nordberg, R., Hamrin, K., Hedman, J., Johansson, G., Bergmark, T., Karlsson, S.-E., Lindgren, I., and Lindberg, B. (1967). "ESCA, Atomic, Molecular and Solid State Structure Studied by Means of Electron Spectroscopy." Almqvist & Wiksells AB, Stockholm.

Siegbahn, K., Nordling, C., Johansson, G., Hedman, J., Heden, P. F., Hamrin, K., Gelius, U., Bergmark, T., Werme, L. O., Manne, R., and Baer, Y. (1969). "ESCA Applied to Free Molecules." North-Holland Publ., Amsterdam.

Smirnov, G. V., Sklyarevskii, V. V., Artem'ev, A. N., and Voscanyan, R. A. (1970). *Phys. Lett.* **32A**, 532 and references therein.

Sternheimer, R. M. (1951). *Phys. Rev.* **84**, 244; (1954). **95**, 736.

Templeton, J. E. and Shirley, D. A. (1967). *Phys. Rev. Lett.* **18**, 240.

Townes, C. H. (1958). "Handbuch der Physik," Vol. 38/1, p. 377. Springer-Verlag, Berlin.

Voitovetskii, V. K., Korsunskii, I. L., Pazhin, Yu. F. and Silakov, R. S. (1970). *JETP Lett.* **12**, 212 and references therein.

Watson, R. E., Gossard, A. C., and Yafet, Y. (1965). *Phys. Rev.* **140**, A375.

Wertheim, G. K., and Guggenheim, H. J. (1968). *In* "Hyperfine Structure and Nuclear Radiations" (E. Matthias and D. A. Shirley, eds.), p. 531. North-Holland Publ., Amsterdam.

Wickman, H. H., Klein, M. P., and Shirley, D. A. (1966). *Phys. Rev.* **152**, 345.

VIII.E   A GUIDE TO NUCLEAR COMPILATIONS

F. Ajzenberg-Selove[†]

DEPARTMENT OF PHYSICS
UNIVERSITY OF PENNSYLVANIA
PHILADELPHIA, PENNSYLVANIA

Unless one has been a nuclear physicist for many years and unless one is very well organized, it is difficult to keep track of the many useful compilations in nuclear physics. The following list of compilations is not a complete one. It is a selected list of those reviews which have been most useful to the author and to a few of her colleagues.

A complete directory did exist in 1958–1960, and the reader is referred to it for early references:

□ "A Directory to Nuclear Data Tabulations," R. C. Gibbs and K. Way, January 1958, and Revisions in 1959 and 1960 *Nuclear Data Tables*. (This directory is out of print.)

In addition, many useful tables are available in

□ "Nuclear Reaction Analysis, Graphs and Tables," J. B. Marion and F. C. Young, North-Holland Publ., Amsterdam, and Wiley, New York (1968).

The reader should consider which values of the fundamental physical constants were used in the tables he consults. The standard contemporary article on these constants is

□ "…The Fundamental Physical Constants," B. N. Taylor, W. H. Parker, and D. N. Langenberg, *Rev. Mod. Phys.* **41**, 375 (1969). (A revision is being prepared by R. L. Cohen and B. N. Taylor.) Reprinted as "The Fundamental Constants and Quantum Electrodynamics," Academic Press, New York (1969).

[†] Work supported in part by the National Science Foundation

Some of the values described in that article are superseded by the measurements reported in

□ "Precision Measurement and Fundamental Constants," D. N. Langenberg and B. N. Taylor (eds.), NBS Special Publ. 343 (1971). (Available from the Superintendent of Documents.[†])

The accepted table of atomic masses is

□ The 1971 Atomic Mass Evaluation, A. H. Wapstra and N. B. Gove, *Nucl. Data Tables* **9**, 265 (1971).

The authors have published Nuclear Reaction Q Values in *Nucl. Data Tables* **11**, 127 (1972). It should be noted that the conversion factor between atomic mass units and MeV is different from that in the *Rev. Mod. Phys.* article of Taylor, Parker, and Langenberg.

The most important of the wall display-type summaries of nuclear data is

□ "Eleventh Edition of the GE Chart of the Nuclides" (1972), N. E. Holden and F. W. Walker. (Available from Educational Relations Department, General Electric Co., Schenectady, N. Y. 12345.)

This is a well-known wall chart, listing among other parameters the best values of isotopic abundances, evaluated by N. E. Holden and P. DeBievre.

Wallet-sized listings of angular momentum, parity, mass excesses,[‡] half lives, and abundances for most known nuclides are available as

□ "Nuclear Wallet Cards," F. Ajzenberg-Selove and C. L. Busch (December 1971). (Available from authors, Dept. of Phys., Univ. of Pennsylvania, Philadelphia, Pennsylvania 19174; no cost.)

As nuclear physics moves toward higher energies readers should be aware of the yearly articles by the Rosenfeld-Berkeley group. The latest is

Review of Particle Properties, A. H. Rosenfeld *et al.*, *Phys. Lett.* **39B**, 1 (1972).

The most portable of the general compendia of nuclear properties is

□ "Table of Isotopes—Sixth Edition," C. M. Lederer, J. M. Hollander, and I. Perlman, Wiley, New York (1967). (This Table is being revised. The seventh edition should appear in 1975–1976.)

[†] U. S. Government Printing Office, Washington, D.C. 20402.

[‡] Masses used are from Wapstra and Gove (see above).

Review articles on the energy levels of various mass chains are listed below:

☐ Energy Levels of Light Nuclei. $A = 3$, S. Fiarman and S. Hanna (being prepared for publication in *Nucl. Phys.* 1974).

☐ Energy Levels of Light Nuclei. $A = 4$, S. Fiarman and W. E. Meyerhof, *Nucl. Phys.* **A206**, 1 (1973).

☐ Energy Levels of Light Nuclei. $A = 5$–10, T. Lauritsen and F. Ajzenberg-Selove, *Nucl. Phys.* **78**, 1 (1966). (Revision in progress: to be published in *Nucl. Phys.* 1974.)

☐ Energy Levels of Light Nuclei. $A = 11$–12, F. Ajzenberg-Selove and T. Lauritsen, *Nucl. Phys.* **A114**, 1 (1968). (Revision in progress: to be published in *Nucl. Phys.* 1975.)

☐ Energy Levels of Light Nuclei. $A = 13$–15, F. Ajzenberg-Selove, *Nucl. Phys.* **A152**, 1 (1970).

☐ Energy Levels of Light Nuclei. $A = 16$–17, F. Ajzenberg-Selove, *Nucl. Phys.* **A166**, 1 (1971).

☐ Energy Levels of Light Nuclei. $A = 18$–20, F. Ajzenberg-Selove, *Nucl. Phys.* **A190**, 1 (1972).

☐ Energy Levels of $A = 21$–44 Nuclei, P. Endt and C. Van der Leun, *Nucl. Phys.* **A214**, 1 (1973).

Evaluated compilations of the energy levels of the heavier nuclei appear in *Nuclear Data Sheets*. For the convenience of the reader the following list shows where reviews of the information on various mass chains may be obtained. The letter R means that these particular mass chains are being revised. These revisions should appear in *Nucl. Data Sheets* for 1973–1975.

A	Nuclear Data Sheets Volume	A	Nuclear Data Sheets Volume	A	Nuclear Data Sheets Volume
45–49	**B4**	113	**B5**	195, 196	**B7, B8**
50–58	**B3**	114–120(R)	**E†**	197	**B7**
59–69	**B2**	121	**B6**	198–200	**B6**
70–83(R)	**B1, B8, B9→B11**	122–141(R)	**B7, B8, E†, B9**	201–204	**B5**
84–87	**B5**	142–148	**B2, B10**	205	**B6**
88–90(R)	**A8**	149–168(R)	**B8, B9, B10, B11**	206	**B7**
91–106(R)	**B7, B8, B10, B11**	169–173	**E†, B10**	207–211	**B5**
107, 108	**B7**	174–181(R)	**E†, B10**	212	**B8**
109	**B6**	182–189	**B1, B10**	213–228(R)	**B1, B10**
110	**B5**	190–193	**B8, E†, B9**	229–242	**B4, B6, B8**
111	**B6**	194	**B7**	243–261	**B3, B8**
112	**B7**				

† **E** = Reprint of Nuclear Data Sheets (1959–1965).

Of interest also is:

□ A Survey of Nonrotational States of Deformed Odd-A Nuclei ($150 < A < 190$), M. E. Bunker and C. W. Reich, *Rev. Mod. Phys.* **43**, 348 (1971); **44**, 126 (1972).

Listings of recent references for nuclei throughout the periodic tables are published at regular intervals in *Nuclear Data Sheets*.

General reviews of nuclear spins, moments, and radii include:

□ "Nuclear Spin-Parity Assignments," N. B. Gove and R. L. Robinson (eds.), Academic Press, New York (1966).
□ Table of Hyperfine Fields, D. A. Shirley, *in* "Hyperfine Structure and Nuclear Radiations" (E. Matthias and D. A. Shirley, eds.), North-Holland Publ., Amsterdam (1968).
□ Table of Nuclear Moments, V. S. Shirley, *in* "Hyperfine Structure and Nuclear Radiations" (E. Matthias and D. A. Shirley, eds.), North-Holland Publ., Amsterdam (1968).
□ Nuclear Spins and Moments, G. H. Fuller and V. W. Cohen, *Nucl. Data* **5A**, 433 (1969).
□ Nuclear Intrinsic Quadrupole Moments and Deformation Parameters, K. E. G. Löbner, M. Vetter, and V. Hönig, *Nucl. Data Tables* **7**, 495 (1970).
□ Nuclear Radii, L. R. B. Elton, R. Hofstadter, and H. R. Collard, *in* "Landolt-Bornstein," Vol. 2, H. Schopper (ed.), Springer-Verlag, Berlin (1967).

The Charged Particle Cross Section Group at Oak Ridge has published a number of compilations. Among these are:

□ "Nuclear Cross Sections for Charged Particle Induced Reactions," F. K. McGowan, W. T. Milner, and H. J. Kim, ORNL-CPX-1 and 2. (Available without charge from Dr. F. K. McGowan, Oak Ridge Nat. Lab., P.O. Box X, Oak Ridge, Tennessee 37830.)
□ Nuclear Cross Sections for Charged-Particle-Induced Reactions, H. J. Kim, W. T. Milner, and F. K. McGowan, *Nucl. Data* **A1**, 203 (1966); **A2**, 1 (1966); **A3**, 123 (1967).
□ Reaction List for Charged-Particle-Induced Nuclear Reactions, F. K. McGowan and W. T. Milner: May 1969–June 1970, *Nucl. Data Tables* **8**, 199 (1970); July 1970– June 1971, *Nucl. Data Tables* **9**, 477 (1971).
□ Reaction List for Coulomb Excitation Data – 1956–1971, F. K. McGowan and W. T. Milner, *Nucl. Data Tables* **9**, 572 (1971). [A cumulated hardcover edition of the McGowan–Milner reaction lists has been published: "Charged-Particle Reaction List 1948–1971," F. K. McGowan and W. T. Milner, Academic Press, New York (1973).]

Range–energy and stopping-power tables include:

□ "Tables of Energy Losses and Ranges of Electrons and Positrons," M. J. Berger and S. M. Seltzer, NASA SP-3012 (1964). (Available from N.T.I.S.†)
□ "Additional Stopping Power and Range Tables for Protons, Mesons, and Electrons," M. J. Berger and S. M. Seltzer, NASA SP-3036 (1966). (Available from N.T.I.S.†)
□ "Tables of Energy Losses and Ranges of Heavy Charged Particles," W. H. Barkas and M. J. Berger, NASA SP-3013 (1964). (Available from N.T.I.S.†)

† National Technical Information Service, Springfield, Va. 22151.

☐ Range and Stopping-Power Tables for Heavy Ions, L. C. Northcliffe and R. F. Schilling, *Nucl. Data Tables* **7A**, 233 (1970).

☐ Theory of Effective Charge and Stopping Power for Heavy Ions, G. D. Sauter and S. D. Bloom, *Phys. Rev. B* **6**, 699 (1972).

☐ Energy Loss, Range and Bremsstrahlung Yield for 10-keV to 100-MeV Electrons in Various Elements and Chemical Compounds, L. Pages, E. Bertel, H. Joffre, and L. Sklavenitis *At. Data* **4**, 1 (1972).

☐ "Calculations of Energy Loss, Range, Pathlength, Straggling, Multiple Scattering, and the Probability of Inelastic Nuclear Collisions for 0.1 to 1000 MeV Protons," J. F. Janni, AFWL-TR-65-150 (1970). (Available from N.T.I.S.[†])

☐ "Tables of Range and Stopping Power of Chemical Elements for Charged Particles of Energy 0.05 to 500 MeV," C. F. Williamson, J.-P. Boujot, and Jean Picard, Rep. CEA-R 3042 (1966). (Available from Documentation Francaise, Secrétariat Général du Gouvernement, Direction de la Documentation, 16, rue Lord Byron, Paris 8, France.)

☐ Multiple Scattering of Charged Particles, J. B. Marion and B. A. Zimmerman, *Nucl. Instrum. Methods* **51**, 93 (1967).

Calibration energies are discussed in

☐ Accelerator Calibration Energies, J. B. Marion, *Rev. Mod. Phys.* **38**, 660 (1966).

☐ Precise Comparison and Measurement of Gamma-Ray Energies with a Ge(Li) Detector, R. C. Greenwood, R. G. Helmer, and R. J. Gehrke, *Nucl. Instrum. Methods* **77**, 141 (1970); R. G. Helmer, R. C. Greenwood, and R. J. Gehrke, *Nucl. Instrum. Methods* **96**, 173 (1971).

Compilations of neutron data are given in

☐ "CINDA 71 (etc.) – An Index to the Literature on Microscopic Neutron Data," (available in U.S.A. from UNIPUB, Inc., P.O. Box 433, New York, N.Y. 10016).

☐ "Neutron Cross Sections," J. R. Stehn, M. D. Goldberg, B. A. Magurno, and R. Wiener-Chasman, BNL 325 (1964). (Available from N.T.I.S.[†])
(A sequel to BNL 325 is in preparation by the National Neutron Cross Section Center at Brookhaven: Vol. 1 contains recommended values for thermal cross sections and resonance parameters — BNL 325, 3rd ed. (June 1973) S. F. Mughabghab and D. I. Garber; Vol. 2 will contain experimental data for cross sections over selected energy ranges.)

☐ "Angular Distributions in Neutron Induced Reactions," (National Neutron Cross Section Center, Brookhaven National Laboratory). D. I. Garber *et al.*, BNL -400 (1970). (Available from N.T.I.S.[†])

☐ Compendium of Thermal-Neutron-Capture γ-Ray Measurements, G. A. Bartholomew, L. V. Groshev *et al.*, Part I: $Z \leqslant 46$, *Nucl. Data* **3A**, 367 (1967); Part II: $Z = 47$ to $Z = 67$, *Nucl. Data* **5A**, 1 (1968); Part III: $Z = 68$ to $Z = 94$, *Nucl. Data* **5A**, 243 (1969).

When working with gamma rays, the reader is referred to

☐ Analysis of Gamma Ray Data, D. H. Wilkinson, *in* "Nuclear Spectroscopy" (F. Ajzenberg-Selove, ed.), Academic Press, New York (1960).

[†] National Technical Information Service, Springfield, Va. 22151.

☐ "Gamma Ray Spectrum Catalogue," R. L. Heath, IDO 16880 (1964). (Available from N.T.I.S.[†])

☐ Catalogue of γ-Rays Emitted by Radionuclides, M. A. Wakat, *Nucl. Data Tables* **8**, 445 (1971).

☐ Photon Cross Sections from 1 keV to 100 MeV for Elements $Z = 1$ to $Z = 100$, E. Storm and H. I. Israel, *Nucl. Data Tables* **7A**, 565 (1970).

☐ "Photon Cross Sections, Attenuation Coefficients, and Energy Absorption Coefficients from 10 keV to 100 GeV," J. H. Hubbell, NSRDS-NBS 29. (Available from Superintendent of Documents.[‡])

☐ "Graphs of the Compton Energy-Angle Relationship and the Klein–Nishina Formula from 10 keV to 500 MeV," Ann T. Nelms, NBS Circ. 542 (1953). (On sale by Superintendent of Documents.[‡])

☐ K X-Ray Transition Probabilities, G. C. Nelson, B. G. Saunders, and S. I. Salem, *At. Data* **1**, 377 (1970).

☐ L X-Ray Transition Probabilities, S. I. Salem and C. W. Schultz, *At. Data* **3**, 215 (1971).

☐ Systematics of Absolute Gamma-Ray Transition Probabilities in Deformed Odd-Mass Nuclei, K. E. G. Löbner and S. G. Malmskog, *Nucl. Phys.* **80**, 505 (1966).

☐ Clebsch–Gordan Coefficients for Nuclear Transition Probabilities for Even-*A* Deformed Nuclei, T. Yamazaki, *Nucl. Data Tables* **A1**, 453 (1966).

☐ Nuclear Transition Probability, $B(E2)$, for $0^+ \rightarrow 2^+$ Transitions and Deformation Parameter, β_2, P. H. Stelson and L. Grodzins, *Nucl. Data* **A1**, 21 (1965).

☐ Table of Electromagnetic M1 Reduced Transition-Probability Matrix Elements between Nilsson States in Odd-A Nuclei, E. Browne and F. R. Femenia, *Nucl. Data Tables* **10**, 81 (1971).

☐ Internal Pair Formation and Multipolarity of Nuclear Transitions, R. J. Lombard, C. F. Perdrisat, and J. H. Brunner *Nucl. Phys.* **A110**, 41 (1968).

☐ Angular Distributions of Gamma Rays in Terms of Phase-Defined Reduced Matrix Elements, H. J. Rose and D. M. Brink, *Rev. Mod. Phys.* **39**, 306 (1967).

☐ Directory to Tables and Reviews of Angular-Momentum and Angular-Correlation Coefficients, K. Way and F. W. Hurley, *Nucl. Data* **1A**, 473 (1966).

☐ Tables of Coefficients for Angular Distributions of Gamma Rays from Aligned Nuclei, T. Yamazaki, *Nucl. Data* **3A**, 1 (1967).

☐ Tables of Coefficients for Angular Correlations of Radiative Transitions from Aligned Nuclei, D. D. Watson and G. I. Harris, *Nucl. Data* **3A**, 25 (1967).

☐ Table of Angular Distribution Coefficients for (Gamma–Particle) and (Particle–Gamma) Reactions, R. W. Carr and J. E. E. Baglin, *Nucl. Data Tables* **10**, 143 (1972).

☐ A Tabulation of Gamma–Gamma Directional Correlation Coefficients, H. W. Taylor, B. Singh, F. S. Prato, and R. McPherson *Nucl. Data Tables* **9**, 1 (1971); P. E. Haustein, H. W. Taylor, R. McPherson, and R. Fairchild, *Nucl. Data Tables* **10**, 321 (1972).

☐ "Angular Correlation Methods in Gamma-Ray Spectroscopy," A. J. Ferguson, North-Holland Publ., Amsterdam (1965).

☐ "Tables of Coefficients for the Analysis of Triple Angular Correlations from Aligned Nuclei," G. Kaye, E. J. O. Read, and J. C. Willmott [Chadwick Phys. Lab., Univ. of Liverpool, England (1963)].

[†] National Technical Information Service, Springfield, Va. 22151.
[‡] U. S. Government Printing Office, Washington, D.C. 20402.

☐ X-Ray Wavelengths, J. A. Bearden, *Rev. Mod. Phys.* **39**, 78 (1967).
☐ X-Ray Atomic Energy Levels, J. A. Bearden and A. F. Burr, *Rev. Mod. Phys.* **39**, 125 (1967).
☐ Internal Conversion Tables, R. S. Hager and E. C. Seltzer, Part I: K-, L-, M-Shell Conversion Coefficients for $Z = 30$ to $Z = 103$, *Nucl. Data* **4A**, 1 (1968); Part II: Directional and Polarization Particle Parameters for $Z = 30$ to $Z = 103$, *Nucl. Data* **4A**, 397 (1968); Part III: Coefficients for the Analysis of Penetration Effects in Internal Conversion and E0 Internal Conversion, *Nucl. Data Tables* **6A**, 1 (1969).
☐ Experimental Values of Internal-Conversion Coefficients of Nuclear Transitions: Total and K-Shell Coefficients and L-Subshell Coefficient Ratios, J. H. Hamilton *et al.*, *Nucl. Data* **1A**, 521 (1966).
☐ Tables of Internal Conversion Coefficients for N-Subshell Electrons, O. Dragoun, H. C. Pauli, and F. Schmutzler, *Nucl. Data Tables* **6A**, 235 (1969).
☐ Contributions of Outer Shells to Total Internal Conversion Coefficients, O. Dragoun, Z. Plajner, and F. Schmutzler, *Nucl. Data Tables* **9**, 119 (1971).

There are several summaries dealing with electromagnetic radiation interactions with nuclei. The following are indexes to the literature of the field (including inelastic electron scattering and inverse capture reactions):

☐ "Photonuclear Data Index (1955–1965)," Photonuclear Data Group, NBS Miscellaneous Publ. 277 (1966). (Available from N.T.I.S.†)
☐ "Photonuclear Data Index (1965–1970)," Photonuclear Data Group, NBS Special Publ. 322 (1970); "Photonuclear Reaction Data," NBS Special Publ. 380 (1973). (Available from Superintendent of Documents.‡)
☐ "Photonuclear Reactions," E. Hayward, NBS Monograph 118 (1970). (Available from Superintendent of Documents.‡)

For work on beta decay and electron capture, the reader is referred to:

☐ Log *f* Tables for Beta Decay, N. B. Gove and M. J. Martin, *Nucl. Data Tables* **A10**, 205 (1971).
☐ Shapes of Beta Spectra, H. Paul, *Nucl. Data* **2A**, 281 (1966).
☐ Shapes of Beta-Ray Spectra, H. Daniel, *Rev. Mod. Phys.* **40**, 659 (1968).
☐ Fermi-Function Integrals for Finding Relative Beta-Group Intensities, G. P. Ford and D. C. Hoffman, *Nucl. Data* **1A**, 411 (1966).
☐ Response of Silicon Detectors to Mono-Energetic Electrons with Energies between 0.15 and 5.0 MeV, M. J. Berger *et al.*, *Nucl. Instrum. Methods* **69**, 181 (1969).
  An expanded version of this paper is: "Tables of Response Functions for Silicon Electron Detectors," M. J. Berger *et al.*, NBS Tech. Note 489 (1969). (Available from Superintendent of Documents.‡)
☐ Nonrelativistic, K-Shell, Auger Rates and Matrix Elements for $4 \leqslant Z \leqslant 54$, D. L. Walters and C. P. Bhalla, *At. Data* **3**, 301 (1971).
☐ Numerical Values of the K-Electron Functions Determining the Probability of Allowed

---

† National Technical Information Service, Springfield, Va. 22151.
‡ U. S. Government Printing Office, Washington, D.C. 20204.

and Forbidden K-Capture, I. M. Band, L. N. Zyrianova, and C. Z. Tsin, *Bull. Acad. Sci. USSR, Phys. Sect.* **20**, 1269 (1956).

☐ Tables of Functions Needed for Determining the Probability of Allowed and Forbidden L-Capture, I. M. Band, L. N. Zyrianova, and Y. P. Suslov, *Bull. Acad. Sci. USSR, Phys. Sect.* **22**, 943 (1958).

Other general tables are:

☐ "Tables for the Transformation between the Laboratory and Center-of-Mass Coordinate Systems and for the Calculation of the Energies of Reaction Products," J. B. Marion, T. I. Arnette, and H. C. Owens, ORNL-2574 (1959). (Available from N.T.I.S.[†])

☐ One-Particle, j–j Coefficients of Fractional Parentage in the Isospin Representation for $n \leqslant 5$ in the $j = \frac{7}{2}$ Shell, L. B. Hubbard, *Nucl. Data Tables* **9**, 85 (1971).

☐ Tables of Fractional Parentage Coefficients in the Isospin Formalism for the $j = \frac{3}{2}$ and $\frac{5}{2}$ Shells, I. S. Towner and J. C. Hardy, *Nucl. Data Tables* **6A**, 153 (1969).

☐ "Tables of Identical-Particle Fractional Parentage Coefficients," B. F. Bayman and A. Lande, PUC-1965-157 (Princeton Univ., 1965).

☐ "Tables of Coefficients," M. Ferentz and N. Rosenzweig, ANL 5324 (1955). (Available from N.T.I.S.[†])

☐ "Numerical Table of the Clebsch–Gordan Coefficients," A. Simon, ORNL-1718 (1954). (Available from N.T.I.S.[†])

☐ "Tables of the Racah Coefficients," A. Simon, J. H. Van der Sluis, and L. C. Biedenharn, ORNL-1679 (1954). (Available from N.T.I.S.[†])

☐ "Tables of Transformation Brackets" (for Nuclear Shell-Model Calculations), 2nd ed., T. A. Brody and M. Moshinsky, Gordon and Breach, New York (1967).

☐ Tables of Single-Particle Reduced Matrix Elements of Spherical Tensors, W. K. Bell and G. R. Satchler, *Nucl. Data Tables* **9**, 147 (1971).

☐ An Introduction to X-Coefficients and a Tabulation of Their Values, P. L. Csonka, T. Haratani, and M. J. Moravcsik, *Nucl. Data Tables* **9**, 235 (1971).

☐ Optical-Model Analysis of Nucleon Scattering from 1p-Shell Nuclei between 10 and 50 MeV, B. A. Watson, P. O. Singh, and R. E. Segel, *Phys. Rev.* **182**, 977 (1969).

For polarization measurements, the reader is referred to a statement of

☐ The Madison Convention (for Polarization Measurements) *in* "Polarization Phenomena in Nuclear Reactions," H. H. Barschall and W. Haeberli (eds.), Univ. of Wisconsin Press, Madison, Wisconsin (1971).

Finally, the reader is referred to these articles and books:

☐ Reduced Widths of Individual Nuclear Energy Levels, A. M. Lane, *Rev. Mod. Phys.* **32**, 519 (1960).

☐ Stripping Reactions and the Structure of Light and Intermediate Nuclei, M. H. Macfarlane and J. B. French, *Rev. Mod. Phys.* **32**, 567 (1960).

☐ "Nuclear Spectroscopy" Vols. A and B, F. Ajzenberg-Selove (ed.), Academic Press, New York (1960).

[†] National Technical Information Service, Springfield, Va. 22151.

□ "Alpha-, Beta- and Gamma-Ray Spectroscopy," K. Siegbahn (ed.), North-Holland Publ., Amsterdam (1965).

□ "Nuclear Structure" Vol. I, A. Bohr and B. R. Mottelson, Benjamin, New York (1969).

□ "Isospin in Nuclear Physics," D. H. Wilkinson (ed.), North-Holland Publ., Amsterdam (1969).

□ "Nuclear Shell Theory," A. de-Shalit and I. Talmi, Academic Press, New York (1963). (The appendices in this book contain many very useful formulae and tables.)

ACKNOWLEDGMENTS

The author is very grateful to her many colleagues who commented on the first draft of this listing. In particular she acknowledges with deep thanks the helpful comments by H. H. Barschall, J. Cerny, E. G. Fuller, E. T. Jurney, C. W. Reich, E. K. Warburton, and K. Way.

SUBJECT INDEX

A

Accelerators
  in atomic spectroscopy, 469-472
  Coulomb excitation and, 13-14
  electron prototype, 384
  energy calibration and, 146-148
Alpha particles, radiative capture of, 168
Alpha radioactivity, 433-434
Amplitude and rise-time compensation
    (ARC) timing, 323-325, 336
Analog states, Coulomb mixing and, 445-451
Angular correlation, 277-305
  azimuthal geometry in, 297-299
  density matrix-statistical tensor method
    in, 280-282
  deorientation effects in, 348-350
  efficiency matrices in, 283
  error minimization in, 303-305
  gram determinant maximization in, 300-303
  matrix element signs in, 287-291
  optimization in, 300-305
  particle-particle gamma reactions in,
    292-296
  spin flip in, 291-292
  techniques in, 291-299
  tensors in, 281-282
  theory of, 275-287
  vacuum recoils in, 299-305
Angular distribution measurements, in cap-
    ture reactions, 179-180
Angular momentum, in heavy-ion reactions,
    197-206
Annihilation photons, 387
Astrophysics, lifetime measurements in,
    494-498
Atomic lifetimes, transition probabilities
    and, 484-494
  see also Lifetime measurements

Atomic spectroscopy
  atomic energy levels in, 480-484
  atomic fine structure and, 498-506
  atomic lifetimes and transition prob-
    abilities in, 484-494
  data analysis in, 473-480
  experimental methods in, 469-473
  field of, 467-468
  foil-excitation mechanisms and, 506-508
  forbidden transitions in, 491-492
  homologous atoms and, 490-491
  in-beam, 467-508
  isoelectronic sequences in, 486-490
  Lamb shift measurements in, 504-506
  Landé g factors in, 503-504
  lifetimes in, 477-480, 493-494
  line identification in, 475-477
  molecular lifetimes in, 492-493
  wavelengths in, 473-475
  zero-field oscillations and, 502-503

B

Beta decay
  vs. deexcitation, 256-260
  theory of, 441-446
Beta-delayed alpha and proton decay,
    435-438
Beta-delayed neutron decay, 438-440
Beta-delayed particle decay, 419
  energy requirements for, 421-424
  occurrence of, 424-431
Beta-delayed-proton precursors, tables of,
    426-430
Beta-neutrino angular correlations, 445-446,
    456-459
Beta-nuclear magnetic resonance perturbed
    angular distribution, 118-121
Biedenharn-Rose convention, in angular cor-
    relation, 290

583

PURE AND APPLIED PHYSICS

A Series of Monographs and Textbooks

Consulting Editors

H. S. W. Massey
University College, London, England

Keith A. Brueckner
University of California, San Diego
La Jolla, California

1. F. H. Field and J. L. Franklin, Electron Impact Phenomena and the Properties of Gaseous Ions. (Revised edition, 1970.)
2. H. Kopfermann, Nuclear Moments. English Version Prepared from the Second German Edition by E. E. Schneider.
3. Walter E. Thirring, Principles of Quantum Electrodynamics. Translated from the German by J. Bernstein. With Corrections and Additions by Walter E. Thirring.
4. U. Fano and G. Racah, Irreducible Tensorial Sets.
5. E. P. Wigner, Group Theory and Its Application to the Quantum Mechanics of Atomic Spectra. Expanded and Improved Edition. Translated from the German by J. J. Griffin.
6. J. Irving and N. Mullineux, Mathematics in Physics and Engineering.
7. Karl F. Herzfeld and Theodore A. Litovitz, Absorption and Dispersion of Ultrasonic Waves.
8. Leon Brillouin, Wave Propagation and Group Velocity.
9. Fay Ajzenberg-Selove (ed.), Nuclear Spectroscopy. Parts A and B.
10. D. R. Bates (ed.), Quantum Theory. In three volumes.
11. D. J. Thouless, The Quantum Mechanics of Many-Body Systems. (Second edition, 1972.)
12. W. S. C. Williams, An Introduction to Elementary Particles. (Second edition, 1971.)
13. D. R. Bates (ed.), Atomic and Molecular Processes.
14. Amos de-Shalit and Igal Talmi, Nuclear Shell Theory.
15. Walter H. Barkas. Nuclear Research Emulsions. Volume I.
                       Nuclear Research Emulsions. Volume II.
16. Joseph Callaway, Energy Band Theory.
17. John M. Blatt, Theory of Superconductivity.
18. F. A. Kaempffer, Concepts in Quantum Mechanics.
19. R. E. Burgess (ed.), Fluctuation Phenomena in Solids.
20. J. M. Daniels, Oriented Nuclei: Polarized Targets and Beams.
21. R. H. Huddlestone and S. L. Leonard (eds.), Plasma Diagnostic Techniques.
22. Amnon Katz, Classical Mechanics, Quantum Mechanics, Field Theory.
23. Warren P. Mason, Crystal Physics in Interaction Processes.
24. F. A. Berezin, The Method of Second Quantization.
25. E. H. S. Burhop (ed.), High Energy Physics. In five volumes.

A 4
B 5
C 6
D 7
E 8
F 9
G 0
H 1
I 2
J 3